Perspectives in Animal Phylogeny and Evolution

Perspectives in Animal Phylogeny and Evolution

Alessandro Minelli

OXFORD

UNIVERSITY PRESS

OXFORD

UNIVERSITY PRESS

Great Clarendon Street, Oxford OX2 6DP

Oxford University Press is a department of the University of Oxford.
It furthers the University's objective of excellence in research, scholarship,
and education by publishing worldwide in

Oxford New York

Auckland Cape Town Dar es Salaam Hong Kong Karachi
Kuala Lumpur Madrid Melbourne Mexico City Nairobi
New Delhi Shanghai Taipei Toronto

With offices in

Argentina Austria Brazil Chile Czech Republic France Greece
Guatemala Hungary Italy Japan Poland Portugal Singapore
South Korea Switzerland Thailand Turkey Ukraine Vietnam

Oxford is a registered trade mark of Oxford University Press
in the UK and in certain other countries

Published in the United States
by Oxford University Press Inc., New York

British Library Cataloguing in Publication Data

Data available

Library of Congress Cataloging in Publication Data

Data available

Typeset by Newgen Imaging Systems (P) Ltd., Chennai, India
Printed in Great Britain
by
CPI Antony Rowe, Chippenham, Wiltshire

ISBN 978–0–19–856620–5 (Hbk)   978–0–19–856621–2 (Pbk.)

10 9 8 7 6 5 4 3 2

Contents

Preface

A recent paper by Dunn *et al.* in *Nature* presents a phylogenetic tree of the metazoans based on a newly increased sample of molecular sequences for 77 representative species. The results support many phylogenetic hypotheses that have been consolidating during the last couple of decades in the context of an increasingly heated debate on the interrelationships among the main animal lineages. For example, the clade of moulting animals, the Ecdysozoa, is supported, which groups together animals as different as arthropods (spiders, crabs, millipedes, and flies) and nematodes (the roundworms, e.g. the filarias and the model species *Caenorhabditis elegans*). Also supported are the Lophotrochozoa, a very diverse assemblage including molluscs, annelids, flatworms, and a diversity of smaller or less familiar groups. Supported too are other groupings the reader will find in this book, such as the Ambulacraria (echinoderms plus hemichordates) and the Tetraconata (crustaceans and insects, to the exclusion of myriapods). However, the phylogeny of Dunn *et al.* (2008) also suggests a number of unexpected and arguably controversial hypotheses. First, the deepest split in the new metazoan tree is one between the Ctenophora (the fragile comb-jellies) and all remaining groups, rather then the traditional contrast between the sponges (or, probably better, a subset of the former Porifera) and all other animal groups. Second, sponges and cnidarians (hydras, medusae, corals, and sea anemones) appear to form, together, a lineage sister to all other animals to the exclusion of the ctenophores. Third, the acoels, a group of small marine worms traditionally classified with the flatworms but recently acknowledged to occupy to a very basal position in the tree, is found in the *Nature* tree in the odd company of little-known groups such as myzostomids and gnathostomulids, in a branch which is deeply nested within the thick crown of the tree.

Will these three phylogenetic hypotheses eventually replace those presented in this book, which have been distilled from the evidence available until last week? And, if so, how robust will they prove to the challenges of future investigations?

No later than June 2007, an international conference on animal phylogeny held at the Royal Society of London left the general impression of a growing convergence of molecular, morphological, developmental, and palaeontological evidence towards a reconstruction of animal interrelationships different from the new tree now presented in *Nature*. Interestingly, some of the 18 authors of the *Nature* paper were among the contributors or the discussants at the Royal Society's conference. This circumstance shows how much our trees are still dependent on a body of evidence that is obviously growing, and thus changing, rapidly, so that we must be cautious before accepting this or that grouping as definitive.

An obvious implication is that a book is not the best place to look for the best tree of the day. On the one hand, phylogenetic research is such a lively branch of science that a serious update on its output can only be found by checking the most recent issues of pertinent journals. On the other hand, at the moment there is, arguably, nothing like a single best tree for the metazoans. Some trees are probably more useful, or more reliable, in delineating the main branching of the animal tree, rather than the internal phylogeny of annelids, molluscs, or chordates, whereas others are more accurate in this respect, but less informative at the level of deeper branchings. Much depends, of course, on the set of representative species included in the analysis. Thus, irrespective of its merits in suggesting the interrelationships between, say, sponges, cnidarians, and bilaterians, the paper by Dunn *et al.* (2008) is probably not the best place to look for the internal phylogeny of molluscs or arthropods, which are represented by only eight and nine species respectively. Therefore, I refrain from expressing my scepticism with respect to the Cephalopoda (squids and octopuses) branching with the shelless, worm-like Chaetodermatida (=Caudofoveata) rather than with some group of shell-bearing molluscs. Correspondingly, I must refrain from manifesting my satisfaction with the suggested branching of myriapods (actually, of the only myriapod in the sample) with the Chelicerata (spiders, scorpions, and mites) rather than with the Tetraconata (insects and crustaceans), in agreement with the phylogenetic relationships suggested in this book.

Thus, taking the expected ephemeral lifespan of so many current hypotheses of phylogenetic interrelationships among animal lineages for granted, what is this book about? As promised by the title, I only aim to offer an articulated set of perspectives on animal phylogeny and evolution. Forty years ago, animal phylogenetics was a Cinderella who could not dream to have its discoveries published in the most important scientific journals. Many causes have contributed to its eventual progress to a much higher status. Some of these causes are environmental rather than structural, but are anyway important; others are more technical and specific. First, whatever the nature of the characters considered and the principles according to which the analysis is performed, reconstructing the tree of life only makes sense in the context of an evolutionary view of life. Developments in phylogenetics have been largely independent from advance in the other provinces of evolutionary biology; nevertheless, the increasingly positive reception of phylogenetic studies has certainly benefited from a generally growing interest in evolution in the community of biologists.

Second, phylogenetic analyses are performed with the help of dedicated software. This is very important, not simply because of the computational power of the machines, but most of all because it demonstrates that phylogenetic inference is obtained by applying a set of explicitly formulated rules. This does not mean that today we must completely dispose of the perceptive but informal insights of specialists with a lifelong experience with a taxonomic group, but their precious experience is required in the selection of the taxa and the characters on which the analysis will be performed, and in the subsequent interpretation of the results, but not in the actual phylogenetic analysis. To be sure, we cannot expect 'objectivity' from a machine-run analysis, which obviously depends on the quality of the input data and the rules embodied in the software, but the very fact that these data and rules are made explicit in every analysis will always allow for that recursive process of challenge and revisitation that is vital to science.

Third, phylogenetics has become molecular. Filling a data matrix with numerical codes to be used for a phylogenetic analysis has proved to be a very critical and often controversial step. This is particularly true in the case of the traditional kind of evidence, mostly from morphology, but also from embryology. Things are generally less controversial with molecular data; that is, essentially, with the nucleotide or amino acid sequences that currently represent the single largest source of data available for phylogenetic analyses. By increasingly moving from morphological to molecular, phylogenetics has not simply obtained a much larger body of evidence. Additionally, it has got a chance of sharing a little bit of the unique visibility of molecular biology, something morphology-based phylogenetics would have never enjoyed. There is a place for phylogenetics even at that frontier of molecular biology, genome sequencing, which has satisfied the media that, similar to other disciplines such as physics and astronomy, biology is able to develop real 'big science' projects.

Thus, phylogenetics has its credentials: it uses algorithms, it analyses molecular (including genomic) data, and it aims to comprehensively reconstruct the whole tree of life. But the next questions are: how successful is this enterprise? And, perhaps more important, what is it actually for? Why are we so keenly interested in reconstructing phylogeny? Is this a target per se, or just a necessary step towards a less immediate, but more interesting and rewarding research target? These questions are introduced in the first two chapters of this book and revisited again in the final chapter. I only anticipate here that I do not see phylogeny as a final aim. Otherwise, I would not have written the last three chapters of this book, those where I dare to express my personal and sometimes idiosyncratic views more extensively.

With Chapter 3 we abandon these mainly conceptual and methodological questions to circumscribe the context within which we will move. The chapter deals with the relationships of the animals (Metazoa) to other lineages of living beings, as well as with the palaeontological evidence on the earliest animals, including some fossils whose animal nature is possible but not certain. Chapters 4 to 6 summarize the main hypotheses of phylogenetic relationships among major animal groups. Some nodes of the tree will deserve a more articulated discussion, especially in a few cases where in recent times alternative hypotheses have been defended that have major consequences for our reconstruction of the evolution of animal morphology and development. Some aspects of this evolution are highlighted in the next two chapters, against the background of the provisional consensus phylogeny emerging from the previous pages. Chapter 7 focuses on evolutionary aspects of reproduction and development, Chapter 8 on the evolution of the most general traits of animal morphology, such as body axes and segmentation, as well as the main organ systems. Finally, Chapter 9 summarizes the main lessons that I have learned from this short walk through animal phylogeny and evolution. The core message is that evolution is not simply one of origin of species or, indeed, of clades, but also one of origin of rules, to such an extent that the very concept of origin becomes problematic.

As often happens with books, *Perspectives on Animal Phylogeny and Evolution* has evolved extensively since I first tentatively drafted its contents, in replying to stimulating input from Ian Sherman, who eventually become my editor at Oxford University Press and, no less important, a good friend of mine. Since then, Ian has been supporting me by all means, including a well-justified strong reaction to my continuous delays in delivering the manuscript. Together with him, Helen Eaton has also let me appreciate how much attention Oxford University Press pays to authors, even the most recalcitrant. More recently, I have experienced also friendly, fruitful interaction with Nik Prowse, the best copy editor I ever met, and with Carol Bestley, sensible and accurate production editor. My most sincere thanks to all of them.

I should add here a list of colleagues and friends from whom I have learned most of what is summarized in these pages. But this list would be too long: to some extent, it would duplicate the list of references. Therefore, I only mention here Ron Jenner (Bath), whose papers have been my single best and most critical source of information about the current phylogenetic debates, and Andreas Schmidt-Rhaesa (Hamburg), to whose excellent book on the evolution of animal organ systems, published by Oxford University Press in 2007, I refer the reader for a subject area I could never dream to cover in this book as accurately as he did. My young research associates at the University of Padova have helped me with healthy discussion and literature search: my sincere thanks to Massimo Bernardi, Lucio Bonato, Leandro Drago, Giuseppe Fusco, Diego Maruzzo, and Marco Uliana. Renzo Mazzaro and Matteo Simonetti have assisted me in preparing or editing the illustrations. Most of the images in this book have been very kindly offered by a huge number of old and new friends worldwide: their names are given individually in the figure legends. To all of them I extend here my warmest thanks for an extremely prompt and sympathetic reply to my shameless request for help.

Padova, 16 April 2008

CHAPTER 1

Reading the history of life

We believe that there has been an evolution of mechanisms as well as of their products. We feel that there is need of a new principle in phylogenetic studies, namely that the *primitive representatives of the older phyla evolved by primitive mechanisms.*

Boyden and Shelswell (1959), p. 116

1.1 Things and processes

In biology, as in other sciences and also in ordinary life, we distinguish between things and processes. A philosopher may dispute whether this distinction has ontological foundation, but being a biologist, rather than a philosopher, I will be content with observing that all living things are subject to obvious and often rapid change. This change is the main research target in virtually all biological disciplines. This is clearly the case for evolutionary biology and developmental biology, but also for physiology and ecology. Change is apparently remote from the subject of morphology and taxonomy, but transformations, either actual or formal, again enter the stage as soon as we make comparisons, search for homologies, or identify parallel or convergent features. Eventually, we may suggest that we recognize things whenever change is small enough as not to attract attention, or slow enough as to escape it. Whenever change is pervasive, as in evolution, or rapid enough, as in development, our attention shifts from things to processes.

The appreciation that everything in nature is subject to change is cognate to many world views, both Eastern and Western, but it conflicts strongly with the kind of questions we often ask, both in science and in everyday life: questions about origins.

Charles Darwin entitled his best-known work *On the Origin of Species*, and a problem of origins (those of humankind) was the reason for most of the strongest objections to evolutionary theory. Cognate to a search for origins is a search for ancestors and, more specifically, the search for the earliest representative, or the founder, of a given lineage. With the neatly delineated branching trees it produces, phylogenetics is often naively perceived as the science of ancestry and, thus, as a means to scientifically discover origins: those of *Homo sapiens*, those of vertebrates, and those of the whole animal kingdom. (Arguably, different tools are needed to reconstruct the origin of life itself.) But this hope, probably far from rare among people with scarce familiarity with the foundations of cladistics, is totally misplaced. Cladograms and phylogenetic trees do not represent ancestor–descendant relationships, as in genealogies. Nodes represent synapomorphies—that is, shared derived characters—and branches embody the historical continuity of sets of characters, punctuated here and there by modifications, the origination of new character states along the way to another node. In unrooted trees or cladograms, the most generalized kind of branching graphs depicting phylogenetic relationships, we cannot even know which way to read the change between two neighbouring nodes; that is, we do not know which character state is the older (plesiomorphic), and which is the derived (apomorphic) one. Polarity of change is only specified as soon as we root the tree, that is, as soon as we determine which is the oldest (most basal) split. By fixing the root, a cladogram becomes a phylogenetic tree. Only following this momentous change does it become, at last, a framework against which we can start reconstructing a segment of evolutionary history; that is, a temporal sequence of character transformations.

If nodes in a phylogenetic tree are not ancestors but simply sets of character states, our hopes to use cladistics as a means to discover the origins of a lineage must be dramatically reconsidered. Imagine two animal groups representing the two branches emerging from a node in the phylogenetic tree: two sister groups, in the language of phylogenetics. Let us consider, in particular, two groups between which we can register a lot of differences, morphological and genetic, as with crocodiles and birds. To be sure, all these differences did not originate in one instant. Three obvious avian innovations—the feathers, the wings, and the virtual tail replacing a long series of vertebrae—evolved sequentially, most probably in that order. However, if we only consider the living representatives of these groups, all these differences, and a great many others, will all collapse on that single node of the phylogenetic tree from which crocodiles and birds split off in the Mesozoic. How should that node be construed? In terms of phylogenetic history, it tells us that crocodiles share a more recent common ancestor with birds than with lizards, or any other living group. The character transformations we have identified as belonging to the tree branch leading to our crocodile/bird node, but more basal than it, suggest a list of characters probably present in the last common ancestor of birds and crocodiles, but it does not put this ancestor, literally, in our hands. However, where this even possible, this would not be one of those ancestors one might hope to find, in what we ordinarily mean by searching for origins. People would like to know the first human, rather than the last common ancestor of humans and chimps; or the first bird, rather than the last common ancestor of crocodiles and birds. But this is exactly the point. Were we able to follow the actual genealogy, rather than the abstractions of the phylogenetic trees, where among the descendants of the last common ancestor of crocodiles and birds would we eventually find the first 'real' bird, or the first

'real' crocodile? Arguably, the answer is in either case 'nowhere'. Strictly speaking, origins would require a quantum leap, from non-crocodile to crocodile, or from non-bird to bird. But what we know about evolution does not offer scope for such quantum leaps. This amounts to saying that, strictly speaking, it is idle to search for origins. Let us look instead at change, its history, and its rules. Let us focus on process rather than on things. If we are searching for evolutionary transitions (Maynard-Smith and Szathmáry 1995) we will probably find them only by cancelling long segments of a continuous history.

1.2 Questions about origins

1.2.1 Uncertain origins

Evolutionary biologists are well aware of the fact that the title of Darwin's major work does not capture the book's contents adequately. Indeed, *On the Origin of Species* is much more about adaptive changes in populations than about the mechanisms by which new species arise. Quibbles about the title apart, to explain why Darwin did not elaborate much on the origin of species we should not perhaps blame the limited amount of evidence available to him, or the inadequate analysis he made of them, but rather the conceptual difficulties any evolutionary biologist would eventually discover with respect to both concepts: the species, of course, and also the origins. Darwin was well aware of the fact that differences between varieties within a species may change insensibly into differences between species, thus revealing that the distinction between the variety and the species is an arbitrary one. But this is only one side of the coin. The other is that a species may not have origins if it were not for arbitrary decisions on how to put species names on a phylogenetic tree.

In evolutionary biology, we would like to unravel the origins of species, adaptations, and novelties. The problems are different, ontologically and epistemologically, with respect to each of these. Speciation involves the change of a pre-existing object into two or more principally independent objects, but the relationship between the original species and those obtained through speciation is generally similar to the relationship between a cell and the two products of its division by mitosis, rather than to the strongly asymmetric division between a parent organism and its offspring, be it produced asexually or sexually. In speciation, as in mitosis, we cannot say, other than by arbitrary convention, than A generates B and C, but only than A changes into B and C (e.g. Bock 2004). With time, a species may eventually become the ancestor of a species-rich lineage, and a cell may become the founder of a structurally and functionally integrated whole that deserves the name of individual, but in many cases obvious, relevant biological systems are not the products of simple and complete lineages: in cladistic parlance, they can be paraphyletic (when corresponding to incomplete lineages) or polyphyletic (when including parts of independent historical origins). However, only monophyletic entities (a 'founder' and all of its descendants) can have origins.

1.2.2 Independent origins

How many times have similar structures evolved independently in two or more distantly related lineages? The many similarities between the animals evolved among the Australian marsupials and those evolved among placental mammals in the other parts of the world are all too popular to deserve be discussed here. But let us consider that convergence in mammals is probably much more extensive than the traditional placental/marsupial parallelism suggests, if we accept an African radiation of placental mammals largely independent from the Eurasian and American branches (see section 6.30.3). There is also no need to give many details about the extraordinary similarity that exists between the anterior (raptorial) legs of praying mantises (close relatives of cockroaches) and those of the very distantly related Mantispidae (close relatives of antlions and lacewings; Figure 1.1). To regard those examples as exceptions would be misleading indeed. It depends, of course, on how exacting are our criteria for inclusion of similar structures in a list of parallel (or convergent) features, but we should not marvel that Bergström and Hou (2003) concluded, from an analysis of Early Cambrian arthropods and arthropod-like animals, that convergence, or parallel evolution, is not exceptional but represents, instead, the general rule. In the lineage on which they focused, parallel trends include conspicuous features like the displacement of the eyes towards the dorsal side, the integration into the cephalic region of an increasing number of postantennal

Fig. 1.1. A mantid lacewing (*Mantispa*), a neuropteran that shows striking convergence in morphology with the very distantly related true mantises. Courtesy of Marcello Romano.

segments, the evolution of a 'bivalve' carapace, and the specialization of head appendages, coupled with abandonment of the original mud-engulfing habits.

Segmentation, morphologically and functionally a major trait of animal organization, is quite certainly polyphyletic (see section 8.1.5). Simonetta (2004) contended that the same is also perhaps true of the development of segmentally articulated appendages in stem-group arthropods.

In this context, it would be perhaps rewarding to critically revisit Sawyer's (1984) curious comparisons of two improbable allies such as leeches and insects, presented by the author in terms of a putative 'arthropodization' of the Hirudinea.

In digeneans, a gross feature of the life cycle such as the presence of a sporocyst generation distinct from the more ancient stage of redia evolved several times independently in different phylogenetic branches (Galaktionov and Dobrovolskij 2003). In molluscs, many shell forms have evolved repeatedly, especially in gastropods. For example, patelliform shell shapes have evolved several times in freshwater limpets (Albrecht *et al.* 2004). Recently, Peel (2006) has shown that even the peculiar tusk-like shell of scaphopods is far from unique: scaphopodization—the evolution of this kind of shell—is a recurrent theme within Palaeozoic benthic molluscs. Within the Lepidoptera, a large pterothoracic ganglion has been formed several times independently by the partial or complete fusion of the two posterior thoracic ganglia (Yack and Homberg 2003).

Carcinization—that is, the evolution of the typical crab form, with a very small pleon folded back-to-back under the pereion—happened several times within the Anomura, but the hypothesis that this applies to the Lithodidae or king crabs evolving from asymmetric hermit crab ancestors (as postulated by Cunningham *et al.* 1992), based on mitochondrial sequences, was rejected by McLaughlin and Lemaitre (1997; see also McLaughlin *et al.* 2004).

Let us add just one more example of parallelism, this time from vertebrates: in a sample of 261 species of squamate reptiles, Wiens *et al.*

(2006) found that a snake-like body form has evolved about 25 times.

The extraordinary discoveries in comparative developmental genetics in the last couple of decades have shown that the homologies between features in different phyla are not necessarily unfathomable: many fundamental traits of metazoan or, at least, bilaterian organization can be traced back to remote common ancestors. This is indeed what several researchers (e.g. Carroll 2005, Carroll *et al.* 2005) regard as the most important result, and message, of the emerging discipline of evolutionary developmental biology. New evidence and new prospects notwithstanding, our current awareness in animal phylogeny suggests that independent origins of the most diverse features of animal organization have evolved many times independently. Specifically, the pervasive occurrence of convergence has been the target of a detailed review by Moore and Willmer (1997). It is also the core idea of Conway Morris' (2003a) thesis of the noncontingent nature of life and evolution.

To discuss this topic seriously, it is not enough to have a good phylogeny and a sound knowledge of the developmental mechanisms by which the features under comparison are shaped in the different lineages. Additionally, we need to be clear on what we mean by 'origins', a tricky question with ramifications in metaphysics and epistemology but also, more specifically, one that must come to terms with a still entangled set of concepts spanning analogy, convergence, parallelism, and homoplasy. Here, as very often in biology, it is not always easy to use in practice the concepts that in a theoretical formulation seem to allow for crisp definitions. At any rate, let us try to fix a few points.

I will restrict the discussion to the concepts of parallelism and convergence. Intuitively appealing as a pair, because of the geometrical metaphor, parallelism and convergence have been often regarded as synonymous, and placed within the broader category of homoplasy together with character reversal, by authors such as Wiley (1981), Patterson (1982), Coddington (1994), Kluge (1999, 2001), and Williams and Ebach (2007). A distinction, however, has been advocated by others (e.g. Desutter-Grandcolas *et al.* 2005, 2007). Willmer has changed her views from neatly distinguishing between convergence and parallelism (Moore and Willmer 1997) to merging all these phenomena into a single broad class (Willmer 2003). In my view, the most sensible theoretical approach to the question has been offered by Powell (2007), who argues that the distinction between parallelism and convergence is one of kind and not merely degree. His starting point is similar to my combinatorial approach to homology (Minelli 1998), in that he advocates a factorial analysis of the homology relationships between two groups to be compared: 'it is the underlying developmental homology *with respect to the generators directly causally responsible for the homoplastic event* that defines parallel evolution and non-arbitrarily distinguishes it from convergence' (Powell 2007, p. 567; italics in the original).

Would it be possible to reliably apply Powell's criterion to the wealth of real cases we have hitherto described as either parallelism or convergence? If so, then it would be possible to understand how much an observed similarity depends on internal constraints such as biased variation (Arthur 2004a, 2004b), and how much instead depends on the strength of natural selection (e.g., Brooks 1996).

To use the language of theoretical morphology (McGhee 1999, 2007), why are morphotypes so often clustered in a small corner of the morphospace? In a study of evolutionary changes in shell shapes among a set of 626 Late Cambrian to Middle Devonian gastropod species, fewer morphotypes than expected were found, given the frequency of change in the corresponding shell features. Not only were the most common morphotypes significantly more species-rich than the others, but they also evolved significantly more times than expected. Wagner and Erwin (2006) interpreted these results as implying a set of architectural attractors, but this is little more than an elegant way to describe the evidence. Are these attractors mainly intrinsic or extrinsic to the evolving molluscs? It is indeed a key question in evolutionary genetics to determine whether, or to which extent,

the independent evolution of similar phenotypes in different lineages depends on shared genetic mechanisms. No general rules should be expected, if we take the lesson of melanization. On one hand, melanism in different populations of rock pocket mice (*Chaetodipus intermedius*) depends on mutation at different genes (Hoekstra and Nachman 2003). On the other hand, melanic phenotypes have developed in two distantly related birds, the lesser snow goose (*Anser caerulescens caerulescens*) and the arctic skua (*Stercorarius parasiticus*), as a consequence of variation in the number of copies of a specific allele of the same gene (*melanocortin-1 receptor*) (Mundy *et al.* 2004).

This kind of investigation requires the highest precision, as provided by the study by Prud'homme *et al.* (2006) of the multiple evolution of a male wing pigmentation pattern in a *Drosophila* clade. These authors analysed two gains and two losses, all involving regulatory changes at the same gene, *yellow*, but two independent gains of wing spots turned out to result from the co-opting of distinct ancestral *cis*-regulatory elements.

At the molecular level (although not necessarily at the organ level), convergence, rather than parallelism, can indeed be demonstrated in cases such as the evolution of crystallins, the proteins that are the main components of the lenses of vertebrate eyes. Co-option in the role of taxon-specific crystallins of different proteins, several of which have roles elsewhere in the body (Fernald 2006), has occurred several times independently (True and Carroll 2002).

In the minuscule eyes of the larva of *Neoheterocotyle rhinobatis* (a monopisthocotylean monogenean flatworm) there are unique lenses of mitochondrial origin (Rohde *et al.* 1999): this is a fatal blow to the hypothesis that the eyes of all bilaterians derive from a single ancestral type of photoreceptor already present in their last common ancestor, the so-called Urbilateria (see section 5.2).

How much animals can be prone to parallelism is shown by the frequent spectacular occurrence of this phenomenon within recently radiated lineages, or, in taxonomic parlance, among species belonging to the same genus or to closely related genera. For example, within the last 750 000 years at most a complex placenta has evolved independently in many clades of the freshwater fish genus *Poeciliopsis* (Reznick *et al.* 2002b).

Multiple parallel changes in life style, sometimes accompanied by conspicuous morphological adaptations, are very often recorded. From the many examples in the recent literature, let us mention the independent shift of several lineages of microhylid frogs of Madagascar from terrestrial to arboreal life, or vice versa (Andreone *et al.* 2005), the multiple evolution of tree-climbing among the grapsoid mangrove crabs (Fratini *et al.* 2005), and also multiple transitions from aquatic to terrestrial larval habitat (with one reversal) among a set of 17 species of the sciomyzid fly genus *Tetanocera*, with corresponding morphological changes (Chapman *et al.* 2006). Less conspicuous, but still interesting, are the multiple adaptations to brackish waters and the distinct but also multiple adaptations to human settlements by different African species of the mosquito genus *Anopheles* (Coluzzi *et al.* 2002).

Multiple loss is possibly the most common process leading to parallel/convergent similarities. This has always been a cause of difficulties in identifying the affinities of lineages within which the phenomenon occurs. Examples involving morphology are extremely numerous, including some classics; for example, the different lineages of miniaturized interstitial 'polychaetes' that in the past where grouped together as the 'Archiannelida'. In the opisthobranch gastropods, a phylogeny based on morphology suggested that the shell has been internalized seven times and lost five times, and broadly similar results (five instances of internalization and six instances of loss) were obtained by mapping these characters onto a molecular phylogeny (Wägele and Klussmann-Kolb 2005). Similarly, a reduction of the anterior appendages (chelifores, palps, and ovigers) evolved independently within each of the two major clades of pycnogonids (Arango 2002).

Wiens *et al.* (2005) have recently demonstrated how misleading morphological characters can be when reconstructing the phylogeny of a group extensively affected by paedomorphosis, as are the caudate amphibians. Paedomorphic salamander families branch as a single clade in the phylogeny of Wiens *et al.* (2005) based on morphological data, right because of multiple misleading effects of paedomorphosis. In addition to the occurrence of shared homoplastic larval traits in paedomorphic adults belonging to distinct lineages, the phylogenetic analysis is also negatively affected by the absence of clade-specific synapomorphies corresponding to traits that develop during metamorphosis in non-paedomorphic taxa as well as by the parallel adaptive changes associated with the aquatic habitat of the larvae.

1.2.3 Mixed origins

There are different ways to describe an animal's body in terms of its components (Minelli and Fusco 1995) and there is little hope that the units eventually recognized by dissecting it into conventional anatomical parts will correspond to parts with largely autonomous developmental identities, or with structural modules with a large degree of evolutionary independence due to their functional integration. As a consequence, it is very often inadequate to frame questions in terms of origin and evolution of an organ or an organ system, such as the brain or the circulatory apparatus. In a later chapter (see section 8.2.1) I will argue that the very notion of organogenesis seldom corresponds to actual phenomena.

As a rule, developmental and evolutionary origins of anatomically distinct body parts are in effect mixed, much more than the uniformity of the resulting pattern or the degree of integration of the product may eventually suggest. This applies, for example, to body segments, for two different reasons at least. First, it can be argued that segments are not archetypal building blocks out of which the body of, say, annelids or arthropods are made. More likely, segments are the result of the ontogenetically and phylogenetically secondary integration (Budd 2001, Minelli and Fusco 2004) of primarily independent serial features, comparable with those found in *Neopilina* (nephridia, foot-retractor muscles, transversal commissures connecting the main longitudinal nerves). Arguably, serially repeated body parts such as nephridia, nervous ganglia, and muscular masses of annelids belong to distinct developmental and evolutionary modules, despite the fact that their units are often integrated in the form of body segments (that is, with a pair of nephridia, a couple of nervous ganglia, etc., per segment). The same can be said of body rings, nervous ganglia, and appendages in arthropods. The second perspective from which a series of body segments can be described as of mixed origins is the likely distinct origin of the most anterior body segments with respect to those that follow along the main body axis. In arthropods, this corresponds to the contrast between the naupliar segments, forming the anterior part of the head, which are not patterned by the expression of the *Hox* genes, and the post-naupliar segments, which form the posterior part of the head and the whole of the trunk (thorax and abdomen) and are patterned by the staggered expression of the *Hox* genes (Minelli 2001).

The mixocoel, the body cavity present in arthropods, has mixed origins, having been traced partly to surviving spaces of blastocoelic origin, and partly to spaces newly opened within the mesodermal derivatives. The origins of an apparently simple and functionally well integrated system such as the tracheal system of insects are also mixed. Here, the finest and most internal branchings are represented by intracellular tracheoles which are structurally and developmentally independent from the larger tracheal pipes, which can be described as tubular invaginations of the epidermis, lined by a thin cuticle. The functional respiratory system of the insect involves connecting together these two components. In the circulatory system of the vertebrates, several distinct developmental modules 'cooperate' in giving rise to the heart (see section 8.2.1) and additional modules are involved in the production of major and minor vascular units.

1.3 Idola theatri

According to Sir Francis Bacon, the idols of the theatre are one of the four classes of prejudices that have biased philosophy for centuries, because the latter failed to adopt a suitable method with which to acquire knowledge. These prejudices, he said, derive from the persisting influence of renowned philosophers of the past, or from religion, and must be removed.

Some *idola theatri* still popular among biologists can exert pernicious influence on our understanding of evolution. It is sensible, therefore, to identify and try to remove them before we start exploring the evolutionary history of metazoans.

1.3.1 The ladder of nature

The first idol is the ladder of nature, or *scala naturae*. The works of the Swiss naturalist and philosopher Charles Bonnet (1720–1793) made this concept popular in the second half of the eighteenth century. Bonnet described nature as a long ladder whose steps lead from the inanimate matter to the simplest forms of living beings and further, up to the flowering plants, the corals, the molluscs, the insects, and finally the vertebrates, from fishes to mammals, with the human species closest to the heavenly angels, the summit of Creation. Today, 150 years on from Darwin's *Origin*, the metaphor of the unbranched ladder has been replaced by the metaphor of the branching tree, but the idea that some living forms are lower and others are higher is still with us. In part, this is due to the fact that a great many biologists do not have an adequate knowledge of the modern comparative method. In particular, they do not appreciate the difference between comparing two sister groups issued from the same node in the tree and comparing a given group with an indeterminately large paraphyletic assemblage together with which it would form a monophyletic group. For example, comparisons are all too often framed in terms of vertebrates versus invertebrates (or, at least, of vertebrates versus invertebrate chordates). This is nothing other than comparing animals before and after the acquisition of a given set of characters, be it the vertebral column or an advanced immune system. But saying 'before' and 'after' an evolutionary change perceived as a structural or functional improvement is tantamount to saying 'below' and 'above' it, along the ladder of nature.

Another reason why many biologists continue to qualify a group as lower or higher with respect to another group is the unwarranted identification of evolution with morphological complexification, or improvement in physiological performance. As a consequence, a mammal is higher than hydra, because it is histologically or anatomically much more complex than it; and a lizard is higher than a frog, because of its adaptations to the terrestrial environment. The obvious anthropocentrism of this perspective does not require further comment.

1.3.2 Morphoclines and Williston's law

Sometimes, this incorrect perspective takes a more subtle and, therefore, more dangerous form. Irrespective of history, it is often possible to arrange in a simple linear series the different states a character can take. For example, in malacostracan crustaceans a variable number of appendages of the pereion (the 'thoracic' region immediately posterior to the head) is often involved in the capture and manipulation of food and has thus a different shape from ordinary walking legs. The number of these specialized appendages (maxillipedes) varies between none and five pairs, according to the group. Thus, it would be easy to arrange them in a progressive series of forms (a morphocline) by aligning, in the order, malacostracans with none, one, two, three, four, and five pairs of maxillipedes. Taking for granted that the most primitive condition, for the whole of the Malacostraca, is the absence of maxillipedes, does it follow that evolution necessarily went through crustaceans with one, two, three, four, and five pairs of maxillipedes in that order? There is no *a priori* reason for taking this conclusion as granted. Eventually, obtaining a phylogeny that suggests otherwise may open

new vistas on the evolvability of this character and thus help in formulating sensible questions as to the way the leg and maxillipede identity are genetically controlled during embryonic development.

Appreciating that the equation simple= primitive=old is unwarranted may also help in selecting new model species for experimental research. Too much has been made of *Hydra*, often regarded as the most basal eumetazoan (Stotz *et al.* 2003), not to mention the colonial ascidian *Botryllus*, which has been recently presented as the ancestral chordate (Stotz *et al.* 2003).

A particular version of the idea that simple is primitive and complex is derived is embodied in the so-called Williston's law. This is the principle according to which the evolution of serial structures (e.g. body segments or vertebrae) starts as a series of many identical elements and proceeds towards a shorter series of increasingly diversified (specialized) elements. This idea is still with us, as shown for example by the recent statement by Bergström and Hou (2003) that arthropods with identical segments are obviously primitive, even when the individual segment may appear complicated. Phylogeny, however, often reveals the opposite. For example, among the centipedes (Chilopoda), the numerous (up to 191) and largely uniform leg-bearing segments of the wormlike geophilomorphs are a secondary feature, the primitive condition for centipedes being arguably a trunk with 15 leg-bearing segments, with some degree of heterogeneity in the size, or the morphological complexity, of the individual segments (Berto *et al.* 1997).

1.3.3 Finalism, adultocentrism, and division of labour

Another *idolum theatri* which is still with us is finalism. This problem is much more serious in developmental biology than in evolutionary biology, but a less than cautious framing of questions can easily lead to missing the opportunity to ask really important questions. To be sure, when Martindale *et al.* (2002) described the medusa as an 'invention' to facilitate dispersal, they did not credit early-Palaeozoic (or late-Proterozoic) polyps with skills worthy of a human designer. Their sentence can be easily read to mean that medusae proved soon to be selectively advantageous and were retained by what eventually evolved into a large and diverse clade. But it is also true that focusing on the selective advantage of a final product does not help us to understand the circumstances under which it first evolved, or the developmental context in which it still takes form today. Still more than in evolutionary biology, latent finalism is arguably more dangerous in developmental biology, where I expect that not too many people would object to the idea that time-measuring devices must exist in the embryo to ensure that developmental processes operate at the right time (Pourquié 2003) or, similarly, that stem cells and early progenitors are deposited in specific sites to ensure organ development throughout embryonic, fetal, and adult life (Thisse and Zon 2002).

This persisting finalism is probably the consequence of the adultocentric perspective from which development (and evolution) are considered. Why should it be otherwise? The adult is, by definition, the reproductive stage and Darwinian fitness is measured in terms of an animal's contribution to the gene pool in the next generation. This requires reproduction, thus the measure of Darwinian fitness is based on the performances of the adult. This seems to justify seeing embryos and juveniles simply as preparatory stages eventually leading to the all-important reproductive stage. But the fact that a gastrula or a trochophore eventually leads to an adult 'polychaete' does not mean that the embryonic or larval stage are there as a means to produce the adult. The gastrula is there because of what the cells in the blastula could do, and the trochophore is there because of what could happen, and actually happened, in the gastrula. To be sure, all these developmental stages have been moulded by evolution and the temporal continuity of the lineage through the millennia has been ensured by a sequence of adults involved in reproduction, but this does not justify adopting a biased adultocentric perspective.

Cognate to finalism is the fashionable concept of the division of labour. For example, Bonner (2003) described the differentiation of stalk cells and spores in cellular slime moulds (*Dictyostelium*) as the result of selection for a division of labour. Similarly, according to Ruiz-Trillo *et al.* (2007), the advent of multicellularity 'created new challenges, including the need for cooperation and communication between cells, and the division of labour among different cell types' (p. 113). I wonder whether this metaphor can help to formulate scientific questions better than we would do by adopting a much more sober language, by speaking, for example, of cells, or cell groups, or body parts, that become involved in local dynamics other than those to which neighbouring cells, cell groups, or body parts participate. This can be framed, for example, in terms of modules, as has recently become fashionable in developmental biology and evolutionary developmental biology (e.g. Schlosser 2002, Schlosser and Wagner 2004). Alternatively, we can use the language of thermodynamics, by speaking of local attractors.

CHAPTER 2

Animal phylogenetics

'Mollusc' is a zoological term, whereas 'mollusc-like' is a paleontological term.

Yochelson (1979), p. 324

In today's practice, to do research in phylogenetics means to fill a matrix with data (molecular more often than morphological) and then to analyse the matrix with the aid of suitable software or, better, with a set of different software packages. The result will be a tree, or a set of trees, representing a hypothesis of relationships among the taxa included in the analysis.

To younger researchers it would be difficult to imagine that over the last few decades progress in this field has not been simply one of tools but also, and perhaps more fundamentally, one of concepts. Phylogenetics has immensely benefited from progress in molecular biology, computing, and microscopy, but unique to this field, and basic to the collection and analysis of any data, is a core of concepts first spelled out explicitly by Willi Hennig in the not too distant past. Fifteen years after the publication of his first book on phylogenetic systematics (Hennig 1950), which went virtually unnoticed, partly because it was published in German, the publication of an American version (Hennig 1966), preceded by an article in the *Annual Review of Entomology* (Hennig 1965), marked the beginning of the modern era in reconstructing phylogenetic relationships.

To be sure, the acceptance of phylogenetic systematics was not immediate and general. For a while, systematists were roughly divided into three schools: the phylogenetic or cladistic school, which was based on Hennig's principles; the phenetic school, the manifesto of which were the books by Sokal and Sneath (1963) and

Sneath and Sokal (1973); and the evolutionary school, with foundations that can be traced in the works of Mayr (e.g. Mayr *et al.* 1953, Mayr 1969, Mayr and Ashlock 1991) and Simpson (e.g. Simpson 1945, 1961). A main point of divergence between the phylogenetic and the evolutionary school was the acceptance by the latter of paraphyletic taxa, such as the Reptilia (for the Recent forms of reptiles to the exclusion of the Aves), whereas the phenetic school rejected phylogenetic reconstructions as dependent on the subjective evaluation of homologies and developed instead numerical methods aimed at generating a robust, repeatable grouping of taxa. Eventually, numerical methods became the main working tools of cladists, while years of debate between schools and currents filled the pages of journals and books. A large part of this history was presented and discussed by Hull (1988) in a well-documented book.

2.1 The Hennigian foundations of phylogenetics

Hennig revisited the traditional concept of homology by introducing the fundamental distinction between plesiomorphy and apomorphy; that is, the respective primitive and derived states of a character involved in an evolutionary transformation. Following this conceptual distinction, Hennig demonstrated that to reconstruct phylogeny we must identify the synapomorphies; that is, the shared derived character states which identify the nodes in the tree or, equivalently, the clades or monophyletic taxa. A clade can indeed be defined as the lineage uniquely sharing one or more synapomorphies, or as an ancestor plus all of its descendants.

A significant part of the subsequent history of systematic theory has been occupied by discussion about how to reconstruct character polarity; that is, how to determine which of two character states, A and B, is the plesiomorphic and which is the apomorphic one. Another problem is what to do in the case—which is the rule rather than the exception, especially in analyses extended over many taxa and many characters—when different characters suggest different trees. To Hennig, who was an entomologist, conflicting characters meant, in particular, larval or adult, but conflicting characters do not necessarily belong to different temporal or spatial partitions. Indeed, conflict was to emerge in the late 1970s and early 1980s as soon as researchers begun comparing trees based on morphological characters with trees based on molecular characters. A hot debate (e.g. Patterson 1987) erupted as to the relative merits and drawbacks of these two kinds of characters used in a phylogenetic analysis.

Another source of conflict was due to the fact that by applying Hennig's phylogenetic method we obtain trees rather than classifications. Thus, how to translate the branching topology of a tree into a nested set of taxa? One aspect of the problem had an immediate solution, at least to the followers of the Hennigian method: all taxa must be clades; that is, monophyletic groups. Other aspects were less obvious. First, shall we name all clades in a tree? Second, how many ranks do we need, in addition to traditional ones such as genus, family, order, class, and phylum? The answer to the first question can be dictated by reasons of opportunity. One the one hand, there is little gain in naming all nodes in a tree, as long as the robustness of the latter is still debatable. That is, one should only name robustly corroborated nodes. On the other hand, to refer to a clade by listing its apomorphies or its included subclades is much less convenient than referring to it by a name. So, how many nodes in a tree deserve to be named will arguably remain a matter of subjective choice.

To the contrary, the question of ranks can be settled in general and definitive terms. There was no need to wait for the origin of phylogenetic

systematics to discover the arbitrary nature of ranks in biological classifications, but this became dramatically clear as soon as researchers began comparing the topology of phylogenetic trees with the nested patterns of taxa in a Linnaean classification. The most sensible use of ranks could arguably be in acknowledging identical rank to the two lineages splitting off the same node, irrespective of the diversity (species number) or disparity (range of forms) in either lineage. But this point is of minor importance. The main difficulty resides in determining whether two clades in distant parts of the phylogenetic tree deserve be acknowledged as deserving the same rank, or not. This is different from assigning a lesser clade with a different rank from a larger clade within which the first is included. The problem is that these two ranks cannot be quietly 'exported' to another set of taxa.

There is arguably some practical advantage in continuing to treat the Mammalia as a class and the Chiroptera as an order, but this perpetuates a common source of error. Are there reasons to treat both the Mammalia and the Amphibia as classes, or the Chiroptera and the Lepidoptera as orders? Or, what makes the Canidae, the Crocodylidae, and the Scarabaeidae, not to mention the Rosaceae and the Ranunculaceae, so accurately comparable that all deserve to be ranked as families? Nothing but tradition is responsible for assigning the same rank to two distantly related groups. This is not the place for in-depth discussion of a question that has long been settled by phylogenetic systematics (e.g. Minelli 2000c), but the point is of consequence whenever we regard two taxa classified at equal rank as directly comparable.

Formal classificatory ranks have disappeared from a few modern zoology textbooks and reference works (e.g. Mickoleit 2004, Westheide and Rieger 2007), but are still widely in use. Unfortunately, lots of calculations are frequently offered, based on the assumed equivalence of taxa classified at the same rank. For example, how many families of bivalves did originate in the Triassic? How many genera of ammonites went extinct during the last 5 million years of the Cretaceous? This kind of computation is very common in palaeontology and a similar way to measure diversity by counting families or other supraspecific taxa is also adopted in ecology and biogeography, although perhaps less frequently. But all this amounts to counting fruit by adding apples and cherries.

Eventually, it may not to be inconvenient to refer to the Gastropoda, the Carnivora, or the Hominidae even without assigning any of these taxa a specific rank, but a long-lasting resistance can be expected if we want to dispose of the rank of phylum. Historically, the phyla of current classifications can be traced back to Cuvier (1816)'s *embranchements*, the four major divisions he recognized in the animal kingdom, each of which corresponded to a body organization radically different from that of the animals referred to the other divisions. In a sense, the members of an *embranchement* (that is, the members of a phylum, in modern parlance) are the species among which organs can be extensively compared in terms of homology, something much more questionable, and often impossible to do, between members of two different *embranchements*, or phyla. This was reasonable in Cuvier's times, before the advent of an evolutionary view of life, but it does not make much sense today. Alternative options are practicable. Were it not for their tadpole larva, would we still classify ascidians as members of the same phylum as the vertebrates? Why not to raise (or, better, maintain) the Vertebrata to the phylum level and regard the Chordata (Vertebrata+Urochordata+Cephalochordata) as a supraphyletic assemblage, as we do, for example, when grouping Annelida and Mollusca as Neotrochozoa? The arbitrary character of any alternative choice should not require additional explanations.

Extending comparisons between groups separated by major morphological gaps is becoming more and more feasible, not simply by eschewing morphology and turning attention to molecular evidence, but also through comparative developmental genetics, which has been revealing unexpected developmental commonalities between morphologically disparate metazoans. This does not mean this will soon definitively solve our search for the phylogenetic relationships of all major metazoan lineages. It means, however, that there is no reason, other than tradition, to single out a number of lineages as 'the metazoan phyla' or, what is basically equivalent, in acknowledging the existence of a well-defined number of alternative animal *Baupläne*, or body plans. As clearly expressed by Scholtz (2004b), phyla and body plans cannot form the basis for evolutionary considerations, contrary to the view still defended by other researchers (e.g. Hughes and Kaufman 2002b, Valentine 2004, Erwin 2006).

2.2 Stem group and crown group

Forty years ago, when still a student, I was amazed to read, in a figure legend in a popular book on evolution, that birds originated in Bavaria. At first, I hypothesized that the author might have conceived that idea in Munich during a lively Oktoberfest, but soon I developed a less humorous explanation. The author must have reasoned as follows: *Archaeopteryx* is the oldest and most primitive of birds; *Archaeopteryx* lived in Bavaria; therefore, the birds originated in Bavaria. There is no need to discuss here the improbable stringency of the final inference, but a part of the initial proposition, that is, that *Archaeopteryx* is the most primitive of birds, deserves be discussed. Why do we classify *Archaeopteryx* with birds, despite the fact that it possessed teeth and a typical reptilian tail? Admittedly, to include this Mesozoic fossil in the Aves we must revise and, in a sense, relax the diagnosis for this taxon, if this was originally formulated in respect to the Recent forms only. This might seem a sensible solution, but this is not true. In the case of groups with an abundant fossil record spanning over an adequate time range, there is the risk of embarking in a never-ending sequence of reformulations of a taxon's diagnosis and, worse, of eventually facing the disappearance of the divide between a less inclusive taxon, like birds, and a more inclusive taxon of which the former is part; for example, the Archosauria (crocodiles plus birds) or the Sauropsida ('reptiles' plus birds). A similar vanishing of boundaries was about to happen many years ago along with the discovery of a series of late-Palaeozoic to early-Mesozoic fossils which share with mammals a more or less extended series of derived characters, but not enough to justify placing them in the Mammalia, as the group can be diagnosed based on its Recent representatives. In terms of phylogeny, all these fossil vertebrates are more closely related to the Mammalia than to any other group, but none of them is phylogenetically closer to a living mammal than are two maximally distantly related living mammals such as, for example, the platypus and the blue whale.

A rational way to accommodate in the classification those 'mammal-like reptiles' (and their equivalents in other parts of the tree of life) was clearly needed and was eventually found in distinguishing crown-group and stem-group members of a phylogenetic lineage, or clade. The crown includes all the descendants, both living and extinct, of the last common ancestor of all Recent members of a clade. It is thus diagnosed by the shared derived traits, or synapomorphies, which identify its basal node in the tree, and includes all living representatives of the clade, plus those extinct forms which branched off from the tree crownwards of the clade-defining node. For example, mammoths and giant sloths are crown-group mammals, as they share a common ancestor with living elephants and living edentates, respectively, more recent than the last common ancestor that the elephants, or the edentates, share with the platypuses or the kangaroos. The crown group closest to the Mammalia is arguably the Sauropsida (the 'reptiles' inclusive of birds). Together, Mammalia and Sauropsida form the crown-group Amniota, to which the extinct 'mammal-like reptiles' certainly belong. But where within the Amniota do they actually belong? Due to the abundant and diverse fossil record they have left, it would be inconvenient to either modify the diagnosis of the Mammalia to include them or to establish a thick series of progressively inclusive new taxa, hierarchically intermediate between the Mammalia and the Amniota, to include, for example, Mammalia+A, then (Mammalia+A)+B, etc., where A, B, etc., are 'mammal-like reptiles' sharing with Recent mammals a decreasing number of apomorphies.

On the other hand, all these extinct forms are phylogenetically closer to the Mammalia than to the Sauropsida. That is, next to the split between the two main amniote lineages, they all belong to the clade of which the Recent Mammalia identify the crown. But those fossils do not belong to the crown. Instead, they are collectively considered as stem-group Mammalia; that is, as forms belonging to the clade whose only extant representatives are the Mammalia, but which branched off the tree more basally than at the node which identifies the crown group. Similarly, if we want to call *Archaeopteryx* a bird, we must add that it belongs to the stem group of Aves, not to their crown group. In the following, we will need sometimes to use this distinction between stem and crown members of a clade, for example when discussing the diverse Cambrian representatives of the arthropodan clade.

2.3 Characters

The long introduction to Adanson's (1763) *Familles des Plantes* contains the oldest analysis of the criteria to be followed in selecting characters to produce a good classification. Eventually, Adanson concluded that there is no single best character for distributing the plants into families; as a consequence, the best taxonomic arrangement is the one based on the most frequently recurrent patterns of grouping emerging from separate comparisons based on a series of different characters. In the essence, this is the same principle adopted 200 years later by Sokal and Sneath (1963) when they launched the programme of phenetic systematics and begun developing numerical methods to accomplish the relevant comparisons in the most effective way. In the long run, however, even among the pheneticists emerged criticisms as to the opportunity to grant all characters equal importance, and new calculation procedures were developed to modulate the weight of individual characters to the way their alternative states are distributed among the sampled taxa.

Collecting fresh data is generally expensive and time-consuming. This applies to both morphological and molecular evidence. Thus, such an effort would be probably more profitable if researchers would make their criteria for character selection explicit, something that does not occur too frequently (Poe and Wiens 2000). Explicit, however, has been the repeated rejection of morphological characters as much less reliable, or at least much less informative, than molecular data (e.g. Scotland *et al.* 2003), but this position has been firmly rejected based on both empirical and conceptual reasons (e.g. Jenner 2004a, 2006). The interested reader is referred to the original literature. Nevertheless, before closing these introductory chapters it seems sensible to summarize some aspects of molecular data that are relevant to understanding their use in phylogenetic reconstruction.

2.4 Phylogenetic analysis of molecular data

2.4.1 Star phylogenies

Homoplasy is not restricted to morphological characters, but can equally plague phylogenetic reconstructions based on molecular evidence. A small amount of homoplasy in nucleotide substitutions, between 2 and 5%, has been predicted to occur by chance (Zhang and Kumar 1997), and simulations (Takezaki *et al.* 2004) have confirmed this. But this is simply the general background level, and above this many different causes can add further amounts of homoplasy. This may cause phylogenetic inference from sequence data not to achieve a satisfactory resolution, even in the presence of abundant evidence.

Phylogenetic inference is particularly difficult when basal internodes are closely spaced in time (Jermiin *et al.* 2004) and very remote from present, to the point that some theoretical calculations (Mossel and Steel 2004, 2005) and some empirical analyses (Ho and Jermiin 2004, Martin *et al.* 2005) suggest that by using DNA sequence data it may be intrinsically impossible to resolve the order of branching in the tree. In this case, a precious phylogenetic signal is eventually found in rare, homoplasy-free genetic events, if any (Rokas *et al.* 2005, Boore 2006, Rokas and Carroll 2006).

Rokas *et al.* (2005) were unable to resolve most nodes of the metazoan tree despite the use of 50 genes (more than 12 000 amino acid positions) from 17 animal species and concluded that the lack of resolution is a positive signal of closely spaced cladogenetic events. Takezaki *et al.* (2004) analysed sequences of 44 nuclear genes at 10 404 positions (out of which ≈300 were informative enough for the parsimony-based analysis to be performed) to resolve the origin of tetrapods, but this large set of sequences did not support conclusively any of the three hypotheses (coelacanth, lungfish, or neither being the sister group of tetrapods). According to Takezaki *et al.* (2004) the coelacanth, lungfish, and tetrapod lineages may have diverged within such a short time interval that the problem is perhaps untreatable. Commenting on this result, Rokas and Carroll (2006) observed that under similar circumstances the main result of the analysis is exactly the appreciation that three strikingly different groups are indeed very strictly related and diverged from their common ancestor within a very short time span; that is, in cladistic jargon, that their relationships reduce, from our perspective, to a 'star phylogeny', with more than two diverging lineages branching off, in practice, from the same node.

2.4.2 Gene and genome duplications

Molecular phylogeneticists have long realized that gene phylogenies are not necessarily identical to species phylogenies. The main point at issue is gene duplication, by effect of which two or multiple (paralogous) copies of the same gene may be present in one genome. If the divergence between two lineages is more recent than the duplication that produced the two paralogous genes, incautious sampling of gene sequences can suggest that the lineages diverged at the time of the gene divergence, rather than at the time of the phyletic split. Gene duplications are all but rare and are sometimes followed by divergent adaptive evolution of the duplicates: the two proteins coded by them may either be specialized in alternative functions within the same broad function of their common ancestor (subfunctionalization), or diverge more extensively with the subsequent involvement of one of the duplicates in a completely new function (neofunctionalization; see Table 2.1). However, functional innovation is not the only reason why multiple gene copies may be retained following duplication. In the ciliate *Paramecium tetraurelia*, most of the nearly 40000 genes in the diploid micronucleus are the product of at least three successive whole-genome duplications. Interestingly, genes involved in the same metabolic pathway have common patterns of gene loss, and there is an excess of highly expressed genes in the set of the multiple-copy genes retained after all duplications. This suggests that many duplicated genes are maintained because of gene-dosage constraints, not because of functional innovations (Aury *et al.* 2006).

Whole-genome duplications have been documented in several major clades of life, including yeasts (Kellis *et al.* 2004) and angiosperms (De Bodt *et al.* 2005). Within metazoans, this phenomenon has been best documented in vertebrates (e.g. Spring 2003, Furlong and Holland 2004). A first duplication did apparently happen at the origin of this lineage (but the precise positioning of this event on the chordate tree is possibly beyond the limits of resolution; Horton *et al.* 2003), followed by a second duplication at the base of the gnathostomes. An additional duplication characterizes the teleosts (e.g. Van de Peer *et al.* 2003, Crow *et al.* 2006). It may be tempting to relate the complexity of vertebrate structure and their eventual evolutionary success to these two or three events of genome duplication, but a careful study of the available evidence, including fossils, provides no support for this hypothesis (Donoghue and Purnell 2005).

2.4.3 Horizontal gene transfer

Horizontal gene transfer was probably common between the genomes of ancient organisms, but in modern metazoans it is likely to be of marginal importance (Kurland *et al.* 2003). Nevertheless, it is still possible. An explanation in terms of horizontal gene transfer, perhaps from a fungus, has been recently made by Rot *et al.* (2006)

Table 2.1 Gene duplication and divergence in the early evolution of vertebrates.

Gene family	Number of gene copies		Divergence of gene copies
	Amphioxus	Vertebrates	
Dll	1 or 2	6	N
En	1	2	N
Hh	1	3	N
HNF-3	1	3	?
Hox	1	4	N
Islet	1	2	N
Mnx	1	2	?
Otx	1	2/3	N
Pax1/9	1	2	N
Pax3/7	1	2	N
Pitx	1	3	S+N
Snail	1	3	N
Sox1/2/3	1	3	N
Wnt1	1	1	?
Wnt3	1	2	?
Wnt5	1	1	?
Wnt7	1	2	?
Wnt8	1	2	S

N, neofunctionalization; S, subfunctionalization.

Main data source: Mazet and Shimeld (2002).

to account for the presence, in the mitochondrial *cox1* gene of the sponge *Tetilla* sp., of an intron similar to the cnidarian mitochondrial introns but without any corresponding example in other sponge orders.

2.4.4 Gene loss and genome compactation

Along the history of a lineage, the eventual fate of a gene is not necessarily its duplication or its modification but also, frequently, its loss. This may cause major difficulties in phylogenetic reconstructions, especially in the case of massive gene loss. The phenomenon affects all major gene families, although some of these are, in a sense, more resistant to gene loss; for example, the *Hox* genes and the collagen genes. But the small number of *Hox* genes in nematodes is likely a secondary feature and examples of lost collagen genes have been found in both insects and nematodes (Spring 2003).

From their analysis of 2301 homeodomain sequences from 11 bilaterian species, Nam and Nei (2005) estimated at least about 88 homeobox genes in the common ancestor of Bilateria. Of these, about 50–60 genes have left at least one descendant gene in all species in the sample, whereas about 30–40 genes were lost in some lineages.

In several groups, particularly small, compact genomes have evolved. This is often associated with very small adult size, but this correlation between body size and genome size is very loose.

According to Gregory (2008), the smallest animal genome known to date belongs to the root-knot nematode *Meloidogyne graminicola*, with a *C* value (the DNA content of a haploid nucleus) of 0.03 pg, less than 1/40000th of the largest animal genome known thus far, that of the marbled lungfish *Protopterus aethiopicus* (132.83 pg). Small genomes are common among nematodes; for example, *C* values between 0.08 and 0.13 pg have been reported in different studies for *Caenorhabditis elegans*. The genomes of several planktonic crustaceans are also very small, like the copepod *Cyclops kolensis* (*C* value 0.14 pg; Grishanin *et al.* 1996) and the cladoceran *Scapholeberis kingii* (*C* value 0.16 pg; Beaton 1988), small-size rhabditophorans like *Stenostomum brevipharyngeum* and *Olisthanella truncula* (*C* value 0.06 and 0.12 pg, respectively; Gregory *et al.* 2000), the tiny annelid *Dinophilus gyrociliatus* (*C* value 0.06 pg; Soldi *et al.* 1994), and the two-spotted spider mite *Tetranychus urticae* (*C* value 0.08 pg; T.R. Gregory in Animal Genome Size Database). One might expect small genomes to be universal among mites, but this is not the case: the Southern cattle tick *Boophilus microplus*—admittedly, a representative of ticks, a clade whose members have a body size far exceeding the average for mites—has a *C* value of 7.5 pg (Ullmann *et al.* 2005) and thus has the largest arachnid genome, larger than the genome of spiders of comparable or even larger size.

In insects, the smallest genome has been recorded in the males of the strepsipteran *Caenocholax fenyesi texensis* (0.105 pg); somewhat bigger is the genome of another strepsipteran, *Xenos vesparum* (male, 125 pg; female, 130 pg). Of similar magnitude is the genome of the Hessian

fly *Mayetiola destructor* (Diptera, Cecidomyiidae) (male, 116 pg; female, 148 pg). It must be noted, however, that endoreduplication is extensive, especially in the females of *C. fenyesi texensis*, with 4, 8, and 16 *C* nuclei (Johnston *et al.* 2004).

Whereas a correlation between genome size and body size is hardly recognizable in insects, a 'signature' of evolutionary relevance is possibly found in the fact that holometabolans, as a whole, have quite small genomes, compared with non-holometabolan insects. It may be worth investigating whether this small genome size, whatever its original cause might have been, may have helped the frequent evolution of life cycles with very short development times, as in *Drosophila* and in many other dipterans, and, perhaps, of endoparasitic habits, through a reduced cell size and/or body size of the infective stages.

Within chordates, small genomes are common in tunicates, with, for example, a *C* value of 0.16 pg in *Ascidia atra* (Mirski and Ris 1951) and 0.20 pg in *Ciona intestinalis* (Atkin and Ohno 1967), and the smallest chordate genome studied to date belongs to the minuscule appendicularian *Oikopleura dioica* (*C* value 0.07 pg; Seo *et al.* 2001). In this case, reduction in size is associated with a very compact genome structure, an effect of which is the disruption of the *Hox* gene cluster (Seo *et al.* 2004) with the consequence that the usual spatiotemporal collinearity of its expression (see section 7.5.2) is also lost.

One more piece of evidence for the strong evolvability of genome size is the fact that among the vertebrates the smallest genome known to date belongs to a bony fish (*Tetraodon nigroviridis*), despite the fact that in the bony-fish lineage an extra whole-genome duplication occurred, in addition to the two events postulated at the origin of vertebrates (Jaillon *et al.* 2004).

2.4.5 Mitochondrial genomes

Mitochondrial genomes have provided a substantial fraction of the molecular data hitherto used in phylogenetic reconstructions. Most animal mitochondrial genomes are of very similar size (for an overview of exceptions see Raimond *et al.* 1999) and contain an equivalent set of genes and other sequences. As a consequence, the alignment of sequences from different species, even of distant ones, is generally easy. The information thus retrieved is not simply one of nucleotide sequences, but also one of gene order. It was indeed the identification of a common change in gene order in the mitochondrial genomes of insects and crustaceans the first convincing evidence suggesting that these arthropods form a clade to the exclusion of the myriapods, to which at the time the insects were regarded to be more strictly related (Boore *et al.* 1998).

However, nontrivial differences in gene order have been repeatedly found among strictly related species, for example in the lancelet genus *Epigonichthys*, where *E. maldivensis* has the same mitochondrial gene order as in two species of *Branchiostoma*, wheres *E. lucayanus* is quite different (Nohara *et al.* 2005); other examples have been recorded in insects (e.g. Dowton and Austin 1999, Shao *et al.* 2001), ticks, birds etc.

Another reason why much care is required in dealing with mitochondrial sequences is the possible dependence of their substitution rate on taxon-specific conditions such as body size and lifestyle: in a sample of arthropod species, Hassanin (2006) found that substitutions in six protein-coding genes was higher in species of diminutive size or with parasitic habits.

A less expected dimension of molecular evolution in animal mitochondrial genome is the occasional divergence from the standard genetic code. For example, several arthropods translate the codon AGG as lysine instead of serine (as in the standard mitochondrial genetic code) or arginine (as in the standard nuclear genetic code) (Abascal *et al.* 2006).

2.4.6 The molecular clock

Next to identifying the sister group of a lineage, a question we may address is the absolute age of its splitting from it; see Table 2.2. Our sources of information are the fossil record and the molecular clock. Both of them have their limits and are best integrated to provide more reliable estimates.

Table 2.2 The molecular clock: some estimates of divergence time.

Age (millions of years)	Node	Reference
251–243	Birds/crocodiles	1
257–252	Birds/lizards	1
310	Living reptiles/mammals	2
360	Amniotes/*Xenopus*	2
419–408	Lungfishes/terapods	2
450	Tetrapods/zebrafish, fugu	1
535	Echinoderms/hemichordates	3
543	Anthozoans/hydrozoans	3
545	Molluscs/annelids	3
546	Arthropods/priapulids	3
549	(Molluscs+annelids)/nemerteans	3
564	Gnathostomes/lampreys	2
579	Protostomes/deuterostomes	3
604	Diploblasts/triploblasts	3
634	Eumetazoans/calcareous sponges	3
664	(Eumetazoans+calcareous sponges)/demosponges	3
751	Vertebrates/Amphioxus	2
993	Vertebrates/*Drosophila*	2
1177	Vertebrates/*Caenorhabditis*	2
1576	Animals/plants/fungi	2

Sources of data: (1) Müller and Reisz (2005); (2) Hedges and Kumar (2003); (3) Peterson and Butterfield (2005).

Donoghue and Benton (2007) listed the following sources of error in estimating the absolute time of origin of a clade from palaeontological evidence: erroneous branching pattern of the reconstructed phylogenetic tree, inadequate sampling of the fossil record, erroneous identification of fossils, wrong absolute dating of rocks, and errors in correlating fossil beds. On the other hand, the molecular clock is far from ticking with the regularity originally suggested by Zuckerkandl and Pauling (1965). Gillooly *et al.* (2005) suggested that the ticking of the molecular clock is not to be expected per unit of time, but in terms of constant substitution rate per unit of mass-specific metabolic energy. Among the evidence they use to support their hypothesis are the higher rate of origination of higher taxa in the tropics (Jablonski 1993) and the faster evolution and higher diversity of smaller rather than larger organisms (Brown *et al.* 1993). Contradictory results have been reported recently. On one hand, a comparison including more than 300 metazoan species has failed to recover a dependence of the molecular clock on metabolic rates (Lanfear *et al.* 2007). On the other hand, a significant slow-down of mutations in mitochondrial genes has been found to correlate with increasing average body size (Fontanillas *et al.* 2007), something that probably does not hold for the invertebrates (Thomas *et al.* 2006).

Thus, it seems fair to say that we must be careful with the use of the molecular clock (Pulquério and Nichols 2007) and that this must be calibrated carefully (Hedges and Kumar 2004). To this effect the fossil record, despite all its gaps and problems of interpretation, can still offer valuable help (Donoghue and Benton 2007).

2.4.7 Controversial results

In a recent survey of the contribution provided by the study of mitochondrial gene arrangements to the resolution of phylogenetic controversies, Boore (2006) listed the following:

• the identification of 'true' plathelminths (von Nickisch-Rosenegk *et al.* 2001) and phoronids (Helfenbein and Boore 2004) as belonging to the Lophotrochozoa;
• the recognition that sipunculans are more closely related to the annelids than to the molluscs (Boore and Straton 2002);
• the rejection of the putative close arthropodan affinities of annelids (Boore and Brown 2000);
• the identification of the close affinities between crustaceans and hexapods to the exclusion of myriapods (and onychophorans) (Boore *et al.* 1995, 1998, Lavrov *et al.* 2004);

• the confirmation of the crustacean nature of the Pentastomida (Lavrov *et al.* 2004).

On the other hand, despite the major efforts produced over the last quarter of century and the availability of an already impressive amount of data, mostly represented by nucleotide or amino acid sequences, we are still very far from the production of a robust and satisfactorily resolved topology of the phylogenetic tree. I have mentioned in the Preface a very recent paper (Dunn *et al.* 2008) suggesting a picture of animal relationships sensibly different from what already seemed to be an accepted, and acceptable, consensus view. Closing up onto the tree, two papers published last year (Rousset *et al.* 2007, Struck *et al.* 2007), both of them based on a very large taxon sampling, have offered two views of annelid phylogeny with very few points in common. But this is just the tip of a huge iceberg. Browsing through the recent literature, we read that a supermatrix including a fragment of the 28S and complete sequences for 18S rDNA, H3 histone, elongation factor-1α, cytochrome oxidases I (COI) and II (COII), and 12 and 16S plus the intervening tRNA, and 170 morphological characters fails to resolve relationships between non-holometabolous neopteran orders (Kjer *et al.* 2006); that mitochondrial genome data alone are not enough to unambiguously resolve the relationships of Entognatha, Insecta, and Crustacea *sensu lato* (Cameron *et al.* 2004); that the position of Strepsiptera is still unresolved (Beutel and Pohl 2006); and that the analysis of mitochondrial genomes has failed to determine the position of the Hymenoptera within the Holometabola (Castro and Dowton 2007). To be sure, there is still a lot of work ahead.

CHAPTER 3

Metazoans enter the stage

The hidden harmony is better than the obvious.

Heraclitus, fragment 54

3.1 Multicellularity

Metazoans are multicellular organisms with gametic meiosis. That is, in metazoans the unicellular haploid phase lasts only one cell generation, being represented by gametes which eventually fuse into a diploid zygote. Gametes are strongly dimorphic, with a female gamete (the egg) being the only surviving cell following meiosis, while all four haploid cells deriving from spermatogenesis differentiate into mature sperm cells (several zoologists are inclined to forget that the products of meiosis have other nature and significance in the life cycles of other organisms, e.g. flowering plants, foraminiferans, or ciliates). The metazoan male gamete has, as a rule, a typical morphology, with an acrosome, a head, peculiar mitochondrial derivatives, and a cilium functioning as a locomotory tail.

Multicellularity is one of the most conspicuous features of animal organization, but is not exclusive to them. Multicellularity has evolved many times along the history of eukaryotes and, apparently, even in geologically recent times. For example, starting from unicellular ancestors the green algae have recently given rise to the multicellular *Volvox* lineage within the last 75 million years (Larson *et al.* 1992).

Arguably, multicellularity evolves easily, provided that cell-adhesion molecules are available, but the eventual long-term stability of a multicellular organism depends on the balance between the benefits eventually obtained by the individual cells being part of a multicellular assemblage and the decrease in individual fitness associated with the same condition. In fact, cooperation among cells will easily benefit the group, but can be costly to the individual cooperating cells. Therefore, we can argue that the stability of multicellular assemblages has been raised by the evolution of genetic modifiers that enhance cooperation among the members of a cell group. In other terms, multicellular organisms are those able to regulate conflicts among the cells of which they are formed (Michod and Roze 2001). But competition among the members of a multicellular assemblage has not been cancelled by a billion years of life together.

As we will see repeatedly in this book, the persistence of cell-to-cell competition helps to explain several key features of metazoan organization, including the origin of tissues, sexuality, cuticles, and others. Common to all these evolutionary events is cell–cell signalling, which can be regarded as a mechanism that allows one cell to gain control of the intracellular signalling of another cell (Blackstone 1998). Terminal differentiation of cells can thus be interpreted as a result of metabolic control under the influence of successful neighbouring cells. In this way, the cells higher in the control hierarchy would have obtained a benefit in the form of increasing availability of resources and increasing replication rate. Above a given size, or number of component cells, a result of this intercellular competition and control would have been the origination of common spaces through which resources can circulate and thus be shared by a multiplicity of cells, in a condition close to equilibrium. These common spaces may correspond to those of simple gastral cavities, but also to those of simple circulatory systems. Eventually, the evolution of control relationships among the cells in a multicellular mass would have brought to the origin of a split between permanently differentiated cells and multipotent cells retaining the ability to divide.

Selection will easily favour cells or cell lineages producing mutants able to manipulate environmental conditions, for example by opening within a multicellular assemblage new channels through which materials can circulate (a kind of primitive angiogenesis). In this model (Blackstone 1998), developmental mechanisms evolve because of the immediate metabolic advantage they may produce, not because of any morphogenetic effect eventually deriving by the operation of the same mechanisms in future generations, something we all too often assume when reconstructing the 'origins' of some feature in modern animals.

Control of morphogenesis through metabolism is arguably much less prominent in modern metazoans than it might have been in the earliest multicellular organisms (Newman 2005), but the regulative role of metabolic gradients is still detectable in processes such as vertebrate angiogenesis and, more extensively, in basal metazoans. Genetic control of morphogenesis has became increasingly prominent during metazoan evolution, especially since the origin of the Bilateria (Davidson 2003, 2006; see also Davidson *et al.* 1995), but, of course, the roots of multicellularity cannot be found in the complexity of differentiation and patterning mechanisms that evolved much later on.

3.2 Competition and synergy, or molecules and organelles

Molecules that mediate cell adhesion are older than the advent of multicellularity. Schneider *et al.* (2003) suggested that in the last common ancestor of metazoans a single β-catenin gene may have fulfilled both adhesion and signalling functions, subsequently performed by diverging molecules deriving from a duplication of the ancestral gene. The multiple convergent acquisition of multicellularity also suggests that

several proteins that regulate transcription and eventually have a role in the body patterning of multicellular organisms such as the metazoans were already present in the common ancestor of eukaryotes (Derelle *et al.* 2007).

The dynamic equilibrium (or, better, the unceasing competition) among the cells in a multicellular assemblage is mediated by cell signalling. This cascade of molecular interactions begins with the binding of a ligand to a transmembrane receptor; this leads, in turn, to the modification of cytoplasmic transducers which activate transcription factors that ultimately alter gene expression (Pires-daSilva and Sommer 2003). The most important signalling pathways found in metazoans are the following: Hedgehog (Hh) (Ingham and McMahon 2001), wingless-related (Wnt) (Cadigan and Nusse 1997, Moon *et al.* 2002), transforming growth factor β (TGF-β) (Massagué and Chen 2000), receptor tyrosine kinase (RTK) (Robinson *et al.* 2000), Notch (Mumm and Kopan 2000), and Janus kinase (JAK)/signal transducer and activator of transcription (STAT) (Castelli-Gair Hombría and Brown 2002).

This complex machinery did not develop following the onset of multicellularity. Iyer *et al.* (2004) have argued that important components of cell–cell signalling in animals originated through horizontal gene transfer from bacteria. They focus on small, rapidly diffusible messengers such as adrenaline (epinephrine), noradrenaline (norepinephrine), dopamine, histamine, acetylcholine, and nitric oxide and suggest that most of the genes encoding enzymes involved in the metabolism of these messengers might derive by horizontal gene transfer from bacteria. Some of these putative transfers would have happened in the most remote history of eukaryotes, but others would have occurred after the divergence of animals from fungi. There is evidence, indeed, of some of these signalling molecules in unicellular organisms and, specifically, in choanoflagellates: receptor tyrosine kinase has been found in *Monosiga brevicollis* (King and Carroll 2001).

In Metazoa, two major classes of molecules contribute to producing and stabilizing the extracellular matrix; these are collagens and fibronectins. These molecules are responsible for cell migration and play an important role in morphogenesis. Different from other molecules involved in cell–cell relationships and morphogenesis in animals, the collagens are not known outside the Metazoa. As a critical oxygen concentration is required for the production of collagens, it has been suggested that metazoans evolved at the time the oxygen concentration in the environment rose to levels comparable with those of today (Müller 2001).

One of the cellular properties to have provided ample opportunity for the evolution of cell differentiation among small and simple multicellular organisms along the stem lineage of the metazoans is the incompatibility between bearing cilia and retaining the ability to divide. This is also true of the choanoflagellates, arguably the unicells most closely related to the metazoans (see below). Different from other unicellular organisms provided with cilia, choanoflagellates do not divide as long as they bear a cilium and the same is true of metazoan cells provided with either cilia or other microtubule-based structures; for example, nervous axons. As a consequence, if a cell is able to induce another cell to produce (or retain) a cilium or to differentiate other microtubule-based structures, the former is in fact impairing the ability of the latter to reproduce. Exclusion from mitotic activity extends to some types of sensory cell, and also to sperm cells (Margulis 1981, Buss 1987, Cavalier-Smith 1991, King 2004). Rieger (2003) regarded monociliated cells as ancestral to all kinds of somatic and germ cells of the Metazoa, but this primitively reversible condition allowing, for example, for the multiplication of choanoflagellates, has become permanent in the sperm cells and in several types of somatic cells, depriving them irreversibly of the possibility of dividing.

3.3 Choanoflagellates, the closest relatives of animals

Similar to the progress of astronomy that progressively removed first the Earth, and later also the Sun from the central position in the universe, our recent advances in phylogenetics have produced a tree of life of which the animals (metazoans), despite their conspicuousness and species diversity, do not represent more than a secondary branch. In a general classification of living beings, animals belong to the clade of the Opisthokonta, together with the Fungi and a number of inconspicuous unicellular forms, among which are our putative closest relatives, the choanoflagellates (e.g. Cavalier-Smith and Chao 2003).

The Opisthokonta can be diagnosed (Adl *et al.* 2005) as eukaryotes, either unicellular or multicellular, with at least one stage in their life cycle represented by cells with a single posterior cilium without mastigonemes (secondarily lost in some lineages), one pair of kinetosomes or centrioles, sometimes modified, and mitochondria with flat cristae in the unicellular stage. Steenkamp *et al.* (2006) have shown that members of all opisthokont protist groups encode a 12-amino acid insertion in a protein called elongation factor-1α, previously found exclusively in animals and fungi. Next to animals and fungi, Opisthokonta include corallochytreans, nucleariids, ministeriids, choanoflagellates, and ichthyosporeans.

Of the four main clades of the Opisthokonta recognized by Adl *et al.* (2005) (see Table 3.1), I provide here a short overview of mesomycetozoans and, especially, of choanomonads, but I will not discuss here the Metazoa, which form the main subject of the whole book, nor the Fungi, of which I will only transcribe the diagnosis provided by Adl *et al.* (2005, p. 405): 'Heterotrophic, not phagotrophic; often with walls and multinucleate hyphae; walls, when present, with β-glucan and usually chitin, at least in spore walls; AAA lysine biosynthesis pathway; mitochondria and peroxisomes present, except in Microsporidia; flattened cristae; plastids and tubular mastigonemes absent.'

The Mesomycetozoa is a quite diverse assemblage of organisms whose life cycle includes at least one stage represented by a spherical cell, posteriorly monoflagellated or amoeboid; the mitochondrial cristae are usually flat. Most members of the Mesomycetozoea, previously known as the DRIP clade (an acronym from

Table 3.1 The system of the Opisthokonta according to Adl *et al.* (2005).

Fungi
Mesomycetozoa
  Aphelidea
  Corallochytrium
  Capsaspora
  Ichthyosporea [=Mesomycetozoea]
    Rhinosporidiaceae[1] [=Dermocystida]
    Ichthyophonae [=Ichthyophonida, Amoebidiidae]
  Ministeria [=Ministeriida]
  Nucleariida
Choanomonada
  Monosigidae [=Codonosigidae]
  Salpingoecidae
  Acanthoecidae
Metazoa
  Porifera [=Parazoa]
  Trichoplax [=Placozoa]
  Mesozoa
    Orthonectida
    Rhombozoa
  Animalia [=Eumetazoa]

[1] Virtually always cited under the original incorrect spelling Rhinosporideacae (Mendoza *et al.* 2001).

their original members *Dermocystidium*, rosette agent, *Ichthyophonus*, and *Psorospermium*), are pathogens of fish, others are parasites of mammals and birds and some are saprotrophic. More detail follows.

• The Aphelidea (*Amoeboaphelidium, Aphelidium, Pseudoaphelidium*) are intracellular parasites of algae with an amoeboid infecting stage but eventually producing flagellate or amoeboid dispersal cells; they have mitochondria with either tubular or lamellar cristae.
• *Corallochytrium limacisporum* is a marine saprotrophic organism found in the coral reefs of the Indian Ocean. Unicellular, its large spherical cell releases numerous elongated amoeboid cells by repeated binary fission.
• *Capsaspora owczarzaki*, a parasite of the sporocysts of the digenean *Schistosoma mansoni*, has been classified with the Mesomycetozoea (e.g. Hertel *et al.* 2002), but more recent molecular phylogenetic analyses have shown it not to be clearly allied with any of the unicellular

opisthokonts, but to represent an independent unicellular lineage closely related to animals and choanoflagellates (Ruiz-Trillo *et al.* 2006).
• The Ichthyosporea, or Mesomycetozoea, is a small but quite heterogeneous group of single-celled organisms (sometimes amoeboid, sometimes forming multinucleated hyphal filaments); they have mitochondria with either flat or tubular cristae, and are without flagella or with one flagellum. Most species in this group are animal parasites, others are free-living and saprotrophic. The position of the Ichthyosporea within the Opisthokonta is still uncertain; comparisons of elongation factor-1α failed to resolve it. Ragan *et al.* (2003) noticed, however, that EF-1α of *Ichthyophonus irregularis* exhibits a two-amino-acid deletion heretofore reported only from fungi. Of the two families generally distinguished within the Ichthyosporea, the Rhinosporideacae or Dermocystida (e.g. *Amphibiocystidium* and *Dermocystidium*) form a heterogeneous group including some animal spherical parasites, whereas other members have a posterior flagellum, and the Ichthyophonida or Amoebidiidae (especially *Amoebidium, Ichthyophonus, Pseudoperkinsus,* and *Sphaeroforma*) are either free-living and saprotrophic, or parasites of fish or arthropods; some species have amoeboid cells; a flagellated stage has been reported from *Pseudoperkinsus tapetis*.
• Ministeriida (*Ministeria*) are imperfectly known marine unicellular organisms with unbranched filopodia radiating from a spherical body; they have mitochondria with flat cristae, and the presence of a flagellum is controversial.
• Nucleariida (*Nuclearia*) are naked amoebae, with a spherical or sometimes flattened body surrounded by radiating filopodia (Amaral Zettler *et al.* 2001); they have mitochondria with flat cristae. Nucleariids are the unicellular lineage most closely related to fungi (Ruiz-Trillo *et al.* 2007).

Trees inferred from multiple concatenated mitochondrial protein sequences demonstrate that animals are specifically affiliated with Choanoflagellata and the fungus-like *Amoebidium parasiticum* (Ichthyosporea). Statistical evaluation of competing evolutionary hypotheses confirms beyond doubt that Choanoflagellata and multicellular animals share a close sister group relationship, as originally proposed more than a century ago on morphological grounds (Fig. 3.1). Lang *et al.* (2002) resolved the phylogenetic position of the Ichthyosporea as basal to Choanoflagellata and Metazoa but after the divergence of Fungi. The close affinity between metazoans and choanoflagellates is supported by many protein sequences (α-tubulin, β-tubulin, elongation factor 2, and heat-shock proteins HSP90 and HSP70; Snell *et al.* 2001, Rokas *et al.* 2003).

The choanoflagellates are heterotrophic marine unicellular organisms, either solitary or colonial, with flat mitochondrial cristae, characteristic for their single flagellum surrounded by a collar of microvilli. Choanoflagellates (or, at least, their best-investigated representative, *Monosiga brevicollis*) possess a number of molecules known to be involved, in metazoans, in cell signalling and cell adhesion. As these phenomena only have a meaning in a multicellular context, the problem remains of what these molecules may eventually do in choanoflagellates. Among the cell-adhesion molecules found in *Monosiga* there are lectins and cadherins (King *et al.* 2003).

The mitochondrial genome of the choanoflagellate *M. brevicollis* has 1.5 times the number of nucleotides of an average mitochondrial genome of bilaterians.

The evolutionary step from the unicellular condition of the flagellate-grade ancestors of metazoans to the multicellular condition of the latter involved extensive loss of genes (Lavrov 2007). Lavrov *et al.* (2005) found several specific insertions and deletions (indels) in mitochondrial protein-coding genes, which are well conserved across the metazoans but are not present in *M. brevicollis*.

Opposite to the mainstream interpretation of the relationships between choanoflagellates and metazoans, Maldonado (2004) and a few other authors suggested that choanoflagellates might be extremely simplified sponges. A phylogenetic position of choanoflagellates within the metazoans was originally proposed by Nielsen (1985), who later came to accept the current scenario of a sister group relationship between

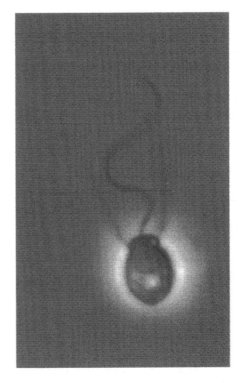

Fig. 3.1. *Proterospongia* sp. (Choanoflagellata); cell body about 5 μm. Courtesy of Nicole King.

choanoflagellates and metazoans. Some support for the hypothesis that choanoflagellates might be secondarily simplified sponges came from a molecular and morphological analysis (Lipscomb *et al.* 1998), whose results were not ruled out by a subsequent phylogenetic analysis based on receptor tyrosine kinase (King and Carroll 2001). However, a difficulty with this hypothesis derives from the choanoflagellate mitochondrial genome (King 2004), which contains several genes also known from other non-metazoan organisms but lacking in all animal mitochondrial genomes sequenced thus far, including those of two sponges (Watkins and Beckenbach 1999).

3.4 Late Precambrian fossils

3.4.1 Vendobionts or Vendian metazoans?

The earliest metazoan fossils (although their animal nature is not universally accepted by palaeontologists) are those from the Vendian period (610–550 million years ago, Ma). Many of these fossils have long been known as the Ediacara fauna after the Ediacara Hills in Flinders Ranges, South Australia, where they were first collected and studied. Other very important sites are along the White Sea coast of northern Russia. The Vendian 'fauna' is characterized by organisms of large size; for example, the medusoid *Ediacaria* is 1 m in diameter, the segmented worm-like *Dickinsonia* is 1 m long, and the sea feather-like *Charnia* and *Charniodiscus* are up to 1.2 m high. These organisms did not possess a skeleton and most of them were flat. In a classic work on the Australian material, Glaessner (1984) referred most of them to the cnidarians, some others to the annelids and arthropods, and left many problematic forms taxonomically unassigned. A very different interpretation was offered by Seilacher (1984, 1989, 1994), who regarded most of the Ediacaran organisms as belonging to a radiation of life (the Vendobionta) distinct from the Metazoa, the latter being

only represented by some coeval trace fossils. According to Seilacher these Vendian organisms were probably multinucleate (plasmodial) rather than multicellular and had no mouth neither digestive or respiratory systems; oxygen and organic matter were most likely absorbed through the body wall. This hypothesis does not find much favour today, at least if applied systematically to the whole set of Vendian macrofossils, but Conway Morris (2003b) believes that Seilacher's interpretation may still apply to some of the more enigmatic taxa such as *Ernietta* and *Pteridinium*.

Fedonkin (1994), who has devoted long studies to the White Sea fossils (Fig. 3.2), suggested that Vendian organisms were broadly similar to Phanerozoic metazoans in body plans and modes of locomotion. Some of these forms show signs of motility, muscular contractions, and behaviours such as epifaunal grazing (Seilacher 1997, Ivantsov and Fedonkin 2001, Waggoner 2003). However, most Vendian organisms do not seem to be in direct morphological

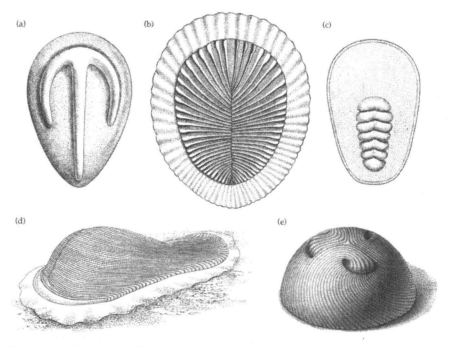

Fig. 3.2. Vendian organisms: (a) *Parvancorina*; (b) *Dickinsonia*; (c) *Onega*; (d) *Andiva*; (e) *Tribrachidium*. Courtesy of Mikhail Fedonkin.

continuity with Phanerozoic metazoans. Fedonkin identified three main lineages of Vendian cnidarians—Cyclozoa, Inordozoa, and Trilobozoa—characterized by different body symmetry, in particular the Inordozoa would have had an 'unstable order of symmetry' while the Trilobozoa show triradial symmetry (something also occurring in Early Cambrian anabaritids or angustiochreids).

Very controversial is the interpretation of flat forms like *Dickinsonia* and *Spriggina*, which have been often regarded as primitive members of the Articulata, because of their 'segmented' structure. However, in these organisms the left and right halves of each 'segment' were out of phase, unlike any modern annelid or arthropod. Additionally, the rejection of the hypothesis of close affinities of annelids and arthropods, as suggested by recent phylogenetic investigations (see section 5.3.10), leaves the interpretation of these Vendian fossils still more uncertain.

3.4.2 Doushantuo embryos

More profitably than in these Vendian faunas we can perhaps identify early stages of metazoan evolution in the many fossil eggs and embryos recently described from the Late Precambrian Doushantuo Formation of south-west China. Compared with eggs and embryos of most modern marine invertebrates, these phosphatized embryos, 580 million years old, are quite large, 300–1000 μm in diameter (Li *et al.* 1998, Xiao *et al.* 1998, 2000, Chen *et al.* 2000, Xiao and Knoll 2000). Some of them present a 'polar lobe' as seen in the embryos of many living molluscs (Chen *et al.* 2006), others have been compared with cnidarian and bilaterian gastrulae (Chen *et al.* 2000). These identifications, if confirmed, would support a long history of bilaterian diversification, as suggested by some models of metazoan evolution (e.g. Davidson *et al.* 1995, Peterson *et al.* 1997, 2000a) as well as by some molecular estimates of the age of the deepest branchings in the metazoan tree.

On the interpretation of these fossils there is absolutely no agreement. According to Hagadorn *et al.* (2006), the embryos from the Doushantuo Formation may belong to stem-group metazoans, as no epithelial organization is recognizable, even in embryos composed of a quite large number of cells. A more radical criticism of the original interpretation has been advanced by Bailey *et al.* (2007), according to which these fossils are clusters of bacterial cells, rather than animal embryos. This objection may apply to some of these fossils, but not to all of them (Donoghue 2007), because there seem to be distinct cell nuclei in the embryos described by Xiao *et al.* (1998) and also because bacteria, when not strictly unicellular, do not go beyond eight-cell clusters, whereas the putative Chinese embryos have up to 128 cells. Still more sceptical about the biological nature of this material was the interpretation suggested by Xiao *et al.* (2000), who regarded the putative 'epithelial layer cells' as simple apatite coatings analogous to those found in other phosphatized fossils (e.g. Yue and Bengtson 1999), whereas Bengtson (2003) suggested that the alleged diversity of these fossils resulted from different *post mortem* alteration and encrustation on globular microfossils. Conway Morris (2006) accepts that these Doushantuo fossils are embryos, but rejects the identification of a blastopore claimed to be present in some of them.

In principle, the presence of fossilized embryos in the Doushantuo Formation is compatible with what we know about the fossilization process. Specifically, Raff *et al.* (2006) have demonstrated experimentally that embryos of marine invertebrates of a size comparable with that of the Doushantuo fossils, if buried when still within their fertilization envelope, have good preservation potential, provided that environmental conditions can prevent autolysis, whereas larvae of similar size, in the same taphonomic conditions, are destroyed rapidly.

A still more controversial fossil from the Doushantuo Formation has been described by Chen *et al.* (2004a) as a minuscule bilaterian with coelomic cavities. This putative animal, called *Vernanimalcula guizhouena*, would have had an anteroposterior body axis 124–178 μm long, a size that would fit an acoelomate larval stage rather than a post-metamorphic coelomate stage, not to say an adult. I am inclined to follow Bengtson and Budd's (2004) and Donoghue *et al.*'s (2006a) interpretations, according to which the putative morphological complexity of these fossils can be simply accounted for by taphonomic and diagenetic processes, a view rejected, but not convincingly, by the original authors (Chen *et al.* 2004b).

Summing up, it seems sensible to repeat here the summary by Briggs and Fortey (2005), in stating that besides the Doushantuo embryos and a few trace fossils not older than about 555 million years (Droser *et al.* 2002), the oldest fossils that can be identified with certainty as metazoans are still no older than the base of the Cambrian.

3.5 The Cambrian 'explosion'

The nature of the fossil record changes dramatically when moving from the Vendian to the Cambrian; that is, when leaving the Proterozoic and entering the Phanerozoic. Indeed, the rocks reveal a real explosion of metazoan forms, most of which can be indisputably assigned to still surviving lineages. However, among the Early Cambrian fossils there are, apparently, a few survivors from the Vendian biota (Conway Morris 1993, Jensen *et al.* 1998, Hagadorn *et al.* 2000). One of these is *Stromatoveris psygmoglena* (Shu *et al.* 2006), a fossil from a Lower Cambrian site that has became very popular in recent years. In the last 20 years this site (Chengjiang, in Yunnan, China) has revealed an extraordinarily diverse and exceptionally well preserved fauna from 525 Ma (for an excellent summary, see Hou *et al.* 2004). Before its discovery, our single best source of information about the early faunas was still the somehow younger Burgess Shales site in British Columbia, Canada, dated to approximately 515 million years before present (BP). Table 3.2 offers a comparative overview of the Cambrian faunas from these two sites.

Many records have been published of Cambrian fossil eggs and embryos, from China, Siberia, Australia, and North America. As in the case of the Doushantuo eggs and embryos, these are also large, from 0.25 to more than 1 mm (e.g.

Steiner *et al.* 2004, Lin *et al.* 2006, Pyle *et al.* 2006). Early Cambrian phosphatized eggs and embryos from Kuanchuanpu, Shaanxi, China, described by Steiner *et al.* (2004), suggest patterns of germ-band formation and irregular 'radial' cleavage types and have been tentatively referred to an 'ancestral arthropodan lineage'. Perhaps counterintuitively, large eggs and embryos may suggest minuscule adult size. At least, this is what results, in Recent animals, when comparing miniaturized forms with their closest relatives of larger size. Arguably, the number of eggs being tightly constrained by body size, producing a small number of large eggs may easily turn out to be better than producing many smaller eggs whose number would anyway be quite reduced. In other terms, the expected survival of a juvenile deriving from a yolk-rich egg can be so high as to overcompensate the drawback of small numbers, whereas the much lower expected survival of individuals derived from very small egg is only sustainable if numbers are very high. Additionally, large eggs suggest direct development, contrary to a commonly shared opinion about 'primary larvae' (see Chapter 7).

It is impossible to guess how much time may be required for what we may describe as a big change in animal organization to evolve. How much time did it take to transform an 'average polychaete' into *Riftia*, the huge tube worm of the oceanic hydrothermal vents, or to produce a worm-like animal like *Buddenbrockia* from arguably polyp-like ancestors (see section 4.6)? When confronted with this kind of question, we must generally content ourselves with estimates of divergence times, other from estimates of the time of first appearance of a specific body feature. From some spectacular evolutionary radiations, such as those of the cichlid fishes in the great African lakes, we know that morphological evolution is sometimes very rapid, while other lineages provide examples of extraordinary morphological stability, as in the case of scolopenders, of which *Mazoscolopendra* from the Carboniferous is virtually identical to some Recent species (Mundel 1979).

Thus, how do we interpret the sudden appearance, in the Cambrian, of a diversity of

Table 3.2 Number of animal genera and species described from Chengjiang, China (≈525 Ma) and the Burgess Shales, Canada (≈515 Ma).

Taxon	Chengjiang		Burgess Shales	
	Genera	Species	Genera	Species
'Porifera': Demospongia			15	28
'Porifera': Silicea	10	13	3	3
'Porifera': Calcarea			2	2
Ctenophora	2	2	1	1
Cnidaria	2	2	4[1]	5
Hyolitha[2]	3	4	1	1
Mollusca			1[3]	1
Annelida	1	1	5	5
Phoronozoa: Phoronidea	1	1		
Phoronozoa: Brachiopoda	5	5	6	6
Halwaxiida[4]			2	2
Nematomorpha	3	3		
Priapulida	6	6	7[5]	7
Lobopodia or Onychophora	6	6	2[6]	2
Anomalocaridida[7]	4	5	4	9
Arthropoda	48	53	44	60
Chaetognatha	1	1	1	1
Echinodermata			4[8]	5
Hemichordata			2[9]	2
Vetulicolia[10]	5	5	1	1
Chordata	3	4	2	2
Enigmatic animals	12[11]	12	12[12]	12

The main sources of information are Hou *et al.* (2004) for Chengjiang and Briggs *et al.* (1994) for the Burgess Shales.

[1] Assignment to Cnidaria largely uncertain. Includes *Thaumatoptilon walcotti*, otherwise regarded as a possible survivor of the late-Proterozoic Ediacara fauna into the Cambrian biota (Conway Morris 1993).

[2] Fossils represented by cone-shaped calcareous shells, known from the whole of the Palaeozoic (Cambrian to Permian), possibly belonging to a lophotrochozoans lineage related to the Mollusca and the Sipunculida (Martí Mus and Bergström 2001).

[3] This refers to the helcionelloid (monoplacophoran) *Scenella amii*; possibly referrable to the Mollusca is also *Odontogryphus omalus* (Conway Morris 1976), tentatively listed by Briggs *et al.* (1994) under '?Superphylum Lophophorata', but see section 5.3.7.

[4] Within the Lophotrochozoa, Conway Morris and Caron (2007) identify a stem-group Halwaxida, including armoured bilaterians with three main zones of sclerites and at least one shell.

[5] Assigment to the Priapulida more or less certain.

[6] These are the well-known *Aysheaia pedunculata* and *Hallucigenia sparsa*.

[7] Perhaps stem-group Arthropoda.

[8] Including *Eldonia ludwigi*, referred here to the Holothuroidea; but see footnote 11.

[9] In Briggs *et al.*'s (1994) list, one of the two forms is tentatively referred to the Enteropneusta, the other, also tentatively, to the Palaeozoic lineage of the Graptolithina.

[10] The first member of the Vetulicolia was described by Hou (1987), but the group was first established by Shu *et al.* (2001) as representing primitive deuterostomes close to the Chordata. According to Lacalli (2002) the vetulicolians are possibly urochordates.

[11] Some of the fossils listed by Hou *et al.* (2004) as 'enigmatic' deserve short comment. *Dinomischus* was described by Conway Morris (1977) as possibly referable to the Entoprocta; Dzik (1991) placed it in its new class Eldonioidea, tentatively referred to a 'lophophorate' complex inclusive of brachiopods and bryozoans. *Eldonia*, after which Dzik established this class, was described by Walcott (1911) as a pelagic holothurian and was listed as such by Briggs *et al.* (1994); a still different interpretation was offered by Sun and Hou (1987) who suggested that *Eldonia* could be a cnidarian medusa. *Allomia* has been variously interpreted as a sponge or an ascidian, or otherwise as a member of the Coeloscleritophora (Bengtson and Hou 2001), a grouping of animals known after their hollow sclerites eventually equivalent, at least in part, to the Halwaxiida, a recently established group assigned to the Lophotrochozoa (Conway Morris and Caron (2007); see section 5.3.7. *Facivermis* was described as an annelid by Hou and Chen (1989) but subsequently interpreted as a lobopodian by Delle Cave and Simonetta (1991), as a lophophorate by Chen and Zhou (1997), while Delle Cave *et al.* (1998) eventually suggested affinities to the Pentastomida. (It is worth noting that up to now the presence of annelids in the Chengjiang fauna is still uncertain, being only based on an undescribed fossil tentatively referred to the 'Polychaeta' by Chen and Zhou 1997). *Rotadiscus* was described as a cnidarian medusa by Sun and Hou (1987), but subsequently interpreted by Dzik (1991) as a member of the Eldonioidea. The most famous of these 'enigmatic taxa' is *Yunnanozoon*, variously referred to the Chordata, the Enteropneusta and, most recently, to the stem-group Deuterostomia (Budd and Jensen 2000) (see section 6.30).

[12] Of the species listed by Briggs *et al.* (1994) as unassignable to a phylum I maintain here the following: *Amiskwia sagittiformis, Dinomischus isolatus, Nectocaris pteryx, Oplatydendron ovale, Pollingeria grandis, Portalia mira,* and *Worthenella cambria.* For five more fossils a tentative assignment can be suggested: *Banffia constricta* as a putative vetulicolian (Caron 2005), *Oesia disjuncta* as possibly a chaetognath (Szaniawski 2005), *Opabinia regalis* as a stem-group arthropod (Briggs and Fortey 2005), and *Wiwaxia corrugata* and possibly also *Chancelloria eros* as halwaxiids. Therefore, in the present table these five species have been removed from the 'unassignable' group and distributed according to the taxonomic assignment suggested for each of them.

animal groups, including many of the 'phyla' traditionally distinguished in the Recent biota? Shall we take it as proof of a sudden radiation of the Metazoa, with an explosive evolution of a diversity of body plans, or simply as a consequence of suddenly changed opportunities for fossilization, such that we can hardly hope to eventually find fossil evidence of animals much older than the Cambrian, despite their existence, perhaps for hundreds of millions of years?

Some palaeontologists regard the fossil record as basically reliable and adequately complete. This means that all the diverse metazoan clades present in the Early Cambrian fossil record must have radiated from a common ancestor within a very short time interval previous to the beginning of the Phanerozoic era. This is the concept of a 'Cambrian explosion' of animal life, which has been championed most conspicuously by Conway Morris (e.g. 1994, 2003a, 2003b, 2006). A fairly recent summary of the palaeontological evidence (Budd 2003b) concluded that the Cambrian fossil record, in its most parsimonious interpretation, represents a broadly accurate picture of the origins of the bilaterian phyla.

Molecules, however, seem to tell a different story, as most estimates of the divergence between protostomes and deuterostomes fall within the range 1000–750 Ma, suggesting a long period of diversification of metazoans before the base of the Cambrian (e.g. Wray *et al.* 1996, Levinton *et al.* 2004, Briggs and Fortey 2005). Molecular phylogenies have often suggested very old times of divergence between protostomes and deuterostomes, or between cnidarians and bilaterians (see Table 2.2). Peterson and Butterfield (2005) have used seven nuclear-encoded protein sequences to calculate the time of divergence between sponges and eumetazoans and between cnidarians and bilaterians: their results bring us to times somewhat deeper, or much deeper, than the base of the Cambrian. A molecular clock based on the so-called principle of minimum evolution offered an estimate for the appearance of the Eumetazoa in the interval between 634 and 604 Ma, whereas the maximum-likelihood method pushed the interval to a range between 867 and 748 Ma.

On the other hand, the intervals between the several lineage splits that originated all major clades represented in the Cambrian faunas (or in the Recent biota) may have been very closely spaced. This may be counterintuitive if we fix on the great differences in the organization of their modern representatives, but these differences are clearly the result of a very long history of divergence. Some molecular phylogenetic studies suggest indeed that many 'phyla' may have originated within a short interval. For example, based on the alignment of 12 060 amino acids encoded by 50 genes from 16 animal taxa, Rokas *et al.* (2005) found a conspicuous polytomy of protostome taxa and another polytomy involving Bilateria, Cnidaria, and Porifera. This serious lack of resolution of the tree, despite the large amount of data used in the analysis, suggest rapid speciation during the early evolution of metazoans and fits well with the implications of the Cambrian explosion hypothesis. This interpretation has been accepted by Jermiin *et al.* (2005), despite some methodological flaws that they found in the original analysis.

Several authors have contended that the Cambrian explosion is, in fact, just an explosion of animals with hard skeletons; that is, of parts which can be easily preserved as fossils. However, this does not explain the extreme scarcity of Precambrian trace fossils, including those left by soft-bodied animals, while the sudden appearance of animals provided with hard skeletons, in the Cambrian, is accompanied by abundant fossils and traces of soft-bodied animals (Vidal 1997, Valentine *et al.* 1999).

The major event associated with the Cambrian sudden surge of animal diversity in the fossil record is the broad-scale acquisition of hard skeletons. These include organic skeletons made of tanned scleroproteins and polysaccharides as well as biomineralized ones. Seilacher (1994) made the important and often overlooked remark that hard skeletons are not simply an addition to pre-existing soft-bodied ancestors. That is, we would not be able to recognize, on their morphology, soft-bodied protomolluscs, protobrachiopods, or protoarthropods, because of the major changes in bodily organization determined by the mode of growth of their skeletons.

What did the first metazoans look like? Several authors (e.g. Fortey *et al.* 1996, Vermeij 1996, Cooper and Fortey 1998) have defended the idea that the protagonists of the so-called Cambrian explosion, or those that represent the Late Precambrian segment of animal diversification, were minute and either planktonic or interstitial. However, features such as arthropod limbs, the annelid parapodia, and the molluscan foot are not adaptations to be expected in minute pelagic organisms (Valentine *et al.* 1999). To the possible exception of the chordates, all major groups, as they first show up in the Cambrian, appear to be conspicuously adapted to a macrobenthic life.

CHAPTER 4

Deep branches of the metazoan tree

Although it may be heresy to say so, it could be argued that knowing that strikingly different groups form a clade and that the time spans between the branching of these groups must have been very short, makes the knowledge of the branching order among groups potentially a secondary concern.

Rokas and Carroll (2006), p. 1903

4.1 The Metazoa, our largest monophylum

The monophyly of the Metazoa was sometimes disputed in the recent past, in the light of Late Precambrian and Early Cambrian fossil evidence (e.g. Walter 1987, Bergström 1991), and a diphyletic origin of the animals was also suggested by early work in molecular phylogenetics (e.g. Field *et al.* 1988), but this hypothesis is no longer at issue: comparative morphology, fossil evidence, and molecular evidence all largely concur to confirm metazoan monophyly (e.g. Lake 1990, Schram 1991, Wainright *et al.* 1993, Raff *et al.* 1994, Willmer 2003). Neither is there any dispute whether sponges (and, possibly, placozoans) should be included in the Metazoa, or not. It has become more and more accepted to use the term Metazoa as a synonym for Animalia, to include all the multicellular heterotrophic opisthokonts without cell walls, with strongly dimorphic gametes (a large egg and a much smaller sperm cell, usually provided with a single cilium), and mostly with an extracellular matrix containing collagens and other fibrous proteins. Less conventional is the choice of Adl *et al.* (2005) to restrict the term Animalia (or Eumetazoa) to embrace what has been otherwise called the Histozoa; that is, the Ctenophora, the Cnidaria,

and the Bilateria. This anyway requires fixing the position of the Placozoa, which were conventionally placed by Ax (1995), together with the Eumetazoa, in a clade Epitheliozoa characterized by the presence of two epithelia (mostly external and internal, but dorsal and ventral, instead, in the Placozoa), with characteristic cell junctions (*zonulae adhaerentes*).

Müller and his associates (e.g. Müller 1998, 2003, 2005, Müller *et al.* 1999, 2004, Müller and Müller 2003a, 2003b) have produced a lot of information about the presence in sponges (especially in the demosponge *Geodia cydonium*) of genes and proteins characteristic of metazoans but often previously regarded to be exclusive of animals with more complex body organization. Integrating their results with those of other researchers, we obtain a list including proteins such as galectin, collagen, myosin, and integrin, all involved in tissue formation; receptor tyrosine kinase, involved in signal transduction; several transcription factors of the following families: POU, LIMHD, Pax, Bar, Prox2, NK-2, T-box, MEF-2, Fox, Sox, Ets, and nuclear hormone receptor (Coutinho *et al.* 2003, Degnan *et al.* 2005, Larroux *et al.* 2006, Leys and Ereskovsky 2006); and proteins involved in immune reactions.

The 'animal' nature of sponges is shown by the role the extracellular matrix plays in their morphogenesis (Wiens *et al.* 2003) but also by the fact that they possess defence pathways and molecules able to prevent adverse effects caused by microbes or parasites (Müller 2003). For example, *Suberites domuncula* and *G. cydonium* can rely on defence systems which involve phagocytosis of bacteria into specific cells, but also mechanisms that actively kill the bacteria.

Sponges are also provided with an interferon-related system. Cytokines have been identified that very probably control gene expression during histocompatibility reactions. *G. cydonium* has several molecules containing domains which are probably homologous to immunoglobulin sequences. In addition, Li *et al.* (2005) reported the presence of substances characteristic of nerve cells, such as neuropeptide Y (NPY), β-endorphin, neuron-specific enolase (NSE), S-100, and 5-hydroxytryptamine (serotonin) in *Spongilla* and regarded sponge collencyte as a primal nerve cell.

All this evidence demonstrates that the evolution of genes precedes the evolution of morphology, a point to which another clade of 'simple' metazoans, the Cnidaria, offers further good examples. Many genes known to be involved in specifying and patterning the nervous system of insects and vertebrates have been found in the genome of the coral *Acropora millepora*, whose nervous system is among the simplest of the metazoans (Kortschak *et al.* 2003, Ball *et al.* 2004).

The presence, in sponges, of molecules characteristically involved in the production or the functioning of structures present in the remaining Metazoa (Epitheliozoa) but not in poriferans is an excellent example of exaptation at the genetic and molecular level. Sponges also possess molecules that have not been found in any other metazoan; for example, silicatein, which polymerizes alkoxide substrates to silica; and an ethylene-responsive protein, also found in yeasts (Müller 2001).

4.1.1 Complexity and phylogeny

Traditional schemes of metazoan phylogeny were strongly biased by a strong inclination to arrange the major lineages (or, at least, the groups supposed to represent early branching monophyla within the Metazoa) in a linear series characterized by growing degrees of complexity. Complexity is a controversial and technically difficult issue in itself (for a useful discussion, see McShea 1991, 1996a, 1996b, 2000, 2002, 2005). An animal's complexity has often been estimated by considering the number

of germ layers recognizable at early stages of embryonic development (but see section 7.3.7), or the number of cell types in the adult organism (Bonner 1988), or the presence and the degree of differentiation of organs specifically involved in performing a given function. In this respect, the complexity, autonomy, and performance of the nervous system have been given a top position, something similar, in different ways, to Lamarck, Cuvier, Oken, and other naturalists of the early nineteenth century. More recently, a series of growing degrees of complexity has been identified in the gene batteries and gene networks involved in developmental events spanning from cell differentiation to body patterning (Davidson 2003, 2006). To be sure, any approach of this kind must come to terms with the frequent, undisputable cases of evolutionary regression, the most conspicuous examples of which are found among the parasites. But the real problem with this approach is not so much the challenge of actually identifying all cases of reversal of the main trend in complexity, but the unwarranted interpretation of increasing complexity levels as steps in the historical sequence of phylogenetic branching.

Sponges provide a very good example of the errors in considering animal evolution from the perspective of an *a priori* selected trend in complexity. Irrespective of the still contentious issue of the phylogenetic relationships between the three or four main sponge lineages (see below, section 4.3.1), zoologists have always been inclined to equate the morphological series (or morphocline) formed by the three kinds of morphological organization and complexity traditionally distinguished among the sponges—the ascon, the sycon, and the leucon—to a phylogenetic sequence.

Zoology textbooks usually present this series of models of sponge organization in order of increasing complexity, suggesting, implicitly at least, that this as an evolutionary sequence. On the one hand, moving from ascon to sycon to leucon would have allowed sponges to reach bigger size (something that some naïve views of evolution suggest as indisputably advantageous); on the other hand, this structural sequence is also found along the development of those sponges (the vast majority, indeed) that eventually reach a leuconoid organization as 'adults'. From pure considerations of developmental evolution, I would suggest that the opposite evolutionary scenario is much more likely. In developmental terms, the diffuse, indefinite, approximately fractal organization of a leuconoid sponge is exactly the kind of structure one should expect in a basal multicellular organism, where local developmental rules are indefinitely repeated in time and space, with the only constraints set by the sustainability of an increasingly large body mass. On the contrary, syconoid and asconoid organization will only result, in an 'adult' sponge, in the presence of further specification; in particular, the specification of an axial polarity, which is not found in the structure of a leuconoid sponge, irrespective of its gross external shape. Interestingly, molecular phylogeny provides strong support for the scenario sketched along these lines. Quite probably, the leuconoid condition is plesiomorphic for the whole of the sponges, whereas leuconoid and asconoid lineages have only emerged within the calcareous sponges: this is an interesting point, in view of the fact that molecular phylogenies suggest that the Calcarea, rather than the Porifera as a whole, are probably the sister group of all remaining Metazoa. I have no reason to expect that exactly the living family of calcareous sponges with asconoid organization (Clathrinidae), rather the whole of the Calcarea, will eventually turn out to be the sister group of the Epitheliozoa; but I believe that the presence within the Calcarea of sponge species with defined axial structure is likely to be the expression of a trend towards forming defined axial structures, a trend that calcareous sponges share with the non-poriferan metazoans. The molecular background and evolution of this 'axial syndrome' deserves closer investigation at this stage of animal evolution.

4.1.2 Germ layers and metazoan phylogeny

Contrary to conventional opinion, according to which no more than two germ layers are present in animals other than the Bilateria, Ruppert (1991a) contended that all metazoans are in fact triploblastic. There are indeed cells between the pinacoderm and the choanoderm of sponges, between the upper and lower epithelium of placozoans and between the epidermis and the gastrodermis of cnidarians; and muscular fibres run through the extracellular matrix between the two main cell layers in ctenophores (Hernandez-Nicaise and Amsellem 1980, Hernandez-Nicaise 1991). Seipel and Schmid (2005) regard the diploblastic condition of cnidarian larvae and polyps as secondary and suggest that that striated muscle-based locomotion coevolved with the nervous and digestive systems in basal metazoans before the branching events that gave rise to the ctenophores, the cnidarians, and the bilaterians. Indeed, cnidarian homologues of bilaterian mesoderm and myogenic regulators have been found in cnidarian larvae and polyps as well as in the medusogenic entocodon and in the medusan striated muscle (see below, section 4.2.3).

4.1.3 Dating early branchings

A careful evaluation of fossil evidence is still regarded by Cartwright and Collins (2007) as the best evidence for the origin of the major metazoan lineages. From the data compiled by Cartwright and Collins (2007), the age of the earliest known representatives of different groups is as follows:

- 560 Ma for crown-group Porifera (increased to 710 Ma if the identification of some Precambrian biomarkers proves to be correct; 750 Ma according to Reitner and Wörheide 2002);
- 500 Ma for several cnidarian lineages (Semaeostomeae and Coronatae among the Scyphozoa, Narcomedusae and Filifera among the Hydrozoa, and Carybdeida among the Cubozoa);
- 540 Ma for stem-group Ctenophora;
- 525 Ma for crown-group Brachiopoda;
- 530 Ma for stem-group Arthropoda;
- 525 Ma for stem-group Chordata.

By putting at 710 Ma the origin of sponges, at 570 Ma the origin of cnidarians, and at 560 Ma the origin of bilaterians, Cartwright and Collins

(2007) estimate at 1147 Ma the origin of Metazoa, at 710 Ma that of the Silicea, at 543, 530, and 537 Ma the origin of protostomes, ecdysozoans, and lophotrochozoans, respectively, and at 540 and 495 Ma that of deuterostomes and chordates. All of this is compatible with the Cambrian explosion hypothesis.

4.2 Metazoan phylogeny

The last quarter of the twentieth century was a time of unprecedented progress in animal phylogenetics, largely thanks to the rapidly increasing availability of DNA, RNA, and protein sequences and the development of increasingly powerful and critically sensible techniques for their analysis. But we must not pass over other precious sources of data and insight. One is the continuing discovery of living metazoans for which no previously defined higher taxon (say, class, or even phylum, to speak in terms of traditional, Linnaean ranks) could provide a natural place. This includes the discovery of the Loricifera (Kristensen 1983), the Cycliophora (Funch and Kristensen 1995), and the Micrognathozoa (Kristensen and Funch, 2000). Another is the enormous growth in our knowledge of Late Precambrian and Early to Middle Cambrian faunas, deriving partly from the discovery and exploration of new sites, such as the Chengjiang Lagerstätte in Yunnan, China, and partly by a fresh revisitation of the fossils from well-known sites such as the Burgess Shales in British Columbia, Canada. Finally, we must acknowledge the standing contribution of comparative morphology and embryology, whose role in phylogenetic analysis is far from being annihilated by the growing availability of molecular data. In recent decades, morphology has profited much from improvements in microscopy and histological (including histochemical) techniques; the same is true of embryology, a field that is benefiting increasingly from the identification of genes involved in the control of developmental events and from the comparative study of their expression. We shall indeed expect interesting results from the incorporation into phylogenetic research of data and concepts deriving from the

new field of evolutionary developmental biology (Minelli 2007, Minelli et al. 2007).

As a consequence of these combined efforts, some of the traditional 'phyla' have been shown not to be monophyletic. Within the sponges, molecular phylogenetics suggests that the Calcarea are the sister group of the non-sponge metazoans (the Epitheliozoa); thus the old taxon Porifera would be paraphyletic. Of the Plathelminthes as traditionally circumscribed, two clades are now generally recognized not to be strictly related to the bulk of the flatworms. A different problem is whether these two clades, the Acoela and the Nemertodermatida, together form a monophyletic group (Acoelomorpha) or not, as currently seems more probable. The remaining flatworms comprise two major clades (Catenulida and Rhabditophora). No morphological synapomorphy has been suggested as uniting them in a single taxon Plathelminthes, but the two lineages still seem to be very closely related anyway. The Acanthocephala have been shown to be nested within the Rotifera, but there is still uncertainty as to their actual sister group. A new name, Syndermata, has been proposed for the whole of the former paraphyletic Rotifera plus the Acanthocephala. The general affinities and the mutual relationships of Endoprocta (=Kamptozoa) and Ectoprocta (=Bryozoa *sensu stricto*) are uncertain. As we will see, some recent work suggests that they should be united again in a clade of Bryozoa *sensu lato*, thus reinstating a nineteenth-century arrangement. Molecular evidence also suggests that the Phoronida are an internal branch of the Brachiopoda; the resulting group is now generally called Phoronozoa.

A large area of unresolved problems is the annelidan clade. Molecular phylogeny has firmly established that the Pogonophora (including the Vestimentifera) are nothing but a specialized polychaete group; less certain are the affinities of the Echiura and the Sipuncula, but at least for the former there are good reasons to suppose that they also belong to the annelidean radiation. Opposite views have been expressed as to the phylogenetic position of the Myzostomida, traditionally regarded as belonging to the Annelida, but recently suggested to have different, perhaps rhabditophoran affinities, There are, however, also good reasons to keep them with the annelids. Provisionally, annelid affinities may also be suggested (Westheide 2007) for two recently described unsegmented 'worms', *Lobatocerebrum psammicola* (Rieger 1980, 1988) and *Jennaria pulchra* (Rieger 1991).

Besides the still intractable problem of circumscribing the Annelida as a monophyletic taxon and, as a consequence, of giving a stable position, in the phylogenetic tree, to some groups whose placement within the annelids is still disputed, the list of taxa whose level of inclusiveness is comparable with that of the traditional phyla is now becoming quite stable. Much more controversy surrounds the topology of the tree of which these taxa are the terminal taxa. This will be discussed in this chapter for the most basal branching of the metazoan tree and in the next chapter for the Bilateria.

Figure 4.1 provides an overview of the higher metazoan taxa recognized in this book. Table 4.1 lists all of them, and many other higher taxa that have been recognized in the past, together with a reference to the work where each of them was proposed, or substantially reinterpreted. The list is not exhaustive, but it covers the large majority of taxonomic or phylogenetic concepts for higher metazoan groups that the reader may expect to find in the current literature. Some major problems of internal phylogeny will be presented briefly, for several 'phylum'-level taxa, either in this chapter (for non-bilaterian clades) or in Chapter 6 (for bilaterian clades).

4.2.1 Epitheliozoa and Eumetazoa

Morphologically simple organisms are often difficult to place in a phylogeny. Indeed, this would hardly be possible without the help of molecular evidence. But even the latter can be misleading, if the morphological simplicity is the result of secondary simplification which has been accompanied by extensive gene loss. This is perhaps the case for *Trichoplax* (that is, for the Placozoa). Nonetheless, a quite basal position for these little animals in the metazoan tree seems to be warranted, more basal indeed than the branching

off of the Bilateria, but their position is still not fixed. For the time being, I would accept that they belong to the sister group of the calcareous sponges—the Epitheliozoa—whereas the Placozoa may be, in turn, the sister group of all remaining non-sponge animals, the Eumetazoa.

In several molecular analyses based on 18S rRNA (Zrzavý *et al.* 1998b, Peterson and Eernisse 2001), 28S rRNA (Lafay *et al.* 1992, Medina *et al.* 2001), and HSP70 (Borchiellini *et al.* 1998) sequences, the eumetazoans were recorded as non-monophyletic, but this was probably due to causes such as inadequate taxon sampling or the limited resolving power of the sequences utilized.

Following Ax (1996), we can list, as eumetazoan apomorphies, the presence of an external epidermis and an internal gastrodermis, of ectodermic and endodermic origin respectively, gap junctions in the epithelia, a gut cavity, monociliate sensory cells in the epidermis, nerve cells at the base of both epithelia, and epithelial muscle cells.

4.2.2 Radiata versus Bilateria? A problem of symmetry and body axes

Within the Eutemetazoa, a contrast has been generally accepted between two clades of diploblastic, radial (or biradial) animals, the Cnidaria and the Ctenophora, collectively grouped as the Coelenterata (or Radiata, in a modern sense of the term), and a remaining clade of triploblastic, bilaterally symmetric animals, the Bilateria. While there seems to be no reason to question the monophyly of the Bilateria, several recent phylogenies reject the hypothesis of coelenterate monophyly, while suggesting that either the Ctenophora or the Cnidaria, rather than the Coelenterata as a whole, may represent the sister group of the Bilateria. This revisitation, however, is not simply based on molecules. Comparative morphology, and especially comparative embryology, may suggest, indeed, that zoologists may have been wrong in regarding the body of cnidarians and ctenophores as being composed of derivatives of only two germ layers, and also in considering these animals as primitively radial.

Silicea
Homoscleromorpha
NN   **Calcarea**
    Epitheliozoa
        Placozoa
        Eumetazoa
            Ctenophora
            NN
                Cnidaria
                Bilateria
                    Acoela
                    NN
                        Nemertodermatida
                      Nephrozoa
                        Protostomia
                          Platyzoa
                              Gastrotricha
                              Gnathifera
                                  Micrognathozoa
                                  Syndermata
                                  Gnathostomulida
                            Catenulida
                            Rhabditophora
                        Spiralia
                            Cycliophora
                            Bryozoa
                                Ectoprocta
                                Endoprocta
                            Mesozoa
                                Orthonectida
                                Rhombozoa
                        Lophotrochozoa
                            Phoronozoa
                            Eutrochozoa
                                Nemertea
                                Neotrochozoa
                                    Mollusca
                                  Annelida incl. Echiura, Myzostomida,
                                                  Sipuncula
                    Ecdysozoa
                      Introverta
                        Scalidophora
                            Vinctiplicata
                                Priapulida
                                Loricifera
                            Kinorhyncha
                      Nematoida
                          Nematoda
                          Nematomorpha
                    Panarthropoda
                      Tardigrada
                      Onychophora
                      Arthropoda
                  Chaetognatha
            Deuterostomia
                Xenambulacraria
                    Xenoturbellida
                    Ambulacraria
                        Echinodermata
                        Hemichordata
            Chordata

Fig. 4.1. A tentative phylogenetic arrangement of the main metazoan lineages. Italics are used for non-terminal taxa; that is, those for which the table gives subordinate taxa. Non-terminal taxa correspond to the phylogenetic hypotheses dealt with mainly in Chapter 5. Bold is used for terminal taxa, which are discussed in Chapters 4 and 6.

Table 4.1 A list of names proposed for major metazoan groups, with references to the works where these names were established and content (with reference to currently accepted 'phyla'). Names are shown in italics when specifying a usage (a circumscription of the group) other than the original usage, but which is somehow of interest and thus deserving of inclusion.

Group	References	Content
Acanthocephala	Rudolphi (1808–1810)	
Acanthognatha	Cavalier-Smith (1998)	Acanthocephala+Gastrotricha+Gnathostomulida+Rotifera
Acoelomata	Hyman (1951a)	Nemertea+Plathelminthes
Acoelomorpha	Baguñà and Riutort (2004a)	Acoela+Nemertodermatida
Acrania	Bleeker (1859)	
Acrosomata	Ax (1995)	Ctenophora+Bilateria
Ambulacralia	Hatschek (1888)	Echinodermata+Hemichordata
		=Ambulacraria
		=Coelomopora
Ambulacraria	Peterson and Eernisse (2001)	Echinodermata+Hemichordata
		=Ambulacralia
		=Coelomopora
Animalia	Linnaeus (1758)	
Animalia	*Sensu* Zrzavý *et al.* (1998b)	Choanoflagellata+Metazoa
Annelida	Lamarck (1809)	
Arthropoda	Latreille (1829)	
Arthropoda	*Sensu* Ax (2000)	Arthropoda+Onychophora
Articulata	Cuvier (1812)	
Articulata sensu lato	*Sensu* Ghiselin (1988)	Annelida+Arthropoda+Brachiopoda+Mollusca+Sipuncula
Articulata	*Sensu* Ax (2000)	Annelida+Arthropoda+Onychophora
Articulata	*Sensu* Nielsen (2001)	Annelida+Arthropoda+Mollusca+Onychophora+Tardigrada
Aschelminthes	Grobben (1910)	
Aschelminthes	*Sensu* Hyman (1951b)	Gastrotricha+Kinorhyncha+Nematoda+Nematomorpha+ Priapulida+ Rotifera
Aschelminthes	*Sensu* Nielsen (1995a)	Chaetognatha+Gastrotricha+Kinorhyncha+Loricifera+ Nematoda+ Nematomorpha+Priapulida+Rotifera
Aschelminthes	*Sensu* Ehlers *et al.* (1996)	Gastrotricha+Kinorhyncha+Loricifera+Nematoda+ Nematomorpha+ Priapulida
		=Nemathelminthes
Aschelminthes	*Sensu* Kristensen and Funch (2000)	Gastrotricha+Gnathostomulida+Kinorhyncha+Loricifera+ Micrognathozoa+Nematoda+Nematomorpha+Priapulida+Rotifera
		=Pseudocoelomata
Bilateria[1]	Hatschek (1888)	=Triploblastica
Brachiopoda	Cuvier (1802)	
Brachiozoa	Cavalier-Smith (1998)	Brachiopoda, incl. Phoronida
		=Phoronozoa
Bryozoa	Ehrenberg (1831)	
Calcarea	Bowerbank (1864)	=Calcispongea
Calcispongea	Johnston (1842)	=Calcarea
Catenulida	von Graff (1905)	=Notandropora
Cellularia	Reiswig and Mackie (1983)	
Cephalorhyncha	Malakhov (1980)	Priapulida+Kinorhyncha+Nematomorpha
Cephalorhyncha	*Sensu* Malakhov and Adrianov (1995)	Priapulida+Kinorhyncha+Nematomorpha+Loricifera
Cephalorhyncha	*Sensu* Zrzavý *et al.* (1998b), Nielsen (2001)	Priapulida+Kinorhyncha+Loricifera
		=Scalidophora
Chaetifera	Zrzavý *et al.* (1998b)	Annelida+Echiura+Pogonophora

Table 4.1 (*Contd*)

Group	References	Content
Chaetodonta	Kasatkina and Buryi (1997)	Chaetognatha+Conodonta
Chaetognatha	Leuckart (1854)	
Choanozoa	Cavalier-Smith (1981)	
Chordata	Kowalewsky (1866)	
Cnidaria	Hatschek (1888)	
Coelenterata[2]	Frey and Leuckart (1847)	
Coelomata	Lankester (1877)	
Coelomopora	Marcus (1958)	Echinodermata+Hemichordata =Ambulacralia =Ambulacraria
Ctenophora	Eschscholtz (1829–1833)	
Cycliophora	Funch and Kristensen (1995)	
Cycloneuralia	Ahlrichs (1995)	Gastrotricha+Nematoida+Scalidophora
Cyrtotreta	Nielsen (1995a)	Chordata+Enteropneusta
Deuterostomia	Grobben (1908)	Chordata+Echinodermata+Hemichordata+Xenoturbellida[3] =Notoneuralia
Dicyemida	Van Beneden (1876)	=Rhombozoa
Ecdysozoa	Aguinaldo *et al.* (1997)	Cephalorhyncha+Nematoida+Panarthropoda
Ecdysozoa	*Sensu* Zrzavý *et al.* (1998b), Peterson and Eernisse (2001)	Cephalorhyncha+Chaetognatha+Nematoida+Panarthropoda
Echinodera	Gosse (1864)	=Kinorhyncha
Echinodermata	Klein (1734)	
Echiura	Sedgwick (1898)	
Ectoprocta	Nitsche (1869)	
Endocnidozoa	Zrzavý and Hypša (2003)	*Polypodium*+Myxozoa
Enterocoela	Hyman (1959)	Chaetognatha (with doubt)+Chordata+Echinodermata+Hemichordata +Pogonophora
Enteropneusta	Gegenbaur (1870)	
Entoprocta	Nitsche (1869)	=Kamptozoa
Epineuralia	Cuénot (1940)	Bryozoa+Brachiopoda+Phoronida+Deuterostomia =Radialia
Epitheliozoa	Ax (1995)	Eumetazoa+Placozoa
Euarticulata	Nielsen (2001)	Annelida+Arthropoda+Onychophora+Tardigrada
Eubilateria	Hennig (1979)	Bilateria to the exclusion of the Acoela and the Nemertodermatida =Eutriploblastica =Nephrozoa
Eucoelomata	Hyman (1940)	Annelida+Arthropoda+Brachiopoda+Bryozoa+Chaetognatha+ Chordata+Echinodermata+Hemichordata+Mollusca+Phoronida+ Priapulida+Sipuncula
Eucoelomata	*Sensu* Zrzavý *et al.* (1998b)	Annelida+Brachiopoda+Bryozoa+Chordata+Cycliophora+ Echinodermata+Entoprocta+Hemichordata+Mollusca+Nemertea+ Phoronida+Sipuncula
Eumetazoa	Bütschli (1910)	Ctenophora+Cnidaria+Bilateria =Histozoa
Euplathelminthes	Bresslau and Reisinger (1928)	=Rhabditophora
Euspiralia	Ax (1995)	Nemertea+Trochozoa
Eutriploblastica	Zrzavý *et al.* (1998b)	Bilateria to the exclusion of the Acoela and the Nemertodermatida =Eubilateria =Nephrozoa
Eutrochozoa	Ghiselin (1988)	Annelida+Brachiopoda+Mollusca+Sipuncula

Table 4.1 (*Contd*)

Group	References	Content
Eutrochozoa	*Sensu* Zrzavý *et al.* (1998b), Peterson and Eernisse (2001)	Annelida+Mollusca+Nemertea+Sipuncula
Gastroneuralia	Ulrich (1951)	Nemathelminthes+Spiralia
Gastrotricha	Metschnikoff (1865)	
Gnathifera	Ahlrichs (1995)	Gnathostomulida+Rotifera
Gnathifera	*Sensu* Sørensen *et al.* (2000)	Gnathostomulida+Micrognathozoa+Rotifera
Gnathifera	*Sensu* Nielsen (2001)	Chaetognatha+Gnathostomulida+Micrognathozoa+Rotifera
Gnathostomulida	Ax (1956)	
Hemichordata	Bateson (1885)	
Histozoa	Ulrich (1950)	=Eumetazoa
Holochordata	von Salvini-Plawen (1989)	Acrania+Craniota
Holozoa	Lang *et al.* (2002)	Choanoflagellata+Ichthyosporea+Metazoa
Homoscleromorpha	Lévi (1973)	
Introverta	Nielsen (1995a)	Kinorhyncha+Loricifera+Nematoda+Nematomorpha+Priapulida
Kamptozoa	Cori (1929)	=Entoprocta
Kinorhyncha	Reinhard (1887)	=Echinodera
Lacunifera	Ax (1999)	Kamptozoa+Mollusca =Sinusoida
Lemniscea	Garey *et al.* (1996)	Bdelloidea+Acanthocephala
Lobatocerebromorpha	Haszprunar *et al.* (1991)	
Lobopoda	Boudreaux (1979b)	Onychophora+Arthropoda
Lobopoda	*Sensu* Cavalier-Smith (1998)	Onychophora+Tardigrada
Lobopodia	Hou and Bergström (1995)	Onychophora+four extinct groups (Archonychophora, Paronychophora, Protonychophora and Scleronychophora)
Lophophorata	Hyman (1959)	Brachiopoda+Ectoprocta+Phoronida
Lophophorata	*Sensu* Peterson and Eernisse (2001)	Brachiopoda+Phoronida
Lophotrochozoa	Halanych *et al.* (1995)	Annelida+Brachiopoda+Bryozoa+Mollusca+Phoronida
Lophotrochozoa	*Sensu* Garey and Schmidt-Rhaesa (1998)	Annelida+Brachiopoda+Bryozoa+Entoprocta+Mollusca+Phoronida+Sipuncula
Lophotrochozoa	*Sensu* Peterson and Eernisse (2001)	Annelida+Brachiopoda+Bryozoa+Cycliophora+Entoprocta+Gnathostomulida+Mollusca+Nemertea+Phoronida+Plathelminthes+Rotifera+Sipuncula
Lophozoa	Cavalier-Smith (1998)	Annelida+Brachiopoda+Bryozoa+Cycliophora+Entoprocta+Mollusca+Nemertea+Phoronida+Sipuncula; more or less coextensive with Lophotrochozoa
Loricifera	Kristensen (1983)	
Malacosporea	Canning *et al.* (2000)	*Buddenbrockia+Tetracapsuloides*
Mastigocystia	Mamkaev (1995)	=Xenoturbellida
Medusozoa	Petersen (1979)	=Tesserazoa
Mesozoa	van Beneden (1876)	
Metazoa	Haeckel (1874)	
Micrognathozoa	Kristensen and Funch (2000)	
Mollusca	Cuvier (1795)	
Monoblastozoa	Blackwelder (1963)	*Salinella*[4]
Monokonta	Cavalier-Smith (1998)	Gastrotricha+Gnathostomulida =Neotrichozoa
Moruloidea	Hartmann (1904)	=Mesozoa
Myxozoa	Grassé (1970)	
Nemathelminthes	Gegenbaur (1859)	Cycloneuralia+Gastrotricha

Table 4.1 (*Contd*)

Group	References	Content
Nemathelminthes	*Sensu* Ehlers *et al.* (1996)	=Aschelminthes
Nemathelminthes	*Sensu* Cavalier-Smith (1998)	Kinorhyncha+Loricifera+Nematoda+Nematomorpha+Priapulida
Nematoda	Rudolphi (1808–1810)	
Nematoida	Schmidt-Rhaesa (1996)	Nematoda+Nematomorpha
		=Nematoidea
		=Nematozoa
Nematoidea	Ehlers *et al.* (1996)	Nematoda+Nematomorpha
		=Nematoda
		=Nematozoa
Nematomorpha	Vejdovsky (1886)	
Nematozoa	Zrzavý *et al.* (1998b)	Nematoda+Nematomorpha
		=Nematoda
		=Nematoidea
Nemertea	de Quatrefages (1846)	=Nemertini
		=Rhynchocoela
Nemertini	Cuvier (1816)	=Nemertea
		=Rhynchocoela
Neodermata	Ehlers (1985)	
Neorenalia	Nielsen (2001)	=Deuterostomia
Neotrichozoa	Zrzavý *et al.* (1998b)	Gastrotricha+Gnathostomulida
		=Monokonta
Neotriploblastica	Zrzavý *et al.* (1998b)	Eucoelomata+Paracoelomata
Neotrochozoa	Peterson and Eernisse (2001)	Annelida+Echiura+Mollusca+Sipuncula
Nephrozoa	Jondelius *et al.* (2002)	Bilateria to the exclusion of the Acoela and the Nemertodermatida
		=Eubilateria
		=Eutriploblastica
Notandropora	Reisinger (1924)	=Catenulida
Notochordata	Nielsen (1995a)	Acrania+Craniota
Notoneuralia	Ulrich (1951)	=Deuterostomia
Olfactores	Delsuc *et al.* (2006)	Urochordata+Vertebrata
Onychophora	Grube (1853)	
Orthonectida	Giard (1877)	
Panarthropoda	Nielsen (1995a)	Arthropoda+Onychophora+Tardigrada
Paracoelomata	Zrzavý *et al.* (1998b)	Ecdysozoa [incl. Chaetognatha!]+Syndermata+Neotrichozoa
Parazoa	Sollas (1884)	=Porifera
Parenchymia	Nielsen (1995a)	Nemertea+Plathelminthes
Phagocytellozoa	Ivanov (1973)	=Placozoa
Pharyngotremata	Schaeffer (1987)	Cephalodiscida+Cyrtotreta
Phoronida	Hatschek (1888)	
Phoronozoa	Zrzavý *et al.* (1998b)	Brachiopoda+Phoronida
		=Brachiozoa
Placozoa	Grell (1971)	=Phagocytellozoa
Plathelminthes	Schneider (1873)	=Platyhelminthes
Platyhelminthes	Gegenbaur (1859)	=Plathelminthes
Plathelminthomorpha	Ax (1984)	Gnathostomulida+Plathelminthes
Platyzoa	Ax (1987)	Gnathifera+Plathelminthes
Platyzoa	*Sensu* Peterson and Eernisse (2001)	Cycliophora+Gnathifera+Plathelminthes
Pogonophora	Johansson (1937)	
Polyzoa	Thompson (1830b)	Bryozoa+Entoprocta

Table 4.1 (*Contd*)

Group	References	Content
Porifera	Grant (1836–1852)	=Parazoa
Priapozoa	Cavalier-Smith (1998)	Loricifera+Priapulida
Priapulida	Delage and Hérouard (1897)	
Proctozoa	Ghiselin (1988)	Bilateria without Plathelminthes
Prosomastigozoa	Zrzavý *et al.* (2001)	Cycliophora+Myzostomata+Syndermata
Protostomia	Grobben (1908)	
Protostomia	*Sensu* Hyman (1940)	Annelida+Arthropoda+Brachiopoda+Bryozoa+Entoprocta+ Gastrotricha+Kinorhyncha+Mollusca+Nematoda+Nematomorpha+ Nemertea+Phoronida+Plathelminthes+Priapulida+Rotifera+ Sipuncula
Protostomia	*Sensu* Nielsen (2001)	Annelida+Arthropoda+Bryozoa+Chaetognatha+Entoprocta+ Gastrotricha+Gnathostimulida+Kinorhyncha+Loricifera+ Micrognathozoa+Mollusca+Nematoda+Nematomorpha+Nemertea+ Onychophora+Plathelminthes+Priapulida+Rotifera+Sipuncula+ Tardigrada
Protostomia	*Sensu* Peterson and Eernisse (2001)	Annelida+Arthropoda+Brachiopoda+Bryozoa+Chaetognatha+ Cycliophora+Entoprocta+Gastrotricha+Gnathostomulida+ Kinorhyncha+Loricifera+Mollusca+Nematoda+Nematomorpha+ Nemertea+Onychophora+Phoronida+Plathelminthes+Priapulida+ Rotifera+Sipuncula+Tardigrada
Pseudocoelomata	Hyman (1951b)	Entoprocta+Gastrotricha+Kinorhyncha+Nematoda+Nematomorpha+ Priapulida+Rotifera
Pseudocoelomata	*Sensu* Kristensen and Funch (2000)	Gastrotricha+Gnathostomulida+Kinorhyncha+Loricifera+ Micrognathozoa+Nematoda+Nematomorpha+Priapulida+Rotifera =Aschelminthes
Pseudocoelomata	*Sensu* Lorenzen (1985)	Gastrotricha+Kinorhyncha+Loricifera+Nematoda+Nematomorpha+ Priapulida+Rotifera
Pterobranchia	Lankester (1877)	
Pulvinifera	Ax (2000)	Articulata+Sipunculida
Radialia	Jefferies (1986)[5]	Brachiopoda+Bryozoa+Phoronida+Deuterostomia =Epineuralia
Radiata	Lamarck (1801)	
Radiata	*Sensu* Hyman (1940)	Cnidaria+Ctenophora =Coelenterata *Auctorum*
Radiata	*Sensu* Cavalier-Smith (1998)	Cnidaria+Ctenophora+Placozoa+Porifera
Rhabditophora	Ehlers (1985)	=Euplathelminthes
Rhombozoa	van Beneden (1882)	=Dicyemida
Rhynchocoela	Schultze (1851)	=Nemertea =Nemertini
Rotatoria	Ehrenberg (1832)	=Rotatoria
Rotifera	Cuvier (1798)	=Rotifera
Scalidophora	Lemburg (1995)	Kinorhyncha+Loricifera+Priapulida =Cephalorhyncha *sensu* Zrzavý *et al.* (1998b), Nielsen (2001) =Scalidorhyncha
Scalidorhyncha	Cavalier-Smith (1998)	Kinorhyncha+Loricifera+Priapulida =Cephalorhyncha *sensu* Zrzavý *et al.* (1998b), Nielsen (2001) =Scalidophora
Schizocoela	Hyman (1940)	Annelida+Arthropoda+Brachiopoda+Bryozoa+Mollusca+ Phoronida+Priapulida+Sipuncula

Table 4.1 (Contd)

Group	References	Content
Schizocoelia	Nielsen (2001)	Annelida+Arthropoda+Mollusca+Onychophora+ Sipuncula+Tardigrada =Teloblastica
Silicea	Bowerbank (1864)	
Sinusoida	Haszprunar (2000)	Kamptozoa+Mollusca =Lacunifera
Sipuncula	Rafinesque (1814)	
Spiralia[6]	Schleip (1929)	
Stomochordata	Ax (2001)	Chordata+Hemichordata
Symplasma	Reiswig and Mackie (1983)	Hexactinellid sponges
Syndermata	Ahlrichs (1995)	=Trochata
Tardigrada	Doyère (1840)	
Teloblastica	Nielsen (2001)	Annelida+Arthropoda+Mollusca+Onychophora+ Sipuncula+Tardigrada =Schizocoelia
Tesserazoa	Salvini-Plawen (1978)	=Medusozoa
Testaria		=Eumollusca
Triploblastica	Lankester (1873)	=Bilateria[7]
Trochata	Cavalier-Smith (1998)	=Syndermata
Trochozoa	Beklemishev (1944)	Annelida+Ectoprocta+Entoprocta+Mollusca+Phoronidea+Sipuncula
Trochozoa	*Sensu* Zrzavý *et al.* (1998b)	Annelida+Cycliophora+Entoprocta+Mollusca+Nemertea+Sipuncula
Trochozoa	*Sensu* Ax (2000)	Annelida+Arthropoda+Entoprocta+Mollusca+Sipuncula
Trochozoa	*Sensu* Peterson and Eernisse (2001)	Annelida+Entoprocta+Mollusca+Nemertea+Sipuncula
Tunicata	Lamarck (1816)	=Urochordata
Urochordata	Lankester (1877)	=Tunicata
Vermizoa	Cavalier-Smith (1998)	Annelida+Nemertea
Vertebrata	Lamarck (1801)	
Vertebrata	*Sensu* Ax (2003)	Acrania+Craniota =Holochordata =Notochordata
Vestimentifera	Webb (1969)	
Vinctiplicata	Lemburg (1999)	Loricifera+Priapulida
Xenambulacraria	Bourlat *et al.* (2006)	Ambulacraria+Xenoturbellida
Xenoturbellida	Bourlat *et al.* (2006)	=Mastigocystia

[1] Nielsen (1995) included the Ctenophora here.

[2] Cavalier-Smith (1998) revived this term for a taxon including Cnidaria and Ctenophora.

[3] This is the current membership, following re-evaluation of the affinities of *Xenoturbella* by Bourlat *et al.* (2003). Authors such as Hyman (1959) also included here the Pogonophora and (with doubt) the Chaetognatha; in this concept, Deuterostomia=Enterocoela. The former Tentaculata were sometimes included in the Deuterostomia and Nielsen (2001) still placed the Brachiopoda and Phoronida here.

[4] It is very unlikely that an animal corresponding to Frenzel's (1892) description of *Salinella salve* actually exists.

[5] Introduced informally: Jefferies (1986), Fig. 2.3.

[6] Besides a core set of taxa represented by Annelida, Entoprocta, Gnathostomulida, Mollusca, Nemertea, Plathelminthes, and Sipuncula, some authors (e.g. Nielsen 1995a, 2001) included in the Spiralia also the Arthropoda, the Onychophora, and the Tardigrada or the Micrognathozoa (Nielsen 2001), the Rotifera (Garey and Schmidt-Rhaesa 1998, Nielsen 2001), and some or all of the 'Tentaculata' and even the Gastrotricha (Garey and Schmidt-Rhaesa 1998) or the Chaetognatha (Nielsen 2001).

[7] Nielsen (2001) extended this taxon's scope to include, additionally, the Ctenophora.

A sweeping revisitation of the traditional views of cnidarians (and ctenophores) as primitively radial metazoans has been prompted by new embryological evidence as well as by data on the expression of body-patterning genes. Based on embryological evidence, Martindale et al. (2002) described cnidarians (and, dubitatively, also ctenophores) as primitively bilaterally symmetrical animals and suggested that the single plane of mirror symmetry of hexacorals might be homologous to the dorsoventral axis of Bilateria. Bilaterality of cnidarians is supported by the expression pattern of multiple *Hox* genes and the transforming growth factor-β gene *decapentaplegic* (*dpp*): in the small anthozoan *Nematostella vectensis*, a recent entry in the set of model organisms in developmental biology, several members of the *Hox* family are expressed in staggered way along the animal's primary axis, whereas *dpp* is expressed in an asymmetric fashion about its secondary body axis (Finnerty et al. 2004). In the same animal, other genes more or less directly involved in establishing the polarity of animal body axes, *chordin*, *Noggin1*, *Gsc*, and *Netrin*, are also expressed asymmetrically along an axis perpendicular to the oro-aboral one (Matus et al. 2006b).

A major difficulty in comparing cnidarian and bilaterian organization is, however, the uncertain equivalence of their body axes. Interpreting the cnidarian body in terms of the expression of *Hox* genes is problematic, because of obvious differences within the phylum. In the anthozoan *Nematostella*, *anthox6*, which is homologous to bilaterian *Hox* genes such as *labial* in *Drosophila*, which are expressed anteriorly along the animal's body axis, is expressed embryonically in the proximity of the blastopore; later, in the solid, mouthless planula larva its expression is found at the end of the body that locomotion identifies as posterior, but in the planula of the hydrozoan *Podocoryne* the homologue of an 'anterior' *Hox* gene is found instead at the functionally anterior end (Yanze et al. 2001, Finnerty et al. 2004). So, which end of a polyp (the oral or the aboral one) is equivalent to the fore end of a bilaterian (provided that the answer is to be actually found between these two alternatives)? The presence in cnidarians of

homologues of the bilaterian *Hox* (and *ParaHox*) genes is not accompanied by comparable patterns of expression; the same is true of other genes involved in anteroposterior patterning such as *orthodenticle*, *empty spiracles*, and *Nkx2.5* or in dorsoventral patterning such as *decapentaplegic/ bone morphogenetic protein* (*BMP*) and *chordin/short gastrulation* (Hayward et al. 2002, Finnerty 2003, Finnerty et al. 2004). This makes a comparison of cnidarian and bilaterian body axes very problematic (Chourrout et al. 2006, De Jong et al. 2006, Kamm et al. 2006, Rentzsch et al. 2006).

Baguñà et al. (2008) have recently provided an insightful analysis of the problem, of which the following is a summary. The planula larva of cnidarians does not have a mouth, and the front end of a swimming larva can be confidently identified as aboral. This is the body end with which the larva will attach to a substrate when transforming into a polyp. But this is where problems begin. If the polarity of the main body axis is conserved from the planula to the polyp stage, then the aboral rather than the oral end of the polyp would correspond to the anterior end of a bilaterian. This would be supported by the similarities in gene-expression patterns between the oral region in cnidarians and the organizer region in chordates (and their equivalent in other bilaterians), which corresponds to either the posterior or the ventral pole of bilaterians (Arendt et al. 2001, Technau 2001). Equating the oral pole of the polyp to the bilaterian anterior pole, as suggested by Martindale (2005), would entail the inversion of the anteroposterior axis between planula and polyp.

Conservation of axis polarity is suggested, however, by *Dickkopf*, a gene expressed in the aboral half of the planula and in the peduncle and basal disc of the polyp. Interestingly, in both cnidarians and bilaterians this gene antagonizes the *Wnt* signalling pathway, and Wnt has been considered a posterior marker in bilaterians. This would thus support the interpretation of the aboral end of the planula and the foot of the polyp as homologous to the anterior region of bilaterians (Meinhardt 2002).

The new scenario, according to which bilateral symmetry was achieved at a cnidarian grade of morphological evolution—that is, among sessile organisms—makes the traditional adaptive explanation unviable: there is no scope for suggesting that bilateral symmetry conferred a mechanical advantage over radial symmetry in facilitating directed locomotion.

4.2.3 Two versus three germ layers

Besides symmetry, another critically important point about which traditional views may require revisitation is the diploblastic nature of cnidarians. Indeed, if polyps are definitely diploblastic, there may be reasons to argue that some medusae are triploblastic and even coelomate (Boero et al. 2005). Seipel and Schmid (2006) argue instead that cnidarians and bilaterians probably share a common Precambrian triploblastic ancestor, with mesodermal muscles but no polyp stage; as a consequence, the diploblastic hydrozoan polyp would be a secondary larval stage.

Two lines of evidence had lead to the suggestion that cnidarians (or, at least, hydrozoans) may be fundamentally triploblastic. One is the presence of sheets of striated muscle cells, repeatedly reported from the subumbrella of

some hydromedusae (e.g. Schmid 1974, Fautin and Mariscal 1991); striated fibres have been also reported from ctenophores (Hernandez-Nicaise 1991).

The other putative evidence in favour of a triploblastic nature of at least some cnidarians is provided by the entocodon (Fig. 4.2). This is a tissue layer intermediate between the ectodermal and the endodermal one, which differentiates during the late development of hydromedusae from their parent polyps. The fact that this endocodon cavitates to produce the muscle of the medusa bell has suggested the idea that hydrozoans are indeed triploblasts (Boero et al. 1998). Seipel and Schmid (2006) regard the entocodon as homologous to the bilaterian mesoderm, because of its origin and position, but also because in transplantation experiments it acts as organizer for the development of the medusa (Reisinger 1957). This behaviour is clearly suggestive of the organizer role of mesoderm of the blastoporal area in many bilaterians. Further support is provided by the several genes involved in the specification of the mesoderm or, more specifically, of the myogenic cell lineage, which are not simply shared by cnidarians and bilaterians, but have been also

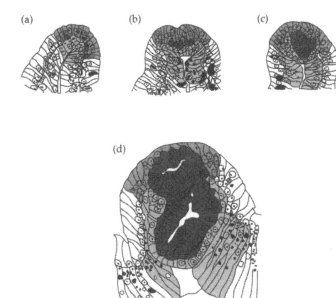

(a)

(b)

(c)

(d)

Fig. 4.2. Four stages in the differentiation of the entocodon in a hydrozoan, eventually forming (d), a tentacular and a subumbrellar cavity between the ectoderm and endoderm. Redrawn from unpublished drawings by Stefano Piraino.

demonstrated to have similar functions in both lineages (Schuchert *et al.* 1993, Gröger *et al.* 1999, Spring *et al.* 2000, 2002, Galliot and Schmid 2002, Seipel and Schmid 2005, 2006).

To investigate the hypothesis that the entocodon of jellyfish is homologous to the mesoderm of bilaterians, a homologue of each of the three gene families *Brachyury, Mef2,* and *Snail* was isolated from then hydrozoan *Podocoryne carnea* and their expression patterns were studied throughout the life cycle and specifically during muscle development. The results demonstrate that all three genes are expressed during myogenic differentiation. Additionally, as is true for their bilaterian cognates, they appear to have other functions as well. The sequence and expression data demonstrate that the genes are structurally and functionally conserved and even more similar to deuterostomes than to protostome model organisms such as *Drosophila* or *Caenorhabditis elegans.* These results further strengthen the hypothesis that the common ancestor of cnidarians and bilaterians already used the same regulatory and structural genes and comparable developmental patterns to build its muscles (Spring 2002).

This interpretation of cnidarians as triploblastic animals has not been adequately discussed to date. Opinions contrary to the putative homology between the entocodon of medusae and the mesoderm of typical triploblastic metazoans have been expressed, for example, by Collins (2002), Collins *et al.* (2006), and Schmidt-Rhaesa (2007). Martindale *et al.* (2002) have also expressed doubts as to the homology between the striated muscles of cnidarians and those of other metazoans. I am inclined to accept the entocodon as a mesoderm equivalent and the striated muscle of some medusae as largely equivalent to the striated muscles of bilaterians but, at the same time, I would be cautious in respect to their phylogenetic significance. On the one hand, the entocodon and the striated muscle are only known from growing or adult medusae of the Medusozoa, but the medusa is quite likely an apomorphy of this clade; a secondary loss of medusa, striated muscle, entocodon, etc. at the origin of the Anthozoa is unlikely. Therefore, the

triploblastic condition of the earliest cnidarians remains questionable. On the other hand, I think that the whole issue of germ layers (their identity, number, and developmental fate) has been generally treated too rigidly. As I will discuss in Chapter 7, we must accept for germ layers the same principle of factorial homology that I will repeatedly advocate in this book (see also Minelli 1998, 2003) as a way to avoid arbitrary classifications of cell lineages and body parts, and developmental processes alike. The two features to be compared will generally deserve to be called homologous according to one or more specified criteria, while failing on other criteria; and to atomize structures and processes to such a degree as to be left with 'units of homology' to which a crude yes-or-no criterion applies would make the whole issue of homology uninteresting, if applicable at all.

To close this account, I will invite those zoologists who are sceptical about the hypothesis of a triploblastic nature of cnidarians to comment on Seipel and Schmid's (2005) remark, that in no bilaterian embryo is there any evidence for the preservation of a diploblastic phase during the blastula–gastrula transition, as should be expected to survive, in some clades at least, if a major diploblastic period had indeed occurred during metazoan evolution.

4.2.4 A cnidarian–deuterostome connection?

The traditional, well-entrenched dichotomy between diploblastic and triploblastic metazoans, together with the 'privileged' position all too often accorded to the deuterostomes (an unconscious consequence of their including the human species or, at least, the vertebrates at large), has shadowed the fact that deuterostomes share with cnidarians, but not with protostomes, a not trivial set of characters. Whether all of these can be regarded as plesiomorphic for the eumetazoans at large may deserve a detailed analysis, which is beyond the scope of this book.

Plesiomorphic, of course, are traits which are not known from any protostome, but are also shared by sponges, as is *Pit-1*, a gene involved in the production of the pituitary gland in vertebrates, and long regarded as exclusively present in chordates, which has been found by Jacobs and Gates (2003) in ctenophores, cnidarians, and sponges. As far back as Beklemishev (1969) we can read about the similarities between the blastulae of deuterostomes (amphioxus, echinoids) and those of sponges and hydroids. Another plesiomorphic trait basal deuterostomes have in common with 'diploblasts' is the epithelial organization of the nervous system, although with some local condensation (Schmidt-Rhaesa 2007). The ganglion of the pterobranchs and the brain structures of tunicates and vertebrates are thus derived features. In other words, a brain may not pertain to the groundplan of the deuterostomes and this would imply multiple independent origins of the brain, contrary to currently widespread opinion (e.g. Kammermeier and Reichert 2001, Reichert 2005).

The draft genome of the anthozoan *Nematostella vectensis* is more similar to the genomes of vertebrates than to those of insects or nematodes (Putnam *et al.* 2007): similarities are found in the overall complexity of the genome, in the gene repertoire, in exon/intron structure, and in large-scale gene linkage. This evidence invites us to credit the eumetazoan common ancestor with a comparably high degree of genome complexity. With respect to non-eumetazoan genomes, some 20% of genes in the genome of *Nematostella* seem to be eumetazoan novelties involved in the control of activities such as cell adhesion, cell signalling, and synaptic transmission.

More generally, from cnidarians we know genes that encode functional domains involved in gene regulation, translational control, signal transduction, apoptosis, extracellular signalling, specification of the myogenic cell lineage, and cell–extracellular matrix interactions (Galliot 2000). The full diversity of metazoan signalling pathways, such as Wnt, TGF-β, hedgehog, Ras-mitogen-activated protein kinase (MAPK), and Notch, has been found in cnidarians (Technau *et al.* 2005).

On the other hand, the divergence between the Cnidaria and the Bilateria must have occurred prior to the development of a *Hox* gene cluster as is found in bilaterians (Kamm *et al.* 2006). At

least, an equivalent of the canonical *Hox* cluster has not been found in cnidarians (Peterson *et al.* 2000a). On the contrary, there is evidence, in *Nematostella vectensis* and *Acropora millepora*, of 'ancient' genes shared by cnidarians with non-metazoan eukaryotes, but never found in bilaterians. While reporting on this discovery, Technau *et al.* (2005) are inclined to reject the hypothesis that this presence is due to recent lateral gene transfer rather than by descent.

4.2.5 The basal branches of the eumetazoan tree

Two of the three possible trees whose branches are Cnidaria, Ctenophora, and Bilateria have enjoyed support during the history of animal phylogenetics (Fig. 4.3). In one of these trees, the two groups traditionally described as diploblastic (Cnidaria and Ctenophora) unite in a clade Coelenterata, as the sister group of the Bilateria. Alternatively, the Cnidaria have been often regarded, in recent works, as the sister group of a clade Acrosomata within which Ctenophora would group with Bilateria. Finally, a third alternative, with Ctenophora branching off first, thus leaving a clade of Cnidaria plus Bilateria, cannot be excluded from consideration, and indeed it has found some support in molecular

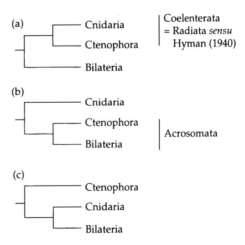

Fig. 4.3. The three possible trees depicting the phylogenetic relationships between Ctenophora, Cnidaria, and Bilateria.

phylogenetic analyses (Wainright *et al.* 1993, Kim *et al.* 1999, Medina *et al.* 2001; Podar *et al.* 2001; all based on 18S rDNA). This is the alternative favoured here.

The fact that this third hypothesis has not been seriously considered before has perhaps its explanation in the difficulty of accepting that seemingly more complex animals (the cteno-phores) may be more distantly related to bilateri-ans than are the seemingly simpler cnidarians. This *scala naturae* perspective, also possibly supported by the additional consideration that the apparent biradial symmetry of ctenophores fits in quite easily between the putative radial organization of cnidarians and the overtly bilat-eral body structure of bilaterians, might be still more justified if ctenophores, as it was some-times suggested, are in fact triploblastic rather than diploblastic. Character sets supporting each one of the three hypotheses are as follows.

1. Cnidaria and Ctenophora have in common unilateral cleavage, the first cleavage furrow establishing the origin of the oral–aboral axis, the second cleavage furrow dividing the embryo into four radial quadrants; the site where the first cleavage furrow is first manifested cor-responding to the future site of gastrulation (Scholtz 2004a).

2. The affinity between Cnidaria and Bilateria is supported by the presence of *Hox* and *ParaHox* genes (Finnerty 1998, Martinez *et al.* 1998, Finnerty and Martindale 1999) and the collin-earity of the *Hox* gene expression (Martindale *et al.* 2002).

3. Common to Ctenophora and Bilateria are the presence of true muscle cells, multiciliate cells, complex sensory organs, a through-gut, and a highly stereotyped cleavage (Martindale and Henry 1997, Martindale *et al.* 2002). Ctenophora and Bilateria have in common, additionally, the following characters (Scholtz 2004a): sperm cell with acrosome (Ehlers 1993) (but this is reduced in several bilaterians; an acrosome is otherwise present in some sponges (Baccetti *et al.* 1986) and in the siphonophoran *Muggiaea kochi* (Carré 1984)); muscular fibres (in the mesogloea of Ctenophora, but also in the largest medusae); and mesoderm (Nielsen 2001), but with the proviso

that genes such as *twist*, *snail*, and *forkhead*, which in the bilaterians are mesoderm markers, are expressed in cnidarian 'endoderm' (Technau and Scholz 2003, Martindale *et al.* 2004).

Molecular phylogenies yield contradictory results: relationships among sponges, placo-zoans, cnidarians, and ctenophores are still uncertain despite comparison of mitochondrial genomes (Haen *et al.* 2007, Wang and Lavrov 2007) and 50 nuclear genes (Rokas *et al.* 2005). A very large database of small-subunit (SSU) rRNA sequences from 528 metazoan taxa, plus outgroups (Wallberg *et al.* 2004), suggests any-way that Ctenophora are the sister group of a Cnidaria+Bilateria clade. I will tentatively accept this topology.

4.3 The sponges: Silicea, Homoscleromorpha, and Calcarea

4.3.1 Sponge paraphyly

Morphological characters such as the presence of choanocytes and their organization into water chambers with inhalant and exhalant pores are the traditional evidence regarded until recently as strong enough as to sup-port sponge monophyly (Böger 1988, Ax 1996, Reitner and Mehl 1996, Nielsen 2001). However, the monophyly of sponges has been rejected by molecular analyses based, for example, on 18S rDNA (Cavalier-Smith *et al.* 1996, Collins 1998, Borchiellini *et al.* 2001, Peterson and Eernisse 2001), protein kinase C (Kruse *et al.* 1998), and a set of seven nuclear-encoded proteins (Peterson and Butterfield 2005). According to these ana-lyses, Calcarea are the sister group of the non-sponge metazoans (Epitheliozoa), a grouping also supported by an ultrastructural character: the striated ciliary rootlets present in the larvae of calcareous sponges (Wollacott and Pinto 1995) and also in the Epitheliozoa (Rieger 1976), but not in the other sponges.

Previous suggestions (e.g. Bergquist 1985) that the sponges may be diphyletic, with the essen-tially syncytial Hexactinellida (Symplasma) and the remaining cellular sponges (Cellularia: the Demospongiae and the Calcarea) representing two distinct lineages, has been rejected.

A newly emerging contentious issue is the position of the Homoscleromorpha, whose placement within the Demospongiae is potentially challenged by their many peculiarities in morphology and development, often shared with non-sponge metazoans (see below).

Summing up, I accept here provisionally the articulation of the former Porifera suggested by Nichols (2005), with three main separate 'phyla', the Silicea (with the Hexactinellida and the Demospongiae *sensu stricto*), the Homoscleromorpha, and the Calcarea.

Embryonic and larval development have been frequently involved in discussions about the metazoan nature and the affinities of sponges. This is an area where much progress has been made recently, with results, however, contributing to a better understanding of developmental processes more than to a better general knowledge of sponge affinities.

In sponges, gastrulation involves cellular reorganizations similar to processes known in other invertebrates (Leys and Degnan 2002, Maldonado and Bergquist 2002, Maldonado 2004). Only the cinctoblastula larva of the Homoscleromorpha forms by a unique process of multipolar egression of cells, which will eventually form the external cell layer of the sponge (Boury-Esnault *et al.* 1995, 2003). Of more general interest is thus the fact that in most groups of sponges gastrulation precedes the developmental transition to the larval stage, but some larvae are instead formed before gastrulation. This diversity of ontogenetic schedules invites a revisitation of the concept of gastrulation (Leys and Ereskovsky 2006). The fact is that the differentiation of two main cell layers and the formation of an internal digestive cavity are not necessarily associated, in sponges, as they generally are in other metazoans. (A dissociation between the formation of the two germ layers and the formation of the gut is also known from several cnidarians; Byrum and Martindale 2004.) This has led some authors (Ereskovsky and Korotkova 1997, Ereskovsky 2005) to deny the existence of gastrulation in sponges, pointing to the fact that sponges do not have cells comparable with those lining the gut of the Eumetazoa and, more generally, denying the homology between the external and internal

cell layers of a sponge and the 'true' germ layers of the other metazoans. Other authors (Lévi 1963, Tuzet 1963, 1973, Brien 1967, Fell 1974), perhaps taking the term gastrulation literally, in the sense of formation of a digestive cavity, identified sponge gastrulation with the time the sponge forms an internal cavity with choanocyte-lined chambers, which are equated with a gut. I prefer to follow a third option, which identifies gastrulation with the formation of two (or three) layers. Specialists of sponge development are still divided in deciding whether this differentiation precedes metamorphosis or is virtually identical to it. Some authors (e.g. Ereskovski and Dondua 2006) have suggested that these layers invert at metamorphosis, with the result that sponge germ layers would be inverted with respect to other animals. Other authors (e.g. Leys and Degnan 2002, Maldonado 2004) regard the changes that occur at metamorphosis as a rearrangement of already differentiated layers.

Under this concept of gastrulation, most sponge larvae are interpreted as post-gastrulation stages consisting of already differentiated cells. This is also supported by the fact that sponge larvae are usually more complex structurally than the adults into which they will metamorphose (Maldonado 2004). An example is provided by the structure of the pinacoderm, the external cell sheet of the adult sponge, which consists of cells that often migrate into the underlying mesenchyme because no basement membrane separates the latter from the pinacoderm (the only exception being provided by the Homoscleromorpha; Boute *et al.* 1996). Larval epithelia have, generally, a less fluid and more complex organization, with different cell types showing a distinct polarity as typical of true epithelia, being often supported, basally, by the presence of abundant collagen (as in the hoplitomella larva of alectonids; Vacelet 1999). More complex than the corresponding features in the adult are also some ultrastructural aspects of the ciliated cells of sponge larvae. At variance with the uniformly uniciliated condition found in adults, biciliated cells are also found in the larvae and the trichimella larva of the hexactinellids has unique epithelial organization with multiciliated cells internal to the surface non-ciliated syncytium, so that their cilia can emerge only by 'piercing' through the latter (Boury-Esnault and Vacelet 1994, Boury-Esnault *et al.* 1999).

4.3.2 Hexactinellida

The main problem to be solved in assessing the phylogenetic position of these sponges is whether, or how, their largely syncytial organization can be derived from the cellular organization found in the other sponges and generally supposed to be present in the earliest metazoans. Accurate ultrastructural research on larval and adult Hexactinellida suggests that their peculiar organization is indeed secondary. Hexactinellid cytoplasm consists of a giant, multinucleated syncytium, which is connected via open and plugged cytoplasmic bridges to archaeocytes, choanoblasts, and other cells. All these components of the sponge are interconnected, and electrical signals can propagate through the animal. The plugged perforate junctions apparently allow tissues to specialize in different ways while maintaining limited cytoplasmic continuity. Larvae of hexactinellid sponges are already largely syncytial since the larval stage, but embryos are cellular until gastrulation, thus hexactinellid sponges may well have evolved from cellular sponges (Leys 2003). Maximum-likelihood trees based on 18 S rRNA sequences indeed suggest that hexactinellid sponges evolved from demosponges (Cavalier-Smith and Chao 2003).

Comparisons involving the nearly complete mitochondrial genome sequence of two hexactinellid sponges (*Iphiteon panicea* and *Sympagella nux*) revealed several genomic features uniquely shared between Hexactinellida and Bilateria, including an Arg → Ser change in the genetic code, a characteristic secondary structure of the serine tRNAs, highly derived tRNA and rRNA genes, and the presence of a single large non-coding region (Haen *et al.* 2007). How this discovery can be accommodated within the emerging consensus that the Calcarea, rather than the Hexactinellida, are the most likely sister group of the Bilateria, is a point that requires further work.

4.3.3 Homoscleromorpha

Uncontentiously included in the Demospongiae in the past, these sponges possess an impressive set of unique characters that have caused their affinities to be seriously questioned. The peculiarities of the Homoscleromorpha include:

- cinctoblastula larvae that form by a peculiar process of multipolar egression (Boury-Esnault *et al.* 1995, 2003);
- a basement membrane underlying both choanoderm and pinacoderm (Boute *et al.* 1996) but also present in the larvae (Boury-Esnault *et al.* 2003). It is sensible to describe these cell sheets as true epithelia, as they also show cell polarization and apical cell junctions (Leys and Ereskovsky 2006);
- aquiferous system and spicules (when present) with distinctive morphology (Gaino *et al.* 1987, Boute *et al.* 1996, Ereskovsky and Boury-Esnault 2002, Muricy and Diaz 2002);
- spermatozoa with acrosome (Boury-Esnault and Jamieson 1999).

Sequences of nuclear genes would suggest that the Homoscleromorpha are quite isolated with respect to the other Demospongiae (Borchiellini *et al.* 2004), but mitochondrial sequences would suggest otherwise. The mitochondrial genome of *Oscarella carmela* (Plakinidae) (Fig. 4.4), with its 44 genes, is the largest known among animals. But this might not be representative of the whole clade, as there are no more than 20 genes in the mitochondrial genome of another plakinid, *Plakortis angulospiculatus*, and mitochondrial DNA sequences place *O. carmela* with other demosponges (Wang and Lavrov 2007), something that would suggest a body-plan simplification in most sponges.

4.3.4 Calcarea

Axial symmetry seems to be primitive for the Calcarea (Manuel *et al.* 2003). Indeed, as mentioned above, the 'simple' ascon kind of sponge organization, long regarded as the most primitive, is found in some calcisponges only. Actually, in developmental terms this is the

Fig. 4.4. Specimen of the homoscleromorphan sponge *Oscarella lobularis* (the dark lobular mass). Courtesy of Giorgio Bavestrello and Carlo Cerrano.

sponge model that requires most patterning cues, as it presents a distinct major axis; therefore, it is more likely that the 'diffuse' leucon model is most primitive and the ascon the most derived.

This agrees with the current views on sponge phylogeny, as in molecular phylogenetic studies, the calcareous sponges have repeatedly emerged as the sister group of non-sponge metazoans (Lafay *et al.* 1992, Cavalier-Smith *et al.* 1996, Collins 1998, Kruse *et al.* 1998, Zrzavý *et al.* 1998b, Adams *et al.* 1999, Borchiellini *et al.* 2001, Medina *et al.* 2001; and other more recent papers).

Morphologically, the monophyly of the calcareous sponges is supported only by the peculiar symmetry of their spicules (Manuel *et al.* 2003), but this hypothesis is also strongly supported by 18 and 28S rDNA sequences (Manuel 2006). An internal phylogeny of calcareous sponges, based on 18S rRNA data, implies that the skeletal architecture and the aquiferous system are highly homoplastic; however, axial symmetry seems to be primitive for all Calcarea (Manuel *et al.* 2003).

4.4 Placozoa

Placozoans lack symmetry, organs, extracellular matrix, muscle cells, and main body axis. There are cell junctions (Rieger 1994a), but no basal lamina. No gametes or, indeed, sexual phenomena, have been reported from placozoans; however, molecular signatures for sex in *Trichoplax* are revealed by comparison of short sequences from the mitochondrial 16S rDNA from different specimens, including sharing of alleles between heterozygous and homozygous individuals, and intergenic recombination (Signorovitch *et al.* 2005). Is the primitive organization of *Trichoplax* a primitive feature or the result of secondary reduction? Structural simplification starting from cnidarian or cnidarian-like ancestors has been suggested. Indeed, maximum-likelihood trees of 18S rRNA suggest that Placozoa are derived from medusozoan Cnidaria (Cavalier-Smith and Chao 2003). However, the circular mitochondrial genome of *Trichoplax* is incompatible at least with the derived, linear organization of the mitochondrial genome of the Medusozoa and the predicted secondary structure of the 16S rRNA molecule is also different between *Trichoplax* and cnidarians, with respect to the number and length of stem and loop regions (Ender and Schierwater 2003).

The first *Hox/ParaHox*-type gene to be identified in *Trichoplax adhaerens* (*Trox-2*) was interpreted as either a candidate *ProtoHox* or a *ParaHox* gene (Jakob *et al.* 2004). It is expressed in a ring around the periphery of *Trichoplax*, and marks the epithelial boundary between the upper and lower epithelial cell layers. If *Trox-2* function is inhibited, growth stops and the animal undergoes binary fission. Further homeobox genes belonging to different families (*Not*, *Dlx*, *Mnx*, and *Hmx*) have been also described from *Trichoplax*. At least two of these genes have spatially restricted expression around the periphery of the animal (Monteiro *et al.* 2006), thus are possibly involved in a minimal degree of body patterning.

Syed and Schierwater (2002) suggested a scenario for the evolution of the Placozoa based on constructional morphology. Placozoa would have been derived from benthic gallertoid pluricellulars, the 'placuloids', and would represent a basal metazoan stem line which separated from the other metazoans earlier than sponges and coelenterates. This is supported by the size and structure of the mitochondrial DNA of *Trichoplax* (Dellaporta *et al.* 2006). While mitochondrial genomes of most metazoans are 15–24 kb circular molecules that encode a nearly universal set of 12–14 proteins for oxidative phosphorylation and 24–25 structural RNAs (16S rRNA, 12S rRNA, and tRNAs) and lack both introns and significant intragenic spacers, the mitochondria of *Trichoplax* contains one of the largest known metazoan mitochondrial DNA genomes (43 079 bp), with numerous intragenic spacers and some introns; moreover, it contains open reading frames of unknown function, and unusually large protein-coding regions. This mitochondrial DNA has traits in common with chytrid fungi and choanoflagellates, but also shares derived features unique to the Metazoa. This strongly suggest a basal placement of Placozoa in the metazoan tree. Although only one placozoan species has been formally described to date, diversity in this group is certainly greater than this (Voigt *et al.* 2004, Signorovitch *et al.* 2006).

4.5 Ctenophora

The contrast between diploblastic and triploblastic organization has been long regarded as a major divide between the Cnidaria and the Ctenophora on one side, and the Bilateria on the other. Occasionally, putative evidence suggesting

the presence of mesoderm in the Ctenophora has been read to indicate that this group may be closer to the Bilateria than are the Cnidaria. Nowadays, there is no doubt anymore as to the presence of endomesoderm (Martindale and Henry 1995, Martindale *et al.* 2002) and of a variety of muscle cell types (Hernandez-Nicaise 1991, Martindale and Henry 1999) in the Ctenophora, but the old argument has lost its force, especially if we accept that the intermediate germ layer is also present in some hydrozoan medusae (see section 4.2.3). Nevertheless, the molecular evidence suggesting that the Cnidaria are phylogenetically closer to the Bilateria than the Ctenophora contrasts with several aspects of ctenophore development, which match the corresponding events in several bilaterian groups better than does the embryonic development of cnidarians: this includes deterministic cleavage and the establishment of equivalence groups of blastomeres sharing a common prospective developmental fate (Henry and Martindale 2004).

The so-called cydippid larva is a kind of 'phylotypic' stage, common to most ctenophore groups, irrespective of their often quite divergent adult morphology, some remaining practically unchanged (many Cydippida), others changing into shapes as different as the cucumber-like *Beroe*, the band-shaped *Cestus*, or the flattened benthic platyctenids. *Stromatoveris psygmoglena* is a recently described fossil from the Lower Cambrian Chengjiang Lagerstätte (Yunnan, China). Despite its manifest similarities to Ediacaran vendobionts, Shu *et al.* (2006) contend that its closely spaced branches were probably ciliated and may represent the precursors of the comb rows of ctenophores.

4.6 Cnidaria

Phylogenetic studies concur with those in developmental genetics in turning upside down the fashionable scenario of a progressive evolution, within the Cnidaria, from radially symmetrical forms to those with a biradial or bilateral symmetry, and from simple, hydra-like polyps to those with a wide, compartmentalized gastral cavity, as in the sea anemones and anthozoans generally. Integral to this changed view of morphological evolution is the current appreciation of the evolution of the cnidarian life cycle. Hyman (1959) championed the idea, recently maintained, for example, by Valentine (2004), according to which the medusa represent the adult organization of the primitive cnidarian, thus requiring a secondary reduction of this stage in several lineages, including the Anthozoa.

Today, a sister-group relationship between the Anthozoa and the remaining Cnidaria (the Medusozoa or Tessarazoa) is well supported by morphology, life history, genome structure, and DNA sequences (Bridge *et al.* 1995, Cavalier-Smith *et al.* 1996, Collins 1998, 2002, Kim *et al.* 1999, Medina *et al.* 2001).

Apomorphies of the Medusozoa (Collins 2002) are the linear (rather than circular) mitochondrial genome, the presence of a medusa, and the rigid cnidocilium. Thus, only the polyp stage would have been present in the earliest cnidarians, a condition retained by the Anthozoa, and the origin of the medusa, characterized by its peculiar shape, a thick mesogloea, a gastrovascular system with radial canals and ring canal, a manubrium, and a nervous concentration along the margin of the umbrella, would be monophyletic (*contra* Ax 1996).

A cladistic analysis of 87 morphological and life-history characters of medusozoan cnidarians, rooted with Anthozoa, resulted (Marques and Collins 2004) in the phylogenetic hypothesis (Anthozoa (Hydrozoa (Scyphozoa (Staurozoa, Cubozoa)))). The Stauromedusae are separated from the Scyphozoa to form, together with the fossil group Conulata, a new clade Staurozoa. The clade Scyphozoa, with the remaining subclades Coronatae, Semaeostomeae, and Rhizostomeae, is redefined as including those medusozoans characterized by strobilation and ephyrae. However, Van Iten *et al.* (2006) place the Conulata with the Coronatae, with which the fossil group (?Vendian to Triassic) shares the presence of a periderm, a trait that appears to be derived within the Scyphozoa. Moreover, the Conulata multiplied by strobilation, as do the Coronatae, Rhizostomeae, and Semaeostomeae—that is, all 'true' Scyphozoa—but not the Stauromedusae.

The isolated position of the Stauromedusae has been confirmed by a new phylogenetic analysis based on nuclear rDNA sequences (18 and 28S) (Collins *et al.* 2006). In this study, the Stauromedusae resulted as the sister group of all remaining Medusozoa, followed by a split between the Hydrozoa and the Acraspeda. The latter clade correspond to the Scyphozoa of the old classifications prior to the separation of the Cubozoa as an independent class, but for the exclusion of the Stauromedusae. The mitochondrial genome of the Cubozoa is the most derived among those of the Medusozoa, as it consists of four distinct linear molecules (Ender and Schierwater 2003).

Discussions of cnidarian phylogeny have increasingly involved two unusual organisms which have very little in common with 'typical' cnidarians, both in morphology and in life history.

One of them is *Buddenbrockia plumatellae*, the other is *Polypodium hydriforme*. But *Buddenbrockia*, in turn, brings with it into the argument the whole group of the Myxozoa.

Myxozoa is a eukaryote taxon with a most incredible history of phylogenetic assignment. The modern period of this history started when Wolf and Markiw (1984) demonstrated that actinosporeans and myxosporeans, previously regarded as members of two separate classes of protists, were nothing other than different stages in the complex life cycles of these parasites. The next step was the demonstration (Siddall *et al.* 1995), on molecular evidence, of the cnidarian affinities of these eukaryotes, now called the Myxozoa, something only vaguely suspected by an earlier researcher, based on the the presence of complex 'spores' somewhat comparable with the cnidarian sting cells, although the filament containing these capsules is eventually used to fix to a new host, rather than to capture prey by use of neurotoxic poison.

More recently, attention has been focused on *Buddenbrockia*, perhaps the most unusual representative of this clade (Fig. 4.5). All myxozoans, as said, are parasites, mostly alternating between a vertebrate and an annelid or bryozoan host. During their life cycle, myxozoans are generally

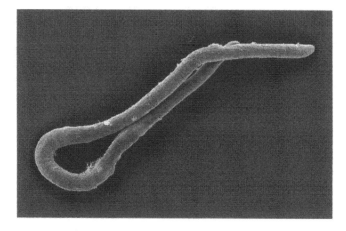

Fig. 4.5. A specimen of *Buddenbrockia plumatellae*. Courtesy of Eva Jimenez Guri.

represented by syncytia or by single-cell stages which do not offer additional morphological clues as to their phylogenetic relationships. The Myxozoa include two classes, the Malacosporea and the Myxosporea. Two genera are referred to the Malacosporea: *Tetracapsuloides*, alternating between freshwater bryozoans and salmonid fishes, and *Buddenbrockia*, which seems not to require a vertebrate host and to complete the cycle within the bryozoan host. Several genera are referred to the better known Myxosporea, which alternate between fish and annelids. *Buddenbrockia* is anatomically more complex than most myxozoans: it is a little worm-like animal, up to 3 mm long, without gut and central nervous system, but with a body wall of two cell layers and four longitudinal muscles topographically arranged as the four muscle bundles of nematodes, a simple organization compatible with a bilaterian ancestry. This was initially supported by some molecular phylogenies, although some studies suggested instead cnidarian affinities; a decisive bilaterian signature would have been the presence of central-class *Hox* genes, reported from a myxozoan by Anderson *et al.* (1998), but this has been found to be due to a contamination by host tissue (Jiménez-Guri *et al.* 2007). Comparisons based on 18S rRNA and the characteristic multicellular reproductive (and infectious) spores maturing in the internal cavity place *Buddenbrockia* firmly with the malacosporean myxozoans.

Buddenbrockia demonstrates that no new organ system has been required to obtain a worm-like parasite from a generalized cnidarian body structure. There is no through-gut, no external sensory organs have evolved, and even the original four-fold symmetry has been maintained, four longitudinal muscles providing the anatomical basis for the locomotion of this 'worm'. Molecular phylogenies based on multiple protein-coding genes have definitely established *B. plumatellae* as a member of the Cnidaria, more precisely of the Medusozoa (Jiménez-Guri *et al.* 2007).

Cnidarian (hydrozoan) affinities have been long suspected in the case of *Polypodium hydriforme*. Its life cycle includes a free-living stage alternating with a stage parasitizing the eggs of sturgeons and other ganoid fishes. From the infectious stage, represented by a binucleate cell, a bilayered stolonal body eventually develops, whose external cell sheet is described as endodermic, the internal one as endodermic, the eventually 'normal' position being obtained by eversion before the stolon fragments into a number of buds, each giving rise to a free globular stage that reproduces asexually by longitudinal fission. The resulting individuals will reproduce sexually (Bouillon *et al.* 2006). Molecular data suggest that *Polypodium* is probably a sister group of the Myxozoa (Siddall *et al.* 1995, Zrzavý *et al.* 1998b, Siddall and Whiting 1999).

CHAPTER 5

The entangled phylogeny of the Bilateria

The need for a solid and well-resolved phylogeny of the Metazoa is now greater than ever.

Jenner (1999), p. 246

5.1 The Bilateria

There seems to be no reason to dispute the monophyly of the Bilateria; nevertheless, listing their morphological and developmental apomorphies (Scholtz 2004a) is not so obvious as one may perhaps expect. Bilateral symmetry belongs to the ground plan of the Bilateria, but even this simple feature of general body organization may eventually turn out not to be a bilaterian novelty, if the relationships among the most basal branches of the phylogenetic tree are different from those summarized in this book's preferred tree, as the organization of the Cnidaria (and, to some extent, also that of the Ctenophora) is primitively bilateral rather than radial. We are on safer ground with listing the triploblastic condition—that is, the presence of mesoderm—as a bilaterian apomorphy. This is hardly challenged by a possible identification of the medusan entocodon as a germ layer different from both ectoderm and endoderm, because the bilaterian mesoderm and the cnidarian entocodon may well have evolved in parallel. Arguably belonging to the bilaterian ground plan is the development from mesodermal derivatives of a musculo-cutaneous body wall including an external layer of circular muscles and an internal system of longitudinal muscles. Additionally, and characteristically, bilaterians have evolved a main body axis distinctly patterned under the control of transcription factors, due to the spatially and temporally patterned expression of a conserved set of genes, of which the *Hox* genes represent the most conspicuous component. The central nervous system was probably the first body feature to be longitudinally patterned in this way. The most anterior part of the body, where there is no *Hox* gene expression, is embryonically patterned by another gene, *orthodenticle*, which fixes the position of another arguably apomorphic bilaterian feature; that is, the brain.

Flatworms have played a major role in shaping our ideas about early bilaterians. To some extent, this is due to the fact that a 'typical' free-living flatworm such as a freshwater planarian is anatomically simpler than many other bilaterians in that it does *not* possess an anus, or a circulatory system, or any kind of body cavity; its nervous system is quite diffuse, rather than centralized, and its sensory organs are simpler than those of many other metazoans. The only major difficulty with presenting a planarian as a lowly worm is the very complex anatomy of its genital system. Ignoring this 'oddity', planarians would play perfectly their role of basal bilaterians preceding the advent of anus and circulatory system, two novelties introduced by the nemerteans (in the pages of textbooks, although not necessarily in evolutionary history).

Within the flatworms, the most perceptive authors, for example Hyman (1951a), have long singled out the Acoela as possibly closest to the organization of the earliest bilaterians. Eventually, molecular phylogenetic studies have shown that the acoels are, arguably, the most basal branch of the bilaterian tree, while the majority of the flatworms (including planarians, flukes, and tapeworms) represent a distinct clade emerging later from within the bilaterian radiation. To be sure, at present we cannot completely rule out the possibility that some traits of the very simple body architecture of acoels are regressive rather than primitive, but there are good reasons for accepting as a primitive feature their nearly total lack of anteroposterior body patterning (section 7.5.2), arguably linked to the very limited number of *Hox* genes in their genome.

Before the advent of molecular phylogenetics, the relationships between the main bilaterian lineages were largely discussed with reference to developmental evidence. The type of cleavage, the fate of the blastopore, the filling of the blastocoel, and the origin of body cavities featured as the main items in this discussion, together with the comparative morphology of larval types.

Spiral and radial cleavage have been long contrasted as irreducibly distinct, not only because of the peculiar spatial arrangement of blastomeres in either model, but also because of the very early commitment of blastomeres in 'mosaic' spiralian embryos to an irrevocably specified fate, whereas the blastomeres in 'regulative' radially cleaving embryos retain for a while a distinct multipotentiality. However, our current knowledge in developmental biology does not allow us regarding this distinction as a reliable divide, because all intermediates exist between a cell-autonomous and an externally controlled behaviour of blastomeres. Nevertheless, the metazoans with spiral cleavage cluster fairly consistently together in phylogenetic trees based on molecular evidence (see below).

Bilaterians have been traditionally divided into Protostomia and Deuterostomia, the names of the two branches mirroring the presumed opposite fate of the blastopore. Protostomes would be the metazoans where the primary embryonic 'mouth' is retained as the animal's definitive mouth, whereas the deuterostomes would be the metazoans where the definitive mouth is something other than the blastopore, the latter either closing definitely or surviving as the animal's anus. However, problems with this distinction have long since been known, especially the infrequent

but phylogenetically scattered cases of amphistomy; that is, the derivation of both mouth and anus from a slit-like blastopore that eventually closes at midlength. Amphistomy is usual in onychophorans (Nielsen 2001), not rare in annelids (Steinmetz *et al.* 2007), and has also been observed in a nematomorph (Inoue 1958) and in enoplid nematodes (Malakhov 2003), and is possibly the ancestral state in the Bilateria (Arendt and Nübler-Jung 1997, Arendt *et al.* 2001).

In most directly developing nemerteans (a typical 'protostome' clade, where the blastopore should give rise to the definitive mouth) the mouth actually forms as secondary invagination at the site of the closed blastopore or close to it. Due to differential growth, however, the mouth moves along the ventral surface; it does not open always externally, because sometimes the oesophagus fuses with the rhynchocoel (Hammersten 1918; Henry and Martindale 1997). Moreover, the fate of the blastopore is characteristically variable within some lineages. For example, within a 'protostome' phylum such as the Arthropoda, in most malacostracan crustaceans the definitive mouth is secondary, as in the Deuterostomia, but this is not true of amphipods, where the blastopore survives as the adult mouth, as expected in the Protostomia (Scholtz and Wolff 2002). A new mouth forms also in some plathelminths, nemerteans and phoronids (Jenner 2004c). Within the Gastropoda there are examples of a mouth arising from the blastopore (as in *Patella*) or near the blastopore (as in *Crepidula*), but also less frequent cases (e.g. *Viviparus*) where the blastopore turns into the anus, while the mouth forms anew at another location, as in a typical deuterostome (Hyman 1967). Arendt *et al.* (2001) regarded the shared expression of two genes (*orthodenticle* and *Brachyury*) as convincing evidence to derive both mouth and anus from the region of the blastopore. But the expression of these genes is not necessarily restricted to these areas and, as noted by Baguñà *et al.* (2008), in the embryos of many molluscs the blastopore does not contribute to the formation of the anus.

Mechanisms of blastopore closure are also variable within a lineage. For example, within the clitellate annelids the blastopore of *Tubifex rivulorum* starts closing at the posterior end, whereas in the enchytraeid *Pachydrilus lineatus* the same process begins at the anterior end (Penners 1922, 1930).

Similar to the traditional phylogenetic interpretation of the blastopore fate, the presence and developmental origin of body cavities has been traditionally used to establish major divisions within the Bilateria, following a typological and progressionist approach. Phylogenetic hypotheses requiring the loss of a general body cavity in this or that lineage have enjoyed little favour. Basically, grouping of bilaterian phyla derived from the following criteria. First, a split between (1) the acoelomates, lacking a general body cavity, and thus parenchymatous, including the flatworms and, tentatively, the nemerteans; (2) the pseudocoelomates, with a body cavity generally regarded as a surviving blastocoel and not lined by an epithelial sheet (a coelothel or mesepithelium); and (3) the coelomates, whose body cavities are formed anew within the mesodermal mass, while the blastocoel generally survives in the form of a circulatory system. Second, within the coelomates, a major distinction was generally made according to the mechanism by which the coelom is produced, by either schizocoely or enterocoely. Schizocoely means the origination of coelomic cavities by dehiscence of tissue within an initially compact mass of mesoderm, whereas enterocoely means that coelomic cavities form as outpouchings of the wall of the archenteron (the embryonic gut). To some extent, the divide between the lineages showing schizocoely and enterocoely, respectively, matches the divide between the protostomes and the deuterostomes. However, if the deuterostomes are all enterocoelic coelomates, the protostomes do not include the schizocoelic coelomates only, but also acoelomate and pseudocoelomate groups. Thus, the monophyly of protostomes and deuterostomes would not support the monophyly of the coelomates, and vice versa. But this is just the most conspicuous problem with using the structure and origin of the main body cavities as a marker of kinship.

In agreement with most traditional scenarios of metazoan phylogeny, Bartolomaeus (1994) argued that the earliest metazoans possessed a compact body without internal cavities, but Rieger (1985) regarded the coelomate condition as plesiomorphic for bilaterians, thus implying that both the acoelomate and the pseudocoelomate condition are derived. In light of more recent phylogenetic reconstructions, the identification of the acoelomate Acoela as the most basal branch of the Bilateria would support the traditional hypothesis that the earliest bilaterians were not coelomate. On the other hand, molecular phylogenies do not suggest that all coelomate lineages share a common coelomate ancestor; that is, a multiple origin of the main body cavities is very likely (e.g. Minelli 1995, Jenner 2004c).

Larvae such as the annelid or mollusc trochophore, or the tornaria of hemichordates have largely played the role of ancestral protostome and deuterostome larva, respectively. In a further effort to derive all ciliate larvae of marine invertebrates from a single ancestral model, Hatschek (1878) elaborated the theory according to which most if not all of the bilaterian phyla derive from a common ancestor developing through a trochophore-like larva. This scenario has been further modified by some authors (e.g. Nielsen 1985, Nielsen and Nørrevang 1985) in a Haeckelian perspective, suggesting that modern trochophore larvae recapitulate the adult condition of early bilaterians. Adult rotifers have been also repeatedly equated to trochophore larvae, in particular by advocating homology of the trochus and the cingulum, two ciliary bands of rotifers, with the prototroch and the metatroch of the trochophore respectively (e.g. Nielsen 1987, 1995a, 2001, Peterson and Eernisse 2001), but this homology has been rejected convincingly (Jenner 2004c).

An opposite view of the evolution of the animal life cycle is emerging today (see section 7.4.5). In many groups, larvae are arguably an addition to a primarily direct development and the striking similarities among larval forms of different clades are probably due to a limited range of body shapes evolvable within the constraints set by the small cell number (often around one thousand) and the small range of available life styles. Parallel independent evolution of larvae is also likely within a single lineage. For example, Müller's larvae and Goette's larvae (collectively called lobophora larvae) are present only in a part of the Polyclada; the non-basal position of this clade within the flatworms (Rhabditophora) argues strongly in favour of an independent origin of these larvae during flatworm evolution, rather than as a trait inherited from a common ancestor with annelids and molluscs (Ehlers 1985, Jenner 2004c). According to the phylogeny of the Nemertea established by Thollesson and Norenburg (2003) based on the nuclear genes for 28 S rRNA and histone H3 and the mitochondrial genes for 16 S rRNA and cytochrome *c* oxidase subunit I, the basal nemertean clades lack a pilidium larva, which thus must have evolved within the phylum, at the base of a clade the authors have established under the name Pilidiophora.

5.2 Urbilateria

Haeckel's (1874) gastraea is probably the most popular among a series of models intended to represent important steps along the early evolution of metazoans. Bütschli's (1884) placula and Metschnikoff's (1886) parenchymella were soon to follow, but the series has continued to grow up to the present. In the current debates on animal phylogeny, prominence is possibly not given to those old models, or to the much more recent trochaea (Nielsen 1985, Nielsen and Nørrevang 1985), as the stage has been solidly occupied by a new, fashionable entry called the Urbilateria (De Robertis and Sasai 1996). This model of the last common ancestor of all living bilaterians has turned out to be not simply popular, but also prolific, as it is currently presented in several alternative versions.

Introducing the Urbilateria was motivated by some of the most unexpected discoveries in developmental genetics: the fact that the genes specifying the dorsal and the ventral side of a vertebrate have their homologues in *Drosophila*, where they are also involved in determining the dorsoventral polarity, but with inverted roles. The *Drosophila* gene *short gastrulation* is expressed ventrally, while its vertebrate homologue *chordin* is expressed dorsally, and the reciprocal is true of the *Drosophila decapentaplegic* gene, which is expressed dorsally, and its vertebrate homologue *bone morphogenetic protein-4* (*BMP4*), which is expressed ventrally. This fuelled the hope of solving one of the most intractable problems in the comparison of animals traditionally assigned to different phyla and thus to trace back in time the main features of animal organization, eventually ending up with a reconstruction of Urbilateria (Fig. 5.1).

However, far from easily leading to a consensus view of the likely organization of the last

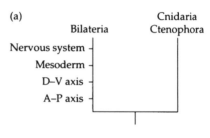

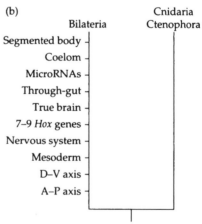

Fig. 5.1. Urbilateria, the last common ancestor of the Bilateria, has been often reconstructed as a quite complex animal (Table 5.1), but this concept may require radical revision, particularly in the light of current ideas about the basal position of the Acoela and Nemertodermatida in the bilaterian tree. A 'simple' Urbilateria (a) implies a limited number of novelties in the bilaterian lineage, with respect to its sister group, whereas a 'complex' Urbilateria would collapse many more novelties onto the corresponding segment of the phylogenetic tree (b). The illustration summarizes ideas developed by Baguñà *et al.* (2008) and others. D–V, dorsoventral; A–P, anteroposterior.

common ancestor of the bilaterians, this effort has resulted in the production of many different reconstructions. Thus, the bilaterian ancestor has been conceived either as small to microscopic (e.g. Baguñà *et al.* 2001, following the identification of acoels as the most basal extant bilaterians; Valentine 2004, based mainly on palaeontology), or as a macroscopic, complex, and possibly colonial (Dewel 2000) or polyp-like animal (Sauer and Kullman 2005).

In the spirit of Jägersten's (1972) model of a pelago-benthic life cycle (with a pelagic larva metamorphosing into a benthic adult) assumed as primitive in the bilaterians, Rieger (1994a, 1994b) suggested that the earliest bilaterians were large animals metamorphosing from a small, free-living larva. I think that there are good arguments to suppose that the early bilaterians were instead small animals with direct development. This is supported by two main lines of evidence.

One of these, following Baguñà *et al.* (2001, 2008; see also Baguñà and Riutort 2004b), is the model provided by the Acoela, a group that molecular evidence strongly suggests to be the earliest branching group among extant bilaterians (see section 6.1). Acoels are small animals (1–12 mm) with direct development. The other argument, from palaeontology, is the large size (mostly, between 0.5 and 1 mm) of all the earliest eggs or embryos putatively referable to bilaterians known from Late Precambrian or Early Cambrian rocks (discussed briefly in Chapter 3). Successful reproduction of very small animals such as the loriciferans or the miniaturized forms that have evolved repeatedly within the annelidan clade requires partitioning the limited resources the animal can invest in reproduction into a very small number of large eggs, thus producing offspring with higher expected average individual success. This is something that, due to small parental size, could not be balanced by very numerous progeny, as the latter would derive from smaller and thus potentially less successful embryos. Thus, odd as it may seem, the production of large eggs is perfectly compatible with the very small size of the animals that produce them.

A problem with most of the suggested reconstructions of Urbilateria (Table 5.1) is the fact that these have not been obtained by applying correct methods of phylogenetic inference and are seldom discussed in relation to real phylogenetic hypotheses. However, even before the now growing consensus about the phylogenetic position of the Acoela, which very strongly argues in favour of a simple Urbilateria, alternative models favouring a high degree of complexity in the common ancestor of Bilateria were distinctly at odds with the mainstream appreciation of metazoan phylogeny. It was not so easy to accept that the earliest bilaterians were segmented and had eyes, a heart, and a complex brain.

On the other hand, it seemed to require too many *ad hoc* hypotheses to explain the common involvement, for example, of the *Pax6* gene homologues in controlling the formation of insect, squid, and vertebrate eyes, or of the *tinman* gene homologues in controlling the formation of the heart in both insects and vertebrates, as simple, repeated instances of convergence. The individual genes, characters, and taxa must be evaluated one by one (several aspects of this question are treated in Chapters 7 and 8), but a general point can be made that largely demolishes the seemingly intractable nature of the problem. As is very often the case in matters of comparative biology, difficulties arise whenever we fail to dissect to an adequate degree the features to be compared into their developmental or historical components, such as, in the present case, the eyes, the brain, and the segments. We have already seen in two instances (choanoflagellates and sponges) that the presence of genes known to be involved in the morphogenesis of structural features only present in more complex metazoans can be generally traced back to much deeper nodes in the phylogenetic tree. The more we know about the expression of these genes, the more evidence we obtain of their involvement in many developmental or functional contexts other than control of the formation of a particular organ, or of a complex anatomical feature. Moreover, the roles in which the products of these genes operate are often quite specific at the cellular level, but also 'generic' at the level of tissues or organs where

Table 5.1 Putative features attributed by recent authors to a complex Urbilateria.

Feature	References
Macroscopic size	Rieger (1986, 1994b), Dewel (2000)
Colonial structure	Dewel (2000)
Bilaterally symmetrical body with conical ends	Ax (1996)
Anteroposterior polarity	De Robertis (1997), Carroll *et al.* (2005)
Dorsoventral patterning	De Robertis (1997)
Cephalization	Finkelstein and Boncinelli (1994), Dewel (2000)
Brain with distinct areas	Arendt and Nübler-Jung (1996), Balavoine and Adoutte (2003)
Segmented body	Kimmel (1996), De Robertis (1997), Holland *et al.* (1997), Palmeirim *et al.* (1997), Christ *et al.* (1998), Holland and Holland (1998), Dewel (2000), Balavoine and Adoutte (2003), Carroll *et al.* (2005), de Rosa *et al.* (2005)
Appendages	De Robertis (1997), Dewel (2000), Dong *et al.* (2001), Balavoine and Adoutte (2003), Carroll *et al.* (2005)
Coelom	Valentine *et al.* (1996) (haemocoel), Dewel (2000) (also gonocoel), Balavoine and Adoutte (2003), Carroll *et al.* (2005)
Contractile blood vessel or heart	De Robertis (1997), Balavoine and Adoutte (2003) , Carroll *et al.* (2005)
Skeleton	Jacobs *et al.* (2000)
Segmented nephridia	Balavoine and Adoutte (2003)
Primitive photoreceptors	Bolker and Raff (1996), De Robertis (1997), Gehring and Ikeo (1999), Neumann and Nüsslein-Volhard (2000), Kumar (2001), Balavoine and Adoutte (2003)
Free discharge of gametes and external fertilization in water	Ax (1996)
Biphasic life cycle	Rieger (1986, 1994b)
Radial cleavage	Ax (1996)
Development through a primary ciliated larva with a tube-shaped gut divided into anterior, middle, and posterior portions	Arendt *et al.* (2001)
Acoelomate or pseudocoelomate larva or juvenile	Rieger (1986, 1994b)
Elongation by posterior addition	Balavoine and Adoutte (2003), de Rosa *et al.* (2005)

they may be involved. We must thus accept that their repeated or consistent involvement in the developmental origin of a specific, complex feature can often be a consequence of their multiple recruitment in ontogenetic processes whose end products are convergently similar because they shared some components, as well as because of the similar selective regimes to which they have been probably exposed.

The simple Urbilateria model supported by phylogenetic considerations invites a revival of one of the traditional scenarios of metazoan evolution, namely the planuloid–acoeloid theory (Hyman 1951, Salvini-Plawen 1978, Baguñà *et al.* 2008). In this hypothesis, an early step in the history of the Metazoa is represented by small,

bilaterally symmetrical organisms with direct development, similar to the planula of most living cnidarians. These animals would have given rise to both the cnidarians and the bilaterians. Along the latter line, the earliest forms would have been not so different from modern acoels. Thus, we must reconstruct Urbilateria as acoelomate, non-segmented, and with a very limited degree of anteroposterior patterning, as mirrored today in the instability of the position of the mouth and the genital organs in living acoels. Comparison with the planula would anyway suggest some degree of body polarity, with sensory cells at the anterior end and a mouth, possibly accompanied by a rudimentary and blind gut, at the opposite end. If modern acoels or nemertodermatids are in effect a good model for reconstructing Urbilateria, the latter was simple, with a compact body (no body cavity, no excretory organs), a blind gut, a nervous system without ganglia, and sensory structures mainly represented by some kind of statocyst (Hejnol and Martindale 2008).

5.3 The main branches

5.3.1 Nephrozoa

Bilateria excluding Acoela and Nemertodermatida
This clade can be established by the possession of the following apomorphies: septate junctions, cerebral ganglion, orthogonal nervous system, protonephridia, spiral quartet cleavage, and fixed cell fate during cleavage. However, this implies at least the loss of protonephridia in large clades such as the Panarthropoda and the Nematoida (Jenner 2004c).

5.3.2 Protostomia

Caution is required in evaluating most of the literature about the monophyly of the Protostomia, because the Acoela and the Nemertodermatida were traditionally assigned to this clade (both of them as subclades of the Plathelminthes), while their much more likely position as the most basal branches of the Bilateria has been recognized only in the last few years (section 6.1). The monophyly of the Protostomia was regarded as doubtful by Jenner (2004c), but this hypothesis is apparently rescued by Philippe *et al.*'s (2005) multigene analysis of a large dataset (146 genes, 35371 positions) from a diverse taxon sample (35 species).

5.3.3 Platyzoa

Acanthocephala+Gastrotricha+Gnathostomulida+ Plathelminthes+Rotifera
Several molecular analyses, including of mitochondrial gene arrangement and the *Hox* gene complement, would suggest that the 'true' plathelminths (Catenulida and Rhabditophora; see section 6.7) have lophotrochozoan affinities

(Balavoine 1997, 1998, Littlewood *et al.* 1998, de Rosa *et al.* 1999, Ruiz-Trillo *et al.* 1999, 2002, Giribet *et al.* 2000, Baguñà *et al.* 2001, Littlewood and Olson 2001, von Nickisch-Rosenegk *et al.* 2001, Zrzavý *et al.* 2001, Jondelius *et al.* 2002). The Platyzoa hypothesis is supported by a molecular analysis based on complete SSU data and nearly complete (>90%) large nuclear ribosomal subunit (LSU or 28S) data for 36 lophotrochozoan species (Passamaneck and Halanych 2006).

5.3.4 Gnathifera

Syndermata+Gnathostomulida+Micrognathozoa
Ahlrichs (1995) introduced a taxon, Gnathifera, to include the Gnathostomulida together with the Syndermata (Rotifera plus Acanthocephala). Supporting apomorphies are the presence of jaw elements with tube-like support rods composed of electron-lucent material surrounding an electron-dense core and the presence of cross-striated pharyngeal muscles that attach to the jaw elements through epithelial cells (see also Jenner 2004c). The subsequently discovered Micrognathozoa (Kristensen and Funch 2000) are easily accommodated within the Gnathifera (Sørensen *et al.* 2000).

In addition to Syndermata, Gnathostomulida, and Micrognathozoa, Gnathifera may include also Myzostomida and Cycliophora, as suggested by some analyses, for example that of Giribet *et al.* (2004b), based on molecular data; other evidence may point towards a close relationship between Cycliophora and Syndermata. But other cladistic studies suggest instead a sister-group relationship between Cycliophora and Entoprocta (Funch *et al.* 2005). The Plathelminthes share with the Gnathifera a lack of mitosis in differentiated somatic cells (Ahlrichs 1995).

5.3.5 Spiralia=Trochozoa

Annelida+Entoprocta+Gnathostomulida+Mollusca+ Nemertea+Plathelminthes+Sipuncula (but see Table 4.1 for the different circumscription of this taxon according to some authors)
This clade is based on the possession of the following apomorphies: spiral quartet cleavage

and a trochophore larva with prototroch, a pair of ocelli, and possibly an apical organ (Ax 1996). Within this large group, a distinction is traditionally made according to putatively different cell patterns during embryonic cleavage stages. The difference would reside in the identity of the blastomeres which appear to be arranged in a cross-like pattern. In the 'molluscan cross', the blastomeres involved in this pattern would be $1a^{12}-1d^{12}$ and $2a^{11}-2d^{11}$ and their derivatives; in the 'annelidan cross', instead, the cross would be formed by $1a^{112}-1d^{112}$ and $1a^2-1d^2$ and their derivatives. But Jenner (2003) has demonstrated that this distinction is an arbitrary one, as both patterns, molluscan and annelid, are readable in embryos of either group, and also of sipunculans. Other aspects of embryonic development are arguably more informative for resolving phylogeny among the spiralians (Jenner 2004c), for example the mechanics of mesentoblast formation (van den Biggelaar et al. 1996, 1997, Guralnick and Lindberg 2001, 2002, Nielsen and Meier 2002). This evidence supports the Eutrochozoa hypothesis; that is, a close relationship between the Nemertea and Mollusca+Annelida (van Loon and van den Biggelaar 1998).

5.3.6 Mesozoa

Orthonectida+Rhombozoa
The earliest molecular phylogenies challenged both the monophyly of the Mesozoa and the idea, embodied in the Greek roots of their name, that these small parasites are somehow intermediate between 'quasi-animals' (the sponges or parazoans, as the poriferans were sometimes called) and the 'true' metazoans. This concept of intermediate organisms along the animal lineage was clearly rooted in a progressionist view of the living world, rather than in 'tree thinking' as implied by an evolutionary world view. At any rate, nowadays if a zoologist thinks of a body organization as 'less than fully metazoan', the model would be offered by placozoans rather than by mesozoans. The often floated idea that their simple anatomy is a consequence of parasitic life seems to be confirmed by recent studies, which suggest a position within the Spiralia. Evidence of spiral

pattern in early cleavage stages of rhombozoans has been reported by Furuya et al. (1992) and a 'spiralian signature' represented by a characteristic sequence of seven amino acids has been reported for a *Hox* gene (*DoxC*) from the rhombozoan *Dicyema* (Kobayashi et al. 1999).

Contrary to earlier results from molecular phylogenies (Hanelt et al. 1996, Pawlowski et al. 1996), the monophyly of the Mesozoa has been confirmed by subsequent studies, mainly based on 18 S rDNA sequences (Winnepenninckx et al. 1998b, Zrzavý et al. 1998b, Siddall and Whiting 1999). The monophyly is also supported by morphological characters, for example the presence of a multiciliated epidermis enclosing a non-digestive gamete-producing tissue, a unique type of ciliary roots (see Slyusarev 1994 for the Orthonectida and Furuya et al. 1997 for the Rhombozoa), and a reduction of many organs and tissue types, including muscular, nervous, and sensory system. Most significant is the absence of *zonulae adhaerentes*, a trait that suggests that the epidermis of the Mesozoa has lost the epithelial nature of the epidermis in the other eumetazoans (Ax 1996). Nevertheless, belt desmosomes (and a cuticle) have been described in orthonectids (Slyusarev 1994, 2000) and gap junctions in rhombozoans (Furuya et al. 1997).

5.3.7 Lophotrochozoa

Annelida (including Echiura, Myzostomida, Sipuncula)+Mollusca+Nemertea+Phoronozoa
The Lophotrochozoa were originally proposed (Halanych et al. 1995) to embrace Annelida+Mollusca and Brachiopoda+Phoronidea+Bryozoa, extended by Garey and Schmidt-Rhaesa (1998) to include also Sipuncula and Entoprocta, and more loosely applied by Peterson and Eernisse (2001) who further extended this term to include, additionally, Cycliophora, Gnathostomulida, Nemertea, and Plathelminthes. The monophyly of the Lophotrochozoa is supported by multigene analysis of a large dataset (146 genes, 35 371 positions) from a diverse sample of animals (35 species) (Philippe et al. 2005).

A close link among all the core lineages of the Lophotrochozoa is possibly represented by a diverse array of Cambrian fossils. One is *Wiwaxia corrugata* from the Burgess Shales, which was dorsally provided with clusters of sclerites that have been considered equivalent to the paleae, a kind of flattened setae present in some modern 'polychaetes' (Butterfield 1990, 2006). Another is *Halkieria evangelista* from the Sirius Passet formation in Greenland, which is generally reconstructed as a flat, slug-like animal bearing two largely separated dorsal shells, one close to the anterior end of the body, the other close to the posterior end. *Halkieria* has been interpreted as somehow related to either the brachiopods (Conway Morris and Peel 1995) or the molluscs (Vinther and Nielsen 2005). Finally, the recently described *Orthrozanclus reburrus* (Conway Morris and Caron 2007) would represent a kind of connecting link between wiwaxiids and halkieriids, all of them being tentatively referred to a newly established clade, Halwaxiida. Several other early-Palaeozoic fossils could fit into the same assemblage. On the whole, these Cambrian forms would represent evidence for a diversified radiation of the Lophotrochozoa, of which the molluscs and the annelids would be the most conspicuous surviving groups. This might still hold even in the case of a revised interpretation of the affinities of some fossil forms. For example, *Wiwaxia* may not be an annelid (Eibye-Jacobsen 2004), but this would not necessarily rule out a placement within the lophotrochozoans.

Another piece in the puzzle is possibly *Odontogriphus* (Conway Morris 1976), a shell-less flat-bodied animal from the Burgess Shales interpreted by Caron *et al.* (2006, 2007) as a stem-group mollusc based on the putative presence of a radula and of respiratory organs (ctenidia) like those of most molluscs. Butterfield (2006) offered an alternative interpretation and regarded *Odontogriphus* as a stem-group lophotrochozoan.

5.3.8 Eutrochozoa

Nemertea+Neotrochozoa
As shown by Jenner (2004c), the most reliable synapomorphies of the Eutrochozoa are the schizocoel and the mesodermal bands. Molecular

evidence for a sister-group relationship of the nemerteans with either the Neotrochozoa or other groups ranging from entoprocts and brachiopods to cycliophorans and rotifers, is still uncertain, but there is no reasonable doubt as to their belonging to the Lophotrochozoa (e.g. Zrzavý *et al.* 1998b, 2001, Ruiz-Trillo *et al.* 1999, 2002, Giribet *et al.* 2000, Peterson and Eernisse 2001, Jondelius *et al.* 2002). The phylogenetic significance of the shared origin of the endomesoderm from the 4d blastomeres (Peterson and Eernisse 2001) is questionable, as the character is also shared by rhabditophorans. No special phylogenetic signal must be read in the molluscan or annelid cross, already mentioned above, of which there is no evidence in nemerteans (Maslakova *et al.* 2004).

5.3.9 Neotrochozoa

Annelida (including Echiura and Sipuncula)+ Mollusca
Close relationships between the molluscs and the annelids has been long suspected, despite the obvious anatomical differences separating a group of unsegmented animals as the molluscs from a lineage of prototypical segmented animals like the annelids. Most conspicuous are the similarities in the highly conserved spiral cleavage pattern, in the origin of apical and cerebral ganglia and in the cell lineage leading to the formation of the prototroch, the ciliary band that represents one of the most characteristic features of their trochophores or trochophore-like larvae (Nielsen 2004) (Figs 5.2 and 5.3). Other putative morphological apomorphies are less certain. In molecular phylogenies, molluscan and annelid taxa cluster regularly together, sometimes even in intermingled way.

5.3.10 Ecdysozoa

Nematoda+Nematomorpha+Kinorhyncha+Loricifera+ Priapulida+Arthropoda+Onychophora+Tardigrada
This very large clade is characterized by the lack of primary ciliated trochophore-type larva, the presence of a cuticle composed of α-chitin which is periodically moulted under the influence of

Fig. 5.2. Trochophore of the 'polychaete' *Serpula*. Courtesy of Claus Nielsen.

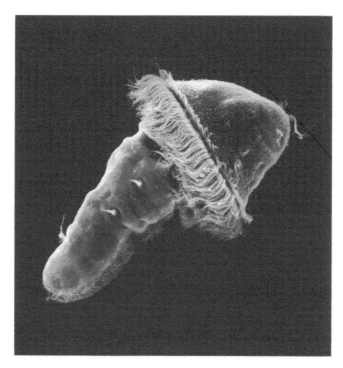

Fig. 5.3. A late larval stage of *Serpula*. Courtesy of Claus Nielsen.

ecdysteroid hormones, the lack of locomotory cilia, and the presence of a haemocoel. Up to the late 1980s, few zoologists objected seriously to the old Cuvierian concept of the Articulata (Cuvier 1816), a major animal group including annelids and arthropods. Apparently, the gross morphology and the internal anatomy provided a satisfactory support for close affinity between the two phyla. Difficulties, admittedly, emerged from embryonic development. Efforts to find evidence of spiral cleavage in arthropod embryos (e.g. Anderson 1973) were hardly convincing and the widespread occurrence among marine annelids of ciliated larvae comparable with those of many marine molluscs provided additional strength to an annelid–molluscan connection already suggested by the similarity of embryonic cleavage patterns.

Even in the context of an Articulata hypothesis it was not difficult to see that many if not most of the idiosyncratic traits of arthropods could be eventually explained as correlated with, or derived from, one major differential trait in respect to the annelids; that is, the presence of a thick cuticle. This cuticle ruled out the presence of ciliated epithelia and required a set of physiological mechanisms allowing for repeated moulting. Very few zoologists, however, were ready to consider the possible phylogenetic significance of the fact that arthropods share a thick cuticle and a development punctuated by moults with the nematodes and other 'worms' that had been classified, over the years, as nemathelminths, aschelminths, or pseudocoelomates (but see Rauther 1909). One of those few unorthodox zoologists, however, went so far as to introduce a supraphyletic taxon Atricha for a grouping of arthropods and nematodes (Colosi 1967), thus largely anticipating later phylogenetic views. More recently, the Articulata hypothesis has been criticized based on developmental evidence against a common origin of segmentation in arthropods and annelids (Minelli and Bortoletto 1988). Soon thereafter, molecular phylogenetic studies (e.g. Eernisse et al. 1992) begun to favour a grouping of arthropods with nematodes rather than with annelids, until Aguinaldo et al. (1997) supported it with more

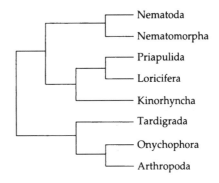

Fig. 5.4. Putative phylogenetic relationships among the main ecdysozoan lineages.

convincing molecular evidence, and eventually introduced the now-fashionable term Ecdysozoa for a grouping of the 'moulting animals' including, besides the Arthropoda and the Nematoda, the Onychophora, Tardigrada, Nematomorpha, Priapulida, Kinorhyncha, and Loricifera (Fig. 5.4). Moulting is obviously an old characteristic on this lineage and has been even directly observed in a soft-bodied Cambrian arthropod fossil (García-Bellido and Collins 2004); other 'moulting sclerites' referrable to the lobopodians *Microdictyon* and *Quadrastopora* have also been described (Zhang and Aldridge 2007).

Over the first few years since the publication of Aguinaldo et al.'s (1997) paper, there has been a lively debate on the relative merits of the two contrasting hypotheses (Articulata versus Ecdysozoa). However, arguments have being changing over time as zoologists with a firm basis in comparative morphology and developmental biology have been progressively overcoming the difficulties originally raised against the Ecdysozoa hypothesis whereas, to the contrary, molecular phylogenetists have been revisiting this hypothesis more critically than at the start of the controversy.

The main arguments initially raised against the Ecdysozoa hypothesis and in defence of the opposite concept of the Articulata (e.g. Wägele et al. 1999, Scholtz 2002, 2003, Pilato et al. 2005) were of morphological nature and included the presence of segments with clear external boundaries and, internally, a pair of coelomic cavities and a pair of metanephridia each; the similar general structure of the central nervous system and the presence of a supraesophageal ganglionic mass; the mushroom bodies (*corpora pedunculata*) in the brain; a segmentally arranged musculature; and a dorsal contractile blood vessel. When discussing segments, Scholtz explicitly referred to the classical morphological definition, for example that of Goodrich (1897), according to which a segment is one of a longitudinally repeated series of body parts with an outer annulus, a pair of coelomic pouches, a pair of ventral ganglia and related commissures, lateral nerves, a set of muscles, and a pair of appendages. As for the homology between annelid and arthropod body cavities, the ultrastructure of their coelomic epithelia is basically similar, but for the less differentiated condition in onychophorans (Bartolomaeus and Ruhberg 1999), but in arthropods these cavities are highly modified and reduced. In the arthropodan line, the onychophorans have metanephridia with a ciliated funnel; however, the origin and differentiation of nephridia in the Onychophora provide no support for the Articulata. In *Epiperipatus biolleyi*, Mayer (2006) found no evidence for 'nephridioblast' cells participating in nephridiogenesis, in contrast to the general mode of nephridial formation in Annelida.

Additional arguments were drawn from development (e.g. Nielsen 2003b). One argument is the sequential production of segments from a posterior growth zone, but even the presence, not to say the precise location and the activity mode, of such a growth zone is all but clear, especially in arthropods (Fusco 2005). Another putative argument in favour of the Articulata would be the presence of embryonic cells with teloblastic activity; that is, of large stem cells located at the posterior tip of the growing embryo, which undergo repeated unequal divisions through which they give off smaller descendants that will eventually give rise to the main cell population of the future segment. But teloblasts occur only in clitellates within the annelids (Dohle 1999) and in malacostracan crustaceans within the arthropods (Dohle and Scholtz 1988), the two occurrences being certainly independently

derived (Scholtz 1998). As a further argument from development, the Articulata hypothesis would be supported by the common occurrence, in the annelidan and the arthropodan lineages, of a kind of segmental ontogenetic repatterning, as the definitive segments, seen in late embryonic and post-embryonic development, would not be identical with the first series of segmental units formed in the embryo. However, we simply have no evidence of the extent to which these units, the so-called parasegments (studied extensively in *Drosophila* in the early years of modern developmental genetics; e.g. Martinez-Arias and Lawrence 1985) actually occur in other 'articulate' clades, and some degree of resegmentation is otherwise known from metazoans other than those that we are discussing here (that is, in vertebrates; see section 8.1.5). The more recent papers in which the Ecdysozoa hypothesis has been rejected based on molecular evidence are often grossly inadequate due to limited range of taxa included in the analysis.

The question has been all too often reduced to discussing whether the Arthropoda are phylogenetically closer to the Nematoda or to the Vertebrata (or the Chordata). Therefore, as the alternative to a minimally conceived Ecdysozoa hypothesis, these papers have not considered an Articulata hypothesis (in several of those papers, no annelid species is taken into account!), but rather a 'Coelomata' hypothesis, where arthropods cluster with the chordates rather than with the 'pseudocoelomate worms': hardly something of relevance in the context of modern animal phylogenetics.

The evidence apparently in favour of this Coleomata hypothesis is, *prima facie*, very impressive, being found in analyses based on data from the 780 single-gene families from 10 completed genomes (Philip *et al.* 2005), to hundreds or thousands of nuclear proteins (Blair *et al.* 2002, Copley *et al.* 2004, Dopazo *et al.* 2004, Wolf *et al.* 2004, Philip *et al.* 2005, Ciccarelli *et al.* 2006), and mitochondrial genomes (Steinauer *et al.* 2005). Similar results have been obtained by Rogozin *et al.* (2007) based on the distribution of rare genomic changes, but their results have been soon rejected based on the analysis of a larger sample of taxa (Irimia *et al.* 2007). Difficulties in evaluating these molecular data, despite the sometimes enormous number of nucleotide or amino acid positions considered, suggest intrinsic problems in the resolution of this phylogeny, possibly due to the small intervals separating the remote events whose sequence we are trying to resolve, so that they practically collapse onto a single, multibranched node (a star phylogeny). Additional problems are due to the extensive loss of genes that seem to characterize the nematode lineage (Copley *et al.* 2004) and perhaps, as often found in molecular phylogenetic studies, to idiosyncratic rates of sequence evolution (Telford 2004, Dopazo and Dopazo 2005).

To a large extent, arguments produced in support of the Ecdysozoa hypothesis are derived indeed from molecular evidence. To the 18S rRNA data used by Aguinaldo *et al.* (1997) to support the proposal of this supraphyletic assemblage, further data have been added, including 28S rRNA (Mallatt and Winchell 2002), *Hox* gene sequences (de Rosa *et al.* 1999, Telford 2000), and elongation factor-1α (de Rosa *et al.* 1999), and, more recently, a comprehensive evaluation of 35371 nucleotide positions from 146 genes (Philippe *et al.* 2005) and mitochondrial genome signatures (Podsiadlowski *et al.* 2008). Other evidence is provided by the cytochemistry of nerve cells (Haase *et al.* 2001), the presence of chitin in the endocuticle (in nematodes, however, limited to the pharynx; Neuhaus *et al.* 1996, 1997a, 1997b); the moulting process triggered by ecdysone (also not exclusive, as this phenomenon has been also reported from the medical leech (Sauber *et al.* 1983), and the presence of active ecdysteroids has been also reported from nemerteans (Porchet *et al.* 1984, Okazaki *et al.* 1998)). Arguments from developmental biology which would support the Ecdysozoa hypothesis have been drawn, as mentioned, from the origin and differentiation of nephridia in Onychophora, which do not match the corresponding events in annelids, and also the expression patterns of *engrailed* (Shankland 2003) and other developmentally important genes (Weisblat *et al.* 2004). Palaeontological arguments have been made by Budd (2003b).

As an implication of the Ecdysozoa hypothesis, Peterson *et al.* (2000a) contended that the ecdysozoan common ancestor must have lost the primary larva that it should otherwise have possessed as a protostome, but the characteristic ciliature of a primary larva is obviously incompatible with the presence of a thick cuticle (Peterson *et al.* 2000a). On the other hand, the very concept of primary larvae is equivocal at best, as we will see in a later chapter (section 7.4.5).

5.3.11 Introverta=Cycloneuralia *sensu* Ahlrichs (1995)

Nematoda+Nematomorpha+Kinorhyncha+ Loricifera+Priapulida
Apomorphies include a trilamellate cuticle and brain in the form of a circumpharyngeal nervous ring. The anterior part of the body, often retractable, is covered by scalids. These are conspicuous, hollow, innervated, and arranged in five rows in the Scalidophora, whereas they are arranged in six rows and reduce to thickenings of the epicuticle in the Nematoida (Schmidt-Rhaesa 1998). This has been regarded as a difficulty in establishing homology between them (Dong *et al.* 2006), but the criticism is probably unwarranted.

5.3.12 Scalidophora=Cephalorhyncha

Priapulida+Loricifera+Kinorhyncha
The body is subdivided into three sections (an evaginable introvert with radially arranged scalids with receptor cells, a neck, and an abdomen). The oesophageal sphincter is represented by a single ring-like cell. Other morphological apomorphies of this clade are the introvert with conspicuous scalids, the reversible foregut, and the presence of tanycytes (tonofibril-containing ectodermal cells in the brain; see Nebelsick 1993).

5.3.13 Vinctiplicata

Priapulida+Loricifera
Apomorphies include late larval stages with a longitudinally plicated lorica and adult protonephridia opening into the gonoducts.

5.3.14 Nematoida

Nematoda+Nematomorpha
Apomorphies include a cuticle with layers of crossing collagenous, rather than chitinous, fibrils, the reduction of the circular body muscles, the epidermal longitudinal nerve cords, the presence of a cloaca, and the tailless spermatozoa (Schmidt-Rhaesa 1998, Nielsen 2001).

Supported by SSU rDNA (Littlewood *et al.* 2004).

5.3.15 Panarthropoda

Arthropoda+Onychophora+Tardigrada
The molecular analysis of Aguinaldo *et al.* (1997), based on 18S rDNA sequences, after which the clade Ecdysozoa was eventually established, failed to resolve the relationships among the major ecdysozoan lineages. Nevertheless, a clade of Panarthropoda has been generally recognized since then (e.g. Halanych 2004, Mallat *et al.* 2004), despite some molecular phylogenies suggesting that tardigrades may be closer to the nematodes than to the arthropods and the onychophorans (see section 6.2.3). At any rate, the relationships of tardigrades, onychophorans, and arthropods are not yet firmly settled.

5.3.16 Deuterostomia

Xenoturbellida+Echinodermata+
Hemichordata+ Chordata
This clade is regularly supported by molecular phylogenetic analyses; traditionally acknowledged apomorphies are bilateral cleavage, dipleurula larvae (e.g. Garstang 1894, Nielsen 1995b, Tagawa *et al.* 2001; but Nezlin and Yushin 2004 suggest convergent evolution of these larvae in echinoderms and hemichordates based on the differences in the way the larval nervous system is organized) and a tripartite coelom (pro-, meso-, and metasome) (Nielsen 2003b).

Cameron (2005) reconstructs the earliest deuterostome as a directly developing benthic vermiform animal with little evidence of regionalization, a straight gut with terminal mouth and anus, and circular and longitudinal muscles. A pharynx with gill slits and collagenous gill bars

may have supported filter-feeding since the origin of the lineage (Cameron 2002).

Modern enteropneusts may provide a blueprint of the ancestral deuterostomes' nervous system. The anteroposterior map of gene expression domains for 42 genes involved in neural patterning is very similar in hemichordates and chordates, despite the fact that in vertebrates this expression is restricted to the neural ectoderm, whereas in hemichordates it encircles the whole body, as these worm-like animals have a diffuse nervous system. According to Gerhart (2006), this map dates back at least to the deuterostome ancestor. Chordates and hemichordates share also the map of dorsoventral expression domains, organized along a developmental axis specified by the expression of *BMP* and *chordin*, as well as conspicuous features of the body syntax such as the placement of the gill slits, the heart, and the post-anal tail. The main differences between the two groups are the lack of regionalization of the epidermis and nervous system in hemichordates, and the position of the mouth, which in hemichordates is on the dorsoventral side marked by *chordin* but on the opposite side in chordates.

The most recent molecular analysis of deuterostome phylogeny (Mallatt and Winchell 2007), based on nearly 4000 informative nucleotide positions corresponding to nearly complete 28 and 18S rRNA from about 65 taxa, supports the Ambulacraria hypothesis (Echinodermata+Hemichordata), previously obtained from an alignment of more than 35 000 homologous amino acids (Bourlat *et al.* 2006) (Fig. 5.5).

5.3.17 Ambulacraria

Echinodermata+Hemichordata
This clade is supported by both morphological and molecular evidence (Turbeville *et al.* 1994, Wada and Satoh 1994, Cameron *et al.* 2000, Peterson and Eernisse 2001, Furlong and Holland 2002, Winchell *et al.* 2002, Peterson 2004, Blair and Hedges 2005, Bourlat *et al.* 2006, Mallatt and Winchell 2007). According to Ruppert (2005), the dorsal hollow nerve cord of vertebrates and the neural cord of enteropneusts are probably

THE ENTANGLED PHYLOGENY OF THE BILATERIA **67**

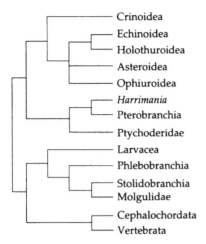

Fig. 5.5. A tree of the Deuterostomia according to Zeng and Swalla (2005).

parallel innovations, relying however on a common pattern of neurulation.

5.4 Other phylogenetic hypotheses

I discuss here briefly some other hypotheses of phylogenetic relationships among major bilaterian lineages. Two of these hypotheses (Lophophorata, Articulata) have been fashionable for decades, but have been progressively abandoned in the light of the alternative patterns of relationships revealed by molecular phylogenies, and eventually supported—although controversially—by a revised evaluation of morphological and embryological evidence. The other hypotheses have been advanced much more recently and some of them are still cherished by a few authors, as is, for example, the Acoelomorpha, the first of the putative clades introduced below. Only a choice of more seriously debated hypotheses is briefly discussed here. A more exhaustive list has been given in Table 4.1.

5.4.1 Acoelomorpha

Acoela+Nemertodermatida
Two groups of 'flatworms' have long been considered as distantly related to the bulk of the

Plathelminthes, and traditionally regarded as basal branches within the latter phylum. These two groups are the Acoela and the more recently described Nemertodermatida. The acoels, in particular, were selected by Hyman (1951a) as a model of ancestral bilaterians, to be derived directly from a metazoan comparable with a cnidarian larva. This is Hyman's planuloid–acoeloid theory of the origin of bilaterians, a long-forgotten theory which is currently finding new favour (e.g. Baguñà *et al.* 2008). During the last decade, very convincing evidence has been accumulated in favour of a basal position within the Bilateria of both Acoela and Nemertodermatida, which must be thus removed from the Plathelminthes. Another question is, whether the Acoela plus the Nemertodermatida form a monophyletic group (for which the name Acoelomorpha is available) or not, with the Acoela as the sister group of the Nemertodermatida plus the Eubilateria or Nephrozoa, this latter clade including all the remaining Bilateria. But let us first consider the arguments in favour of removing the acoels and the nemertodermatids from the plathelminths. These arguments (Ruiz-Trillo *et al.* 1999, 2002, Henry *et al.* 2000, Telford *et al.* 2000, 2003, Baguñà *et al.* 2001, 2008, Jondelius *et al.* 2002, Zrzavý 2003, Baguñà and Riutort 2004 a, 2004b, Jenner 2004c) are molecular, developmental, and morphological. The molecular evidence includes, besides the more usual 18 and 28S rRNA sequences, other nuclear genes and the mitochondrial genome. Embryological evidence mainly refers to the very characteristic spiral cleavage, during which duets, rather than the usual quartets, are produced progressively. Morphological evidence refers to some ultrastructural traits as the peculiarities of the rootlet systems of epidermal cilia (Ax 1996), but is largely negative, including the lack of musculature, of gonoducts, of a nervous orthogon, the characteristic grid-like pattern of nerve fibres typically encircling the body of flatworms (Rieger *et al.* 1991, Raikova *et al.* 2001) and, often, of mouth and gut. While some molecular analyses seemed, for a while, to support the monophyly of the Acoelomorpha, this hypothesis has progressively been abandoned and finally rejected by Wallberg *et al.*'s (2007) analysis, based 18 and 28S rRNA sequences, in favour of the alternative view of the Acoela and Nemertodermatida as independent lineages of basal Bilateria.

5.4.2 Plathelminthomorpha

Plathelminthes+Gnathostomulida
Ax (1984, 1996) grouped flatworms and gnathostomulids together, based on the following characters: thread-like sperm without a distinct acrosomal complex, hermaphroditism, internal fertilization, and—with some uncertainty—the inability of differentiated somatic cells to divide. This grouping was formulated in the context of the traditional interpretation of flatworms as basal within the Bilateria. This caused the lack of anus to be considered a plesiomorphy of these two groups, whereas it acquires the value of a possible synapomorphy of the Plathelminthes and the Gnathostomulida if these, as current phylogeny suggest, are deeply nested within the Protostomia. There is some uncertainty, however, as to the actual lack of anus in the gnathostomulids (Jenner 2004c). Peterson and Eernisse (2001) reported the lack of cilia in the intestinal wall as another possible apomorphy of the Plathelminthomorpha, but this trait is also shared by the Micrognathozoa (Kristensen and Funch 2000).

5.4.3 Cycloneuralia

Gastrotricha+Nematoda+Nematomorpha+Kinorhyncha+Loricifera+Priapulida
The group corrisponds to the old nemathelminths after removal of the Gnathifera. Apomorphies of the Cycloneuralia would be the loss of mesodermal segmentation, haemal system, and metanephridia, and a radially symmetrical mouth surrounded by rings of spines or jaws (Nielsen 2003b).

5.4.4 Parenchymia

Plathelminthes+Nemertea
One hundred years ago, the nemerteans were classified with the flatworms; thus this tentative grouping of Plathelminthes and Nemertea is little more than a revisitation of an old view. Since they were removed from the Plathelminthes and granted the status of independent phylum, the nemerteans have been often regarded as a kind of 'advanced flatworms' provided not simply with a unique structure (the rhynchocoel) but, more importantly, with two widespread features that the flatworms do not possess; that is, the anus and a circulatory system. On balance, however, the supposed similarities between the two groups mainly refer to their acoelomate organization and to the abundant parenchyma filling the space between body wall and internal organs (Nielsen 1985), but the lateral vessels and the rhynchocoel of the nemerteans are currently recognized as coeloms (e.g. Struck and Fisse 2008) and the other characters are likely to have evolved independently in these phyla (Rieger 1985, Turbeville 1996). Finally, the similarity between the pelagic larvae of many polyclad flatworms (Müller's larva and Goette's larva) and the pelagic larvae of many nemerteans (e.g. the pilidium) cannot be called to support the Parenchymia hypothesis, as these larvae are unlikely to be plesiomorphic within the respective phyla.

5.4.5 Vermizoa

Nemertea+Annelida
This putative clade was recognized by Cavalier-Smith (1998), based on potential apomorphies such as the presence of closed blood vessels and the central nervous system with two ventrolateral cords. The Neotrochozoa hypothesis (Mollusca+Annelida) is, however, much better supported by molecular analyses, morphology, and embryonic and larval development.

5.4.6 Schizocoelia=Teloblastica

Annelida+Arthropoda+Mollusca+Onychophora+Sipuncula+Tardigrada
The Schizocoelia would be defined by the following apomorphies: coelomic sacs developing from mesodermal bands, metanephridia, spiral cleavage, trochophora larvae, and a collagenous cuticle (Nielsen 2003b).

5.4.7 Lophophorata

Brachiopoda+Phoronida+Ectoprocta

The presence of a lophophore supporting the crown of tentacles encircling the mouth, together with the largely erroneous belief that these groups share three pairs of coelomic pouches, lead to the establishment of this once fashionable group, for which deuterostome rather than protostome affinities were also often claimed. Molecular analyses have shown that the Lophophorata is not monophyletic and that its two component groups (Brachiopoda+Phoronidea=Phoronozoa, and Ectoprocta) are protostomes, with a closer (Phoronozoa) or more distant (Ectoprocta) relationship to the Annelida and Mollusca (e.g. Halanych *et al.* 1995, Passamaneck and Halanych 2006).

5.4.8 Articulata

Annelida+Onychophora+Tardigrada+Arthropoda

This hypothesis has already been mentioned when discussing the Ecdysozoa (section 5.3.10). The putative apomorphies (segmented body, primarily with segmentally arranged pairs of coelomic cavities) are mainly based on a strong belief in the common origin of annelid and arthropod segmentation from their last common ancestor, but the recent studies of segmentation, especially at the level of developmental genetics, have failed to support it (Weisblat *et al.* 2004).

Particularly informative are the differences in the expression pattern of the *engrailed* gene, which in arthropods such as *Drosophila* marks transversal rows of cells immediately anterior to the future segmental margin, and remains expressed during both embryonic and larval life, whereas in annelids such as the leech *Helobdella engrailed* is first expressed at a time the segments are already visible as morphological units, and fades away very soon (Wedeen and Weisblat 1991, Lans *et al.* 1993, Shankland 2003).

In a very well documented and balanced comparison of the Ecdysozoa and Articulata hypotheses, Giribet (2003) accepts the close similarities in the nervous system of some annelids, arthropods, and tardigrades, and also notes that, curiously, when typical ganglia are not present, a similar cellular architecture develops both in annelids (Sternaspidae and Flabelligeridae; see Schmidt-Rhaesa *et al.* 1998) and onychophorans, with nerve cells having their nuclei scattered along the ventral nerve cord. However, immunohistochemical investigations (using anti-5-hydroxytryptamine (serotonin), FMRFamide, and acetylated α-tubulin antibodies) in combination with laser scanning microscopy demonstrate that, in contrast to the arthropod pattern, the annelidan central nervous system is highly variable in terms of the number and position of connectives and the number of commissures per segment (M.C.M. Müller 2006).

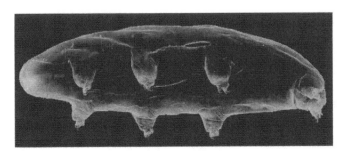

Fig. 5.6. The tardigrade *Macrobiotus macrocalyx*, Italy. courtesy of Roberto Bertolani.

On the other hand, tardigrades, arthropods, and kinorhynchs share a sphincter between the midgut and rectum, and the sensilla of nematodes, tardigrades, and arthropods are extremely similar (Giribet 2003).

A difficulty with the Ecdysozoa hypothesis would be the difference in the position of the mouth, which is terminal in most Introverta, but not so in the Arthropoda and their closest relatives. However, a terminal, jawless mouth was present in some Cambrian lobopods, such as *Aysheaia* and *Paucipodia*, whose affinities to extant onychophorans are generally accepted. Moreover, a terminal mouth is also present in some extant tardigrades (Fig. 5.6) and even in an arthropod clade, the pycnogonids, as well as in some Cambrian arthropod-like animals (e.g. *Kerygmachela*).

Nielsen (2003b) has tried to combine the Ecdysozoa hypothesis with the principally alternative concept of a close relationship between Arthropoda and Annelida. Nielsen has thus proposed an enlarged concept of the Articulata, within which the Ecdysozoa would be the sister group of the Annelida. In turn, Nielsen's Articulata would be the sister group of the Mollusca, to form an unnamed clade characterized by the presence of a haemal circulatory system. This would save the idea of a segmented common ancestor of arthropods and annelids, but would also imply a complete loss of segmentation in phyla such as Nematoda and Priapula. For this revised Articulata hypothesis there is no support from the available molecular analyses.

5.4.9 Cyrtotreta

Chordata+Enteropneusta
Some morphological characters (a branchial gut with U-shaped gill slits, the lack of ciliated tentacles and the presence of both uniciliate and multiciliate cells) would support this grouping. Against this there is, however, consistent molecular evidence supporting the Ambulacraria hypothesis.

CHAPTER 6

A gallery of the major bilaterian clades

The existence of an acron has never been directly shown but rather it is an inference based on the assumption of a close relationship between annelids and arthropods, the Articulata hypothesis.

Scholtz and Edgecombe (2006), p. 396

The case of the mythical arthropodan acron demonstrates that there is a two-way relationship between the way we describe organisms and the hypotheses of phylogenetic relationships that we derive by comparing these descriptions. If we start by carelessly accepting homology between the antennae, or the body cavities, or the trocophore-like larvae of two animals, or with carelessly aligned nucleotide sequences, we cannot hope to eventually obtain a viable phylogenetic hypothesis. Vice versa, by starting with unwarranted hypotheses of phylogenetic relationships we may see in a lineage evidence of something that in fact only occurs in other groups, wrongly considered to be related strictly to the taxon we are considering. No acron, arguably, in the Arthropoda. There is no need to search desperately for the origin of arthropod segments if it is independent from the origin of the segments of annelids. The idea of an arthropod acron has been killed off by the Ecdysozoa hypothesis. In due time, the newly emerging animal phylogeny will probably leave further victims on the battlefield.

6.1 Acoela

The isolated position of the Acoela in relation to the flatworms, to which they were traditionally assigned, has long been established, based on the characteristic spiral cleavage where duets rather than tetrads of blastomeres are produced, and also on morphological characters such as the presence in the sperm cells of two cilia not emerging from the cell body, the lack of glandular cells in the often indistinct digestive system, the complete lack of an extracellular matrix, and other fine structural or ultrastructural characters concerning the statocyst and the ciliary rootlets (Ax 1996). But in the last few years there has been a growing consensus on the hypothesis that acoels are not members of the Plathelminthes but rather represent (either alone or together with the nemertodermatids) the most basal branch of the Bilateria. This basal position of the Acoela (Fig. 6.1) is confirmed by newly addressed aspects of developmental genetics. From investigations based on *Symsagittifera roscoffensis* and *Paratomella rubra* we know the presence in acoels of three *Hox* genes (one anterior, one central-class, and one posterior) whose sequences support a basal position within the Bilateria (Cook *et al.* 2004), but their expression does not fit within the 'zootype' model shared by the eubilaterians (see section 7.5.2). Similarly, the acoel homologue of the *pax6* gene is not involved in producing eyes as in the other bilaterians (Hejnol and Martindale 2008).

In two acoel species, *Childia* sp. and *S. roscoffensis*, Sempere *et al.* (2006, 2007) found a subset of the microRNAs (miRNAs) (see section 7.5.2) shared by protostomes and deuterostomes, whereas the 'true flatworms' (polyclads and triclads) possess, in addition, protostome- and lophotrochozoan-specific miRNAs.

Fig. 6.2. The nemertodermatid *Flagellophora apelti.*, Mediterranean Sea. Courtesy of Ulf Jondelius.

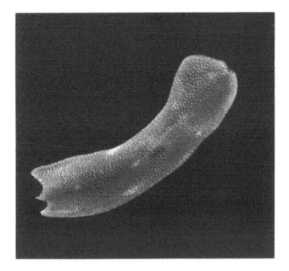

Fig. 6.1. The acoel *Convolutriloba longifissura.* Courtesy of Andreas Hejnol.

6.2 Nemertodermatida

From the traditional perspective that Nemerto-dermatida (Fig. 6.2), as well as Acoela, are members of the Plathelminthes rather than inde-pendent basal branches of the Bilateria, Tyler (2002) regarded as plesiomorphic with respect to the Acoela some characters (statocyst with two statoliths, gut with epithelial lining) that on the contrary turn out to be apomorphic, with respect to the Acoela, in the phylogenetic scen-ario accepted here.

Mitochondrial and 18S rDNA sequence data support their position as basal bilaterians branch-ing off immediately after the acoels as the sister group to the remaining Bilateria (Baguñà *et al.* 2001, Jondelius *et al.* 2002, Telford *et al.* 2003).

6.3 Gastrotricha

Todaro *et al.* (2006) included SSU rDNA sequences from 15 species of Chaetonotida (Fig. 6.3) and 28 species of Macrodasyida (26 genera) in an alignment of 50 metazoan taxa representing 26 phyla. Gastrotricha clustered robustly together with Micrognathozoa, Rotifera and Cycliophora within a monophyletic clade of Lophotrochozoa, with the Gnathostomulida as

Fig. 6.3. The marine gastrotrich *Chaetonotus neptuni.* Courtesy of Antonio Todaro.

sister to that clade. Other molecular phylogenies suggest affinities to Acanthocephala. Combined analyses of morphological and molecular evi-dence suggest that gastrotrichs are the sister group to either Gnathostomulida (Zrzavý *et al.* 1998b) or Plathelminthes (Giribet *et al.* 2000) or even place them within Ecdysozoa (Peterson and Eernisse 2001). By bringing together evi-dence from morphological, developmental, non-sequence molecular, and ecological characters, together with the conserved regions of SSU rRNA genes, Zrzavý (2003) obtains support for a sister group relationships between Gastrotricha and Ecdysozoa (the Cycloneuralia hypothesis). If confirmed, this grouping would help in the for-mulation of hypotheses about the early evolution of the ecdysozoan cuticle. A thin cuticle is present in the Gastrotricha, uniquely associated with long epidermal cilia. There are no moults, but also no post-embryonic mitoses.

6.4 Micrognathozoa

Only one species of micrognathozoan is known, called *Limnognathia maerski*, which was originally described from specimens collected in running freshwater habitats in Disko Island, Greenland (Kristensen and Funch 2000), and subsequently recorded from the sub-Antarctic region. This animal has been assigned to the Gnathifera because of the structure of the jaws, which are formed of special cuticularized rods, similar to those of gnathostomulids and rotifers. The fine structure of the very complex jaw apparatus of *Limnognathia maerski* suggests that Micrognathozoa is more closely related to the Rotifera than to the Gnathostomulida (Sørensen 2003).

Analysis of three nuclear genes suggests relationships with the Syndermata and Cycliophora. However, adding the sequence of the mitochondrial gene *cytochrome c oxidase subunit I* favours a relationship of the Micrognathozoa with the Entoprocta (Giribet *et al.* 2004b).

6.5 Syndermata

The molecular phylogenetic evidence showing the large, parasitic acanthocephalans as a lineage branching off from within a paraphyletic Rotifera found initial resistance, but is now widely accepted. The hypothesis is supported by phylogenetic analyses of SSU and LSU rDNA and *cox1* sequences (Near *et al.* 1998, García-Varela and Nadler 2006, Passamaneck and Halanych 2006).

Morphological apomorphies supporting this clade are the syncytial integument with an internal skeletal lamina (a character shared with the Micrognathozoa), the sperm cells with anteriorly directed cilia (also in Myzostomida), the loss of cilia in the protonephridial canals, and, possibly, the complex jaws (Fig. 6.4; a character lost, within the Syndermata, in the Acanthocephala,

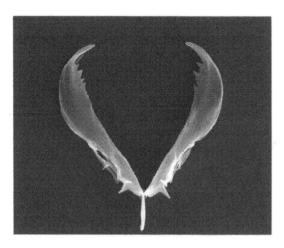

Fig. 6.4. Trophi (mouthparts) of the rotifer *Asplanchna priodonta*. Courtesy of Giulio Melone.

but also present in the Gnathostomulida and Micrognathozoa) (Wallace *et al.* 1996, Zrzavý 2001, Littlewood *et al.* 2004).

The internal relationships among the main syndermatan lineages (Acanthocephala and the three main rotiferan groups: Bdelloidea, Monogononta, *Seison*) are still unsettled. An initial suggestion that the Acanthocephala are the sister group of the Bdelloidea (Acanthocephala+ Bdelloidea=Lemniscea), advanced by Garey *et al.* (1996) based on a limited SSU rDNA evidence, was questioned by Ricci (1998) on morphological evidence, but it has been revived by the combined SSU and LSU rDNA and *cox1* data analysed by García-Varela and Nadler (2006).

Mark Welch (2005) reported a sister-group relationship between the Acanthocephala and the Eurotatoria (=Monogononta+Bdelloidea), with *Seison* (Fig. 6.5) sister to the Acanthocephala+ Eurotatoria.

Finally, a sister-group relationship between the Acanthocephala and *Seison* (Acanthocephala+ Seisonida=Pararotatoria) is supported by the ultrastructure of the sperm cell (Ferraguti and Melone 1999), by the presence of characteristic fibre bundles in the epidermis (Ahlrichs 1997, 1998) and also by the molecular phylogeny of Herlyn *et al.* (2003). In a study based on 18S rDNA sequences from 22 species of Syndermata, the

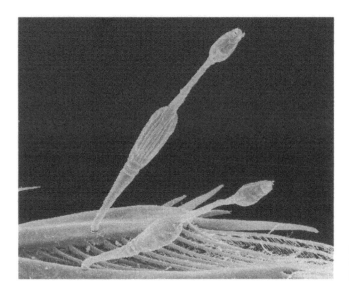

Fig. 6.5. The syndermatan *Seison nebaliae*, epizoic on the leptostracan crustacean *Nebalia bipes*. Courtesy of Giulio Melone.

latter authors found evidence for a monophyletic clade of Eurotatoria (Monogononta+Bdelloidea), whereas *Seison* appears as the sister group of the Acanthocephala. These relationships would suggest a scenario for the origin of endoparasitism in the latter group, starting from a common ancestor of *Seison* and Acanthocephala that lived epizoically on mandibulate arthropods, followed by a change to an endoparasitic lifestyle by some acanthocephalan stem species. Vertebrates would have become part of the parasite's life cycle at a later stage. Herlyn *et al.* (2003) further suggested that a terrestrial life cycle evolved in the stem line of the Archiacanthocephala, with new intermediate hosts being provided by tracheate arthropods.

6.6 Gnathostomulida

Still regarded as Protostomia *incertae sedis* by Littlewood *et al.* (2004), Gnathostomulida have been otherwise regarded as the sister group of Plathelminthes (in the traditional sense, i.e. including also the Acoela and the Nemertodermatida), to form a clade Plathelminthomorpha, sister to all remaining Bilateria (see section 5.4.2) (Ax 1996). A later morphological analysis (Sørensen *et al.* 2000) suggested instead relationships

to the Syndermata within a clade related to Lophotrochozoa. Jaw ultrastructure suggests indeed affinities to Rotifera and to the newly described taxon Micrognathozoa (Sørensen and Sterrer 2002). According to Jenner (2004c), a sister group relationship of the gnathostomulids and syndermatans appears, provisionally, to be the best supported. However, molecular evidence has pointed to different affinities, initially indicating ecdysozoan relationships (Littlewood *et al.* 1998), but further combined analysis of SSU rDNA and morphology has instead suggested affinities to Syndermata, Cycliophora, and Gastrotricha, and all these taxa have been placed in a new clade, the Platyzoa (Giribet *et al.* 2000).

6.7 Catenulida and Rhabditophora

The monophyly of the flatworms, once the Acoela and the Nemertodermatida have been removed from the traditional phylum Plathelminthes, is sometimes recovered by molecular phylogenies, but no morphological synapomorphy is known for the two main clades recognizable in this assemblage; that is, the Catenulida and the Rhabditophora. The vast majority of the flatworms (marine, freshwater, and terrestrial), including the planarians (the fashionable handbook model of free-living flatworms) belong to the latter, together with the large groups of parasitic flatworms (Monogenea, Cestoda, and Digenea). The phylogenetic analysis of Peterson and Eernisse (2001) indicates the lophotrochozoan affinities of Catenulida+Rhabditophora. Zrzavý (2003) identifies Catenulida+Rhabditophora as sister to a clade including nemerteans, neotrochozoans, entoprocts, and the recently described *Lobatocerebrum* (see section 6.16). In general terms, this is the part of the metazoan tree where the true flatworms reasonably belong, but the identity of their actual sister group is still uncertain (Jenner 2004c), as is the monophyly of Catenulida+Rhabditophora.

6.7.1 Catenulida

Apomorphies of this group (Fig. 6.6) are the unpaired protonephridium, the dorsally located male genital porus, and the sperm cell without cilia (Ehlers 1985). A new investigation of the trochophore-like larva (uniquely provided with multiple pre-oral bands) described by Reisinger (1924) for the catenulid *Rhynchoscolex* would be rewarding (Jenner 2004c).

6.7.2 Rhabditophora

The Rhabditophora includes two main clades. One is the Macrostomorpha (=Haplopharyngida+ Macrostomida) whose apomorphies, according to Doe (1986) and Rieger (2001), are the duo-gland adhesive organs emerging in a collar of modified microvilli, the peculiar kind of pharynx (*pharynx simplex coronatus*), and the sperm cells without cilium. The other is an unnamed clade including the Polyclada, the Lecithoepitheliata, and a large clade with the remaining groups formerly classified in the polyphyletic 'Turbellaria', together with the Neodermata, which includes the major groups of parasitic flatworms (digeneans, monogeneans, and cestodes).

The molecular phylogeny of Lockyer *et al.* (2003) of the Rhabditophora based on nearly complete sequences of LSU and SSU rDNA from 32 flatworm species confirmed the monophyly of the Monogenea, of the large 'turbellarian'

Fig. 6.6. The catenulid *Stenostomum sieboldi*. After von Graff (1882), courtesy of Tim Littlewood.

clade of the Proseriata and of the parasitic Neodermata. Within the Neodermata, the Cercomeromorpha (Cestoda+Monogenea) was not supported, whereas there was evidence suggesting a clade of Cestoda+Trematoda.

6.8 Cycliophora

When describing *Symbion pandora*, the first representative of the Cycliophora (Fig. 6.7), Funch and Kristensen (1995) discussed possible

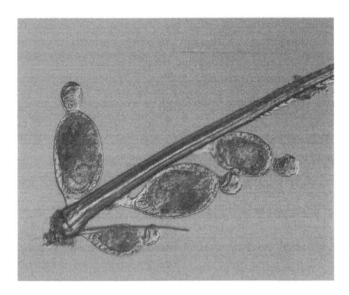

Fig. 6.7. Four young specimens, about 300 μm long, of the feeding stage of *Symbion pandora* (Cycliophora) attached on a seta from the mouthparts of the Norway lobster *Nephrops norvegicus*. Courtesy of Peter Funch.

affinities to Entoprocta and Ectoprocta. Later phylogenetic analyses widened the range of the alternatives to be explored. Based on morphology, Sørensen *et al.* (2000) suggested that this group is possibly sister to the Entoprocta. This is supported by a recent comparative analysis of complete small (SSU or 18S) and nearly complete large (LSU or 28S) nuclear ribosomal subunits for 36 lophotrochozoans (Passamaneck and Halanych 2006). Alternatively, the Cicliophora would belong in the Platyzoa, with Catenulida, Rhabditophora, Gnathostomulida, Gastrotricha, Myzostomida, and Syndermata. Evidence supporting affinity to the Syndermata can be found both in morphology (Funch and Kristensen 1995, Sørensen *et al.* 2000) and in molecular data (SSU rDNA; Winnepenninckx *et al.* 1998a, Giribet *et al.* 2000). Zrzavý *et al.* (2001) unite Cycliophora, Myzostomida, and Syndermata in a clade Prosomastigozoa characterized by sperm cells with anterior flagella. The unique life cycle with distinct sexual and asexual phases and very peculiar larval forms, the chordoid larva and the pandora larva (Funch 1996), does not help to identify the sister group of the Cycliophora, but in the newly described *Symbion americanus* a prometheus larva has been found, which is provided with 'toes' similar to those of rotifers (Obst *et al.* 2006).

6.9 Ectoprocta

A recent analysis (Hausdorf *et al.* 2007) based on sequences of 79 ribosomal proteins of 38 metazoans suggests resurrecting the Bryozoa in the traditional sense (Ectoprocta+Endoprocta) and places this lineage closer to the Neotrochozoa than to the Syndermata. If so, in cleavage patterns, coelomic cavities, gut architecture, and body segmentation there is more homoplasy than usually accepted.

Irrespective of their possible closest affinities to the Endoprocta, the lophotrochozoan affinities of the Ectoprocta are supported by evidence from *Hox* gene sequences (Passamaneck and Halanych 2004). Different affinities have been suggested in the past, including the now definitely discarded opinion that this group, together with the brachiopods and the phoronids, would branch with the deuterostomes rather than the protostomes. The way the coelom is formed in these three groups (ectoprocts, brachiopods, and phoronids) was the main argument indicated to suggest deuterostome affinities, but bryozoans do not show typical enterocoely, their body cavities forming in different and sometimes idiosyncratic ways. Even schizocoely, a typical protostome feature, occurs as a mechanism giving rise to the ectoproct tentacle coelom (Jenner 2004c).

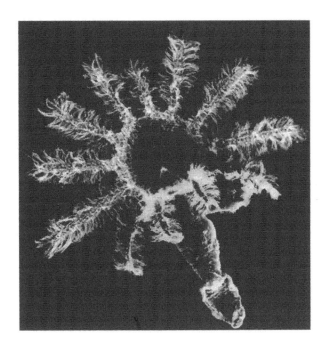

Fig. 6.8. The endoproct *Loxosomella elegans*. Courtesy of Claus Nielsen.

The ectoprocts are an invertebrate group with a very good fossil record. The oldest forms known to date are from the Late Tremadocian (Early Ordovician) of the East Yangtze Gorges in China (Xia *et al.* 2007). Thus, according to palaeontological evidence, their origin may well be more recent than the origin of most other 'phyla'.

6.10 Entoprocta

The newly obtained molecular evidence suggesting closest relationships to the Ectoprocta (Hausdorf *et al.* 2007) has just been mentioned in the previous section. The old idea of a monophyletic lineage of Bryozoa including both Ectoprocta and Endoprocta (Fig. 6.8) had been already revived by Nielsen (1971, 1987) (note, however, that this author's views on the affinities of the entoprocts have later turned much more cautious; see Nielsen 2001, 2002), based on shared larval characters such as the structure of larval eyes and some aspects of metamorphosis. At any rate, the problem is far from settled. The spiral quartet cleavage, the origin of the

mesoderm from a 4d micromere, and trochophore larvae would suggest close relationships to the neotrochozoans (Mollusca and Annelida *sensu lato*), and the nemerteans (Jenner 2004c). Haszprunar and Wanninger (2008) have recently described the creeping larva of the entoproct *Loxosomella murmanica*. In its advanced phase, this larva exhibits large cerebral ganglia from which two paired nerve cords project backwards, closely resembling the tetraneuralian pattern of basal molluscs. The foot sole shares similarities with that of the putatively basal clades of the Mollusca: the Solenogastres especially. This evidence adds to previous findings, in particular the chitinous, non-moulted cuticle, the sinus circulatory system, and several neural features, supporting a sister-group relationship between Mollusca and Entoprocta.

Other affinities have been suggested, including the Cycliophora (Zrzavý *et al.* 2001) and Rotifera+Gnathostomulida (Nielsen 2001). The unique Cambrian sessile metazoan *Dinomischus isolatus* (Conway Morris 1977) may belong to the ectoproct lineage.

6.11 Orthonectida

In the orthonectid *Intoshia variabili*, Slyusarev and Kristensen (2003) found that the external cell layer, composed of alternating rings of ciliated and non-ciliated cells, is covered by a thin cuticle. Two types of cell contact, spot desmosomes and tight junctions, are present between these cells. The presence of a true cuticle and fibrous bands in the epithelial cells provides some support to the hypothesis that Orthonectida are derived Annelida or at least derived Trochozoa, rather than basal metazoans ('Mesozoa') as long accepted in the past.

6.12 Rhombozoa

The view that rhombozoans (=dicyemids) are simplified bilaterians rather than basal metazoans has gained support from molecular phylogenies based on 18S rDNA sequences (Winnepenninckx *et al.* 1998b, Zrzavý *et al.* 1998b, Siddall and Whiting 1999) and the discovery in this group of parasites of a *Hox*-type gene with a sequence containing a lophotrochozoan 'signature' (Kobayashi *et al.* (1999). Nevertheless, dicyemids have many protistan-like characters such as tubular mitochondrial cristae, a body cover with endocytic ability, and a lack of collagenous tissue. Noto and Endoh (2004) suggest that this contradictory evidence could be explained by a substantial amount of lateral gene transfer from a bilaterian (lophotrochozoan) host to a prostist-like parasite.

6.13 Nemertea

The monophyly of this clade has been robustly confirmed by the molecular analysis of Thollesson and Norenburg (2003). Traditionally, the nemerteans were regarded as acoelomates close to flatworms (e.g. Nielsen 2001), but their ultrastructure is more like that of typical coelomates (circulatory system and rhynchocoel comparable with the annelid coelom; Turbeville and Ruppert 1985) and molecular evidence (SSU rRNA, LSU rRNA) places them broadly in a phyletic assemblage including both flatworms and annelids+molluscs.

Their blood vessels, lined by a cellular epithelium rather than by the basal membrane, are unique among the invertebrates (Ruppert and Carle 1983). Developmental and, especially, ultrastructural evidence suggests convincingly that these cavities must be regarded as true coelomic spaces (Steinböck 1963), rather than as vessels homologous to the blood vascular system of metazoans such as, for example, the annelids (Ruppert and Carle 1983, Turbeville and Ruppert 1985, Turbeville 1986, Jenner 2004c).

A phylogeny of the Nemertea based on sequence data from two nuclear genes (28S rRNA and *histone H3*), and two mitochondrial genes (16S rRNA and *cytochrome c oxidase subunit I*) (Thollesson and Norenburg 2003) suggests that the pilidium, the characteristic larva known from many nemertean genera, was secondarily acquired during the history of the phylum as it is an apomorphy of a nemertean subclade, hence called the Pilidiophora.

6.14 Phoronozoa

Affinities between the Brachiopoda and the Phoronidea have long been recognized. With the frequently changing perspectives on their position in the metazoan tree, these groups have been moved frequently, often accompanied by the Ectoprocta, from the protostomes to the deuterostomes and back to the protostomes. Nowadays, their protostomian nature and their lophotrochozoan affinities seem to be settled, especially because the putative report of enterocoely in brachiopods has not been confirmed by recent reinvestigations (Freeman 1993, Lüter 2000). Coelom formation by amoeboid wandering of mesodermal cells may be unique to this phylum (Zimmer 1997). A related problem is the origin of the mesoderm. In phoronids, Freeman and Martindale (2002) have shown that the mesoderm derives from the boundary between ectoderm and endoderm and that the ectomesoderm is the major source of larval muscles; this supports the phylogenetic affinities of phoronids (and brachiopods) with the spiralians within the Lophotrochozoa (Halanych 2004, Jenner 2006). A nearly complete DNA sequence of the mitochondrial genome of *Phoronis archi-tecta* showed a gene order remarkably similar to that of the chiton *Katharina tunicata* (Helfenbein and Boore 2004).

Molecular phylogenies (e.g. Cohen 2000, Cohen and Weydmann 2005) recover phoronids as a lineage internal to the brachiopods; two different but equivalent names (Phoronozoa, Zrzavý *et al.* 1998b; Brachiozoa, Cavalier-Smith 1998) have been provided for the whole group.

Important questions are still open as to the interpretation of some major features of these animals' morphology, but embryological and palaeontological evidence seem to converge towards new, convincing scenarios. For example, the two parts of the brachiopod shells have been traditionally described as dorsal and ventral, but Nielsen (1991) demonstrated that in *Novocrania anomala* the 'ventral' shell of the adult actually derives from the posterior dorsal side of the larva. This is compatible with the scenario suggested by Conway Morris and Peel (1995) for a derivation of brachiopods from animals like the Cambrian halkieriids, which possessed two dorsal cap-shaped shells, one close to the front end, the other close to the rear end of the body.

6.15 Mollusca

Mollusc monophyly is arguably supported by several morphological apomorphies, including a coelomic space, or pericardium, enclosing the heart (in several taxa its fluid content is processed by ultrafiltration) and also in relation to the gonads, as gonopericardium; a tripartite mantle edge divisible into outer, middle, and inner folds; a contrast between ventral locomotory foot and dorsal protective mantle with cuticle and calcareous bodies (Lindberg *et al.* 2004, Steiner 2004). Nevertheless, molecular phylogenies including a number of neotrochozoan or lophotrochozoan representatives have often failed to recover the Mollusca as monophyletic, conspicuous outliers often being the bivalves or the chitons. However, the Mollusca were recovered recently as monophyletic by Giribet *et al.* (2006) in their analysis based on a large dataset of nuclear and mitochondrial sequences.

Molluscan affinities are also uncertain (Lindberg *et al.* 2004). Contenders in the role of molluscan sister group include the Brachiopoda (or the Phoronozoa), as suggested by the 28S sequences (Mallat and Winchell 2002), but Haszprunar (1996) suggested the Ectoprocta based on morphological and developmental data. Strict affinities to the Sipuncula were suggested by Scheltema (1993, 1996) based on developmental and larval characters.

Fossil evidence potentially relevant to an understanding of molluscan affinities is controversial. The Early Cambrian *Halkieria evangelista*, originally interpreted as a stem-group brachiopod (see the previous section) is regarded by Vinther and Nielsen (2005) as a mollusc, to be placed, however, in a separate class, the Diplacophora.

Acaenoplax hayae (Fig. 6.9) is a vermiform animal from the Herefordshire Lagerstätte of the Silurian of England (about 425 Ma) with 18 dorsal ridges, each of which is provided with spines, and seven dorsal valves. It was originally interpreted as an aplacophoran (Sutton *et al.* 2001), although this reconstruction has been disputed, favouring instead an annelidan identity (Steiner and Salvini-Plawen 2001).

The internal phylogeny of the Mollusca (Fig. 6.10) is still in flux. The molecular analyses hitherto produced have reported contrasting results. Analyses of molluscan phylogeny using LSU and SSU nuclear rRNA sequences (Passamaneck *et al.*

Fig. 6.9. The Silurian lophotrochozoan *Acaenoplax hayae* from the Herefordshire Lagerstätte. Courtesy of Derek Siveter.

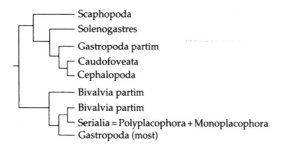

Fig. 6.10. Phylogenetic relationships of the main molluscan groups according to Giribet *et al.* (2006). Note that Serialia is a term proposed in that paper.

2004) support the monophyly of Polyplacophora, Gastropoda, and Cephalopoda, but do not recover the Bivalvia as a monophyletic clade. The Scaphopoda would be more likely related to Gastropoda and Cephalopoda than to Bivalvia.

A recent study (Giribet *et al.* 2006) based on complete sequences of 18S rRNA, a 3 kb fragment of 28S rRNA, the protein-coding nuclear gene histone H3 and two mitochondrial gene fragments for cytochrome *c* oxidase subunit I, and 16S rRNA from an extensive taxon sampling (101 mollusc species including two Caudofoveata, two Solenogastres, 13 Polyplacophora, one Monoplacophora, nine Scaphopoda, 32 Gastropoda, 24 Bivalvia, and 18 Cephalopoda) obtained an interesting grouping of the mollusc clades with serially repeated structures (the Polyplacophora, with their dorsal shell, which is articulated into—usually—eight plates, and the Monoplacophora, with their serially repeated foot-retractor muscles, nephridia, and nervous commissures). This putative clade (Serialia) is deeply nested within the molluscan tree; therefore, its recognition does not raise again the old question concerning the derivation of molluscs from segmented animals. Interestingly, of all molluscs, only the polyplacophorans have an unganglionated nervous system (Nielsen 2008).

It makes sense to suggest a single origin of serially repeated structures in molluscs, together with their independence from other more or less overt instances of segmentation in other phyla. It must be remarked, however, that other aspects of the tree of Giribet *et al.* (2006) are quite

problematic. At variance with more traditional phylogenies, where the Chaetodermomorpha (or Caudofoveata) and the Neomeniamorpha (or Solenogastres) feature, either as successively branching basal lineages (e.g. von Salvini-Plawen 2003), or as a single, basal clade (Aplacophora), this trees suggests that the Solenogastres are only basal to Caudofoveata+Cephalopoda, while the Scaphopoda branch off earlier than the Solenogastres from one of the two main molluscan clades. The other clade includes, in turn, a paraphyletic Bivalvia, the vast majority of the Gastropoda (but a couple of species end up with the Solenogastres), and finally the Polyplacophora, within which the only sampled monoplacophoran is retrieved.

Bivalve monophyly, to the contrary, was found by Giribet and Wheeler (2002) based on combined morphology and DNA sequence data including two nuclear ribosomal genes (18 and 28S rRNA) and one mitochondrial gene (*cytochrome c oxidase subunit I*). Contrasting results are also obtained within the Bivalvia, as Giribet and Wheeler (2002) report protobranchiate bivalves as paraphyletic, whereas Zardus and Martel (2002) were inclined to confirm their monophyly, mainly based on morphological and developmental traits, while not excluding that the Protobranchia, with its unique larval form (Drew 1899, 1901, Gustafson and Lutz 1992), may represent a lineage distinct from the Bivalvia, which split off early in molluscan evolution.

As for gastropods, significant advances have been recently obtained in understanding the nature and origin of the torsion, an ontogenetic process which is, at the same time, a phylogenetic signature of the Gastropoda and a developmental event that dramatically affects their adult organization.

Textbook interpretations of the torsion in gastropods have been dominated by Garstang's hypothesis that the larval retractor muscles power the morphogenetic movement of ontogenetic torsion in all basal gastropods. However, if the retractor muscles are prevented from establishing attachment onto the protoconch by experimental intervention in embryos of the trochid *Calliostoma ligatum* or the keyhole limpet

Diodora aspera soon after fertilization, the larvae of these species nevertheless accomplish a complete or almost complete torsion (Page 2002).

Page (2006b) argues that the rotation hypothesis of the origin of gastropod body plan is a tautological argument, as the conserved aspect of gastropod development is, actually, not the synchronous rotation of the whole complex of mantle and visceral mass relative to the head and foot, but an anatomical organization with the developing mantle cavity on the right but the shell coil posterior; that is, in endogastric orientation. Thus, Page (2003, 2006b) proposes an alternative 'asymmetry hypothesis', according to which the gastropod mantle cavity originated from one side only of an originally bilateral set of mantle cavities. This does not require an exogastric orientation (with a coil over the head) in the monoplacophoran-like gastropod ancestor. Additionally, Page's (2006a) further studies on the early differentiation of neurons in larval abalone (*Haliotis kamtschatkana*) reveal that, contrary to widespread belief, the crossing of the pleurovisceral nerve cords is not necessarily the effect of ontogenetic torsion. Using an early differentiating neuron as a marker to delineate the trajectory of the pleurovisceral nerve cords, Page found that the two neurites of this cell did not cross over during ontogenetic torsion because the soma of this neuron shifted in the same direction as the rotating head and foot. Only during later development, as the mantle cavity deepened and expanded leftward, did the crossing of the pleurovisceral cords eventually occur.

The internal phylogeny of the Gastropoda has undergone recent extensive investigation, with wide-scope studies dealing with the Euthyneura (Grande *et al.* 2004a), the Opisthobranchia (Grande *et al.* 2004b, Vonnemann *et al.* 2005), and the land snails and slugs (Pulmonata, Stylommatophora) (Wade *et al.* 2006). Molecular studies have confirmed the monophyly of the Caenogastropoda, but the monophyly of the putative subclade Neogastropoda is not confirmed (Colgan *et al.* 2007).

Analysis of the complete 18S (SSU) rRNA as well as partial 28S (LSU) rRNA and partial mitochondrial cytochrome oxidase I (*COI*) sequences

(Klussmann-Kolb and Dinapoli 2006) suggest the monophyly of the pelagic Pteropoda, a traditionally recognized group whose two main members, the Thecosomata and the Gymnosomata, were recently regarded as not strictly related and only similar due to convergent adaptation to pelagic life.

Repeated examples of convergence, but also changes from one life form to a quite different one, are demonstrated by a molecular phylogeny of the Stylommatophora. On the one side, shelless slugs have evolved repeatedly within this clade; on the other, a genus of typical slugs (*Tandonia*) is shown to be very closely related to a genus of typical snails with flat, disc-shaped, many-whorl shells (*Vitrea*) (Wade *et al.* 2006).

Among gastropods, parallel trends involving a reduction in the complexity of many characters are indeed frequent (Haszprunar 1988, Ponder and Lindberg 1997, Lindberg *et al.* 2004). Besides the shell, these trends involve the number of radular teeth, the renopericardial system, the ctenidia and associated features of the circulatory and nervous system, the coiling of the hindgut, and the complexity of the stomach.

Lindberg and Guralnick (2003) exploited the phylogenetic signal present in cell-lineage patterns of different gastropod groups. In most groups, cell lineages show characteristic trends of acceleration or retardation relative to the outgroup and more basal ingroup taxa. Cladograms based on these heterochronies are in agreement with the fossil record of the timing and sequence of gastropod subclade origination. 18S rDNA sequences support a Scaphopoda+Cephalopoda clade (Steiner and Dreyer 2003). Morphological characters associated with the burrowing lifestyle of the Bivalvia and Scaphopoda (body enclosed into mantle and shell, burrowing foot with true pedal ganglia) are convergent, whereas the prominent and tilted dorsoventral body axis, the ring of cephalic tentacles, and a ring-shaped attachment of the muscle on the shell are probably synapomorphies shared by the Scaphopoda and Cephalopoda. In a combined approach to the phylogeny of the Cephalopoda based on three nuclear SSU rRNAs, fragments of 28S rRNA and histone H3, and one mitochondrial *cytochrome oxidase subunit I* combined with 101 morphological characters (Lindgren *et al.* 2004), most analyses support the monophyly of Cephalopoda, Nautiloidea, Coleoidea, and Decabrachia; however, the monophyly of Octobrachia was refuted (but is supported by mitochondrial *cytochrome c oxidase subunit I* and a partial 16S rRNA gene; Lin *et al.* 2004).

6.16 Annelida

One of the major controversies in metazoan phylogeny and classification is the composition of the Annelida. Problems derive in part from objective difficulties in determining the relationships among the clades that have been traditionally classified as annelids, such as earthworms, leeches, and the *bona fide* 'polychaetes', and groups such as myzostomids, echiurans, and sipunculans, whose inclusion in the Annelida is still, to some extent, a subject of debate. In part, however, difficulties with circumscribing annelids are a consequence of the typological habit of setting boundaries between major groups in respect to conspicuous morphological traits, such as segmented or unsegmented body. Finally, difficulties derive from the widespread and often parallel character losses occurring in these animals. According to a recent review by Bleidorn (2007), this includes loss of segmentation, chetae, nuchal organs, blood vascular system, branchiae, jaws, and other traits. Character loss may extend even to the coelom, as in *Microphthalmus* (and in the putative annelid *Lobatocerebrum*; see below), perhaps as an effect of progenesis (Westheide 1967, Smith *et al.* 1986, Rieger and Purschke 2005). *Microphthalmus aberrans* and *Microphthalmus carolinensis* lack also a blood vascular system (Smith and Ruppert 1988). For some character, in the absence of a very robust internal phylogeny (see below) it is difficult to determine whether the absence of an otherwise widespread feature is primitive or derived. This is the case for circular muscles, which have been reported as absent in many families (Opheliidae, Protodrilidae, Spionidae, Oweniidae, Aphroditidae, Acoetidae, Polynoidae, Sigalionidae, Phyllodocidae, Nephtyidae, Pisionidae, and Nerillidae; Tzetlin

et al. 2002). Tzetlin and Filippova (2005) suggest that absence of circular muscles is possibly plesiomorphic in the Annelida. I would argue anyway that the condition found in the serpulid *Pomatoceros lamarckii*, in which circular muscles are first formed very late during post-larval development, is secondary with respect to the usual annelid condition with circular muscles present from much earlier stages.

At present, we can be quite confident in regarding the pogonophorans (including the vestimentiferans) of previous classifications as 'polychaetes' with numerous peculiar apomorphies. The affinities of echiurans and myzostomids are somehow less certain, but I will summarize below arguments that reasonably suggest that they should also be included in the annelids. Less obvious is the position of the sipunculans.

At the same time as the external boundaries of the Annelida have been subjected to revision, conspicuous changes have also affected the arrangement of the major lineages within this large clade. The monophyly of the Clitellata is not disputed, besides the question of the actual affinities of a few families often referred to this clade (Aeolosomatidae, Potamodrilidae, Parergodrilidae), but clitellates are now clearly emerging as a lineage internal to a paraphyletic assemblage of 'polychaetes': this is inclusive of the former pogonophorans and, perhaps, of the echiurans and the myzostomids, if not also of sipunculans.

Uncertainty as to the relationship between these smaller groups and the core annelids troubles all discussions of annelid monophyly. But even leaving 'minor' groups aside, we are left with problems. No support for annelid monophyly was found in the molecular phylogeny of Bleidorn *et al.* (2003a). Nevertheless, Bartolomaeus *et al.* (2005) have argued that, if the Ecdysozoa are a monophyletic taxon, Annelida would emerge as one of the best supported monophyletic taxa within the Lophotrochozoa, within which they possess a unique combination of characters such as segmental organization and the generation of additional segments from a posterior sub-terminal growth zone. Whether additional characters like nuchal organs, head appendages, and parapodia can be assumed to represent apomorphies of Annelida, this depends on the internal relationships within the group (Purschke 2000).

As to the internal phylogeny of annelids, the Clitellata, originally established based on morphological traits such as the presence of a clitellum, a restriction of gonads to some anterior segments, hermaphroditism, sperm ultrastructure, and also developmental features such as lecitotrophic eggs, direct development, and the production of cocoons (e.g. Rouse and Fauchald 1997, Bartolomaeus *et al.* 2005, Rousset *et al.* 2007), has been confirmed by molecular studies, whereas the 'Polychaeta' seem to be firmly established as paraphyletic in respect to the Clitellata (e.g. Joerdens *et al.* 2004, Siddall *et al.* 2004, Rousset *et al.* 2007, Struck *et al.* 2007).

A 18S rDNA phylogeny of Clitellata analysed using 18S rDNA sequences (Erséus and Kallersjö 2004) supports an earlier hypothesis based on morphology that the Capilloventridae represents a basal clade of Clitellata. The remaining clitellate taxa form a basal dichotomy, one clade containing Tubificidae (including the former 'Naididae'), Phreodrilidae, Haplotaxidae, and Propappidae, the other clade with two subgroups: (1) Lumbriculidae together with all leechlike taxa (Acanthobdellida, Branchiobdellida, and Hirudinida), and (2) Enchytraeidae together with a monophyletic group of all earthworms included in the study. This phylogeny supports the hypothesis that the first clitellates were aquatic.

Within the Clitellata, the Hirudinea are consistently monophyletic, their apomorphies including the fixed number of 32 segments plus prostomium and peristomium, the involvement of the seven most posterior segments in forming a posterior sucker, the conspicuous development of the oblique musculature, the reduction of mesenteries and coelomic spaces, and the separation of the nephrostome from the nephridial duct (Bartolomaeus *et al.* 2005). Recent phylogenetic assessments suggest that leeches originated from freshwater ectoparasites feeding on vertebrate blood, with several convergent losses of sanguivory for a carnivorous way of life (Borda and Siddall 2004b). Within the leeches, a phylogeny based on nuclear 18 and 28S rDNA, mitochondrial 12S rDNA, and *cytochrome c oxidase subunit I* sequence data, as well as 24 morphological characters, conflicts with most of the traditional classification schemes of the Arhynchobdellida (Borda and Siddall 2004a). Although bloodsucking leeches such as the medicinal leech live in freshwater habitats, this phylogeny implies that the ancestral hirudiniform was terrestrial. This is nicely in agreement with a morphofunctional trait whose phylogenetic relevance has been largely overlooked; that is, that the medicinal leech and its closest relatives possess a long copulatory organ and have internal fertilization (Siddall *et al.* 2006). Contrary to the Hirudinea, the oligochaetes are likely to be paraphyletic; not a single apomorphy is known to suggest the contrary (Bartolomaeus *et al.* 2005, Erséus 2005).

Molecular phylogenetic studies and ultrastructural investigations have addressed the affinities of several species-poor but morphologically interesting taxa like *Parergodrilus* (Fig. 6.11), *Stygocapitella*, *Hrabeiella*, *Potamodrilus*, and the Aeolosomatidae, whose clitellate nature, or affinities, have been often affirmed in the past.

On morphological characters, a distinct 'higher' taxon (Aphanoneura; Timm 1981) was created to include the Aeolosomatidae and *Potamodrilus*. Sequences of the nuclear 18S rDNA and the mitochondrial COI gene support a sister-group relationship of Aeolosomatidae and Potamodrilidae; however, their close affinities to the Clitellata are not supported (Struck and Purschke 2005), despite some obvious morphological traits shared with clitellates (at least by *Potamodrilus*; Purschke and Hessling 2002).

The terrestrial polychaete species *Hrabeiella periglandulata* has many characters in common with the Clitellata and also with *Parergodrilus*, but characters such as the lack of a typical clitellum, the different structure and position of the genital organs, the different ultrastructure of spermatozoa and chaetae, and the presence of nuchal organs would discourage placing it with the clitellates. However, the ultrastructure of the dorsal

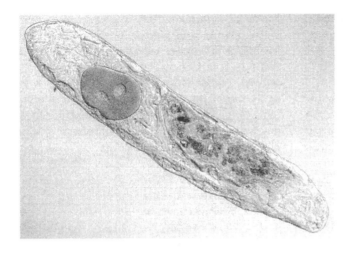

Fig. 6.11. An annelid with uncertain affinities, *Parergodrilus heideri.* Courtesy of Emilia Rota.

pharynx of *H. periglandulata* is very similar to that of the clitellate species *Enchytraeus minutus*. This similarity, together with comparable structure of cerebral sense organs and the central nervous system, still makes a sister-group relationship between *H. periglandulata* and the Clitellata conceivable (Purschke 2003). 18 and 28S rDNA and *COI* sequences place Parergodrilidae close to the orbiniid 'polychaetes' (Joerdens *et al.* 2004).

In the recent molecular phylogeny of annelids produced by Rousset *et al.* (2007), the Clitellata allied with the Dinophilidae, close to 'typical polychaetes' such as Sabellidae and Spionidae; the Parergodrilidae are in the same branch as other 'polychaetes' such as Nereidae, Pilargidae, and Scalibregmatidae; the Aeolosomatidae and *Hrabeiella* with Nephtyidae, Hesionidae, and Lumbrinereidae among the traditional 'polychaetes', and the Polygordiidae among those clades of borderline position with respect to the Annelida. As to the traditional 'Polychaeta', three morphological characters would support their monophyly: the parapodia, the nuchal organs as pits and grooves, and the mixonephridia (Rouse and Fauchald 1997), but this hypothesis has no molecular support (Bleidorn *et al.* 2003a).

The position of the cerebral ganglion and the structure of the circumoesophageal connectives offer a way to distinguish 'polychaetes' from clitellates (Orrhage 1995, Purschke 1999, 2002, M.C.M. Müller 2004, 2006, Müller and Henning

2004). The adult nervous system of clitellates may well have been derived from the organization found in 'polychaetes' (Hessling and Westheide 1999, M.C.M. Müller 2004, 2006).

Molecular phylogenies have caused two lineages previously treated as independent phyla, the Pogonophora and the Vestimentifera, to be reduced to components of one annelid family, the Siboglinidae, which is deeply nested within the 'polychaetes'. More precise affinities, however, are controversial. Different analyses report siboglinids as a member of a clade including other sedentary (sabellimorphs) 'polychaetes' (Struck *et al.* 2007), or branching with Oweniidae (Rousset *et al.* 2004), or as a quite isolated annelid clade (Rousset *et al.* 2007).

Morphological (e.g. Ax 2000) and molecular traits (e.g. Halanych *et al.* 2002) have been ooted for years in support of interpreting the Echiura as a derived annelid clade. This is confirmed by more recent 18S rDNA analyses, which additionally suggest close relationship to the Capitellidae (Bleidorn *et al.* 2003a, 2003b, 2006), with which they share the absence of a circulatory system, the ventral part of the coelom acting as a haemocoel, and the presence of haemoglobin in the coelomic 'corpuscles'.

Studies of the development of the echiuran nervous system have revealed evidence of metamerical organization (Hessling 2002, 2003, Hessling and Westheide 2002). Using antibodies against different neurotransmitters, Hessling (2003) identified discrete repetitive units of perikarya in the ventral nerve cord of *Bonellia viridis*. Evidence of serial organization was additionally provided by histochemical studies of the distribution of peripheral nerves, shown to be serially arranged, as these nerves are clearly paired and evenly distributed, corresponding to the serial units of serotoninergic neurons in the ventral cord. The serial organization of the nervous system of *Bonellia viridis*, at the level of cellular specificity, is in agreement with the segmental arrangement of ganglia in most typical annelids and thus supporting the hypothesis that the Echiura are derived from segmented ancestors.

Another clade of controversial affinities is the Myzostomida (Fig. 6.12). In a phylogeny of the 'polychaetes' by Rouse and Fauchald (1997), myzostomids grouped within the phyllodocids, with which they share traits such as a specialized muscular proboscis and the presence of metameric protonephridia in the adult. In myzostomids, organs of ectodermal origins, such as the protonephridia (Pietsch and Westheide 1987) and the nervous system (Müller and Westheide 2000), are segmentally arranged. Morphology and post-embryonic development (five pairs of parapodia with chaetae, and a larva similar to a trochophore, but without a prototroch, thus not fitting Rouse's (1999) restricted definition of a trocophore) indeed suggest annelid affinities, but the segmentation of myzostomids is incomplete, and there is no coelom. Alternative affinities have been thus suggested. According to Zrzavý (2001), myzostomids are 'platyzoans that exhibit a complicated mixture of annelid-like, basal spiralian and platyzoan/prosomastigozoan characters.' A molecular phylogeny based on sequences of 5S rRNA and *elongation factor 1α* suggested that Myzostomida are the sister group of flatworms (Eeckhaut *et al.* 2000). This result was criticized by Littlewood *et al.* (2001) as well as by Zrzavý (2001), who suggested instead affinities to the Cycliophora. Jenner (2003, 2004c) has stressed the fact that myzostomid affinities represent a textbook example of a phylogenetic problem where results obtained

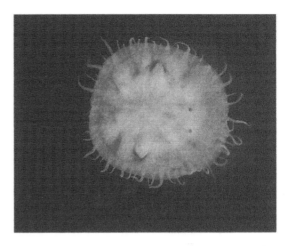

Fig. 6.12. The myzostomid *Myzostoma polycyclus*, a parasite of the feather star *Capillaster multiradiata*, Madagascar. Courtesy of Igor Eeckhaut.

from molecular evidence only can be grossly wrong, as the data matrices deliberately ignore conspicuous morphological characters consistently pointing to a different interpretation. In this case, myzostomids share with several 'polychaete' families characters such as the nectochaeta (a post-metamorphic stage characterized by bundles of long chaetae), the aciculae (specialized chaetae functioning as struts for the parapodia), and the putatively sensory cirri with their peculiar innervation pattern.

At present, hypotheses of plathelminth or cycliophoran affinities of the Myzostomida have been discarded, as has Mattei and Marchand's (1987) grouping of the Myzostomida with the Acanthocephala in a taxon called Procoelomata, only based on similarities in sperm cells. Furthermore, despite renewed rejection by Eeckhaut and Lanterbecq (2005) of a hypothesis of annelid affinities, myzostomids should be reasonably regarded as belonging within the annelid radiation following the phylogenetic analysis by Bleidorn *et al.* (2007) based on 18 and 28S rDNA, *myosin II*, and *elongation factor 1α*, plus the nearly complete mitochondrial genome. In particular, the *myosin II* sequence and mitochondrial gene order place myzostomids with annelids, whereas the *elongation factor 1α* sequences suggest less probable nematode affinities.

Two fairly recently described genera must be cited before leaving the annelids. Both of them are broadly similar to very elongated free-living flatworms, rather than to annelids, but it has been suggested that they are probably specialized annelids showing a high degree of anatomical simplification in relation to their adaptation to interstitial life. Neither of them is segmented. *Lobatocerebrum psammicola* (Rieger 1980, 1988) is a thin worm about 4 mm long, distributed worldwide in sublittoral sandy shore environments. Mouth and anus occupy the two body ends. The epidermal cells are multiciliated, with a thin cuticle. The nervous system includes differentiated ganglia. An external layer of longitudinal musculature encloses an inner layer of circular fibres. The presence of an anus rules out the interpretation of *Lobatocerebrum* as a flatworm. The lack of a coelom is unusual in annelids, but would not be unique. If it is not an annelid, it might represent a very specialized mollusc lineage. Our knowledge of *Jennaria pulchra* (Rieger 1991), a 1–2 mm-long worm from the interstitial water of the Atlantic coast of the USA, is more fragmentary. *Jennaria* (Fig. 6.13) lacks circular muscles and a ciliated epithelium covers the two terminal parts of the body, whereas a long midbody region is simply covered by a cuticle comparable with that of the annelids.

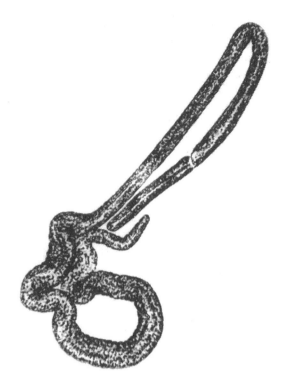

Fig. 6.13. Habitus of *Jennaria pulchra*. After an unpublished drawing of Reinhard Rieger; courtesy of Gunde Rieger.

6.17 Sipuncula

Close relationships between the Sipuncula and Annelida (including the Echiura) are supported by sequences of mitochondrial nucleotide and amino acid sequences and gene order (Boore and Straton 2002, Staton 2003, Jennings and Halanych 2005), to the point that Bleidorn *et al.* (2006) do not hesitate in placing the Sipuncula within the Annelida, perhaps close to the Orbiniida.

A characteristic 'polychaete' trait, the nuchal organs, is present in Sipuncula (Purschke *et al.* 1997). This seems to be a good argument in favour of the hypothesis that sipunculans are specialized annelids, and also in view of the fact that nuchal organs have been lost by clitellates and by other derived 'polychaete' groups

(Purschke 1997). In addition, the pharynx in the pelagosphaera larva of sipunculans has many similarities to the pharynx of polychaetes (Tzetlin and Purschke 2006). On the other hand, at variance with the newly demonstrated evidence of segmentation in echiurans, no trace of segmentation was found in their musculature and nervous system even when using modern immunocytochemistry techniques (Wanninger *et al.* 2005).

6.18 Priapulida

Despite the very small number of Recent species of Priapulida described to date, there is conspicuous variation among them, even with respect to major traits such as body cavities. One species only, *Meiopriapulus fijiensis*, is known to possess a true coelom lined by a mioepithelium, this also being the only example in all cycloneuralians (Storch *et al.* 1989).

Several Cambrian 'worms' have been interpreted as belonging to the Priapulida, but this is not firmly established. In the case of the Palaeoscolecida, the resemblance of many of these forms to the priapulids has been regarded by Conway Morris (1997) as strongly suggestive of their affinities, but Budd (2003a) flags the possibility that this might simply be plesiomorphic and only showing, at most, membership in the Cycloneuralia.

6.19 Loricifera

The ecdysozoan and scalidophoran affinities of this recently discovered group (Kristensen 1983) do not seem to be problematic. Molecular evidence, nevertheless, will be welcome.

6.20 Kinorhyncha

The main evolutionary issue to be discussed about kinorhynchs is still perhaps the traditional question about the homology between their zonites and arthropod segments. Since the acceptance of both groups as members of the Ecdysozoa, this hypothesis looks much less odd than before. It must be stressed that in the Kinorhyncha, as in the Arthropoda, the articulation of the trunk into segments or zonites is not limited to the epidermis and the associated cuticle, but extends to the musculature, thus forming an integrated functional unit (Clark 1964), and also to the central nervous system. This is not enough to conclude that this whole set of serial units can be traced back to a corresponding feature in the last common ancestor of Arthropoda and Kinorhyncha (this would imply, indeed, a lot of parallel loss of segmentation in other ecdysozoan lineages); nevertheless, it is probable that both arthropod segments and kinorhynch zonites evolved from a common background of mechanisms producing serial structures in remote ecdysozoan ancestors. Individual features may have been added up to this original core, and the integration of segmented body rings, myomeres, neuromeres, etc., is likely to have been obtained in the two groups independently.

6.21 Nematoda

Despite their robust cuticle, which would suggest a fossilization potential much higher than that of a ctenophore or a medusa, nematodes have left a very limited fossil record. In part, this is due to the lifestyle of many nematode species, for example the parasites, which is unlikely to allow their bodies to be buried in conditions that would favour preservation. However, we would not expect fossil free-living nematodes to be as rare as hitherto recorded. Valentine (2004) mentions a few dubitative examples from the Carboniferous, while the oldest reliable nematode would be a Jurassic fossil from Osteno, Italy (Arduini *et al.* 1983). To be sure, cladistic relationships suggest that the nematode clade must be older, but the putative sister group of the Nematoda, the Nematomorpha, does no offer a reliable temporal reference, as the Cambrian Chengjiang fossils referred by Hou *et al.* (2004) to the Nematomorpha may belong instead to the Priapulida, thus leaving an Eocene fossil as the oldest recorded member of the group (Voigt 1938). Thus, the Nematoda are a remarkable case of major animal lineage of apparently young geological age.

Apart from a small number of genera that depart more or less dramatically from the structure of the average nematode, morphology has provided limited help in generating hypotheses of phylogenetic relationships within the Nematoda, but to date more than 200 taxa have been already considered in molecular studies (De Ley and Blaxter 2004, Holterman *et al.* 2006, Meldal *et al.* 2007). Deeply nested within a long series of more basally branching clades is a set of 'core' clades, all together corresponding to the Secernentea of traditional classifications, which includes both *Caenorhabditis elegans* and most plant and animal parasites among the nematodes. In these internal clades, parasitic lifestyle is associated with higher rates of nucleotide substitutions (Holterman *et al.* 2006).

6.22 Nematomorpha

Most of what we know about this small group of ecdysozoans with parasitic larvae has been

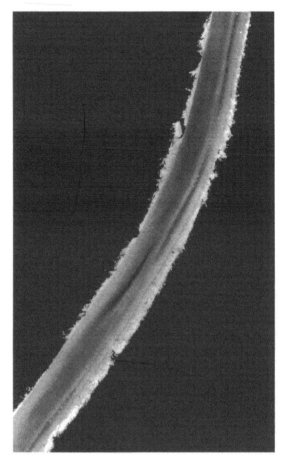

Fig. 6.14. A part of the body of the marine nematomorph *Nectonema*. Courtesy of Andreas Schmidt-Rhaesa.

presence of four pairs of appendages, usually provided with claws, would seem a good argument in favour of this. In fact, most phylogenetic discussions involving the tardigrades have revolved around the mutual relationships of the three groups of Panarthropoda, the Tardigrada being more often regarded as the sister group of Arthropoda+Onychophora. The new phylogenies have placed the tardigrades firmly within the Ecdysozoa, but their closest affinities are less certain than before, from the perspectives of both palaeontologists (Dzik and Krumbiegel 1989, Dzik 1993, Waggoner 1996, Budd 2003a) and molecular phylogeneticists. Some recent analyses have indeed recorded tardigrades as branching with the nematodes rather than with the arthropods (Marlétaz *et al.* 2006, Jiménez-Guri *et al.* 2007, Dunn *et al.* 2008); thus, we must regard the problem, at present, as still open. If confirmed, the nematode affinities of tardigrades would add new evidence to the emerging idea of multiple parallel evolution of a 'segmented' body within the Ecdysozoa, as suggested above when discussing the Kinorhyncha.

6.24 Onychophora

Between the late 1980s and the early 1990s, the discovery (or the reinterpretation) of several Cambrian metazoans provided with serial pairs of lobopods—that is, flexible but not truly articulated appendages similar to those of the Recent onychophorans (Ramsköld and Hou 1991, Hou and Bergström 1995, Ramsköld and Chen 1998, Bergström and Hou 2001, Hou *et al.* 2004)—provided abundant material for phylogenetic speculation (e.g. Dzik and Krumbiegel 1989, Dzik 1993). Cambrian lobopods, all marine, were much more diverse than their Recent terrestrial relatives. Fossils (a couple of Carboniferous species and some Tertiary amber forms) do not help much in bridging the gap between the former and the latter, as these post-Cambrian forms are all too similar to modern velvet worms. The study of the Cambrian lobopods raised some hope in clarifying the earliest steps in the history of the Arthropoda, but those fossils are too recent, being coeval to several stem-group and even

obtained by studying their larger subclade, the Gordioidea. More investigations are required about the marine Nectonematoidea (Fig. 6.14). However, despite the sensible morphological gap between the two groups and the great differences in their ecology and life cycles, the monophyly of the Nematomorpha is probably warranted.

6.23 Tardigrada

Most zoologists have traditionally regarded the tardigrades (Fig. 6.15) as strictly related to the arthropods and the onychophorans. Besides the cuticle and the periodic moults, the

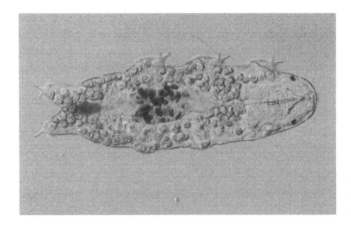

Fig. 6.15. The tardigrade *Hypsibius dujardini*; note the mouth in the apical position. Courtesy of Roberto Bertolani.

crown-group arthropods (see below) and possibly also too specialized to provide much help.

Extensive ultrastructural and developmental research of the last few years has provided interesting but inconclusive evidence as to the precise affinities of the onychophorans. On the one hand, close affinities to the Arthropoda are suggested by brain organization (Strausfeld *et al.* 2006), by the confirmed presence of a mixocoel (Mayer *et al.* 2004), and by the ultrastructure and development of nephridia in the antennal segment (Mayer and Koch 2005). Suggestive of arthropodan affinities is also the presence of the blue pigment haemocyanin in the haemolymph of *Epiperipatus* (Kusche *et al.* 2002). Other evidence, however, would not suggest such close affinities between onychophorans and arthropods. In Recent arthropods, for example, there is no evidence, embryological or anatomical, of their derivation from ancestors with a terminal, frontally directed mouth (which is the condition found in arthropod-like Cambrian fossils like *Anomalocaris* and *Opabinia*; see below), whereas this is evident in modern onychophorans, as witnessed by embryology and by the innervation patterns in the cephalic area (Eriksson 2003). The homology of eyes and antennae in the two groups is also disputed.

6.25 Arthropoda

The increasing number of stem-group arthropods described by palaeontologists causes the set of potential apomorphies of the crown Arthropoda (Euarthropoda) to be more and more restricted. Waloszek (2003a) suggested, as a euarthropodan apomorphy, the presence of a head with one pair of antennae plus three pairs of biramous limbs. Harzsch (2006) adds a few traits of the nervous system, such as the three brain neuromeres, a pair of lateral compound eyes with a growth zone that gives off new elements throughout the post-embryonic development, and the presence of a pre-oral frontal commissure composed of deutocerebral and tritocerebral fibres, innervating hypostome, oesophagus, and the anterior part of the gut.

During the 1970s, the monophyletic nature of the Arthropoda was rejected by many zoologists, mainly based on the arguments of Manton (e.g. Manton 1973, 1977, Manton and Anderson 1979), drawn from morphology and embryology, but never articulated in a cladistic argumentation scheme. Despite this methodological flaw, Manton's views enjoyed, for a while, an extraordinary success. In particular, it became fashionable to group together the hexapods, the myriapods, and also the onychophorans in a phylum Uniramia, while the chelicerates and the crustaceans were regarded as two separate phyla. As sudden and clamorous as its rise in favour was also this arrangement's fall into disrepute, one of the latest survivors among the defenders of arthropod polyphyletism being Fryer (1987). These events of the recent past are recalled here as a background onto which to evaluate the more recent migration of the myriapods from their traditional place in the closest proximity of the insects (see Dohle 1988) to a position in the arthropod tree which is still problematic but nevertheless external to the hexapod+crustacean clade.

During the last two decades or so the monophyly of the Arthropoda, in its traditional circumscription, has never been seriously questioned, but the name has also sometimes been used (perhaps with a *sensu lato* qualification) to embrace, additionally, the Onychophora and the Tardigrada. Panarthropoda is a more current name for this larger clade, while the Arthropoda *sensu stricto* are often called Euarthropoda. Alternative views about the mutual relationships between (Eu)arthropoda, Onychophora, and Tardigrada were discussed briefly in section 5.3.15.

The fossil evidence of interest with respect to reconstructing the origin of the (Eu)arthropoda and the relationships among their main lineages is abundant but, in many ways, confusing. On the one hand, the earliest Cambrian representatives of the crown-group Euarthropoda are approximately of the same age as forms mainly regarded by palaeontologists as stem-group (eu) arthropods or even as stem-group panarthropods. As mentioned above in relation to the Onychophora, many of these forms well deserve the name of lobopods, rather than arthropods, but even those that seem to share more characters with (eu)arthropods than the modern velvet worms do not throw much light on arthropod origins. On the other hand, there are forms such as the anomalocaridids (see below) that are distinctly more similar to the (eu)arthropods than onychophorans, but these similarities are not indisputable proof of a closer relationship. All these forms were poorly provided with rigid cuticular sclerites, and characters such as the development of segmentally repeated, paired appendages, together with the presence of a pair of peculiar sensory or grasping appendages at the front end of the animal, may have evolved more than once in this panarthropodan lineage.

Another difficult question is the interpretation of head structure and the evolution of arthropod cephalization. Indisputably, many of the oldest panarthropodan and even euarthropodan forms possessed a head region involving a number of segments less than that of the head of, say, a modern insect or crustacean, but there is no reason to accept that all the steps of progressive incorporation of further segments into the cephalic region (e.g. from two to three, or from three to four segments) happened only once. That is, arthropod (or arthropod-like) forms with two, three, and four segments are easily grades rather than clades. Additionally, as we will see further in the following, the traditional views about the segmental composition of the head in the individual arthropod lineages (Recent as well as extinct) are still far from universally agreed. For example, the presence in Cambrian stem-group arthropods of an extra, eye-bearing tergite in front of the cephalic shield may well suggest the presence of a pre-antennal ocular segment which does not fit into the canonical segmentation scheme for the arthropod head (Waloszek 2003a).

6.25.1 Cambrian arthropod-like animals and stem-group Euarthropoda

Our knowledge of Cambrian arthropods and arthropod-like animals has increased tremendously in the last couple of decades, partly because of the systematic exploration of new sites, especially in China, and partly because of a careful re-evaluation of already described but poorly understood taxa. Problems, however, remain.

The first problem is the position, with respect to the Arthropoda, of Cambrian forms such as *Anomalocaris*, *Cassubia*, *Hurdia*, *Opabinia*, and *Proboscicaris*. These animals were clearly segmented (at least dorsally) and provided with appendages. Problems are, however, the terminal (rather than ventral) position of the mouth, the poor, or lacking, skeletonization, and the unsegmented, flap-like nature of the paired appendages borne on their trunk segments (e.g. Hou and Bergström 1997). In the case of *Opabinia*, for example, the evidence for leg-like ventral appendages remains equivocal (Briggs and Fortey 2005). These animals are probably not true arthropods (Bergström and Hou 2003, Briggs and Fortey 2005, Hou *et al.* 2006), but it is not possible, in the face of the available evidence, to decide whether any of their arthropod-like traits was uniquely derived from a common ancestor they share with the (Eu)arthropoda, or not. Otherwise, we should revert to the hypothesis of 'convergent arthropodization', a view that some authors (e.g. Bergström and Hou 2003) consider seriously.

Other Cambrian forms (*Fuxianhuia protensa*, *Chengjiangocaris longiformis*, and *Shankouia zhenghei*, all from the Chinese Chengjiang deposits), which possessed a distinct two-segment head, a condition we can interpret as an early step towards cephalization involving an ever-increasing number of segments, are interpreted as stem-lineage (Eu)arthropoda (Waloszek *et al.* 2007). The 'unfinished' nature of arthropod cephalization is shown, indeed, by the widespread occurrence of taxa where one or more segments that comparative morphology would indicate as belonging to the trunk rather than to the head are more or less distinctly incorporated into the cephalic region (but not necessarily 'fused' to form a more extended cephalic capsule), especially because their appendages are largely integrated, morphologically and functionally, with those of the segments we conventionally describe as cephalic. This is not limited to all those crustacean lineages where

one or more pairs of 'post-cephalic' appendages are developed as maxillipedes, but also applies to the Chilopoda and, within insects, to the Mantodea (praying mantises), whose poison claws and raptory legs, respectively, well deserve the name of maxillipedes, not least than the one to five pairs of appendages of this name in crabs, shrimps, and many other malacostracans. Besides the three genera mentioned above, the variegated world of stem Euarthropoda includes also *Canadaspis* (Fig. 6.16), *Cindarella*, *Ercaia*, *Isoxus*, *Primicaris*, and *Xandarella*, and also the bradoriids, tiny arthropods similar to the ostracod crustaceans (Waloszek *et al.* 2007). The legs of these Cambrian arthropods were articulated into a large number of podomeres, all quite similar; for example about 20 in *Fuxianhuia* and about 13 in *Canadaspis* (Bergström and Hou 2003).

6.25.2 Problems with segments and eyes

Before presenting some major points of arthropod phylogeny (see Table 6.1 for a list of recently proposed or debated taxa), it is worth discussing three aspects of comparative morphology whose better understanding is of consequence for phylogenetic studies.

The first point relates to arthropod affinities rather than to the internal phylogeny of arthropods. It is due to Valentine (2004), who makes the good point that, although arthropods are currently described as segmented, they could

Fig. 6.16. A specimen of the Cambrian stem euarthropod *Canadaspis* from the Chengjiang formation, Yunnan, China. Courtesy of Jerzy Dzik.

Table 6.1 A list of names proposed for major arthropodan groups, with references to the works where these names were established and content (with reference to currently accepted clades).

Group	References	Content
Artiopoda	Hou and Bergström (1997)	Chelicerata plus the extinct groups Nectopleura, Conciliterga, Petalopleura, Trilobita, Xenopoda, and Aglaspidida
Chelicerophora	Dubinin (1959)	Pantopoda+Chelicerata
Collifera	Boudreaux (1979a)	Diplopoda+Pauropoda
Cormogonida	Zrzavý et al. (1998a)	Arthropoda less Pycnogonida =Euchelicerata+Eumandibulata
Ellipura	Börner (1910)	Protura+Collembola =Parainsecta
Ichthyostraca	Zrzavý et al. (1998)	Pentastomida+Branchiura
Myriochelata	Pisani et al. (2004)	Chelicerata+Myriapoda =Paradoxopoda
Nonoculata	Luan et al. (2005)	Protura+Diplura
Pancrustacea	Zrzavý and Štys (1997)	Crustacea+Hexapoda =Tetraconata
Paradoxopoda	Mallatt et al. (2004)	Chelicerata+Myriapoda =Myriochelata
Parainsecta	Kukalová-Peck (1987)	Protura+Collembola =Ellipura
Tetraconata	Dohle (2001)	Crustacea+Hexapoda =Pancrustacea

be better seen as jointed, as there is no evidence that the body of arthropods was ever subdivided into compartments with a morphological individuality and a functional autonomy comparable to those seen in annelid segments. In terms of segmentation versus articulation, a look at trilobite development offers useful insights. In the post-embryonic development of these Palaeozoic arthropods, new trunk units went through two distinct phases. Non-articulated 'segments' made their first appearance in the posterior part of the last section of the trunk (the so-called 'provisional pygidium'); a few moults later these serial units were individually released, as fully articulated segments, from the anterior part of the same region of the trunk, until a residual set of unarticulated segments eventually remained as the definitive pygidium even in the more advanced (holaspid) instars. Thus, the individualization of segmental units and their release as articulated elements are, in principle, two

different processes; their decoupling offered trilobites the scope for a diversity of post-embryonic developmental schedules, because the release of the last articulated segments could either precede, or accompany, or follow the end of the production within the pygidium of new inarticulate units (Hughes et al. 2006).

Second, of the units identified by morphologists in the arthropod body, special attention was traditionally reserved for the acron, the putative 'pre-segmental' anterior region of the body. Its very existence in arthropods has been revisited in the context of the Ecdysozoa/Articulata controversy. An acron is indeed recognizable in the annelids and its existence in the arthropods would be an argument in favour of the Articulata. But renewed efforts have failed to provide any evidence (embryological, anatomical, or palaeontological) in favour of the existence of an acron in either the Onychophora or the Arthropoda (Scholtz 1998, 2002, Eriksson and Budd 2000, Budd 2003a, Waloszek 2003a, Scholtz and Edgecombe 2006).

Third, another contentious problem is the homology of the compound eyes throughout the Arthropoda. The question has been revisited by Richter (2002) and Oakley (2003) in the face of recent evidence suggesting that the compound eyes of myodocopid ostracods perhaps do not share homology with those of other crustaceans (and arthropods in general). Despite these difficulties, the cited authors did not reject the hypothesis of a monophyletic origin of this structure.

6.25.3 Pycnogonida

The affinities of Pycnogonida are still uncertain, the majority of the phylogenetic analyses recovering them either as the sister group of the Euchelicerata or as sister to all remaining crown Arthropoda (Dunlop and Arango 2005). However, in a phylogeny based on nearly complete 28 and 18S rRNA genes of 35 ecdysozoan taxa, analysed with Bayesian inference, pycnogonids are grouped within the chelicerates or are included in the Paradoxopoda (Chelicerata+Myriapoda; see section 6.25.5), but are never recovered as basal arthropods (Mallatt et al. 2004). The putative affinities between Pycnogonida and Acari found in phylogenetic analyses using mitochondrial genes (Hassanin 2006) are explained by Podsiadlowski and Braband (2006) as due to long-branch attraction and convergence in nucleotide composition and amino acid frequency,

A problem in evaluating pycnogonid affinities is the questionable homology of their anterior segments. The chelicera of pycnogonids have been traditionally regarded as homologous to those of Euchelicerata, both in the sense of special homology and in the sense of positional homology. A few authors (e.g. King 1973) were not convinced, but this homology has been recently reaffirmed, mainly based on embryological evidence (see Dunlop 2002). Comparative embryology (Mittmann and Scholtz 2003) and the expression patterns of Hox genes (Telford and Thomas 1998 on the oribatid mite Archegozetes; Damen 2002 on the spider Cupiennius) have been

used to establish positional homology between the chelicerae and the first (crustaceans) or only (hexapods, myriapods) antennae of other arthropod groups: all these appendages belong to the deutocerebral segment. Expression patterns of two Hox genes (labial and Deformed) in the protonymphon larva of Endeis spinosa suggest homology between their chelicerae and the first antennae of the Mandibulata (Manuel et al. 2006), confirming Vilpoux and Waloszek's (2003) suggestion based on comparative morphology. If so, the adult pantopod head (the cephalosoma) would correspond to the euarthropod head and the protonymph with three appendage-bearing segments may represent a phylogenetically older larval type than the euarthropod 'head larva' with four pairs of appendages. If segments in pycnogonids can be legitimately compared, in a one-to-one correspondence, with those of other arthropods, than the fourth walking legs of Pycnogonida would correspond to the chilaria of euchelicerates.

As of Palaeozoic fossils, Haliestes dasos (Silurian, 425 Ma) was provided with large chelate first appendages, consistent with a chelicerate affinity of the pycnogonids. Siveter et al. (2004) place it near the base of crown-group pycnogonids.

6.25.4 The main arthropod clades

Leaving the Pycnogonida aside, the next phylogenetic set of problems posed by crown-group Arthropoda (Fig. 6.17) revolves around the mutual relationships of the four major groups (Chelicerata, Myriapoda, Crustacea, and Hexapoda), with the additional difficulty that the monophyly of these groups is also far from certain. Let us summarize here the two main points of the recent and current debate:

• are the Hexapoda more closely related to the Crustacea (the Tetraconata or Pancrustacea hypothesis) or to the Myriapoda (the traditional Atelocerata hypothesis)?
• if the Atelocerata hypothesis is rejected, is the Mandibulata hypothesis (perhaps as (Myriapoda (Crustacea+Hexapoda)) still tenable? Or, alternatively, are the Myriapoda closer to the Chelicerata (the Myriochelata or Paradoxopoda hypothesis)

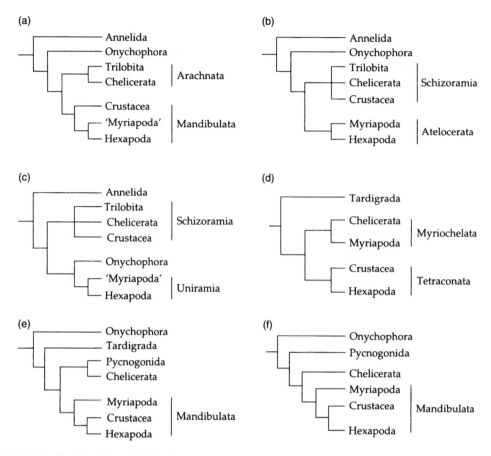

Fig. 6.17. A choice of traditional (a–c; all of them developed within the frame of the Articulata hypothesis) and modern (d–f) views of the relationships among the major arthropod lineages. The trees summarize the views of the following authors: (a) Snodgrass (1938); (b) Cisne (1974); (c) Tiegs and Manton (1958); (d) Friedrich and Tautz (1995) and Giribet *et al.* (1996); (e) Zrzavý *et al.* (1998a); and (f) Giribet and Ribera (1998).

than to the Tetraconata? An additional possibility, namely that the Myriapoda are a sister group to all remaining arthropods (pycnogonids excluded), does not seem to deserve serious scrutiny.

6.25.5 Pancrustacea=Tetraconata versus Paradoxopoda=Myriochelata

During the last few years, only a few students of arthropod phylogeny (e.g. Klass and Kristensen 2001, Kraus 2001, Willmann 2003) have supported a clade of hexapods and myriapods (Tracheata or Atelocerata), in favour of which there would

be indeed some putative morphological apomorphies; in particular is the lack of appendages on the segment that in crustaceans bears the second antennae (i.e. the intercalary segment of the Tracheata), a condition mirrored by the segmental expression of *engrailed* (Kettle *et al.* 2003). Less compelling (Richter and Wirkner 2004) are other putative apomorphies of the Tracheata, for example the tracheae, some sensory organs (the temporal organs), and the excretory organs (the Malpighian tubules). Tracheae, however, are quite diverse in structure and position, suggesting multiple origins, as already argued by Ripper (1931) and more recently demonstrated by Dohle (1988), Kraus and Kraus (1994), and Hilken (1998).

The monophyly of Pancrustacea (Zrzavý and Štys 1997) or Tetraconata (Dohle 2001) is well supported by molecular evidence: mitochondrial gene order (Boore *et al.* 1998), mitochondrial protein coding genes (Hwang *et al.* 2001), mitochondrial ribosomal genes (Ballard *et al.* 1992), nuclear ribosomal genes (Friedrich and Tautz 1995, 2001, Giribet and Ribera 2000, Mallatt *et al.* 2004), and nuclear protein coding genes (Regier and Shultz 2001, Kusche *et al.* 2002, 2003). But there are also putative morphological apomorphies, such as the structure of the ommatidium, with two corneagenous cells, four crystalline cone cells, eight retinula cells, and eight pigment cells (Dohle 2001, Harzsch 2006), and the origin of neuroblasts and of particular pioneer neurons (Stollewerk 2008). A moderate support for the Tetraconata hypothesis is found by Harzsch (2004) in the arrangement of serotonin-immunoreactive neurons in the ventral nerve cord, as the pattern found in Chelicerata, Chilopoda, and Diplopoda differs from the pattern present in Hexapoda and Crustacea, and the hexapodan pattern most closely resembles that of the Crustacea.

No molecular study supports the alternative Atelocerata hypothesis, but the Mandibulata hypothesis, irrespective of the internal phylogeny of the main lineages belonging to it, still has many defenders.

The monophyly of the Mandibulata has been challenged by many molecular analyses (Cook *et al.* 2001, Hwang *et al.* 2001, Mallatt *et al.* 2004, Negrisolo *et al.* 2004). However, it must be acknowledged that the morphological evidence in support of a sister-group relationship between Chelicerata and Myriapoda is very limited, despite some aspects of embryonic neurogenesis that have been cited in support (Dove and Stollewerk 2003, Kadner and Stollewerk 2004).

A clade of Mandibulata is still accepted by several researchers (e.g. Willmann 2001). There are, indeed, molecular or combined phylogenies supporting it (e.g. Giribet *et al.* 2001, based on combined morphology, plus 18 and 28S rDNA sequences), but the hypothesis is mainly defended for morphological reasons, which are not restricted to the mandibular nature of the appendages of the second post-antennal segment. Studies of gene expression would support homology of mandibles throughout crustaceans, hexapods, and myriapods (e.g. Akam 2000). Additional arguments would be the structure of the ommatidia with crystalline cone consisting of four cone cells (recently found in the compound eye of a myriapod, the centipede *Scutigera*; Müller *et al.* 2003), and the common structure of the mandible, with a distinct *pars molaris*.

Harzsch *et al.* (2005) have produced a review of mechanisms of neurogenesis, morphology of serotonergic interneurons, number of motoneurons, and cellular features and development of the lateral eyes in arthropods. These authors find that in Chelicerata, as compared with the Mandibulata, the numbers of neurons of the different cell classes examined are much higher and in many cases are not fixed but variable. These characters argue against a sister-group relationship of Myriapoda and Chelicerata, but instead provide evidence in favour of the Mandibulata concept. Several lines of molecular evidence, however, suggest that myriapods, rather than belonging to a clade Mandibulata, together with crustaceans and hexapods, are instead more closely related to the Chelicerata (Friedrich and Tautz 1995, Cook *et al.* 2001, Hwang *et al.* 2001). This putative clade of Chelicerata+Myriapoda has been called Paradoxopoda (Mallatt *et al.* 2004) or Myriochelata (Pisani *et al.* 2004). Molecular evidence in favour of this clade derives from mitochondrial gene sequences (Friedrich and Tautz 1995, 2001, Hwang *et al.* 2001, Negrisolo *et al.* 2004), *Hox* gene sequences (Cook *et al.* 2001), and complete 18 and 28S rDNA sequences (Mallatt *et al.* 2004, Mallatt and Giribet 2006).

Neurogenesis in the millipede *Glomeris marginata* (Dove and Stollewerk 2003) and in the centipede *Lithobius forficatus* (Kadner and Stollewerk 2004) suggests more similarities to chelicerates than to insects, but the comparative framework is still inadequate to evaluate whether these similarities represent arthropod plesiomorphies or apomorphies of the Myriochelata. The genetic network involved in the recruitment and specification of neural precursors is probably conserved among euarthropods, but two characters distinguish insects+crustaceans from myriapods+chelicerates (Stollewerk and Chipman 2006). First, in pancrustaceans the neuroectoderm gives rise to epidermal and neural cells, whereas in chelicerates and myriapods only neural cells are produced in the central area of the neuroectoderm. Possibly more important, in insects and crustaceans neural cells arise by stem-cell-like divisions of neuroblasts, whereas in chelicerates and myriapods there is recruitment of mainly post-mitotic neural precursors. This developmental difference is arguably a support for the Myriochelata hypothesis.

In addition to a weak molecular suggestion in the same direction, Negrisolo *et al.* (2004) listed a few possible non-molecular synapomorphies uniting Chilopoda+Chelicerata (thus negating, by implication, myriapod monophyly). A first aspect is the feeding mechanisms, as the vast majority of chelicerates and centipedes feed on fluid food, mostly of animal origin and digested pre-orally or extra-orally, whereas this is not true of millipedes. Second, the excretory system of arachnids is represented by coxal glands that mostly open behind legs I and III, whereas centipedes have a maxillary rein opening on the second maxillary segment: this distribution is perhaps an instance of positional homology, if both the third leg pair of arachnids and the second maxillae of the centipedes are segmentally homologous to the insect labium (Kraus and Kraus 1994, Telford and Thomas 1998).

6.25.6 Trilobita and Euthycarcinoida

In a recent insightful review on the phylogenetic position of the Trilobita, Scholtz and Edgecombe (2006) have challenged the mainstream opinion that places this extinct group within the Arachnomorpha, together with the Chelicerata (e.g. Bergström 1979, 1980, 1992), and revive instead the opinion of Boudreaux (1979b) that trilobites are instead stem-group Mandibulata. This would be suggested by head tagmosis. Scholtz and Edgecombe (2006) move from the consideration that the antennae of extant mandibulates are deutocerebral and post-ocular,

and the insertion of trilobite antennae at the anterolateral margin of the hypostome would suggest for these appendages an identical segmental identity to mandibulates, rather than a pre-ocular position and a protocerebral connection as in the case of the 'primary antenna' of the onychophorans. Scholtz and Edgecombe regard the loss of the primary antenna and the specialization of the deutocerebral appendages as sensory structures as an apomorphy of the Mandibulata. Placing the trilobites in the Arachnomorpha would imply massive character reversal in the arachnids (and in pycnogonids if these actually belong here).

Irrespective of a decision between the opposite Mandibulata and Myriochelata hypotheses, I dare to suggest that the affinities between trilobites (and minor trilobitomorph groups) and millipedes (plus some Palaeozoic myriapod lineages) deserve closer study. There are, indeed, morphological and developmental traits that trilobites share with millipedes, especially with a group which is likely basal within the Diplopoda like the pill millipedes or Glomerida. I will only mention here the post-embryonic progression in the production and release of fully articulated 'segments' and the contrast between a sclerotized (indeed, strongly calcified) dorsal side and a very weakly sclerotized ventral side. Only careful but also open-minded study will reveal whether any of these traits may represent an apomorphy for a putative lineage whose main members are trilobites and millipedes.

The phylogenetic position of the Euthycarcinoidea (a little more than a dozen species ranging in age from Late Cambrian to Middle Triassic) is also uncertain. Affinities to crustaceans have been claimed as well as, alternatively, a position within a myriapod–hexapod assemblage, or a basal position in the Euarthropoda, or even in the Hexapoda. A Cambrian species recently described by Vaccari *et al.* (2004) demonstrates that the general morphology of these arthropods was strictly conserved in the transition from marine to freshwater life. A morphology-based phylogeny resolves euthycarcinoids as stem-group Mandibulata. What will remain of these views if the Mandibulata must be rejected?

6.25.7 Chelicerata

The monophyly of the crown Chelicerata is well founded on molecular evidence (e.g. Giribet and Ribera 2000, Giribet *et al.* 2001). If we accept that Pycnogonida are the sister group of the (Eu)chelicerata, the latter turn out to be characterized by a developmental apomorphy; that is, by hatching at a more advanced stage than their likely plesiotypic sister group (Dunlop 2002).

Several arthropods from the Early to Middle Cambrian possessed uniramous 'great appendages', a characteristic pair of pre-oral limbs most likely used in prey capture. Besides this common trait, these arthropods exhibit quite remarkable morphological differences, to the point that multiple independent evolution of the great appendage has been suggested. However, Chen *et al.* (2004c) remarked that all 'great appendage' arthropods possessed a head provided with three pairs of biramous limbs, in addition to the great appendage itself. They also described *Haikoucaris ercaiensis* from the Lower Cambrian Maotianshan Shale of southern China, whose short, spiky 'great appendages' resemble the chelicera of the Chelicerata to the point that this particular type of 'great appendage' is hypothesized by them to be the actual precursor of the chelicerae. A position of these arthropods within the Arachnomorpha is supported by Cotton and Braddy (2004) who list the loss of the first cephalic appendages (the antennae), the loss of the exopods of the second cephalic appendages, and modification of the endopods of these appendages into spinose grasping organs as potential synapomorphies. To this group of great appendage (or macrocheiran) arthropods belong also genera such as *Alalcomenaeus, Fortiforceps, Leanchoilia, Jianfengia, Mafengocaris*, and *Yohoia*. The legs of these forms were still subdivided into a number of uniform podomeres, about 15 in *Fortiforceps*.

The internal phylogeny of the crown-group Chelicerata is still problematic (e.g. Weygoldt 1998, Wheeler and Hayashi 1998, Giribet *et al.* 2002, Paulus 2004, Shultz 2007). An interesting result of recent investigation is the rejection of traditional views that regarded scorpions as basally branching within the Arachnida, if not

even as a chelicerate lineage not belonging to the Arachnida; scorpions, instead, seem to be well settled as an internal branch of the arachnids, perhaps closest to the Opiliones.

6.25.8 Myriapoda

Supporters of the Mandibulata hypothesis have repeatedly suggested myriapod paraphyly, mainly using morphological data (Borucki 1996, Kraus 1998, Edgecombe *et al.* 2000, Giribet and Ribera 2000). Otherwise, molecular or total evidence suggests a monophyletic Myriapoda as a basal branch of the Mandibulata (Giribet *et al.* 2001), whereas Regier and Shultz (2001) obtained a monophyletic Myriapoda as a basal clade of all Arthropoda in a tree based on the nuclear gene *elongation factor 2*, and Loesel *et al.* (2002) found a paraphyletic Myriapoda as a basal arthropodan clade, based on immunocytochemical and neuroanatomical data.

As mentioned above, several molecular and developmental studies (Boore *et al.* 1995, 1998, Friedrich and Tautz 1995, Burke *et al.* 1998, Cook *et al.* 2001, Hwang *et al.* 2001, Dove and Stollewerk 2003, Negrisolo *et al.* 2004) support a close relationship between Myriapoda and Chelicerata and Negrisolo *et al.* (2004) go even further, by favouring a paraphyletic myriapodan clade, with the Chilopoda as sister group of the Chelicerata, although this conclusion would require more stringent evidence.

Putative morphological apomorphies for a monophyletic Myriapoda include the lack of median eye(s), the acrosome of sperm cells lacking a perforatorium (Baccetti and Dallai 1978), as well as characters of the head endoskeleton and the mandible (Edgecombe and Giribet 2002). Other traits are provided by Koch's (2003) detailed morphological study of head morphology, but even this is not conclusive. Edgecombe (2004) defended myriapod monophyly within the Mandibulata, while a phylogenetic analysis of six mitochondrial protein-coding genes retrieved a Myriapoda clade within the Paradoxopoda (Hassanin 2006).

The monophyly of the Chilopoda is founded on several morphological characters (Dohle 1985,

Edgecombe and Giribet 2002, Wirkner and Pass 2002), the most obvious being the transformation of the appendages of the first trunk segment into a pair of poison claws. The internal phylogeny seems to be reasonably settled (Edgecombe and Giribet 2007), by confirming the sister-group relationship between the Geophilomorpha and the Scolopendromorpha and, at a more basal level, the split between the Scutigeromorpha (Notostigmophora) and all remaining centipedes (Pleurostigmophora). The most conspicuous advance produced by combined analysis of morphological and molecular evidence is the recognition of two large clades supported by behavioural evidence, in addition to the morphological and molecular characters. One is the Phylactometria (Edgecombe and Giribet 2004), grouping together all centipedes (the Epimorpha and the monogeneric Craterostigmomorpha) whose females provide extended parental care to their brood; the other is the Adesmata, which groups all those geophilomorphs where the brooding females coil their body around the brood in such a way as to expose to the outside the ventral surface, which is provided with glands producing defensive exudates (Bonato and Minelli 2002).

For the millipedes (Diplopoda) most modern analyses, based on either molecular or total evidence, confirm the basal split between the soft-bodied Pselaphognatha or Polyxenida and the generally strongly calcified Chilognatha. The latter, in turn, are basally divided into Pentazonia (pill millipedes and relatives) and Helminthimorpha, whose internal arrangement is still much in flux (Sierwald et al. 2003, Sierwald and Bond 2007).

Of the two lesser myriapod groups, the Pauropoda has traditionally been regarded as the sister group of the Diplopoda, together forming the Dignatha, characterized by the presence of one pair of post-mandibular appendages rather than two, as in the other mandibulate arthropods. The position of the other group, the Symphyla, is more problematic, but there is some agreement in placing them with the Dignatha in a clade Progoneata, putatively characterized by the anterior position of the genital opening, but also by the peculiar shape of the trichobothria (a kind of sensory structures), and the origin of gonoducts as ectodermal invaginations (Dohle 1980, 1988, Ax 1999). Eleven mitochondrial protein-coding genes would provide a weak support for the affinity between Symphyla and Chilopoda (Podsiadlowski et al. 2007).

6.25.9 Crustacea

Morphology supports the monophyly of the Crustacea (Lauterbach 1983, Schram and Hof 1998, Waloszek 1999, Edgecombe et al. 2000), with characters such as the presence of a naupliar eye, no more than two pairs (antennal, maxillary) of nephridia, a poorly sclerotized labrum with posteriorly located gland openings, and a telson with a furca. Further arguments for the monophyly of the Crustacea have been reviewed by Richter (2002). Disputable is the phylogenetic significance of the presence of a naupliar stage. On the one hand, Scholtz (2000) has convincingly argued for an independent origin of the malacostracan nauplius (only present in the Euphausiacea and in the dendrobranchiate Decapoda) with respect to the nauplius of other crustacean groups such as copepods, cirripeds, and branchiopods; on the other hand, free-living lecithotrophic nauplii comparable with those in Euphausiacea and Dendrobranchiata have been recently found in another very isolated crustacean lineage, the Remipedia (Koenemann et al. 2007).

Waloszek (2003a) lists morphological characters for the ground pattern of the Crustacea and those for the Eucrustacea (Fig. 6.18); the same author also groups all non-malacostracan crustacean in the traditional group Entomostraca, for which he also provides a list of ground pattern characters. Molecular evidence, nevertheless, suggests crustacean paraphyly (e.g. Wilson et al. 2000, Cook et al. 2001, Nardi et al. 2001), but there is no agreement on which crustacean lineage should represent the sister group of the Hexapoda (provided, additionally, that the latter are monophyletic; see below). Early molecular phylogenies tentatively suggested that this group could be the Malacostraca, but more recent studies suggest otherwise.

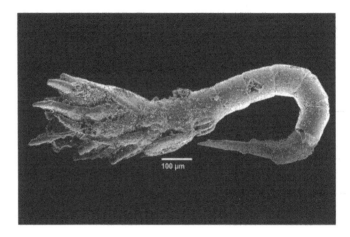

Fig. 6.18. *Martinssonia elongata*, a crustacean-like euarthropod from the upper Cambrian 'Orsten' limestones of Sweden. *Composite figure, the anterior part being the holotype specimen UB 748, whereas the posterior part is specimen UB 104. Both specimens belong to the collections of the Palaeontological Institute, University of Bonn* (courtesy of Dieter Waloszek and Andreas Maas).

Based on brain anatomy across the Mandibulata, Fanenbruck and Harzsch (2005) reject the hypothesis that the Remipedia are basal crustaceans. Instead, they suggest that the Remipedia belong to a clade including also Malacostraca and Hexapoda, the only mandibulate arthropods with a brain complexity comparable with that of remipeds.

Analyses of pancrustacean phylogeny based on complete or near-complete mitochondrial genomes suggest that crustaceans and hexapods are mutually paraphyletic (Cook *et al.* 2005). In this study, Malacostraca+Branchiopoda emerges as sister to the Insecta *sensu stricto*, while the Collembola groups with the Maxillopoda. Another phylogenetic analysis, based on three protein-coding nuclear genes from 62 arthropods and lobopods (Onychophora and Tardigrada) suggests that Hexapoda is most closely related to the Branchiopoda and Cephalocarida+Remipedia, thereby making hexapods terrestrial crustaceans and the traditionally defined Crustacea paraphyletic; additionally, Malacostraca unites with Cirripedia and this clade groups in turn with the Copepoda, making the Maxillopoda paraphyletic (Regier *et al.* 2005). A more recent analysis of a large matrix of 28 and 18S rRNA sequences provides strong evidence that the sister group of the Hexapoda is limited to the Branchiopoda (Mallatt and Giribet 2006).

Problems in resolving the deepest branches of the crustacean (or pancrustacean) phylogeny may appear less surprising if we consider the probable age of these events, although we do not know the length of the time span within which all the deepest branchings in which we are interested actually occurred. At any rate, crown-group eucrustaceans were present in the Lower Cambrian (Zhang *et al.* 2007), while a Bayesian statistical estimate of divergence times suggests a Precambrian origin for the Pancrustacea (600 Ma or more) (Regier *et al.* 2005).

Irrespective of the monophyly of the Crustacea as a whole, four crustacean clades seem well founded, both on morphological and molecular evidence. These are the Branchiopoda (Waloßek 1993, Giribet and Ribera 2000, Giribet *et al.* 2001; but this opinion is not shared by Ax 1999) and the Malacostraca (Spears and Abele 1998, Shultz and Regier 2000, Richter and Scholtz 2001) as well as the two small, homogeneous groups, the Cephalocarida and the Remipedia. In the past, a fifth group, the Maxillopoda, was often regarded as monophyletic, but some molecular studies suggest otherwise. In a phylogenetic analysis of three protein-coding nuclear genes, the Copepoda appear as the sister group of the Malacostraca plus the Cirripedia, thus making the Maxillopoda paraphyletic (Regier *et al.* 2005). The affinities of some groups within the Malacostraca have also been challenged. In particular, a combined analysis of nearly complete 28 and 18S rRNA gene sequences places the stomatopods with an euphausiacean rather than at the base of the Eumalacostraca, as currently accepted (Mallatt and Giribet 2006). Membership of the Peracarida is also disputed, as nuclear SSU rDNA indicated a monophyletic Peracarida that excludes the Mysida and includes instead the Thermosbaenacea (Spears *et al.* 2005).

As an appendix to the crustaceans, let us add a few lines about the pentastomids. The phylogenetic position of these parasitic, vermiform animals has been long disputed, but several lines of evidence may place them within the Crustacea. In our times dominated by molecular phylogenetics, it is easy to ignore that the currently debated hypothesis of maxillopodan affinities of the Pentastomida was originally advanced based on morphological evidence. This was indeed one the earliest important contributions of electron microscopy to the progress of animal phylogenetics. Wingstrand (1972) based his suggestion on the strong, putatively synapomorphic similarities between the sperm cells of the pentastomid *Raillietiella* and the branchiuran crustacean *Argulus* (Fig. 6.19). More recently, analyses of the mitochondrial DNA gene arrangements and sequences confirm that pentastomids are crustaceans, probably related, indeed, to branchiurans (Lavrov *et al.* 2004). A placement near the branchiurans is also supported by 18 and 28S rDNA data (Mallatt and Giribet 2006). This conclusion, however, is not undisputed.

According to Maas and Waloszek (2001) (see also Waloszek 2003b, Waloszek *et al.* 2006), pentastomids are not close relatives of the branchiurans and, probably, do not belong to the crustacean lineage; even their (eu)arthropodan nature can be disputed. Pentastomid characters stressed by these authors (see also Walossek and Müller 1994), as generally indicative of a pre-euarthropodan condition, are the lack of arthrodization of the body, the frontal position and the segmental composition of the head, the morphology of limbs, the topography of the nervous system, epimorphic development with pseudometamerism of the 'caudal end', and the cuticular β-chitin. One may question, however, how far these characters must be regarded as primitive traits rather than as structural simplifications correlated with the parasitic lifestyle of pentastomids. But this, in turn, is questionable, as pentastomids were already present in the Cambrian (Fig. 6.20), long before the kind of animals (vertebrates) they parasitize today (Walossek and Müller 1994, Walossek *et al.* 1994, Waloszek *et al.* 2006). Were the Cambrian pentastomids also parasitic and, if so, did they perhaps attack early representatives of the chordate lineage such as the conodont animals?

6.25.10 Hexapoda

In zoological nomenclature, the changing fate of the term Insecta is one of progressively narrowing scope. In the pre-Linnean literature, the notion of insect was hardly more restrictive than what later it became usual to call an invertebrate.

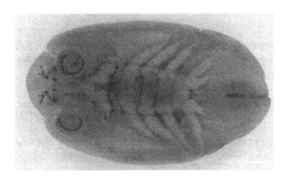

Fig. 6.19. Ventral view of *Argulus* sp. (Crustacea, Branchiura). Courtesy of Marco Uliana.

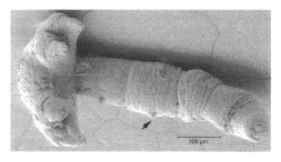

Fig. 6.20. Hammer-head larva of an undescribed Cambrian pentastomid from Västergötland, Sweden. Arrow, rudimentary or vestigial limbs. Specimen UB W 359, Palaeontological Institute, University of Bonn (courtesy of Dieter Waloszek and Andreas Mass).

The Insecta of Linnaeus (1758) correspond to our Arthropoda, but by the early nineteenth century the concept had been further restricted, within zoological circles, to the six-legged arthropods only. Things changed again with the earliest cladistics analyses (e.g. Hennig 1969), which forced a definite move away from the old dichotomy between Apterygota (in the traditional sense, thus including Protura, Collembola, Diplura, and Thysanura) and Pterygota, provisionally replacing it with a basal split between the Entognatha (Protura, Collembola, and Diplura) and the Ectognatha, the latter embracing the Pterygota together with the paraphyletic group Thysanura, eventually split into the Archaeognatha or Microcoryphia (jumping bristletails) and the Zygentoma (silverfish and allies). The Zygentoma were soon recognized as the sister group of the Pterygota and grouped together with them to form a clade Dicondylia, based on the mandible provided with two articular condyles rathen than one as in the Archaeognatha. Eventually, both terms, Entognatha and Ectognatha, became less and less in use as names for clades, whereas the term Insecta was increasingly applied to the Ectognatha only, its former synonym Hexapoda being now employed for the whole of the six-legged tracheate arthropods. Uncertainties as to the relationship between the 'true' insects and the entognathan groups, sometimes coupled with doubts as to the monophyletic nature of the whole hexapodan lineage (see below), eventually led to the stabilization of the modern use of both terms, Hexapoda and Insecta.

A clade Hexapoda is, apparently, well founded on morphological characters (Hennig 1969, Kristensen 1998, Willmann 2003) and also several molecular phylogenies have recovered a monophyletic Hexapoda (e.g. Regier and Shultz 1997, based on *elongation factor 1α* and *RNA polymerase II*; Wheeler 1998a, 1998b, Zrzavý *et al.* 1998a, based on a combined analysis of molecular and morphological data), but other studies tell otherwise, based both on mitochondrial genes (Nardi *et al.* 2003) and nuclear ribosomal genes (Giribet *et al.* 1996, Spears and Abele 1998, Wheeler 1998a, 1998b, Giribet and Ribera 2000). For example, comparisons of complete mitochondrial genomes have suggested that the Collembola constitute a separate evolutionary lineage that branched much earlier than the separation of many crustacean lineages and insects and independently adapted to life on land (Nardi *et al.* 2003), but other molecular analyses suggest instead that Collembola belong to a monophyletic Hexapoda (Bitsch *et al.* 2004).

A newly revised analysis of nearly complete 28 and 18S rRNA gene sequences suggests the subdivision of a monophyletic clade of the Hexapoda into Insecta (=Ectognatha) and Entognatha, with the latter divided in turn into Collembola and Nonoculata (Protura+Diplura) (Mallatt and Giribet 2006) rather than in Diplura and Ellipura (Collembola+Protura) (e.g. Hennig 1969). The monophyly of the Entognatha, however, is not supported by comparative embryological evidence (Machida 2006). But the relationships among the three main groups of non-insect Hexapoda (Collembola, Protura, and Diplura) are still hotly debated (another recent account is Carapelli *et al.* 2006), and even the monophyly of the Diplura has been questioned.

Within the 'true insects' (Insecta), there is still universal consensus for a sister-group relationship between the Zygentoma and the Pterygota, and between the Archaeognatha and Zygentoma+Pterygota, but the monophyly of the Zygentoma is sometimes disputed because of the problematic position of *Tricholepidion*, either within the Zygentoma or as sister to the remaining Zygentoma plus the Pterygota (Giribet *et al.* 2004a).

Future phylogenetic studies will hopefully see a re-evaluation of some morphological characters about which current research in developmental genetics is challenging the traditional interpretations. A newly re-opened question rounds on the segmental composition of the insect head. The question cannot be easily dismissed as long settled, even if we disregard more radical criticisms in respect to the traditional views of segments as 'given' building blocks of arthropod body structure (see section 8.1.5). For example, Boyan *et al.* (2002) have questioned the nature of the insect labrum and, at variance with the traditional opposite views, regard it as articulated and representing the appendage pair of the so-called intercalary segment; that is, the third head metamere, innervated by the third brain neuromere (the tritocerebrum). Evidence for the appendicular nature of the labrum would be provided by the local expression of a stripe of the transcript of the *engrailed* genes, as is typical of all appendages in the early embryo, as well as by the innervation pattern. In a recent re-evaluation of the problem from a morphological perspective the putative equivalence of the labrum to a pair of limbs is not decided (Scholtz and Edgecombe 2006).

Of the internal phylogeny of the Pterygota I will mention here a few conspicuous (and generally contentious) points only. The first question is the relationships between mayflies (Ephemeroptera) and dragonflies (Odonata) to each other and with respect to the remaining winged insects (the Neoptera). Some molecular studies would support a Palaeoptera clade (Ephemeroptera+Odonata) as sister to the Neoptera, but other analyses favour a sister-group relationship between Ephemeroptera and Odonata+Neoptera; and even the remaining possible tree, with Odonata as sister to Ephemeroptera+Neoptera, has been also suggested (Kjer 2004). A clade of Chiastomyaria (=Ephemeroptera+Neoptera) has been also recovered, although not robustly, by the molecular analysis of Misof et al. (2007) based on 18S sequences.

Within the non-holometabolous insects, many relationships are still uncertain and even the monophyly of a few traditional orders has been questioned. Within a monophyletic Dictyoptera, the cockroaches are likely paraphyletic with respect to the termites, whereas the mantises are likely to be a sister group to all cockroaches and termites. The molecular analyses of Terry and Whiting (2005) recovered as monophyletic clades those such as Orthopteroidea (=Orthoptera+Phasmatodea+Embioptera), Eukinolabia (=Embioptera+Phasmatodea), Haplocercata (=Dermaptera+Zoraptera), and Xenonomia (=Grylloblattodea+Mantophasmatodea). The Mantophasmatodea is the only 'order' of living insects discovered in the last three-quarters of a century (Klass et al. 2002).

Early cladistic analyses revealed that the traditional subdivision of the Hemiptera into Homoptera and Heteroptera is unwarranted, as the latter are a derived hemipteran lineage whose sister group is the Auchenorrhyncha (cicadas, leafhoppers, and relatives) or, perhaps, the auchenorrhynchan clade Fulgoromorpha (Wheeler et al. 1993, Bourgoin et al. 1997, Grimaldi and Engel 2005). Very recent is the suggestion (Misof et al. 2007), based on a molecular phylogeny using 18S rDNA sequences, that the thripses (Thysanoptera) may also belong to the hemipteran radiation.

There seems to be no doubt as to the monophyly of the Holometabola (e.g. Kjer 2004). Their sister group is probably the Paraneoptera or Acercaria (Psocoptera, Phthiraptera, Thysanoptera, and Hemiptera) (Kjer 2004) or the Paraneoptera plus the Zoraptera (Beutel and Weide 2005).

Interestingly, the imaginal discs, out of which the holometabolous insects develop the adult eyes, antennae, legs, wings, and genitalia, are perhaps not an apomorphy of the group. It is also questionable whether the pupa is an apomorphy of this clade (Beutel and Pohl 2006).

Within the Holometabola, particularly contentious are the phylogenetic position of the Strepsiptera and the status of the Mecoptera. As a candidate sister group to the Strepsiptera, Coleoptera are newly gaining favour with respect to another candidate, the Hymenoptera. The alternative suggestion, that the Strepsiptera are sister to the Diptera (Whiting and Wheeler 1994, Wheeler et al. 2001), has been rapidly discarded. Mecoptera are almost certainly non-monophyletic (Beutel and Pohl 2006). In particular, molecular evidence shows the brachypterous, jumping Boreidae to be closer to the fleas (Siphonaptera=Aphaniptera) than to the remaining traditional Mecoptera (Whiting et al. 1997, Whiting 2001).

6.26 Chaetognatha

In terms of phylogeny, the arrow-worms are still possibly the most problematic of the major clades. Represented today by about 100 marine species, this group has diverged from its closest relatives in the distant past, as documented by the presence, in the Cambrian deposits of the Burgess Shales, of a form (*Oesia*) referrable to the Chaetognatha (Szaniawski 2005). Long classified with the deuterostomes because of their radial cleavage, the secondary origin of the mouth and a presumed enterocoely (for a recently revised interpretation see section 8.2.6), based on new palaeontological evidence, chaetognaths have been regarded (Szaniawski 2002) as deriving from protoconodonts, characteristic microfossils of the early Cambrian skeletal faunas (Bengtson 1976, Qian *et al.* 2004). But this hypothesis, even if proved, will not lead us far, as we know much less about protoconodonts than about chaetognaths themselves.

A first molecular study lead to the identification in *Spadella cephaloptera* of six *Hox* genes including a 'mosaic' gene with features of both median and posterior *Hox* genes, two classes that are distinct in nearly all remaining bilaterians (Papillon *et al.* 2003). Interpreting this gene as representing a condition antecedent to the splitting of the median and posterior classes of the *Hox* genes would suggest that the divergence of the chaetognath lineage would predate the deuterostome/protostome split.

All other recent investigations, morphological and molecular alike, point towards the identification of the chaetognaths as a protostome lineage. This is suggested by the spiral cleavage configuration in the four-cell embryo (Shimotori and Goto 2001), by brain ultrastructure (Rehkämper and Welsch 1985), and by several details of neuroanatomy (Harzsch and Müller 2007). That Chaetognatha are protostomes is supported by the analysis of the mitochondrial DNA genome of *Spadella cephaloptera* (Papillon *et al.* 2004). This is the smallest metazoan mitochondrial genome known. Similarly, the complete mitochondrial DNA sequence of *Paraspadella gotoi* revealed the lack of 23 of the genes commonly found in animal mitochondrial DNAs, and only contains 14 genes. Comparisons of amino acid sequences from mitochondrially encoded proteins supports a position of Chaetognatha as sister to all the protostomes included in the study (Helfenbein *et al.* 2004). Similar results were obtained by a

phylogenetic analysis based on ribosomal proteins (Marlétaz *et al.* 2006) whereas a parallel study involving a larger dataset (SSU and LSU rRNAs plus the complete mitochondrial genome) suggested that the chaetognaths are either lophotrochozoans, possibly close to the molluscs, or at least a sister group to the lophotrochozoans (Matus *et al.* 2006a).

6.27 Xenoturbellida

The phylogenetic position of these marine worms (Fig. 6.21) remained problematic until the advent of molecular phylogeny, because of their extremely simple anatomy. The very simple gastral cavity lacking an anus suggested affinities to the flatworms (perhaps sister to the Acoela; Hyman 1951, Franzén and Afzelius 1987), or a very basal position within the bilaterians (e.g. Ehlers and Sopott-Ehlers 1997). The first

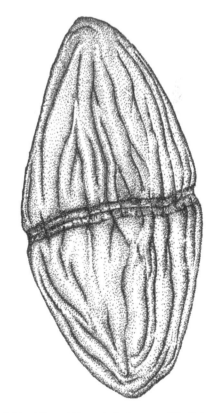

Fig. 6.21. Habitus of *Xenoturbella*. Courtesy of Tim Littlewood.

molecular investigation suggested otherwise, pointing to a very close affinity of *Xenoturbella* to the nuculid bivalves (Israelsson 1997, 1999, Norén and Jondelius 1997). Nuculid bivalves, however, are only a favourite food of *Xenoturbella* and this worm's putative sequences were nothing other than a contamination. New investigations have shown that *Xenoturbella* is in fact a deuterostome. The original study (Bourlat *et al.* 2003) suggested that *Xenoturbella* is probably a sister to the Ambulacria, a placement supported by further nuclear and mitochondrial data (Bourlat *et al.* 2006), but the most recent study based on the analysis of the mitochondrial DNA of *Xenoturbella* (Perseke *et al.* 2007) suggests a more basal branching of this lineage within the deuterostomes.

6.28 Echinodermata

We are accustomed to regarding the echinoderms, alongside amphioxus, as prototypical deuterostomes. Indeed, much of the classical deuterostome embryology has been based on their study. Echinoderm embryonic development, however, is far from uniform and this must be taken into account when determining the weight to be accorded to this kind of evidence in a phylogenetic analysis. Some echinoderms are indeed schizocoelic rather than enterocoelic, and some are, literally, protostomous (Fell 1948).

The Palaeozoic fossil record includes many extinct lineages, some of them to be referred to the stem-group Echinodermata, with a general organization very different from all Recent forms. Among them are the Stylophora, within which the ancestry of the vertebrates would be found according to Jefferies's (1986) long-debated Calcichordata hypothesis. It has proven impossible to single out any univerally shared morphological trait of any importance: the calcitic skeleton is lacking in some Holothuroidea, the water-vascular system was not present in some fossil lineages, the five-ray radial symmetry is changed into secondary bilaterial symmetry in several lineages and into radial symmetry with more than five rays in several Asteroidea, not to mention the numerous extinct echinoderm taxa without any evident symmetry (Pawson 2007).

Based on current molecular phylogeny of deuterostomes, Smith (2005) sketches the morphology of the earliest echinoderms at their split from the hemichordates in this way: presence of gill slits, absence of a notochord, adult with bilaterally symmetrical body, and, tentatively, development characterized by torsion. Presence and evolution of these traits can be traced through several Lower Palaeozoic groups of deuterostomes, collectively known as carpoids, whose relationships to crown echinoderms are not always clear. In the Cincta, a large atrial opening was present, in addition to mouth and anus, leading to the interpretation of these animals as pharyngeal basket feeders. Paired grooves associated with the mouth are interpreted as a proof of the presence of a hydrovascular system, but this was not necessarily built from just the left hydrocoel as in (crown) echinoderms. In the Stylophora (Cornuta+Mitrata) the bilateral organization was obscured by torsion; these animals were provided with a locomotory appendage and with gills, either unpaired and external, or paired and internal, opening into an atrial cavity. The presence of a hydrovascular system homologous to the water-vascular system of echinoderms is doubtful in the stylophorans, but certain for the remaining group, the Soluta, which are the closest to crown-group Echinodermata, possessing an ambulacral system with a hydropore and no pharyngeal gill openings. The relationships of the Recent classes are still somewhat controversial, besides general agreement regarding the Crinoidea as the sister group of the remaining Recent classes and the Echinoidea as the sister group of the Holothuroidea. The Ophiuroidea are sister either to the Asteroidea or to the Echinoidea+Holothuroidea. The first hypothesis is better supported by morphology (Mooi and David 1997) and by gene order in the mitochondrial genome (Smith *et al.* 1993), but molecular evidence is generally in favour of the alternative topology (Littlewood *et al.* 1997).

6.29 Hemichordata

Recent ultrastructural research on the stomochord of *Rhabdopleura compacta* (Pterobranchia) fails to provide a support for the hypothesis that the notochord of the chordates and the stomochord of the hemichordates are homologous structures (Mayer and Bartolomaeus 2003). There are major differences between the two groups traditionally assigned to the Hemichordata, the infaunal, solitary worm-like Enteropneusta, and the sessile and often colonial Pterobranchia. Morphology suggests that both groups are monophyletic (recently confirmed by Cameron 2005), but molecular analyses based on 18S rRNA sequences have suggested otherwise. The Enteropneusta would be paraphyletic, with their family Harrimaniidae sister to the Pterobranchia (Halanych 1995, Cameron *et al.* 2000), but this is probably an artefact due to unequal rates of changes in the homologous molecules of the different lineages (Winchell *et al.* 2002).

6.30 Chordata

Molecular phylogenetic analyses have uniformly confirmed the monophyly of the Chordata. This hypothesis has been only questioned recently by Raineri (2006), based on embryological evidence, but without offering an explicit alternative phylogenetic hypothesis. The choice of Zeng and Swalla (2005) to consider the Tunicata as a separate phylum rather than a subphylum of Chordata because of their unique (apomorphic) adult morphology is cladistically irrelevant. During the last two decades there has been important progress in chordate palaeontology, as a consequence of a fresh interpretation of long-known fossils, such as conodonts, and with the description of new forms that probably belong in part to the stem-group chordates, in part to one or the other of the three main chordate lineages, the cephalochordates, the urochordates, and the vertebrates.

Conodonts, small teeth-like fossils (Fig. 6.22), known from most of the Palaeozoic, and long regarded as being problematic to assign taxonomically, are now recognized as chordates since Aldridge *et al.* (1986) associated them with fossils preserving traces of soft parts that suggested their nature. Cladistic analyses of relevant taxa by Donoghue *et al.* (1998, 2000) suggest that conodonts are, specifically, vertebrates (see also Aldridge *et al.* 1993, Aldridge and Purnell 1996). The chief competing hypothesis (Kasatkina and Buryi, 1996, 1997, 1999) regards conodonts as representing feeding structures of extinct chaetognaths.

Another 'popular' chordate fossil is *Pikaia*, now tentatively referred to the cephalochordates rather than to the vertebrates, as previously suggested (Briggs and Fortey 2005). Its morphology, however, is ambiguous as its myomeres are not clearly chevron-shaped as in amphioxus and there is no clear evidence of gill slits (Schubert *et al.* 2006).

Fig. 6.22. Conodonts. Courtesy of Phil Donoghue.

Of the more recently described forms, the Vetulicolia from the Cambrian Chengjiang site have a body divided into an anterior part covered by a kind of carapace and a tail-like posterior part. These animals were described by Hou (1987) as arthropods. Chen and Zhou (1997) created for them the new class Vetulicolia and referred them to stem-group Arthropoda, to include also *Banffia* from the Burgess Shales. Shu *et al.* (2001) raised the Vetulicolia to the rank of phylum and regarded them as primitive deuterostomes; the perforations on the 'carapace' were interpreted as gill slits. Gee (2001) compared these fossils with tunicate tadpole larvae and suggested them to be possibly the sister group of the Chordata. According to Lacalli (2002), the Vetulicolia are perhaps stem tunicates, while Butterfield (2003) still accepts that they are more likely to be arthropods because of the presence of a cuticle.

The Vetulocystida (Shu *et al.* 2001) are another group of fossils form Chinese Lower Cambrian localities, also with bipartite body. Two conical projections found on the globose, putatively anterior part have been tentatively compared to ascidian siphons. The other body part has been described as a tail, but this is not compatible with the presumed presence of a posterior gut running throughout its length.

Myllokunmingia (possibly identical with *Haikouichthys*; but this is rejected by Conway Morris 2006) is regarded by Hou *et al.* (2002) as the earliest vertebrate because of its V-shaped myomeres and the presence of gill filaments and a dorsal fin. In this fish-like fossil the head seems also to be provided with large eyes and paired nostrils, possibly also paired olfactory organs (Shu *et al.* 2003, Schubert *et al.* 2006). *Yunnanozoon* (Fig. 6.23), originally described by Hou *et al.* (1991) as a 'problematicum', has been subsequently interpreted as a cephalochordate (Chen *et al.* 1995) because of the presence of notochord and myomeres; alternatively it has been described as a stem-group chordate (Dzik 1995) and as a primitive chordate (Chen and Li 1997). Possibly closely related to it is *Haikouella* (Chen *et al.* 1999), which is more craniate-like than cephalochordate-like (Briggs and Fortey 2005).

6.30.1 Cephalochordata

The mutual relationships of the three main clades of Recent chordates have been the subject of renewed attention since molecular phylogenetic analyses suggested that vertebrates are possibly closer to the urochordates than to the cephalochordates. This was first suggested by the structure of the cadherin gene (Oda *et al.* 2002) and subsequently supported by several multigene phylogenies (Blair and Hedges 2005, Philippe *et al.* 2005, Delsuc *et al.* 2006). In addition, Wada (2001a, 2001b, Wada *et al.* 2006) has

Fig. 6.23. The Chengjiang chordate *Yunnanozoon.* Courtesy of Jerzy Dzik.

pointed out that some of the genes involved in neural development show similar expression patterns in ascidians and vertebrates, but are distinct from their amphioxus counterparts.

6.30.2 Urochordata

An analysis of the morphological, life-history, and biochemical characters of 57 tunicate species (Stach and Turbeville 2002) recovered both the Thaliacea and the Larvacea as lineages nested within the 'Ascidiacea'. The same result has been obtained by Zeng and Swalla (2005), based on 18S rRNA sequences. If confirmed, this would invite a dissection of the developmental and genetic changes involved in the dramatic transitions from the ascidian organization to that of a *Salpa* or an *Oikopleura*. In the latter case, profound changes in the whole organization of the genome might have been involved (see section 2.4.4).

6.30.3 Vertebrata

It would be impossible to account adequately in this book for all major recent advances in vertebrate phylogenetics and evolution to which many different disciplines have contributed, from phylogenetics (based on molecules and morphology of Recent forms) to palaeontology, from developmental genetics to comparative physiology. Synthetic reviews are available for many groups or problems and I will not attempt to summarize them here. However, before closing these chapters devoted to sketching what is going on in animal phylogenetics I will pick up three items that may be exemplary of the inextricable mix of old problems and new vistas with which we are confronted daily in these studies. Other examples, such as the water-to-land transition in the history of vertebrates (see section 9.1.1), are mentioned elsewhere in this book in a different context.

A first point is that we still have inadequate knowledge of morphological and developmental issues that we would routinely regard as settled. For example, the evolutionary history of coelom formation in vertebrates is still controversial

(Jenner 2004c). We might assume that vertebrates, as deuterostomes, form the coelom by enterocoely, but this is mainly based on the blueprint provided by amphioxus although, admittedly, the anterior coelomic cavities of the lampreys seem also to form this way (Goodrich 1958, Schaeffer 1987, Presley *et al.* 1996). But some authors (e.g. Ghiara, 1995) deny the occurrence of enterocoely in vertebrates.

Second is the origin of snakes. Limb loss has occurred many times among the squamate reptiles and this has probably determined a great deal of homoplasy that has caused problems in reconstructing phylogenies from morphological evidence. This may help to explain some divergence between analyses based on morphological and molecular evidence. In a recent phylogenetic analysis of the Squamata based on 248 osteological, 133 soft anatomical, and 18 ecological traits (Lee 2005), morphology suggests that snakes are nested within the anguimorph lizards (the group to which the familiar European slow worm, *Anguis fragilis*, belongs). But a molecular phylogenetic study of 69 squamate species using 2876 parsimony-informative base pairs from nuclear and mitochondrial genes (Townsend *et al.* 2004) suggests that all the main groups of limbless squamates evolved independently: the amphisbaenians (worm lizards) would be the sister group of lacertids, whereas the dibamids diverged early in squamate evolutionary history and snakes are grouped with iguanians, lacertiforms, and anguimorphs, but do not emerge from within the latter as in the previously cited morphological analysis. Cladistic hypothesis may help to resolve a key question in snake evolution. One hypothesis suggests a close relationship to terrestrial lizards, and a terrestrial scenario for limb reduction, whereas an alternative hypothesis suggests that snakes are most closely related to Cretaceous marine lizards, such as mosasaurs (Caprette *et al.* 2004). In the latter case, limb reduction would have taken place in the water. As monitor lizards (Varanidae) are believed to be close relatives of the extinct mosasaurs, the relationship between varanids and snakes are critically important in assessing the validity of the alternative hypotheses. DNA sequence

Table 6.2 A list of names proposed for 'superordinal' groups of mammals, with references to the works where these names were established and content (with reference to currently accepted clades).

Group	References	Content
Afroinsectiphilia	Waddell et al. (2001)	Tubulidentata+Afroinsectivora
Afroinsectivora	Waddell et al. (2001)	Macroscelidea+Afrosoricida
Afrosoricida	Stanhope et al. (1998)	Chrysochloridae +Tenrecidae
Afrotheria	Springer et al. (1997)	Proboscidea+Sirenia+Hyracoidea+Tubulidentata+Macros celidea+Chrysochloridae+Tenrecidae
Archonta	Gregory (1910)	Scandentia+Dermoptera+Chiroptera+Primates
Atlantogenata	Waddell et al. (1999b)	Xenarthra+Afrotheria
Boreotheria	Waddell et al. (2001)	Laurasiatheria+Supraprimates
Cetartiodactyla	Montgelard et al. (1997)	Artiodactyla+Cetae
Cetungulata	Irwin and Wilson (1993)	Cetartiodactyla+Perissodactyla
Eparctocyona	McKenna (1975)	Artiodactyla+Cetae plus the extinct clades Procreodi, Condylarthra, and Arctostylopida
Epitheria	McKenna (1975)	Placentalia less Xenarthra
Euarchonta	Waddell et al. (1999c)	Primates+Scandentia+Dermoptera
Eulipotyphla	Waddell et al. (1999c)	Lipotyphla less Afrisoricida
Eurchontoglires	Murphy et al. (2001)	Euarchonta+Glires =Supraprimates
Euungulata	Waddell et al. (2001)	Cetartiodactyla+Perissodactyla
Exafroplacentalia	Waddell et al. (2001)	Boreotheria+Xenarthra
Ferae	*Sensu* Waddell et al. (2001)	Pholidota+Carnivora
Fereuungulata (*sic*)	Waddell et al. (1999a)	Pholidota+Carnivora+Cetartiodactyla+Perissodactyla
Hyracoproboscidea	Waddell et al. (2001)	Proboscidea+Hyracoidea
Laurasiatheria	Waddell et al. (1999c)	Cetartiodactyla+Perissodactyla+Carnivora+Pholidota+ Chiroptera+Eulipotyphla
Lipotyphla[1]	Haeckel (1866)	
Pegasoferae	Nishihara et al. (2006)	Chiroptera+Perissodactyla+Carnivora+Pholidota
Preptotheria	McKenna (1975)	Crown Epitheria (=Epitheria less the extinct Leptictida)
Primatomorpha	Kalandadze and Rautian (1992)	Primates+Rodentia
Primatomorpha	Beard in Szalay et al. (1993)	Dermoptera+Primates
Pseudoungulata	Waddell et al. (1999c)	Proboscidea+Sirenia+Hyracoidea+Tubuilidentata
Scrotifera	Waddell et al. (1999b)	Pegasoferae+Cetartiodactyla
Sundatheria	Olson et al. (2005)	Scandentia+Dermoptera
Supraprimates	Waddell et al. (2001)	Euarchonta+Glires =Euarchontoglires
Tethytheria	McKenna (1975)	Sirenia+Proboscidea
Uranotheria	McKenna and Bell (1997)	Hyracoidea+Tethytheria (plus the extinct Embrithopoda)
Volitantia	Illiger (1811)	Chiroptera+Dermoptera
Zooamata	Waddell et al. (1999c)	Pholidota+Carnivora+Perissodactyla

[1]Lipotyphla Haeckel, 1866 corresponds to the 'true' Insectivora; that is, to the exclusion of both Macroscelidea and Scandentia, often classified with the Insectivora in early times.

evidence does not support a close relationship between snakes and monitor lizards (Vidal and Hedges 2004); thus a terrestrial origin of snakes seems more likely, but the question is possibly still open.

The third and more popular point is the radically new interpretation of the interrelationships among the main lineages of mammals that emerge from molecular phylogeny. This can be easily appreciated from the long

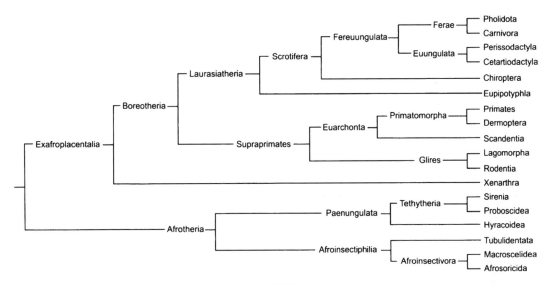

Fig. 6.24. The phylogeny of mammals according to Waddell *et al.* (2001).

list of mammalian 'superordinal' clades given in Table 6.2 and the phylogeny presented in Fig. 6.24. Of all the novelties emerging from these studies the most conspicuous, and controversial, is the hypothesis of a very diverse clade of mammals comprising a diversity of lineages endemic to Africa or, at least, originated in that continent. This lineage, the Afrotheria, would include mammals as different as elephants, sirenians, hyraxes, elephant shrews, the aardvark, and also a couple of families traditionally classified with the insectivores. Until now, no support has been found in morphology for the Afrotheria

(Asher 1999, Whidden 2002, Asher *et al.* 2003, Zack *et al.* 2005), but the signal deriving from molecular evidence is strong and consistent. It is possible that this phylogenetic hypothesis will eventually become accepted by mammalogists, as is already happening for the Cetartiodactyla, a clade corresponding to the cetaceans plus the artiodactyls of the old classifications, with the hippopotamuses as the sister group of the cetaceans. In this case, convincing palaeontological evidence (e.g. Thewissen *et al.* 1996, Thewissen and Madar 1999) has been found that supports the relationships suggested by molecules.

CHAPTER 7

The life cycle and its evolution

I believe that ontogeny has only a very limited value for phylogenetic questions.

Steinböck (1963), p. 49

A genuine comparative morphology must be based on a comparison of life cycles, and an adequate phylogeny, on the phylogenesis of life cycles.

Beklemishev (1963), p. 234

During the last few decades of the nineteenth century, the rush to collect the 'proof of evolution' was often coupled with a search for evidence to be used in drawing the first tentative versions of the tree of life. Largely under the influence of Ernst Haeckel, the most vigorous German apostle of Darwin, zoologists invested much effort in the study of the embryonic and larval development of a rapidly increasing number of animal species, ready to take similarities in the morphology of early developmental stages as solid proof of phylogenetic relatedness (Gould 1977). This is the cultural milieu from which Kowalewski's (1866) discovery of the affinities between vertebrates and ascidians eventually emerged. Indeed, even before the publication of the *Origin of Species*, developmental evidence had proved critically important in determining the affinities of animals of otherwise uncertain position. A good example is the barnacles, from whose typical nauplius larvae Thompson (1830a) inferred the crustacean nature of these sessile, shelled invertebrates. However, the enthusiasm with the phylogenetic importance of developmental data was far from universal, and an increasing number of zoologists started paying attention to Garstang's view (e.g. Garstang 1922), who provided convincing examples of the independent and likely recent

evolution of many larval adaptations and, more generally, strongly rejected any simple relationship between ontogeny and phylogeny.

A new era in the contrasted history of the relationships between the study of animal development and the reconstruction of phylogeny has taken form, initially in the context of the revisitation of the comparative method, which accompanied the advent of cladistics, and more recently under the aegis of those advances in comparative developmental genetics that have been the main force behind the birth of evolutionary developmental biology (in short, evo-devo). For comprehensive and partly contrasting views on the nature and aims of this discipline, useful book-size accounts are provided by Hall (1998a), Wilkins (2002), Hall and Olson (2003), Minelli (2003), Carroll *et al.* (2005), and Minelli and Fusco (2008); excellent article-size overviews are Arthur (2002) and Müller (2008). The relevance of evo-devo studies for systematic and phylogenetic research is discussed in Minelli (2007) and Minelli *et al.* (2007).

Basically, the role of developmental evidence in the context of cladistic analysis amounts to an appreciation of the value of ontogenetic information in the formulation of hypotheses about homology. This point was very hotly debated in the late 1970s. Nelson (1978) introduced the so-called ontogenetic criterion of homology in the following terms: 'given an ontogenetic character transformation, from a character observed to be more general to a character observed to be less general, the more general character is primitive and the less general advanced' (p. 327). Other cladists (e.g. Rieppel 1979) were critical, however, either of the actual importance of ontogenetic information or of its independence from the evidence acquired by the application to morphological data of the outgroup comparison method. A very informative study was that by Mabee (1989) on the ossification schedule in some centrarchid fishes. She demonstrated that ontogeny indeed provides data of use in phylogenetic reconstruction, but the ontogenetic criterion per se is not an accurate predictor of character polarity. Similarly, in the developing limbs of the marbled newt *Triturus marmoratus* the order in which the chondrogenetic condensations appear is not the same in which these cell condensations will eventually chondrify (Blanco and Alberch 1992). Haeckel's, or indeed Nelson's, recapitulatory scenarios capture only one of the possible relationships between the temporal schedule of ontogeny and the temporal schedule of evolutionary events.

At around the same time, a debate on the notion of homology was taking place in terms of developmental mechanisms (with, e.g., Wagner's 1989 biological concept of homology) rather than in terms of character distribution and transformation. These efforts were intended to identify shared integrated modules of developmental processes as an explanation for the occurrence of homologies between different lineages rather than at the evaluation of nested patterns of shared character states as foundation for phylogenetic hypotheses. Whether the concepts of homology around which phylogeneticists and developmental biologists have been discussing since then have enough in common to justify the use of the same term by both communities of researchers is something that philosophers of biology will hopefully assess in the near future. Whatever the answer, I dare to say that these efforts have helped to improve the link between phylogenetics and developmental biology.

Today, mainly as a consequence of a growing body of knowledge in comparative developmental biology, we are forced to ask new questions, some of which will be considered further in this chapter. Many of these questions revolve around the involvement of homologous genes in the production of more or less similar organs in animals belonging to remotely related lineages. This is no more simply a question about the phylogenetic

signals contained in the nucleotide sequences of the homologous genes selected from different species as these, basically, are nothing other than morphology at the molecular level rather than at that of the cell or whole organism. The new question we can (and must) address today is whether the shared involvement of these homologous genes in producing, say, the eyes of two different animals, A and B, can be taken as a 'proof' of the homology between the eye of A and the eye of B. The answer is not simple, for various reasons. The most serious problem is probably that the actual role of a given gene in the ontogenetic production of an organ is not necessarily specific to the body part we are considering. This is what we are learning increasingly, the more we know about the functional roles of gene products. Thus, the list of what we may eventually call, specifically, 'eye genes' or 'heart genes' is not that long, and quite probably fated to decrease rather than to increase with the progress of knowledge in this field. On the other hand, a gene can still have a very specific role in determining the eventual organization of a body feature because of precise spatial and temporal restrictions in its expression (at least at one of the several times, possibly an early one, at which it is expressed during development). In turn, these spatial and temporal restrictions in the expression of a gene are the consequence of precise control, in which the products of other genes are also involved. Thus, for us to state the homology between, say, two eyes is it not enough to demonstrate the common involvement of homologous copies of the same gene, provided that its expression is necessarily required to produce both kinds of eye? Alternatively, must we also obtain proof of similar expression patterns, or even of the involvement of homologous copies of identical genes in the upstream control of the expression of the gene in question? I believe that this chain of questions can only be terminated by acknowledging the inadequacy of the traditional all-or-nothing concept, according to which two features being compared either are, or are not, homologous. As often repeated in these pages, this concept of homology only works in full if the comparison is one between two minimal items, such as two molecular sequences, but does not apply very strictly to more complex and interesting items, within which individually homologous components may have been coupled and uncoupled many times and in many different ways. To accept as homologous only features with a uniquely derived *whole set* of components would render the notion of homology virtually useless. The workable alternative is a factorial approach, by which the homology between the features we are comparing is only suggested, or negated, for specific sets of descriptors we should try to circumscribe in each case as carefully as possible (Minelli 1998).

Summing up, shall we adopt Steinböck's critical attitude with respect to the value of developmental evidence in the reconstruction of phylogeny, or Beklemishev's claim that a sound understanding of phylogeny must be based on a reconstruction of the evolution of the whole life cycle? In general terms, my personal position is closer to Beklemishev's than to Steinböck's, but with an important qualification, in the same vein as my advocacy of a factorial approach to homology. The dynamic continuity linking all the stages in a life cycle, the usually strong integration among the parts of the organisms at any time along this cycle, and the obvious constraints that every stage poses to the eventual deployment of the next stage should not lead us to adopt a holistic view, which would obscure the steady entangling and disentangling of individual features with respect to the remaining components of the animal body and their ontogenetic dynamics. Additionally, and no less importantly, the concept of a life cycle should be taken literally, by granting similar status to all developmental stages, rather than regarding eggs, embryos, larvae, and juveniles simply as preparatory stages whose eventual meaning is to be found in the adult, for which they are paving the way. Let us try to describe and explain a gastrula, a trochophore, and even an insect pupa with all its dramatic changes during metamorphosis for what they are, with their specific anatomy and with their peculiar dynamics. Let us try to understand how their peculiar traits relate to those of the preceding stage, on which they

depend, rather than 'justifying' their structural and functional properties in terms of how these stages eventually contribute to making a sexually mature adult.

Removing ourselves from both holism and adultocentrism will help us to frame many questions concerning animal structure and development in a fresh way. This suggested revisitation is not easy, so entrenched is the traditional perspective. In this chapter I will try to articulate a number of points concerning reproduction and development, about which many blueprints, commonly accepted without much criticism, may obscure the understanding of key aspects of the evolutionary history of the Metazoa.

7.1 Cells and tissues

7.1.1 The cell as a building block of the animal body

The cell is the most obvious, universal building block of the animal body. Its evolution deserves much more attention than it has received from this perspective up until now. We know a lot about the specific patterns of gene expression that characterize many cell types in a few model species, but the problem of the evolution of specific cell-differentiation pathways has only been addressed with respect to a few cell types and, especially, within very limited and phylogenetically poorly informative comparisons, such as between mouse and fruit fly. Something like an essay on comparative cytology was attempted by Willmer (1960), but his approach suffered too much from a vertebratocentric bias, and from a *scala naturae* vision, to offer stimulating reading today.

The cell can be often regarded as the most elementary building block of the animal body, but not always. Problems with taking the cell as the morphological and functional unit of the animal body have nothing to do with the beginning of animal evolution (section 3.1). Indeed, Hadži's (1963) idea, that animals derive from multinucleate, ciliate-like ancestors, has definitively fallen into disrepute, at least since metazoans have found a place in molecular phylogenies close to unicellular organisms like the choanoflagellates, which have conventional uninucleate cells. Rather, problems derive from the fact that the individuality of the textbook animal cell has been challenged repeatedly in evolution.

The issue of individuality—that is, of what actually defines the spatiotemporal boundaries and the historical continuity of an individual—emerges time and again in the analysis of living beings, at different structural levels. In zoology, the most obvious grey area is multinucleate units without cytoplasmic boundaries, whereas morphological (junctions) and functional (electrical coupling, exchange of molecules) ties between neighbouring epithelial cells are less permissive than the links provided by the plasmodesmata in plant cells, to the extent that in plants the very notion of cells is sometimes disputed (Lucas *et al.* 1993). With respect to cell individuality, metazoans are usually less controversial than plants and many protists. Events such as the temporary presence in the cytoplasm of *Elysia* sea slugs of functional chloroplasts stolen from the green algae upon which they prey do not seem to have opened significant avenues towards evolutionary innovations, despite the recent demonstration that a gene encoding a chloroplast protein has been transferred from a siphonaceous alga to the nuclear genome of *Elysia crispata* (Pierce *et al.* 2003). New discoveries in this area, however, are perhaps waiting around the corner. Karyoklepty—retaining prey nuclei after feeding on the cells to which they belong—has been very recently discovered in the marine ciliate *Myrionecta rubra* which, in addition to stealing chloroplasts from its prey, the cryptophyte *Geminigera cryophila*, also accumulates the prey's nuclei, which remain transcriptionally active for several weeks and continue to serve the requirements of the stolen cryptophyte plastid cells (Johnson *et al.* 2007).

The plasticity of animal cells is also shown by the unique topology evolved among the rhombozoans, with cells (assoblasts) developing inside other cells (the axial cells) (see Ax 1996).

Thus, rather than taking the cell, typologically, as a standard unit of morphological organization, it seems advisable to regard it as a unit of function, integrating a complex network of local dynamics, whose independence, as a distinct module of form, is not always granted, as the widespread occurrence of syncytia amply demonstrates.

7.1.2 Syncytia

Syncytial organization as a condition ensuring strict morphofunctional coupling beyond the range of a conventional uninucleate cell has evolved independently many times, through different mechanisms. A distinction must be drawn between multinucleate structures deriving from the fusion of previously independent cells and those deriving from incomplete cell division. In 1934, Studnička proposed a terminology that attempted to combine origin and organization of these multinucleate structures, using 'symplasma' for multinucleate systems whose cytoplasmic continuity results from incomplete cytokinesis but which otherwise remain independent, 'syncytia' for multinucleated structures whose cytoplasm is not organized around centrioles, and 'plasmodia' for multinucleated tissues formed by fusion of separate cells or by division of nuclei in a growing cell. Irrespective of the merits (and applicability) of those distinctions, current literature has increasingly adopted the term 'syncytium' for all these structures (Mackie and Singla 1983).

With respect to a cellular organization, syncytia exhibit enhanced intercellular transport by way of cytoplasmic continuity. But syncytia can be compartmentalized: for example, within the coenocyst, a germline multinucleate structure surviving throughout prophase I in the oogenesis of the urochordates *Oikopleura*, entry of the individual nuclei into meiosis is asynchronous (Ganot *et al.* 2007).

The origin of a syncytial organization is open to several alternative explanations. In so far as the selective filter determined by plasma membranes is removed, molecules move more freely and over larger distances, thus facilitating the spread of local dynamics. But in some instances a syncytial organization may reduce the strength of external constraints on the cell, or cell sheet, as it happens with cuticles, whose bonds are likely to prevent cells from undergoing mitosis and, as a consequence, to deprive the epidermis of many ecdysozoans of the chance to grow in size during the intermoult. In nematodes, where most of the epidermis is formed by a single syncytium, size does not depend exclusively on the numbers of the intervening fused cells, but also on the extent of its acellular growth, and is partly associated with the endoreduplication of hypodermal nuclei (Flemming *et al.* 2000, Knight *et al.* 2002).

As recently summarized by Kontani *et al.* (2005), cell–cell fusion initiates fertilization, sculpts tissues during animal development, and changes the eventual state into which the progeny of stem cells can differentiate, not to mention being a key step in cancer progression.

Syncytia are often a transitory embryonic feature, because of retarded cytodieresis, as in the blastoderm of *Drosophila* and teleost fishes (Kishimoto *et al.* 1996), or in the embryos of *Patiriella* sea stars (Cerra and Byrne 2004), or as the result of the fusion of epidermal cells surrounding the wound plug during epidermal wound healing in *Drosophila* larvae (Galko and Krasnow 2004).

Syncytial structures are present virtually everywhere; that is, in the most diverse body parts of the most different animal groups. Syncytia are known, in fact, from groups as diverse as Silicea, Placozoa, Cnidaria, Acoela, Gastrotricha, Syndermata, Gnathostomulida, Rhabditophora, Nematoda, Arthropoda, Echinodermata, and Chordata. Clearly, this condition has been obtained independently a great many times.

The organization of the hexactinellid sponges is unique; their syncytial organization was revisited by Leys (2003). These sponges are usually described as fully syncytial, but this is not strictly correct. Their cytoplasm consists of a giant, multinucleated trabecular syncytium, which is connected via open and plugged cytoplasmic bridges to cells such as archaeocytes, choanoblasts, and cells with spherical inclusions. The whole sponge is cytoplasmically interconnected, so electrical signals can propagate throughout the whole animal. Similar to plant cells, which could be considered syncytial because they are connected by plasmodesmata, but which are nevertheless considered by most plant biologists to be cells because of their large functional independence, in hexactinellids there are uninucleate components still termed cells, although all components are connected cytoplasmically (Leys 2003). The cytoplasmic bridges between uninucleate and multinucleate components are generally plugged by intrasyncytial proteinaceous junctions (Mackie and Singla 1983), which probably act selectively on transport between 'cellular' units. Important for phylogeny, this plug is a specialized cytoplasmic structure, different from the true cell junctions (desmosomes, gap junctions, or septate junctions) common to the remaining metazoans, which are formed by pairwise apposed membranes. The larva of hexactinellids does not resemble either the amphiblastula of the Calcarea and some demosponges, or the parenchymella larva of most demosponges (Boury-Esnault *et al.* 1999). Apparently, in hexactinellids the earliest developmental stages until gastrulation are cellular, as in the other sponges. According to Leys (2003; also Leys *et al.* 2006) it seems likely that in the early stages of development blastomeres are connected by cytoplasmic bridges, as otherwise seen in the development of squid embryos (e.g. Arnold 1974), and that subsequently a proteinaceous barrier, the perforate plug, is inserted between the two daughter cells.

In many cases, the syncytial organization is restricted to epithelia and two metazoan groups have been named after this trait. One group is the Syndermata; that is, the clade including the probably paraphyletic, mostly free-living Rotifera and the parasitic Acanthocephala (e.g. Kristensen 2002). Three distinct syncytia (frontal, central, and epidermal) are identified in the acanthocephalan larva, the acanthor (Albrecht *et al.* 1997). The other clade of animals named after their conspicuous syncytia is the Neodermata, the large group of parasitic flatworms comprising the digeneans (flukes), the monogeneans, and the cestodes (tapeworms). Their neodermis is a continuous syncytial sheet that covers the whole body and is connected to perikarya sunk below the body wall musculature (Tyler and Hoodge 2004). In a flatworm group closely related to the Neodermata, the epizoic Temnocephalida, the epidermis comprises several distinct syncytia, separated by lateral membranes. For example, in *Diceratocephala boschmai*, Joffe *et al.* (1995) demonstrated a regional differentiation of the six syncytia differing for the presence or absence of locomotory cilia and many traits of cytoplasmic organization, including the development of a cytoskleteton and the abundance of hemidesmosomes linking these syncytia to the basal membrane. An extended hypodermal syncytium is found in nematodes. For example, in *Caenorhabditis elegans* about one-third of the 959 somatic cells fuse to form 44 multinucleated syncytia (Chen and Olson 2005).

In *Hydra*, nematoblasts in the process of differentiating into nematocytes form a syncytium (Bode 1996).

Muscles are often syncytial, as in the well-known case of the voluntary muscles of vertebrates, following fusion of embryonic myoblasts. Similarly, insect muscles are syncytial, arising from the fusion of mesodermal cells (Paululat *et al.* 1999). The circular visceral muscles of *Drosophila* have a peculiar origin, with binuclear syncytia arising from the fusion of a circular visceral founder cell with a visceral fusion-competent myoblast (Schröter 2006). A tendency towards syncytial organization is also found in gnathostomulids like *Rastrognathia* and *Gnathostomula* (Lammert 1991), especially in the pharynx (Herlyn and Ehlers 1997); but a syncytial pharynx is also present in nematodes like *C. elegans* (Vinogradova *et al.* 2006).

The secretory cells involved in the production and storage of poison in the serous cutaneous glands of several anurans (Bombinatoridae, Discoglossidae, Ranidae, Hylidae, Pseudidae, and Leptodactylidae) are also organized as syncytia (Delfino *et al.* 2001a, 2001b, Alvarez *et al.* 2005).

Syncytia, however, are far from restricted to epithelia and muscles. A syncytial mesenchyme is comprised between the cellular dorsal and ventral epithelia in Placozoa (Buchholz and Ruthmann 1995) and a skeletogenic mesenchyme

forms in sea urchin larvae by incomplete division of the progeny of 32 precursors migrated into the blastocoel (Hodor and Ettensohn 1998, Ingersoll and Wilt 1998, Treccani *et al.* 2003).

A syncytial organization is also frequently found in the wall of gonads, in the clusters of growing germ cell, or in the trophocytes that accompany them. For example, the wall of the testis of the Rotifera is a thin syncytium (Gilbert 1983) and in some nematodes the male spermatogonia form from the syncytial wall of the testis (Soliman 2003), whereas in tapeworms syncytial organization can be found in the interstitial tissue of the ovary (Poddubnaya *et al.* 2005). The germ line is syncytial in the oikopleurid appendicularians (Ganot *et al.* 2006). Syncytial spermiogenesis has been described in arthropods such as the Xiphosura (Hong and Huang 1999) and *Drosophila* (Fabrizio *et al.* 1998). A syncytium combining numerous nurse cells has been often described in mites (Alberti *et al.* 1999, Di Palma and Alberti 2001) and insects, such as Raphidioptera (Jedrzejowska and Kubrakiewicz 2002) and beetles (Büning 2005), but also in bony fishes (Diaz *et al.* 2002). Syncytia formed by nurse cells and maturing oocytes are found in Heteroptera (Dittmann 1998) and Diptera (Dobens and Raftery 2000).

The Acoela lack a gut, but their body is filled by a syncytial tissue where food is digested in large vacuoles (Rieger *et al.* 1991, Tyler 2001). A digestive cavity lined by a conventional epithelium is present, however, in the Nemertodermatida, and a somewhat intermediate condition has been described in the acoel *Paratomella rubra* (Smith and Tyler 1985). Three distinct syncytia are found in each of the two protonephridial systems of *Seison* (Ahlrichs 1993), a marine 'rotifer' that occupies a distinct position within the syndermatan radiation (section 6.5).

In the nervous system, the giant neurons of annelids (e.g. Nicol 1952) and squids (e.g. Young 1939, Williamson and Chrachri 2004) are multinucleate following fusion of originally separate cells. An astroglial syncytium is present in the cerebral cortex of mammals (e.g. Colombo *et al.* 2000) and in sensory cells, as the horizontal cells of the vertebrate retina (e.g. salamanders; Hare and Owen 1998, Bornstein *et al.* 2002) or the glial cell syncytium in the retina of the honeybee (Marcaggi *et al.* 2004).

Syncytial organization may be induced by parasites in the tissues of their host, as caused by the infective cells of the Orthonectida. Similarly, syncytia are induced in the peritoneal lining of the metacoelom of the freshwater bryozoan *Plumatella* by the microsporidian *Schroedera airthreyi* (Morris *et al.* 2005). The structures hosting intracellular symbiotic bacteria in insects are also often syncytial (e.g. Fukatsu and Nikoh 1998).

7.1.3 Cell division

In biology, defining default conditions or processes is often dangerous, but I think we can accept Soto and Sonnenschein's (2004) suggestion that proliferation is the default mid-term activity of all cells. If so, we must reject Martin Raff's opposite view, according to which death would be the default fate of all metazoan cells, and survival would be conditional on the steady supply of specific external rescue signals (Raff 1992, Raff *et al.* 1993, 1998, Meier *et al.* 2000). I have already criticized this interpretation as adultocentric (Minelli 2003). I might add here that it also would hardly apply to metazoans like *Hydra*, where the cells steadily lost from the tentacles (and also steadily renewed by proliferation of interstitial cells) are hardly the victims of programmed cell death. (In *Hydra*, apoptosis does probably occur in the nurse cells which are eventually phagocytosed by the growing oocyte; Technau *et al.* 2003.)

If anything like a metazoan prototypic cell ever existed, it was certainly not an apoptotic cell (otherwise, no metazoan would be around nowadays), nor was it a totipotent but terribly specialized cell as is a typical egg. Rather, as I suggested elsewhere (Minelli 2003), a possible example would be one of those blastomeres in 'regulative' embryos which are able to generate a new complete embryo if isolated during an early cleavage stage. Still better, to avoid speaking of blastomeres, which are ontogenetically close to the egg cell, we could consider as a better

prototype of the metazoan cell one of those neoblasts that are responsible for regeneration of lost parts, or, more generally, a stem cell.

If the ability to undergo mitosis, rather than to fall victim to programmed death, is a likely default attribute of metazoan cells, it is also true that an early stop to mitotic activity is a specific feature of many selected taxa, developmental stages, or body parts. Unfortunately, no comprehensive review is available and many potentially interesting taxa, stages, and body parts have never been investigated in this respect. I will thus give here a few examples suggesting the need for a detailed, comprehensive overview of this phenomenon.

In the flatworms there seems to be no mitosis in the epidermis and, possibly, in all differentiated cells generally (Littlewood *et al.* 1999), but it would be rewarding to extend investigations to a larger number of taxa than those studied to date. During the early development of the polyclad *Notoplana humilis*, mitoses are already rare in the ectoderm when organogenesis begins and no mitosis is observed in the epidermis of advanced embryonic stages or in the larvae. However, a generalization would not be safe, as in another polyclad, *Cycloporus japonicus*, mitoses were also observed in the ectoderm at the larval stage. In the embryos and larvae of both worms, mitoses were observed in the mesoderm but not in the endoderm (Drobysheva 2003). Mitosis of the epidermal cells is also probably lacking in the post-embryonic stages of chaetognaths (Shinn 1997). As a rule, no post-embryonic mitosis occurs in nematodes, chaetonotid gastrotrichs, and mites, but exceptions can be expected.

As a rule, cell divisions are downregulated or shut off before the initiation of large-scale morphogenetic changes. For example, the complex changes in cell shape that occur in *Drosophila* during gastrulation are incompatible with cell division (Foe *et al.* 1993). In other instances, however, cell divisions contribute to morphogenesis. For example, cell proliferation is required for the development of the vertebrate kidney (Davies *et al.* 1995), whereas other tubular structures such as the tracheae of *Drosophila* form without any cell division (Manning and Krasnow 1993).

An early stop to mitosis is possibly an important factor in determining eutely; that is, the constancy in cell number observed (but sometimes only supposed to occur) in several metazoans, especially among those of smaller size, as rotifers and some nematodes. Cell numbers around 1000 seem to be critical in this respect, as this is the order of magnitude of the total number of cells in a small nematode such as *C. elegans*, but also in several embryos of marine invertebrates at the time they turn into larvae. A bit higher (around 2600) is the number of cells in the tadpole larva of the ascidian *Ciona* (Lemaire 2006), where eutely has been recorded, in particular, for the larval central nervous system (Meinertzhagen 2005). Eutely, however, is not so strict as sometimes presumed in the past, and as is still recorded in textbooks. Indeed, variation in cell number is long known even for putatively paradigmatic groups like rotifers and nematodes (Shull 1918, Cunha *et al.* 1999). On the other hand, cell number is quite fixed for selected structures, such as arthropod ganglia, in animals that globally are far from eutelic.

7.1.4 Apoptosis

Apoptosis is very unlikely to be the default fate of metazoan cells; nevertheless, this phenomenon is involved in many morphogenetic processes in the most diverse metazoans. A well-known example is the role of apoptosis in sculpting the fingers of most tetrapods, otherwise bound by an interdigital membrane as seen in waterfowl. Less known, but perhaps more impressive in its selectivity, is the involvement of apoptosis in producing the asymmetric mouthparts of the Thysanoptera, where the anlage of the right mandibular appendage degenerates during embryonic development (Heming 1980).

An evolutionary approach to this cellular phenomenon must begin with some speculation about its origin. Accepting that apoptosis is not a default feature of metazoan cells, and avoiding *post factum* 'explanations' of its occurrence in purely adaptive terms, it is perhaps sensible to consider the suggestion by Biella *et al.* (2002) that programmed cell death first appeared in animals, fungi, and plants as a defence response against pathogens. Alternatively, it has been suggested that apoptosis evolved as a means by which a multicellular organism would fight against 'renegade' cancerous cells (Krakauer and Plotkin 2002, 2005). In these terms, the interpretation is dramatically finalistic, but it could be easily rephrased in terms of competition: non-cancerous cells would be given an advantage by inducing apoptosis as this would eliminate competition from cancerous cell lineages. This makes sense in terms of Darwinian competition among cells within a multicellular body. Cells able to induce apoptosis in their neighbours may effectively raise a barrier against further invasion by viruses or other enemies, including their nearest neighbours; that is, other cells in a multicellular organism. By exaptation, this ability would have been eventually exploited in sculpting multicellular organisms in ways that turn out to be advantageous to the surviving cells in other ways, either directly (perhaps by reducing competition for metabolic resources) or indirectly, because of a global selective advantage to the organism of which the apoptosis-inducing cells are part. Such a scenario does not necessarily imply mechanisms of apoptosis universally shared across eukaryotes, or a unique origin of this cellular phenomenon. At any rate, a revisitation of the phylogenetic history of apoptosis would be well deserved.

Programmed cell death is sometimes the fate of a large percentage of the cells in an embryo, or an organ anlage. For example, in *C. elegans* 131 of the 1090 somatic cells of the hermaphrodite individuals normally die by apoptosis, whereas in the retina of the cat more than 80% of the ganglion cells die shortly after they are born (Meier *et al.* 2000). The most massive occurrence of apoptosis, however, is found in colonial ascidians such as *Botryllus schlosseri*. Here, the whole parental generation of zooids in a colony is fated to die, while the next generation, which is issued asexually from them, reaches maturity within 1 week. This massive programmed cell death of zooids is accompanied by programmed removal of cell corpses by blood phagocytes (Lauzon *et al.* 2007).

Apoptosis can be involved in the extensive loss of larval structures, or in that dramatic reworking of body structure we usually describe as metamorphosis. For example, in the metamorphosis of the hydrozoan *Hydractinia echinata* the first evidence of apoptosis occurs very early, approximately 20 min after the induction of metamorphosis, and is followed by phagocytosis of a large number of larval cells (Seipp *et al.* 2001). A thorough comparative study across all metazoan groups with extensive metamorphosis would be useful.

Cell death also helps to create joints between the future elements of the vertebrate limb. This requires a precise localization of the cells eventually affected by the process. Future joints become morphologically detectable by a change in shape and density of chondrocytes in the prospective joint-forming region, while the local differentiation of cartilage is inhibited. Mariani and Martin (2003) demonstrated the crucial role in joint formation of two genes, *Gdf5* and *Nog*, which encode a bone morphogenetic protein (BMP)-related protein and a BMP antagonist, respectively. Programmed cell death has also a role in regenerating deer antlers (Colitti *et al.* 2005).

7.1.5 Tissues

In principle, competition among cells is universal, and a cell's neighbours are not simply competitors for a common pool of limited resources, as they themselves can turn into resources. For example, the embryo of the Plagiorchiidae (Digenea) consists of 70–80 cells, whereas the miracidium larva into which it develops contains only 10–15 cells: degenerating cells are probably used as nutrients for the embryo (Galaktionov and Dobrovolskij 2003). The same has been suggested for free-living flatworms (Ivanov and Mamkaev 1973) and equivalent examples can be found in other groups.

But competition can be reduced, avoided, or balanced. This may happen in three different ways. One is avoiding, or destroying, cell–cell boundaries, either by incomplete cytodieresis (in the case of the so-called plasmodia, see above),

or by more or less extensive fusion, either limited to the cytoplasm (as in syncytia), or additionally involving karyogamy (as in fertilization). Competition is obviously reduced, and often completely avoided, if cells are very distant in the body, or belong to metabolically distinct compartments, or become independent from a multicellular mass. Of course, this will work unless hormones, or pheromones, are involved. But the most interesting and less obvious way in which cells can escape from disruptive competition is by acquiring metabolic or differentiation conditions very similar to those of their nearest neighbours. In such a context, competition will be obviously strong, but will probably be stable, especially if their mitotic activity is small or zero.

Old theories of the metabolic control of development have long been discounted (Mitman and Fausto-Sterling 1992), but new evidence has lead to a resurrection of these ideas (e.g. Nijhout 1990, Blackstone 1998). For example, maternally inherited differences in the distribution of mitochondria are supposed to influence, in some embryos, the eventual fate of differentially located blastomeres or to be involved in the establishment of the embryonic axes (Dumollard *et al.* 2007).

There are several ways in which neighbouring cells can end up being very similar, structurally and metabolically. They may become similar because they are all subjected to identical differentiation cues. Otherwise, they may all have inherited the epigenetic state of their mother cells. This amounts to saying that cells acquire instructions either from their contemporaries or from their ancestors (Gurdon 1992) or, better, that a cell can inherit predispositions (genetic states) from its mother, or take cues (through diffusible ligands) from neighbours, or both (Held 2002).

Perhaps more interesting, however, are two other mechanisms by which a set of cells can take its identity as a distinct tissue. One of these mechanisms is self-sorting, the other is a self-sustained interaction among neighbouring cells (a kind of chorus effect). The actual relevance of self-sorting in natural conditions remained uncertain for a long while after the phenomenon was discovered early in the twentieth century in reaggregating sponge cells that had been previously disassociated (Wilson 1907). But in more recent times (Minelli 1975) we have become increasingly aware of its importance. Self-sorting explains the lack of mixing between cells forming neighbouring compartments, such as polyclones, half a body segment in length, that contribute to building the body of *Drosophila* larvae (Garcia-Bellido *et al.* 1973). Self-sorting (more precisely, the lack of miscibility of cells from even-numbered rhombomeres with those from odd-numbered rhombomeres) explains the origin and stability of the segmented structure of the vertebrate rhombencephalon (Guthrie and Lumsden 1991, Guthrie *et al.* 1993). But the most surprising finding is perhaps the recent discovery that the whole embryo of *C. elegans* is involved in a process of global self-sorting through which cells with similar identity, first originated at scattered locations, eventually gather together to form tissues and organs at what will eventually become their final place (Bischoff and Schnabel 2006, Schnabel *et al.* 2006).

Parallel differentiation, cell lineage coupled with epigenetic effects and, finally, cell self-sorting are, possibly in that order, established ways by which a tissue may take its origin. But there is possibly a fourth mechanism, not necessarily alternative to the others or wholly independent from them. This is the sustained, two-way exchange of chemical or electrical inputs between neighbouring cells, a phenomenon studied extensively in many developmental and adult models. It is quite possible that this cell–cell coupling eventually produces effects of metabolic uniformity comparable to the wide-ranging synchronizing effects observed in both nature and human life when comparable links develop between neighbours. A similar chorus effect is the synchronization of 'olas' in the stadion, other examples are the synchronous mating calls of the frogs in a pool, or the synchronous flashing of a million fireflies as often observed in the tropics, and discussed in a theoretical context by Strogatz (2003). A developmental system that is suggestive of this mechanism is the insect imaginal wing disc, within which small clusters of cells within the disc progress

through the cell cycle together, tightly coupled to their neighbours, irrespective of their clonal origin (Johnston and Gallant 2002).

7.2 Reproduction and sexuality

7.2.1 Convergence and co-option

A comparative study of the phenomena of reproduction and sexuality throughout the Metazoa reveals abundant evidence of the pervasivity of convergence and co-option, as well as of functionally and adaptively dramatic changes occurring even within phylogenetically restricted lineages, and often open to a possibility of reversal.

Let us begin with the roots of metazoan sexuality. No evidence of sexuality has been found to date in the choanoflagellates, the putative sister group of the Metazoa. Due to the uncertainties regarding still deeper branchings in the eukaryote tree, this may leave open two possibilities: either a loss of sexuality in the choanoflagellate lineage, or a fresh origin of sexuality in the metazoan branch, an alternative I would guess to be more likely than the former. A polyphyletic origin of sexuality, as such, is indeed likely (Boyden and Shelswell 1959), although every time it has occurred it has involved many of the same cellular structures and mechanisms. Polyphyletism means, however, that we must be extremely cautious when using the same terms to describe features of the sexual cycle of distantly related organisms. For example, are differences in size and mobility sufficient to call a gamete either an egg or a sperm cell in whatever eukaryote lineage they may occur? The following examples will show that the domain of sexuality and reproduction phenomena is one where a factorial approach to homology (Minelli 1998) is most conspicuously needed.

The concept can be introduced through the following example. The clitellate annelid *Enchytraeus japonensis* can reproduce both sexually and asexually. Following asexual reproduction, the new individual deriving from a parent's body part where no gonads were present will produce new primordial germ cells from a set of stem cells. Accordingly, the germ cells in this individual have a different origin with respect to the ordinary germ cells, derived from a cell lineage of early embryonic origin, as found in other individuals of the same species. Thus, the two sets of germ cells will be homologous under many, but not all, of the criteria that may apply in other animals. The fact that the stem cells entitled to produce new primordial germ cells are distinct from the ordinary somatic stem cells in the same worm, as only the first ones express the gene *piwi* (Weisblat 2006), illuminates a further component of the complex set of factors we may be interested to disentangle before declaring homology between two cells (as in this case), or organs, or processes.

To move from an intraspecific comparison to one involving two closely related species, it is known that no Y-linked gene is shared by *Drosophila pseudoobscura* and *Drosophila melanogaster* (Carvalho and Clark 2005). In other terms, the Y chromosome of the former species is not homologous to the Y chromosome of the other, but only functionally analogous to it. Of course, this does not rule out homology between more or less extended parts of the Y chromosomes of other species pairs, even of distantly related ones.

The very mechanisms by which germ cells are specified have a checkered phylogenetic distribution, which clearly suggests a lot of change from 'preformation' to 'epigenesis' or the other way around. Extavour and Akam (2003) have provided an excellent summary of the distribution of preformation and epigenesis in the origin of germ cells in the main animal groups (Table 7.1). This detailed analysis shakes the empirical foundation of the conventional view that preformation—with the germ line being determined during embryonic development—is plesiomorphic in the Bilateria (Blackstone and Jasker 2003) and rather offers arguments in favour of the alternative. A close look at the basal branches of the metazoan tree shows lineages where populations of endodermally derived pluripotent stem cells (the archaeocytes of sponges, the interstitial cells of cnidarians, and the neoblasts of the acoels) acquire their fate in early to mid-embryogenesis, and give rise later to both somatic cell types and gametes (Extavour 2008). But this implies that, if epigenesis was the mechanism by which the Urbilateria specified its germ line, then preformation must have evolved convergently several times during the bilaterian radiation (Extavour 2008).

Repeatedly evolving variegated mechanisms for the origin of germ cells is perhaps less puzzling if we consider that germ cells and stem cells are very similar at the levels of morphology and gene expression (Extavour 2008). This allows for some degree of reversibility, even in 'modern' metazoans like *Drosophila*, where germ cells already advanced in oogenesis can be induced to revert back to a germline stem-cell state (Kai and Spradling 2004).

Convergence, again, can be demonstrated in the primordial germ cells of the holothurian *Synaptula hydriformis*, which has epithelial ciliated cells with a collar of microvilli surrounding the cilium. This is the morphological equivalent of a sponge choanocyte (Frick and Ruppert 1996).

If convergence is one side of the coin, the other is the ease with which there can be evolution from one condition to another. Amphibians provide a good example, as anuran germ cells are predetermined by germ plasm, whereas in urodeles they are specified by inducing signals (Johnson *et al.* 2003).

There can be a market, however, even in the absence of money. Here this means that the very criteria by which we currently identify germ cells are not necessarily obvious. A notoriously difficult case is provided by digeneans. The very nature of the reproduction process by which sporocysts or rediae produce other sporocysts, rediae, or the minuscule cercariae fated to grow into adults of the marita generation (which is parasitic in vertebrates) is still far from clear. No evidence of meiosis has been found in either rediae or sporocysts (Galaktionov and Dobrovolskij 2003). This is generally assumed to imply apomictic and probably diploid parthenogenesis, but is this conclusion warranted and, if so, does it apply to all digeneans? For example, in the rediae of paramphistome digeneans the germinal cells undergo little multiplication so

Table 7.1 Origin of germ cells in the different animal groups. Data mainly after Extavour and Akam (2003).

Germ cells formed by epigenesis
 'Porifera'
 Ctenophora
 Cnidaria
 Acoela
 Gastrotricha
 Rhabditophora exclusive of Digenea
 Bryozoa
 Nemertea
 Phoronozoa
 Sipuncula
 Mollusca *partim*: Aplacophora, Polyplacophora
 Kinorhyncha
 Arthropoda *partim*: Myriapoda
 Xenoturbellida
 Hemichordata
 Echinodermata *partim*: Crinoidea, Asteroidea, Holothuroidea
 Vertebrata *partim*: Agnatha, Dipnoi, Caudata, Lepidosauria, Testudines, Mammalia
Germ cells formed either by epigenesis or by preformation
 Mollusca *partim*: Gastropoda
 Annelida (preformation only in Hirudinea; epigenesis in Echiura)
 Onychophora
 Arthropoda *partim*: Insecta, Crustacea, Chelicerata
 Echinodermata *partim*: Echinoidea
 Cephalochordata
 Urochordata
 Vertebrata *partim*: Chondrichthyes, Actinopterygii
Germ cells formed by preformation
 Syndermata
 Rhabditophora *partim*: Digenea
 Mollusca *partim*: Cephalopoda, Bivalvia
 Nematoda
 Arthropoda *partim*: Collembola
 Chaetognatha
 Vertebrata *partim*: Anura, Archosauria

that no true germinal masses are formed and there is no persistent centre of multiplication. Is this enough to grant the nature of 'eggs' to the cells from which the offspring will be generated? As far as I understand from the specialist literature, these cells are not different from somatic cells, not only in that they are not the product of meiosis, but also because nothing like vitellogenesis happens in them or, as expected in a neoophoran flatworm, in distinct yolk cells

by which the 'egg' would be eventually surrounded. Therefore, in every respect, these cells have fewer properties of conventional female gametes than the diploid 'eggs' of amictic parthenogens. This indeed blurs the boundaries between sexual and asexual reproduction and, as a consequence, between embryonic and blastogenetic development, a topic to which I will return below.

Before leaving this point on whether there is any unique set of necessary and sufficient characters by which an egg can be defined, let us consider the diversity of mechanisms by which vitellogenesis can be accomplished and the instances of convergence in this phenomenon. In some sponges, the embryos apparently ingest parental cells and/or cellular products as sources of nutrition (Fell 1997). In *Hydra*, an aggregate of germ cells differentiates into one oocyte and thousands of nurse cells, which are eventually phagocytosed (Technau *et al.* 2003). In other hydrozoans (*Tubularia*), when an egg has completed its growth, a few oogonia are left in the gonophore; these fuse to give rise to one oocyte that will eventually survive and differentiate (Van de Vyver 1993). In some 'polychaetes' (*Phragmatopoma lapidosa*, *Streblospio benedicti*, *Spio setosa*, and *Leitoscoloplos fragilis*) nutrients are transferred directly from the blood to the oocytes (Eckelbarger 2005). In the case of flatworms, the anatomical and functional arrangements involved in the production and delivery of yolk are so different in different groups that they were used in the past as a key to divide the 'Turbellaria' into the Archoophora, whose germovitellaria produce conventional endolecithic eggs, and the Neoophora, where morphologically, topographically, and functionally distinct organs, the vitellaria, produce yolk cells that will be eventually added to the egg cells, produced by the ovary, to form compound esolecithic eggs.

In female insects, a diversity of ovariole types is known, where vitelline cells have different genealogical, topographical, and functional relationships to the developing oocytes, but the phylogenetic distribution of this ovariole diversity is not so clear-cut as is the egg/yolk relationship in flatworms. This suggests that evolutionary transitions in the way insects perform vitellogenesis are easier than the corresponding changes in flatworms.

But let's turn from easy change to frequent convergence. Just to offer one example, compound eggs are not exclusive to neoophoran Rhabditophora. An envelope of somatic cells is found, for example, around the ovulated eggs of some oviparous sponges, and in some cases at least it is known that these somatic cells are internalized by the developing embryos and very likely serve a nutritional function (Fell 1997). A functional equivalent of the yolk cells in neoophorans are the so-called food-ova of some nemerteans, crinoids, 'polychaetes' (e.g. *Pygospio elegans*), and gastropods (e.g. *Nucella lapillus*) (Beklemishev 1969).

Metazoan history has also witnessed a number of parallel changes from gonochorism to hermaphroditism and less numerous changes from hermaphroditism to gonochorism. Sometimes, these kinds of transition identify clades of some consistency and age, but in other cases there are differences within a family or a genus, sometimes even within a species. For instance, a change from gonochorism to hermaphroditism occurred many times independently in sabellid 'polychaetes' (Rouse and Fitzhugh 1994), as well as in several groups of bony fishes (e.g. Sparidae), sea stars, etc. This is probably less unexpected if we consider that a change from parthenogenetic reproduction to simultaneous hermaphroditism can even occur during the life of a single individual, as has been shown for chaetonotid gastrotrichs (Weiss 2001).

7.2.2 The one-cell bottleneck

I cannot avoid the feeling that sexual reproduction is perceived by most biologists, and no less by lay people, as the 'true', legitimate way an animal is expected to reproduce. On the other hand asexual reproduction has, at most, a subsidiary role, in case sexual reproduction is not possible any more for a particular species. Corollaries of this assumption are two more debatable opinions. One is the primacy of embryonic over blastogenetic development, the latter being perceived

as a clumsy proxy of the former. The other is the privileged role of the fertilized egg (or, anyway, of the egg) as the authentic starting point of a new individual's life cycle.

I will argue that none of these widespread opinions is warranted. Some arguments will be presented here, others are deferred to a subsequent section on animal development (section 7.4).

Let me start by discussing the opinion that selection must favour development from a single cell, rather than from a cluster of cells, to move subsequently to the related opinion that only development from the egg, fertilized or not, can satisfy this condition. To begin with, I will note that only in some specific instances are we obliged to acknowledge a multicellular origin to the 'bud' that will eventually develop, asexually, into a new individual. This may apply indeed to those sponges where the bud is formed by condensation of mesenchymal cells through a mechanism which is probably identical to regeneration (Brien 1973). Generally speaking, however, to say whether a bud has a unicellular or a multicellular origin would require the ability to fix a precise point in time at which a cell, or a group of cells, will deserve to be called a bud, either because of a largely irreversible commitment to that fate, or at least because of the overt manifestation of some morphological or molecular singularity. But even so, the worst we may find is that the cells forming the bud are the equivalent of what in cladistic language is a paraphyletic assemblage. That is, all the cells in the bud will be traceable to a single ancestor cell, whose offspring, however, is not completely included in the bud. But how would this be different, say, from the conditions in an adult sea urchin, whose cells derive from a small fraction of the initial descendants of the zygote, most of which formed instead the larval body, which is largely discarded at metamorphosis? Thus, I do not think that the expected genetic heterogeneity among the founders of a bud out of which a new individual will develop asexually is arguably larger than the heterogeneity found among the cells of an embryo, larva, or juvenile deriving from an egg after several cell divisions.

Nevertheless, several biologists (e.g. Grosberg and Strathmann 1998, Wolpert *et al.* 2007) have insisted on the risks derived from starting development from more than one cell, as one or more of these may have been affected by mutation. Additionally, or alternatively, to focusing on the potential heterogeneity of the cells eventually contributing to the new individual, some authors stress the importance that the offspring takes their origin from a cell or cells that possess the genetic characteristics of the adult from which they originate. This is possibly the point Gerhart and Kirschner (1997) had in mind when they argued that the only way for the genome to be fully tested is to have only one line of germ cells, a problem solved by the early sequestration of a dedicated germ line (Extavour 2008). Indeed, the germ line is thus protected from somatic mutation by exhibiting reduced mitotic activity, reduced transcriptional activity, and reduced mobility of transposable elements until gametogenesis begins (Extavour 2008), but these attributes are not necessarily exclusive to them. Even if these circumstances contribute some minor advantage to the generalized (but not universal) fixation and early sequestration of the germ line, this does not offer any perspective on how this may have originated early in metazoan history; nor does it protect the cells deriving clonally from a zygote from the strongest forms of competition between cell clones which may differ by differential gene expression rather than by mutation, up to the point of eliciting phenomena like apoptosis or phagocytosis.

7.2.3 Free cells, or unicellular metazoans

Everybody is familiar with two kinds of 'free' animal cell; that is, gametes and blood cells. Less familiar, but closely comparable with blood cells, are the coelomocytes of annelids and other metazoans. The question is, how and under which circumstances a single cell may acquire independence from the multicellular assemblage, and how long it can survive in the unicellular state. Flagellated sperm cells probably have the least promising future, as they are denied any possibility of mitosis. Blood cells, coelomocytes, and

the like form a very diverse set, at one extreme including anucleate 'cells' like mature mammalian erythrocytes, which cannot reproduce but also cannot get food by endocytosis, and at the other extreme an array of active phagocytic cells that are not simply able to divide, but may also give rise to gametes, as in the case of the ascidian blood cells.

One may object that blood cells and coelomocytes remain within the body of the multicellular animal that gives rise to them and are thus functionally integrated within it. In this respect, truly independent unicellular stages would be found only among the gametes. But equating free, unicellular stages of metazoans with gametes would be, again, an unjustified generalization. On one hand, the female gamete does not need to be free at the time it is fertilized. In some sponges and in some hydrozoans, for example, the eggs remain fixed even when mature and eventually fertilized (see below). On the other hand, a unicellular free stage is not necessarily a gamete. In the rhizocephalan *Loxothylacus panopaei*, a crustacean parasite of crabs, the multicellular vermiform stage injected into the host following the attachment of a cyprid larva splits soon into approximately 25 cells that move about independently in the host's haemolymph (Glenner and Høeg 1995). Although detailed evidence is lacking in this respect, it is pretty certain that, conditions permitting, each of these cells would be able to give rise eventually to a female multicellular adult.

7.2.4 Asexual reproduction and regeneration

Before returning to sexual reproduction and embryonic development, let us consider some aspects of asexual reproduction, where we are likely to find conditions that may help us determine a scenario for the early steps of sexual reproduction and embryonic development in metazoans. Remember that sexuality, as a way to put together genetic material originally found in two distinct nuclei, has no necessary relation to reproduction. To be sure, reproduction without sexuality is a familiar occurrence, whereas sexuality without reproduction is not that evident, in animals at least. Indeed, to provide a textbook case of 'pure' sexuality, it is sensible to use the familiar example of the ciliates, whose conjugation phenomena start and end with the same number of individuals (two), the intervening change being essentially a nuclear affair that leaves virtually unchanged the complex cortical architecture of the two cells. Thus there are no males and females, no gametes, no parents, and no offspring, just nuclear exchange: that is, pure sexuality. Reproduction is a completely different affair, not only in ciliates, but universally. Reproduction is multiplication of independent systems (individuals), provided that this expression makes sense, which may be questionable in many marine invertebrates that produce loosely integrated colonies, and in many representatives of other kingdoms, including green plants and fungi. Restricting attention to the metazoans, we can practically ignore reproduction among unicellular organisms, and thus we are left with a range of possibilities between the two extremes, with a multicellular organism that either splits into two multicellular units of a complexity comparable to that of the parent organism, or which gives off single cells, each of them being eventually the founder of a new multicellular organism. Examples closer to the first case are widespread among metazoans with relatively simple organization—for example sponges and cnidarians, but also catenulids, rhabditophorans, annelids, and echinoderms—and are even better represented by phenomena currently classified as regeneration, rather than reproduction. The phylogenetically scattered distribution of all these phenomena (Table 7.2) and the diversity of mechanisms involved (sometimes within a single species; more generally, among smaller or larger clades) suggests that these do not represent uniquely evolved new mechanisms, but are instead multifarious expression of what still survives of properties of early multicellulars whose evolution begun on the edge between opposite morphofunctional conditions and selective regimes. A conspicuous, indeed precious leftover from those times is still with us. This is true, in particular, of the most basic alternatives a metazoan cell may confront besides the one between

life and death: the alternative between remaining a part of a multicellular body, or becoming separated from it. We will come back to the issue of 'free' cells.

Let us consider briefly phenomena in the often fuzzy area between asexual reproduction and regeneration. In this respect, the most interesting phenomena are those for which Wagner (1890) proposed the term paratomy, which starts with fragmentation of the parent animal, followed by extensive growth and differentiation of new parts, more or less distinctly preceded by the dedifferentiation of existing parts. In the enteropneust *Balanoglossus australiensis* young individuals (2–10 mm long) are produced by vegetative division through fragments splitting off from the hind end of the anterior half of a worm. The small fragments possess all structures originally present at the level of the parent body from which they split. Development requires redifferentiation of proboscis, collar, gills, hepatic caeca, etc. (Packard 1968). From the point of view of the mechanisms involved, this is not different from the pre-ordered loss of an appendage (autotomy) followed by regeneration. Possible distinctive criteria are only the conspicuousness of the fact, and its eventual consequences: reproduction versus reconstruction of lost parts.

In the case of another phenomenon, architomy, the individual features of the offspring are largely delineated before it frees itself from what remains of the 'parent' individual, but there is no reason to regard this condition, per se, as very different from paratomy. Basically, the difference resides in a different balance between the autonomy of local morphogenetic processes and the timing of fragmentation of the parent organism. That is, a distinction is largely matter of heterochrony, as heterochrony is the difference between releasing either gametes as free cells or embryos following fertilization of egg cells still bound to the mother organism.

From a phylogenetic perspective, how shall we interpret these ways to produce a new individual? As advocated repeatedly in these pages, I think that also in this case we must carefully dissect the phenomena to disentangle old from new features. To simply classify architomy or

Table 7.2 Occurrence of asexual reproduction and/or regeneration in representatives of the main animal groups (but mostly only in a subset of the members of the clade). The extent of regeneration is quite variable within each clade where the phenomenon has been recorded.

Taxon	Asexual reproduction	Regeneration
'Porifera'	+	+
Placozoa	+[1]	?
Ctenophora	+[2]	+
Cnidaria	+	+
Acoela	+	+
Nemertodermatida	–	?
Gastrotricha	–	+[3]
Micrognathozoa	–	?
Syndermata	–	–
Gnathostomulida	–	?
Catenulida	+	+
Rhabditophora	+[4]	+[5]
Cycliophora	+	?
Bryozoa	+[6]	+
Entoprocta	+	+
Dicyemida	+	?
Orthonectida	–	?
Nemertea	+	+
Phoronozoa	–	+
Mollusca	–	+[7]
Annelida	+[8]	+[9]
Sipuncula	+[10]	+[11]
Priapulida	–	–
Loricifera	?	–
Kinorhyncha	–	–
Nematoda	–	–
Nematomorpha	–	–
Tardigrada	–	–
Onychophora	–	–
Arthropoda	+[12]	+[13]
Chaetognatha	–	–
Xenoturbellida	?	?
Echinodermata	+	+
Enteropneusta	+[14]	+
Pterobranchia	+	?
Cephalochordata	–	–
Urochordata	+[15]	+
Vertebrata	+[16]	+

[1] Asexual reproduction as the dominant (only?) form of reproduction; mostly by binary fission, but also by budding off of small (40–60 μm) hollow, spherical propagules.

[2] Only in the benthic genera *Ctenoplana* and *Coeloplana*; the process is comparable with laceration of the pedal disc in sea anemones.

[3] Only known in *Turbanella*.

Table 7.2 (*Contd*)

[4] A phase with asexual reproduction is a regular trait of the life cycle of the Digenea and the Cestoda Cyclophyllidea. There are asexual forms in Macrostomorpha and Tricladida; there are no asexual forms in Polycladida, Lecithoepitheliata, Prolecithophora, Proseriata, Bothrioplanida, or Rhabdocoela.

[5] Excellent regeneration capacity in Macrostomorpha and Tricladida; poor in Rhabdocoela.

[6] Budding; statoblast production. Polymebryony in the Stenolaemata.

[7] For example, part of mantle, the arms of cephalopod, and the cerata of nudibranchs.

[8] Not in the Hirudinea and in several other groups.

[9] Not in the Hirudinea and in several other groups.

[10] Fission and budding in two species: in *Sipunculus robustus* (only recorded from laboratory animals), lateral budding from the posterior end and transverse fission in the posterior half of the trunk; in *Aspidosiphon elegans*, transverse fission with portions of the internal organs within the posterior region serving as anlage for the regeneration of new organs (Rice and Pilger 1993).

[11] Regeneration of tentacles and of the anterior introvert.

[12] Only as polyembryony.

[13] Appendages only.

[14] At least in *Balanoglossus australiensis*.

[15] At least in the colonial ascidians.

[16] Only as polyembryony.

paratomy as primitive or derived, as recapitulative or coenogenetic, would just mean blurring matters. Let us ignore, for a moment, the widespread entangling of sexuality and reproduction. Reproduction, indeed, is fundamentally a question of increasing the number of physically independent and eventually autonomous units (individuals). This only requires (1) the existence of a mechanism ensuring the fragmentation of a parent organism or the detachment from it of single cells or groups of cells and (2) the continuation or the reactivation of developmental processes by the effect of which at least a subset of the physically separated parts will acquire or re-acquire the structural complexity and the specific organization of the adult individuals of the species. This means that:

• neither development nor reproduction implies the presence of gametes per se, despite the prevalence of sexual reproduction and development through embryonic stages;

• hypothetical early multicellular organisms that we would classify as stem-group metazoans may have possessed asexual reproduction only;
• development from the egg is likely to have inherited features or mechanisms already evolved in the context of a regeneration-like development as an integral part of some form of asexual reproduction.

On the other hand, all Recent metazoans with asexual reproduction are crown metazoans, even the most simple ('primitive') of which belong to lineages that have also experienced sexual reproduction and embryonic development. Therefore, the mechanisms and the possible outcome of asexual reproduction (and regeneration) in Recent animals have evolved alongside the mechanisms of sexual reproduction and embryonic development of the respective lineages as they are dependent (perhaps increasingly so) on each lineage's idiosyncratic cellular and genetic properties. This is an important point, because a common dependence of embryogenesis, asexual reproduction, and regeneration on a common set of genetic factors and cellular properties suggests that to study the evolution of regeneration, or asexual reproduction, in isolation from embryogenesis actually makes little sense, other than within fairly strict phylogenetic limits. In other terms, it will make little sense to ask what a newt and a stick insect have in common, given that both of them, but not many of either's close relatives, can regenerate a lost leg. On the other hand, I think that there is a phylogenetic signal in some idiosyncratic behaviours with respect to regeneration and related phenomena, such as the lack of autotomy in the Myriochelata (curiously associated with a virtually total lack of directional asymmetry in external body features) and its likely apomorphic presence in the Pancrustacea, albeit with secondary loss in some lineages, for example the Hemipteroidea and the endopterygote insects (Maruzzo *et al.* 2005). The common background of embryogenesis, asexual reproduction, and regeneration has been shown abundantly by recent studies comparing morphogenetic events in embryogenesis with those associated with asexual reproduction

or regeneration in animals where two or three of these classes of phenomena occur.

During the embryonic development of triclads (planarians and relatives) there is no clear evidence of germ layers or of well-defined organ primordia. This condition makes planarian embryos more comparable with regenerating blastemas than other embryos would be. Thus, Cardona *et al.* (2005) compared embryogenesis with regeneration in these animals. The onset of the two processes differs because of the different origin and spatial distribution of the involved cells, but also because in regeneration axial patterning cues are associated with already differentiated structures, while relying on maternal determinants only in the embryo. But strict similarities will show up soon with the formation, in either case, of a provisional epidermis, followed by the differentiation from proliferating neoblasts of definitive epidermal cells which replace the provisional epidermal cells by intercalation, similar to normal cell renewal in the adult. More precise similarities between the two processes are found in the muscle tissue, in the formation of ventral nerve cord pioneers that differentiate into ventral nerve cords independent of any brain structures, and in the processes of the brain precursors, such as condensation and nerve-cord development.

In the freshwater oligochaete *Pristina leidyi* Bely and Wray (2001) have shown extensive similarities in the expression patterns during regeneration and fission, respectively, of one homologue of *engrailed* and two homologues of *orthodenticle*. Moreover, by comparing their data with evidence on the expression patterns of the same genes in the embryos of the leech *Helobdella* (an annelid with neither regeneration nor asexual reproduction) they found evidence that embryonic processes are re-deployed during regeneration and fission.

Of course, at the finer level of regulation we must expect some difference between development through an embryo and regeneration or development associated with asexual reproduction. For example, fin regeneration in bony fishes involves the expression of several genes that are also expressed during the embryonic development of the fins, but the function of some of these genes in regeneration is closer to their role in larval than in embryonic fin development (Akimenko *et al.* 2003). Similarly, in the newt and the axolotl, the expression of the genes *Tbx4* and *Tbx5* that code for two transcription factors implicated in controlling the identity of vertebrate legs and arms, is regulated differently during embryogenesis and regeneration (Khan *et al.* 2002).

A peculiar behaviour observed in some regenerating flatworms is that they can live longer than an intact conspecific. A specimen of *Macrostomum lignano* (Fig. 7.1) regenerated 29 times, following as many amputations, eventually living for 1 year; that is, 2 months longer than the life span of an untreated specimen (Egger *et al.* 2006). One might suggest different interpretations of this fact, for example in terms of relaxed competition

Fig. 7.1. A living specimen of *Macrostomum lignano* (Rhabditophora Macrostomida). Courtesy of Bernhard Egger.

to which the surviving cells may have been subjected following amputation or, possibly better, in terms of metabolic rejuvenation due to the repeated injection of freshly divided, temporarily undifferentiated cells. In an adult planarian neoblasts can comprise up to 30% of the total number of cells (Ellis and Fausto-Sterling 1997). These cells are released from 'dormancy' when their neighbours die or when wound 'hormones' signal colonizable empty space.

7.2.5 Gametes

A gamete is a cell fated to fuse to another gamete of the same species (and one of the opposite type, if in that species macrogametes and microgametes, or eggs and sperm cells, are distinguishable). We must acknowledge that this definition is intrinsically weak. First, it is not framed in terms of intrinsic characteristics of the cells that are called gametes, but in terms of potential behaviour (if not of finality). Second, it does not clearly identify a biological entity of which we can trace the origin and evolution. Therefore, we should possibly look for other defining criteria, even if eventually it turns out that the most obvious behaviour of the cells in question is in fact their participation in an event of cell fusion followed by karyogamy. A good starting point is perhaps the realization that most animal gametes, and especially the female ones, are all but primitive cells. This may fuel the hope of finding some specific markers of gamete identity.

Animal eggs and sperm cells are usually the product of meiosis, but this is not an unequivocal signature of gamete nature, as shown by gamete production by haploid parthenogens. The expected fusion to a complementary gamete is equivocal, not just because of the failure of a vast majority of gametes to eventually fulfil this fate, but also because of the reproductive (or developmental?) success of cells we usually define as parthenogenetically developing eggs. These are often produced by processes involving regular meiosis (perhaps preceded by a duplication of the chromosomal complement, or followed by 'restitution'; that is, by the fusion of the egg's nucleus to a polar body nucleus) and yolk is regularly stored in its cytoplasm, as in the vitel-logenesis of a 'normal' egg. As for the sperm cell, we know that its involvement in the production of an individual of the next generation is some-times limited to the activation of the otherwise quiescent female gamete, the male nucleus even-tually being discarded (gynogenesis; e.g. Schultz 1967). The opposite may also happen, with the male haploid nucleus replacing the female hap-loid nucleus, which is discarded, and the fol-lowing events resemble parthenogenesis. This process strictly parallels gynogenesis, differing from it only because of the paternal origin of the genetic material in the 'egg' and thus in the cells of the future embryo, so in fact it deserves to be called androgenesis (e.g. Mantovani and Scali 1992, Komaru *et al.* 1998).

On the other hand, it is quite possible that all the traits we find today in the gametes of a majority of metazoans (some of which have a parallel in eukaryotes other than animals) may have originated in contexts other than sexuality, although the involvement of these cells in fer-tilization will have obviously contributed in a powerful way to their subsequent evolution. If so, scenarios of cell–cell competition come eas-ily to mind: competition between these cells and other cells in what eventually becomes their parent organism, but also, eventually, competi-tion with cells of different origin, usually from another conspecific organism.

Gamete release—that is, the severing of phys-ical ties with the parent organism—is a process by which competition is settled, temporarily at least. By asking whether this is contributed more by the gamete or the parent organism we may be tempted to argue in terms of which one of the two competitors will eventually profit most from gamete release, but this is not the only way to frame the question. One issue is the differ-ing nuclear status of the parent and gamete—diploid multicellular organism and individual haploid cell—but more important is the fact that a typical male gamete looks like a cell that has been induced to stop mitosis, possibly forever. Its centriole will not be able to organize a mitotic spindle, as long as it acts as the cilium's basal body; this drawback is additionally coupled to its

meagre energetic resources and thus limited life expectancy. Conditions with respect to the par-ent organism may look more favourable in the case of the egg, but it is hard to decide how much the huge accumulation of maternal molecules in the growing oocyte or in the maturing egg sig-nals a winning cell that removes resources (or life) from the surrounding cells and how much, to the contrary, the surrounding cells hamper the future egg's proliferation by inducing the nuclear and cytoplasmic changes that occur in oogenesis. Significantly, the physical release of the egg from the parent individual is not a pre-requisite for the egg's fertilization or even for the start of embryogenesis. Thus, the egg is usu-ally a product of meiosis in whose cytoplasm some amount of yolk has been stored, but nei-ther of these two features is shared universally by the so-called female gametes. In amictic par-thenogenesis, embryonic development starts with a cell that has got yolk, but which has not gone through meiosis. On the other hand, in some sponges, a meiotic product that other-wise behaves like an egg, as it is fertilized by a sperm cell, thus starting embryonic develop-ment, is barely distinguishable from a somatic cell in terms of yolk production or storage, as the macromolecules and/or the organelles pro-duced by nurse cells (trophocytes) are supplied to the developing embryo only after fertilization (Fell 1997).

And what about the sperm cell? We might say that spermatogenesis transforms the sperm cell into a ciliated cell without a future of reproduc-tion, and thus into a cell without a future, at the point when it becomes physically independent of the parent organism. If so, we can perhaps suggest that by fusing to a conspecific haploid cell the sperm cell can be rescued, in a sense, from its gloomy prospects. This amounts, more or less, to regarding the male gamete as a kind of parasite of its eventual partner. If we see the male gamete as a parasite, exploiting the meta-bolic resources and synthetic machinery of the egg, then the recently described 'polychaetes' of the genus *Osedax* exemplify an extreme develop-ment of this trend. These worms live on the sea floor, on the bones of large marine mammals, and exhibit a strong sexual dimorphism. The female is a tube-forming sedentary 'polychaete' with a lifestyle comparable with that of the other siboglinids (the pogonophorans of former classifications), whereas the smaller, paedomorphic males live in the tube of the female and only feed on yolk; that is, on the resources originally stored in the egg (Rouse *et al.* 2004).

But this entire story is about sexuality, not reproduction. Reproduction does not go necessarily, or even primarily, through the cells eventually evolving into gametes. This amounts to saying that the perpetuation of multicellularity does not necessarily require gametes, fertilization, or embryonic development. There is no foundation in equating the fertilized egg to the unicellular ancestor of the metazoans. The latter might well have lacked sexuality, or at least gametes, as is possibly true of today's choanoflagellates, and the egg is indeed one of the most specialized (and successful) cell types evolved in the animal lineage (Boyden and Shelswell 1959). This is not simply shown by the peculiar shape and size of a typical mature egg, but also by the specialized set of genes which are expressed during oogenesis. In the oocytes of the sea urchin *Strongylocentrotus purpuratus*, Song *et al.* (2006) have identified the expression of genes required for the egg's growth, meiotic recombination and division, the storage of nutrients, and fertilization. Clearly, this does not suggest a 'generic' metazoan cell.

Thus, we should probably remove both the egg and the sexually reproductive adult from the list of the *compulsory* components of an animal's basic developmental cycle. Both of them only result from grafting on to multicellular development both sexuality and those components of reproduction that, in most metazoans, have become, in turn, tightly bound to sexuality. We can imagine that in a hypothetical early metazoan, where the composing cells were still quite loosely integrated, all cells were more or less likely to turn into reproductive cells. Therefore, the evolutionary innovation that gave rise to the germ line was not the establishment of a gametogenic lineage, but rather the loss of gametogenic potential from the majority of cells

in the organism (Extavour 2008). In modern animals, most cells are not able to produce gametes, even in cases (as in non-bilaterian metazoans) where a dedicated gametogenic cell population may not exist at all (Extavour and Akam 2003, Extavour 2008). Interestingly, even in modern animals there is no systematic restriction of germ-cell production to any of the germ layers. For example, in most Hydrozoa gametes are ectodermal, but they are endodermal in several hydrozoan clades or species (*Polypodium*, Actinulidae, *Protohydra leuckarti*, *Nannocoryne mammylia*, *Pegantha clara*, and *Solmaris flavescens*; Bouillon *et al.* 2006), and insect germ cells can derive from any of the three embryonic germ layers (Corley and Lavine 2006).

The evolution of a germ line was made possible by the evolution of specific molecular markers, as are the products of the genes *vasa* and *nanos*. This happened quite early in animal evolution. In the anthozoan *Nematostella vectensis*, expression of genes belonging to the *vasa* and *nanos* families is not limited to presumptive primary germ cells, but during early embryogenesis it occurs also in multiple somatic cell types (Extavour *et al.* 2005). The expression of *nanos* in the different stages of the life cycle of the hydrozoan *Podocoryne* shows that the germ line differentiates exclusively during medusa development, but not in the polyp. This is part of an active process of local cell proliferation and tissue differentiation that produces the medusa by differentiating tissue types not present in the polyp (Torras *et al.* 2004). The specificity of *nanos* expression in the presumptive germ cells should not be overstated, however. In *Drosophila*, *nanos* functions as a posterior determinant essential for the formation of the fly abdomen (Hülskamp *et al.* 1989) and in *Podocoryne* Torras *et al.* (2004) found *nanos* expression at the posterior pole of developing embryos, not related to germline formation, raising the possibility that *nos* embryonic expression could be involved in establishing axial polarity, at least in cnidarians. I wonder, however, whether the role of *nanos* in germ-cell specification may represent a case of exaptation, from a role as positional marker to a role as specifier of a cell lineage with characteristic (posterior) original localization (see also section 7.5.3). Anyway, the function of nanos proteins is largely conserved among invertebrates and vertebrates (Tsuda *et al.* 2003).

7.3 Development

7.3.1 What is development?

The evolution of the metazoans was one of most important events in the history of multicellular life, and in the history of development. Understanding animal evolution thus requires us to adopt a sensible view of development. Unfortunately, despite the many facts we know about development, both in terms of descriptive accounts of ontogenetic changes in animal species and in terms of molecular mechanisms and their genetic control using a few model species, we still have no satisfactory and comprehensive theory of development. In my opinion, the biggest impediment to obtaining a sound theory of development is caused by the near-universal adultocentric attitude adopted until now. By adultocentric I mean the consideration of all developmental instars and developmental processes as steps, or means, required to eventually obtaining an adult. This attitude, most explicit in the seventeenth-century preformist views of development, has quietly survived the collapse of the now naïve-looking models in which the future adult (or the whole series of future generations) was imagined to be present in the ovum or sperm cell (ovist and animalculist preformism, respectively). It is still alive in the modern view according to which development is simply the deployment of a genetic programme targeted to eventually produce an adult. That the adult represents the end point of development is perhaps a given even in the minds of many biologists, and philosophers of biology, who energetically object to the notion of a genetic programme and do not hesitate to accept epigenetic factors as additional causal agents of development. Viewing the adult as the end point of development is, however, one of those elements that shapes what I regard as an evolutionarily incorrect view of development. Why do we take development as

being, in principle, the largely programmed way by which an egg is turned into an embryo that will undergo a largely predictable and to some extent stereotyped series of changes after which, following additional more or less conspicuous post-embryonic modifications, it will eventually give rise to the final reproductive stage, the adult? There are indeed many reasons to dispute this description. To begin with, in the case of vegetative reproduction, development does not start from an egg. Nevertheless, a polyp of *Hydra* produced by vegetative budding is identical to those *Hydra* polyps that are sometimes produced from fertilized eggs. To counter the objection that building a simple cnidarian polyp is easy, thus likely to be accomplished in different ways, an additional, more sophisticated example can be offered, that of colonial ascidians (e.g. *Botryllus*). Here, individual zooids are produced either by embryogenesis, starting with a fertilized egg, or by blastogenesis, from groups of cells budding off from the parent zooid, and development goes on in the two instances through surprisingly overlapping stages, to end up forming zooids with very similar organization (Fig. 7.2).

Another phenomenon that challenges the current perspectives on development is polyembryony, by which more than one embryo is obtained from a single fertilized egg. In the tiny parasitic wasp *Copidosoma*, polyembryony starts with the splitting of a large morula-like embryo into a huge number of partial morulae, each of which will eventually develop into a larva (Fig. 7.3). In a sense, polyembryony is a kind of vegetative reproduction occurring at an early, perhaps very early, embryonic stage, rather than post-embryonically as, for example, in *Hydra*, in catenulids, or in many 'polychaetes'.

We do not need to discuss all these points in detail in this book. Rather, I will argue here that the current view of development as the sequence of processes changing an egg into an adult (1) does not fit well with what we can tentatively reconstruct as early stages of animal multicellular life, (2) embodies strong, unwarranted residuals of Haeckelian recapitulationist views and, especially, (3) involves a confusing mix of phenomena pertaining to sexuality with those

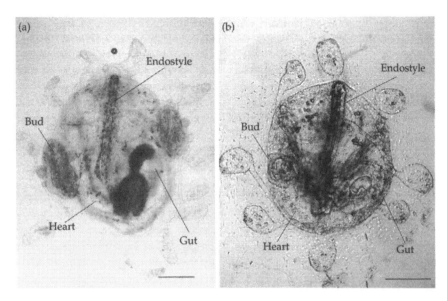

Fig. 7.2. Blastozooid (a) and oozooid (b) of the colonial ascidian *Botryllus schlosseri*. Scale bar, 350 μm. Courtesy of Paolo Burighel and Lucia Manni.

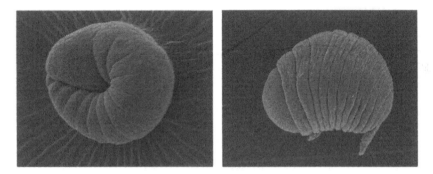

Fig. 7.3. The two kinds of embryos and larvae that derive by polyembryony from a single zygote of the tiny wasp *Copidosoma floridanum*. Precocious larvae (left) die inside the host without pupation, whereas the reproductive larvae (right) complete development into adults. Courtesy of Miodrag Grbic.

properly pertaining to development. The latter point deserves comment.

Let me start by explaining in what sense the current concept of development embraces biological phenomena that actually belong within sexuality and, in part, reproduction rather than within development proper. This requires some preliminary considerations about the relationships between sexuality and reproduction. If we want to understand the nature of either

phenomenon, most metazoans are indeed a poor starting point, because sexuality and reproduction are usually intertwined in what we call sexual reproduction. Things are much simpler in those living beings where the phenomena of sexuality have nothing to do with those of reproduction, and vice versa. In this respect, organisms with asexual reproduction only provide partial help, as in these we find pure reproduction without sexuality, but not sexuality without reproduction, as sexuality is either completely absent (if asexual reproduction is obligate) or still occurs—albeit occasionally—together with reproduction (if asexual reproduction is facultative). Thus, let us focus on other organisms, for example on ciliates. In these protozoans, conjugation is a purely sexual phenomenon, without any consequence for reproduction, as it only involves two partners exchanging nuclei, ready to separate again after this exchange: that is, two individuals are involved in conjugation, and two genetically changed, but somatically unchanged, individuals are recovered after conjugation. On the other hand, ciliate reproduction occurs by cell division, without involvement of sexual phenomena.

Basically, sexuality is nothing other than putting together genetic material originally belonging to the genomes of different nuclei, or different cells. As such, it has nothing to do with reproduction. To the contrary, as it implies the coming together of genetic material originally belonging to different nuclei (most commonly, of whole nuclei originally belonging to different cells), no wonder that sexuality mostly results in exactly the opposite to reproduction; that is, in the fusion of two unicellular individuals into a single zygote. At most, the original number of individuals involved in a sexual exchange can be maintained, exactly as in gamontogamy (the process exemplified by ciliate conjugation, where the two partners are not two gametes fated to fuse together, but gamonts each of which contributes a nucleus, but not the cytoplasm, to the exchange).

On the other hand, if reproduction is an increase in the number of individuals, in unicellular organisms it is obviously coincident with cell division, whereas in pluricellular organisms it corresponds to the breaking down of the original individual into two or more parts, perhaps by the detachment from it of smaller fragments, or of individual cells. Read literally, this definition of reproduction would lead to the idea that the core event in sexual reproduction is spawning, whereas the fertilization of an egg by a sperm cell (the sexual event) is a kind of accident superimposed on reproduction. Disturbing as it

perhaps is, this view of metazoan biology can help us to put into a clearer framework the evolution of the relationships between reproduction and sexuality at the origin of metazoans and during subsequent phases of their history.

A first fact that this view of sexuality and reproduction helps put into a reasonable historical context is that the conventional sequence by which so-called sexual reproduction starts with setting gametes free from the parent's body, to be followed by fertilization and then by embryonic development, is quite probably not the most primitive among pluricellular organisms. Indeed, in sponges the eggs are often fertilized when still deeply embedded in the parent's mesenchyme and reproduction (in the true sense of the term; that is, the detachment of a part from a parent's body) only happens with the release of a multicellular 'embryo' (Fig. 7.4). Interestingly, the same is also true of green plants: at the time they are fertilized, the 'female' gamete (the oosphere) of mosses, liverworts, ferns, and horsetails and the 'female' gametic nucleus of the flowering plants are still part of the parent gametophyte and the sporophyte that will eventually form from the zygote will still remain attached to the latter (most conspicuously in mosses), if this does not wither away.

We may now speculate what would happen to a simple multicellular organism in the absence

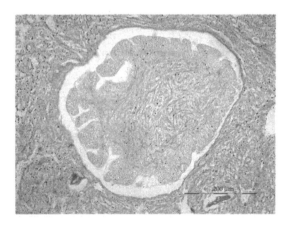

Fig. 7.4. Embryo of the sponge *Crambe crambe* developing inside the parent sponge. Courtesy of Manuel Maldonado.

of either reproduction or sexuality, or both. Let us start with reproduction. In this respect, the very existence of multicellular organisms is a consequence of delayed or limited reproduction with respect to what happens to unicellular organisms, where cell division and reproduction are one and the same thing. In principle, one could imagine a multicellular organism undergoing indefinite growth, until the whole or parts of it eventually collapse because of a failure to gain adequate access to the resources required to support metabolism, or to get rid of increasing amounts of metabolic waste. Of course, the upper limit that such a multicellular organism may reach will depend strongly on its specific shape and organization, as well as on the spatial distribution and accessibility of the resources on which it depends. In this respect, be autotrophic or heterotrophic will make a difference, as will having a massive, flat, or branching shape. At any rate, a simple and thus likely way to increase in size would be to go on repeating time and again, on different scales, the same basic elements of organization. This will end up, for example, with a green plant provided with repeatedly branched branches, or—back to animals—with a sponge provided with repeatedly branched systems of canals leading into loosely interconnected flagellate chambers. This is the so-called leuconoid model of sponge organization, which is apparently more complex than those with only one layer of flagellated chambers surrounding a central spacious cavity (syconoid model) or, even simpler, with just a single flagellated central cavity (asconoid model); but see section 4.1.1 for a different evolutionary interpretation.

7.3.2 Everything everywhere, or a principle of developmental inertia

Scale-invariance is a concept that applies across the most diverse areas of science and often stimulates a burgeoning of models and research programmes in fields far remote from the core area where they originated. Fractal patterns were long known to mathematicians, and also had a place in the popularization of mathematics long before Mandelbrot identified them as a field worthy of research, for both their intrinsic conceptual interest and potential applications (e.g. Mandelbrot 1982). In biology, fractal models have been applied to objects as different as the sutural lines of the ammonite shell (e.g. García-Ruiz et al. 1990, Long 2005) and the branching pattern of the respiratory and vascular systems of vertebrates (e.g. Glenny and Robertson 1990, Zamir 2001), not to mention the distribution of lower taxa among higher taxa in classification (e.g. Burlando 1990, Minelli et al. 1991).

Restricting here the focus to the frequent recursive patterns in animal morphology, we cannot limit our horizon to an appreciation of the fractal geometry of these features. Rather, we should ask what these patterns suggest in terms of the mechanisms that generate them. The main question is whether the recursive similarity of these patterns on differing size scales is actually produced by the recursive operation of the same morphogenetic processes or not. In other terms, are these patterns approximating the fractal geometries so easily produced, for example on a computer screen by the application of often simple recursive rules, actually the way the branching pattern of insect tracheae, or the finely dissected pattern of the compound leaves of many ferns, are produced? We must be cautious. The more we know about development, the more we must be prepared to discover that nature is less 'elegant' than we would expect. I will mention below the case of the segmentation of the *Drosophila* embryo, where the primary striped pattern is produced in a complex series of steps where the origination of each transverse stripe is controlled separately from the other, eventually equivalent stripes. In the case of the branching pattern of the insect trachea, genes are known whose expression determines the branching of an otherwise elongating tube, but the first-, second-, and third-order branching (three recursive, self-similar steps in a description in terms of fractal geometry) are actually under distinct genetic control (Samakovlis et al. 1996). Fractal-like geometry is, thus, no guarantee of simple recursive application of a small set of elementary morphogenetic processes.

Nevertheless, we must not discard the idea that such developmental control specifically directed at successive levels of an otherwise simply recursive process are probably burdened by an excess of determination. That is, largely similar patterns would probably be obtained even with less stringent and specific control, as we might have expected from our computer simulations of fractal patterns. Additionally, we must be aware of a fundamental difference between fractal models and the quasi-fractal patterns found in nature. In the former, the self-similarity applies *ad infinitum*, whereas in nature (and specifically, in our case, in animal morphology) the pattern cannot be repeated more than a very few times. As a rule, a macroscopic, supracellular pattern cannot be perpetuated down the scale as a microscopic, intracellular pattern. I am aware of only one noticeable exception to this rule, that is, the branching of the intracellular tracheoles that adds one more step to the branching pattern of the multicellular insect tracheae.

Eventually, the specific mechanisms involved in the production of a fractal-like morphology must be studied case-by-case. However, we already know enough of patterns other than the branching tracheae of *Drosophila* to suggest that the 'inelegant' (but probably adaptive) way this process is accomplished does not rule out the idea that self-similar patterns do not require, in principle, such a step-specific genetic control. For example, this seems to broadly apply to the branching pattern of the vertebrate circulatory apparatus, which is largely determined by hydrodynamic forces; that is, by the differential resistance met here and there by the blood that starts moving through the growing channels in the embryo (e.g. Nekka *et al.* 1996, Yashiro *et al.* 2007). This reminds us of the branching pattern of a river system, which is a classic in applied fractal geometry.

The possible origination of quasi-fractal patterns by recursive application of a fixed and possibly simple set of 'rules' may help to explain the evolution of body features such as the indeterminately branched organization of chambers and channels in many sponges, or the indeterminate arborescent structures of many colonial hydroids and corals. But I venture to say that these patterns, and the developmental mechanisms by which they are possibly originated, tell us something of much more fundamental importance with respect to development. They suggest that development is based on an indeterminate, inertial self-replication of developmental modules of different size, complexity, and robustness.

Developmental modules can be often identified with morphological units which are the actors of specific dynamics, such as dividing cells, or branching multicellular tubes, but in other cases it is more sensible to identify developmental modules with processes such as the periodic oscillations in gene expression whose clock-like behaviour controls the regular progression of somite production in an elongating vertebrate embryo (see section 8.1.5). Of course, this inertial behaviour cannot go on forever, because it is sooner or later challenged by the limited availability of resources, or by the control operated on it by other developmental modules. Thus, if the origin of development is in the inertial propagation of existing dynamics, its temporal progression, and its evolution, is found in the interactions internal to the developing system, by which local dynamics can be stopped or merged with those of neighbouring modules, while other dynamics may be started; that is, new developmental modules may originate.

Evidence of continuing, locally focused development is provided by the process of regeneration (see section 7.2.4), but also by many mechanisms of segmentation (see section 8.1.5). Developing segments can be described as parallel worksites (Minelli 2009) where similar or even virtually identical sequences of events give rise, either simultaneously or in regular temporal progression, to similar or virtually identical body parts. But this happens frequently (and, arguably, as a primary, default condition) even in the absence of an overt serial arrangement; that is, even in the absence of segmentation. For example, the motoneurons in the ventral nerve cord of nematodes are produced by parallel, quasi-identical patterns of cell division starting from 13 regularly spaced precursors (Walthall 1995). Again, everything happens wherever nothing forbids its perpetuation. This is *developmental inertia*, something to be explained by looking backwards, to the system's past states, rather than forward, to a hypothetical target (the adult, for example) towards which the developing system is often supposed to proceed.

Diffuse, isometric growth can be thus read as a consequence of this principle of developmental inertia. Intercalary growth, as in the epidermis of a nematode following the last moult, is another. Developmental inertia and lateral inhibition offer a kind of null hypothesis for the production of regularly spaced features which grow in number while conserving an average distance between two elements, as observed during the post-embryonic growth of mecistocephalid centipedes: the coxae of their last pair of legs are covered by pores the size of which does not vary with age. However, their number varies, new pores being formed at each moult, scattered over the whole surface of the coxa, in such a way as to roughly conserve the same pore number per surface unit.

7.3.3 Development for the development

To speak of developmental inertia means to use words closer to those of physics rather than to the adultocentric (if not finalistic) language so current in biology. A living being goes on doing whatever its current status and, thus, its recent history allow it to do. It is only in the textbooks that the polyps of Scyphozoa are obliged to give off medusae. In the case of *Aurelia*, the developmental behaviour of the scyphistome will be determined by environmental conditions, according to which its development will be directed. To give rise to ephyrae eventually growing into adult medusae is only one of the options, while other avenues open to it are much closer to a demonstration of its developmental inertia. The polyp, indeed, can either bud off new polyps, or produce stolons out of which new scyphistomes will be generated, but it can also produce larvae that will later turn into scyphistomes (Franc 1993). The fact that cnidarian specialists call these larvae pseudoplanulae

is clearly due to the fact that the 'true' planula is, by definition, the 'legitime' larva deriving through embryonic development from a generally fertilized egg produced by a medusa.

Accepting a principle of developmental inertia does not mean that we can easily identify an archetypal developmental module whose indeterminate iterativity can be accepted as the definitive starting point of our reconstruction of the evolution of development. Our closest approximation would arguably be a mitotically active cell, but what cell? Certainly not an egg, which is very specialized, but rather, as argued above, something comparable to a stem cell. But this is, in turn, a somewhat idealized condition, as no animal consists of stem cells only, outside a species-, stage-, and often site-specific context. We are thus obliged to allow for some degree of indeterminacy in this model.

By focusing on cells undergoing indeterminate mitotic activity as a convenient starting point for our reconstruction, the following history will be largely one of the challenges imposed by a cell, or a group of cells, on another cell or another group of cells. A most obvious effect of these interactions will be retarding or prohibiting mitosis of other cell(s), by inducing them to produce cilia (something incompatible with the production of the mitotic spindle), or by inducing them to produce a cuticle or a mineralized matrix that can provide a mechanical impediment to cell division (see section 7.3.10), or by filling them with mitostatic molecules. All these altered conditions of the cells whose mitotic activity is thus challenged by means of their neighbours and competitors can eventually turn into the starting point of phenomena of exaptation by effect of which new developmental modules and new body features will eventually result. But there is no need to describe these changes in terms of how they will eventually contribute to giving rise to the organization of the adult. That is, there is no need to use finalistic explanations, not even in that domesticated form that has become fashionable in biology under the name of teleonomy. Development can be self-explanatory in terms of local dynamics, developmental inertia, and competition. Development is a world of change, but processes may last enough, locally, to justify our description in terms of developmental modules. With the fading away of the corresponding dynamics, these may eventually disappear, their products becoming integrated into different anatomical structures. This is the case of the neural crest of vertebrates, which deserves to be recognized as an independent germ layer, but eventually gives rise to elements as different as skeleton and pigmentary cells (see section 9.1.2). Other developmental modules, however, can leave a permanent mark on the final structure of the animal, as is often (but not always) the case with serial (segmented) structures.

Stopping mitosis is not the only possible effect of cell–cell competition: other expected outcomes are apoptosis (see section 7.1.4), differential cell growth, and the induction of differential transcription, hence cell differentiation. Irrespective of their eventual consequence for the structure of the animal at later developmental stages and also of the selective advantage eventually deriving to the whole organism, all these phenomena must be first explained in terms of inertial and interactive properties of the cells and other modules involved in development: *hic et nunc*, here and now, rather than looking into an ontogenetic or phylogenetic future. This will require a revised reading of ontogenetic processes, from which I am confident of predicting substantial progress in our understanding of development and evolution alike. Let us try to explain the blastopore for what it can do in a gastrulating embryo, rather than seeing it only, or foremost, as the prospective opening of the future digestive tract. Let us look for a possible meaning of the presence of a cuticle in an advanced ecdysozoan embryo, when and where it cannot function as a site of muscle attachment, or as a way to protect the animal's soft parts, two jobs that are reserved for a later time.

7.3.4 Generic mechanisms of development

As soon as we start describing development in terms of actually ongoing processes rather than as a long chain of events eventually culminating in the adult, the value of the current metaphor of a developmental programme written in the genome is challenged. Of course, developmental events are controlled by gene expression, but this is not proof that nothing would happen in its absence. Outside the domain of life, the physical and chemical properties of matter are able to produce the regular shapes of the Earth and its fellow planets, the patterned sequences of the waves on the surface of sea, and the wonderful geometry of crystals. Why should living matter not being able to make spheres or tubes even in the absence of dedicated gene expression?

The robustness of many biological processes and of the biological structures eventually produced by them, with respect to mutations in genes whose expression is anyway involved in those processes is a proof of sovradetermination. That is, many genes are involved, but the process would nevertheless go on, more or less the same way, even with less genetic control. Extrapolating from this kind of evidence, why not to imagine that some simple developmental event might 'spontaneously' proceed, only depending on the intrinsic physicochemical properties of living matter? This is the notion of 'generic', as opposed to genetically controlled, developmental processes suggested by Newman and Müller (2000; see also Müller and Newman 2003, Newman 2003, 2005) as pivotal in the transition from the unicellular to the multicellular condition that accompanied the emergence of the metazoan lineage. To take this hypothesis seriously, we must be able to consider that life's phenomena are not simply those that go on in the familiar model systems of *Caenorhabditis*, *Drosophila*, and *Arabidopsis*. Evolutionary history, indeed, is not simply the production of new species, or new life forms, by effect of the steady operation of the same old rules, but also involves changes in these very rules, including, arguably, those that relate phenotypes to gene expression.

To be sure, reconstructing the nature of genotype–phenotype mapping in ancestral eukaryotes, or in hypothetical stem-lineage metazoans, would be even more problematic than inferring the presence, in the same 'ancestors', of a homologue of this or that gene. However, even if we will hardly be able to reconstruct in adequate detail the history of the changes in which we are interested, we can nevertheless obtain solid proof that those changes are possible and have indeed occurred. This is something we can often obtain by looking at branches of phylogenetic trees other than the one on which we originally focused.

In our case, something of value for an appreciation of how stringent the genotype–phenotype mapping must be can be learned from a non-metazoan clade where complex morphological features are produced and reproduced without the intervention of specific gene expression. These organisms are the ciliates, whose often very sophisticated patterns of cilia are a self-replicating system independent of gene expression. A complex self-replicating system is something that cannot be built anew from scratch, if nothing like its description, or a set of generative rules, is preserved somewhere. The future—that is, the replication of the pattern—may depend on the survival of a minimum set of bits and pieces of the existing pattern, around which additional elements will be made, or recruited, eventually completing a copy of the original pattern. This requirement is obtained when a ciliate cell divides, each daughter cell inheriting a part of the parent cell's pattern of ciliature. But, much more interestingly, this pattern is also saved when two ciliates are involved in a sexual exchange. Ciliates such as *Paramecium* or *Stentor* do not produce zygotes by merging of two gametes, but only exchange gametic nuclei that eventually fuse, producing zygotic nuclei, while the somatic individuality of the two sexual partners (the gamonts, or conjugating cells) is kept virtually intact during the whole process, and the original pattern of ciliature is saved.

In the late 1950s, Vance Tartar performed a wonderful series of experiments (summarized in Tartar 1961) on a very large ciliate, *Stentor*, in which he manipulated the cellular cortex, mechanically, in such a way as to obtain gross changes in the arrangement of the ciliary rows. At mitosis, the new ciliary pattern was faithfully inherited by the two daughter cells, and a morphologically stable 'monstrous' clone was eventually obtained.

In metazoans, cyclically passing through gametes and zygote does not leave much scope for correspondingly major effects of similar epigenetic changes, but an interesting example is provided by asexually reproducing colonial ascidians (*Botryllus*). In these colonies, each bud shows bilateral asymmetry (Fig. 7.5). Usually, the heart is on the right side of the organism and the digestive tract on the left side. However, if the asymmetry is experimentally reversed, the altered pattern is transmitted to the future asexual zooids (Sabbadin *et al.* 1975). Also relevant in this context is that in sipunculans portions of the internal organs within the posterior portion are used as anlagen for the regeneration of new organs (Pilger 1997).

To find examples of morphogenetic processes where physical forces play a major role we do not need to concentrate on asexually reproducing animals. For example, in the repeated branching process that gives rise to the bronchial tree

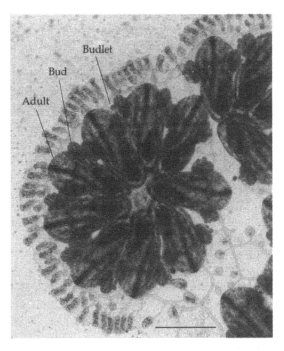

Fig. 7.5. A living colony of the ascidian *Botryllus schlosseri* showing three blastogenetic generations (adult, bud, and budlet, respectively); scale bar, 1.4 mm. Courtesy of Paolo Burighel and Lucia Manni.

of terrestrial vertebrates, the pattern of at least the most internal branchings is probably determined by physiological conditions such as local oxygen and nutrient concentration rather than by gene expression (Chuang and McMahon 2003). A complex interplay of gene-controlled specifications and physical (hydrodynamic) forces has been elegantly demonstrated in the development of the aortic arch in the mouse embryo (Yashiro *et al.* 2007).

Physical forces obviously prevail upon genetic control in the self-sorting of cells, something that has been studied *in vitro* since the beginning of the last century, but which was eventually shown to be important in embryogenesis. In *C. elegans*, the whole embryo is involved in a cell-sorting process (called cell focusing), eventually resulting in pattern formation, as the sorting cells assemble into coherent regions according to their prospective fate before the beginning of morphogenesis (Schnabel *et al.* 2006; see also Bischoff and Schnabel 2006). Newman (2003) suggests a non-genetic scenario for the origin of the organizer, the cluster of cells that embryologists recognize in many embryos as a centre controlling the fate of the other cells by sending chemical messages that will affect their mitotic activity and their eventual differentiation. This is another question where insights from a non-metazoan model can help in the reconstruction of a putative ancestral condition in animals.

Here, useful evidence is provided by *Dictyostelium*, a representative of a group traditionally known as the cellular slime moulds. These spend most of their existence as amoeboid unicells, but under unfavourable conditions hundreds of these amoebae may aggregate and eventually turn into a multicellular 'fruiting body'. Aggregation is mediated by chemical communication. A lot of effort was devoted for a while in the search for what (say, the expression of a particular gene, or a particular metabolic condition) might turn a 'normal' *Dictyostelium* amoeba into the founder of a multicellular group. But it was eventually realized that no founder cell may actually exist (Keller 2002). Rather, aggregation starts wherever two amoebae touch, and this happens simply because they were close enough

to have a better chance of one to be hit by the chemical signal produced by the other. That is, no founder, no project. Additionally, the chemical signal to which a *Dictyostelium* amoeba responds with moving towards another amoeba is the same molecule (cAMP) that signals the presence of the bacterial cells on which these amoebae feed. This suggests that the aggregation behaviour originates, likely at a very low cost, from the locomotory activity ancestrally used in the search for food. The evolutionary novelties are found in what happens after a bunch of amoebae have gathered together, rather than in the mechanism that brings them together.

Let us now go back to the organizer. Why does a localized group of cells in an embryo acquire this peculiar role in embryogenesis? A group of organizer cells affect the behaviour and eventual fate of other embryonic areas by releasing some chemical at a higher rate than the neighbouring cells, but this may have different causes (Newman 2003) including spontaneous activity, stochastic effects, or induction by an external cue. The divide between the cells belonging to the organizer and those that are controlled by it becomes deep if an effect of the chemical produced by the organizer cells is to prevent the surrounding cells from making the same. Maternal determinants may fix the position where the organizer will eventually form, but this predetermination is not a necessary requirement. Rather, those determinants may cause the positional (and, perhaps, temporal) canalization of a rupture of symmetry that even in their absence would nevertheless appear among the growing mass of embryonic cells (Newman 2003).

If we now move back in phylogenetic rather than ontogenetic time, we are confronted with early stages in the history of multicellularity, in the lineage that became the Metazoa. It has been suggested that multicellular organisms passed through a pre-Mendelian phase where context-dependent properties of self-organization were much more important than the canalization or determination of processes derived from gene expression, as generally happens in today's multicellular organisms (Newman 2005). According to this hypothesis, the morphological features of the earliest multicellulars were mainly the outcome of epigenetic processes. According to Newman and Müller (2000; see also Müller and Newman 2003), in a first phase these processes were simply manifestations of the physics of condensed, excitable media represented by primitive cell aggregates. With the evolution of more complex and integrated multicellular systems, there was also the emergence of conditional responses of tissues to each other and to external forces. In this scenario, there was little place for gene sequence variation and gene expression as the primary agents of biological form, something that would have progressively evolved, with a sovradetermination of form by the added expression of genes able to fix in either space or time the production of these forms. Shapes produced without genetic control might have been tubes, spheres, and also serially repeated structures. Physical forces, unaided by spatially restricted gene expression, would even be able to cause a kind of gastrulation, provided that a cell aggregate included two different non-mixing kinds of tissue (Newman 2003).

This is certainly a hypothetical scenario, rather than the result of a historical reconstruction, but its plausibility is indirectly supported by the occurrence, even in modern animals, of developmental processes that are not strictly canalized by gene expression. West-Eberhard (2003) has recently devoted a whole book to presenting and discussing this evidence. Therefore, I will not attempt to summarize it here, but I will focus on the fact that all kind of cells, whether embryonic, stem, or finally differentiated, retain the ability to modulate most of their functions. It has been poignantly remarked that even cells devoid of a nucleus, as are mammalian red blood cells, can still regulate their behaviour as a function of their environmental context (Bissel *et al.* 2003).

This brings us to the more general question, of whether the role of genes in development is instructive or permissive. To a large extent, the answer to this question will depend on the aspect of the phenotype on which we want to focus. There is ample scope to acknowledge that the role of genes in controlling phenotype is quite

often a permissive one, the specificity being given by external influences which can instructively alter gene expression (Gilbert 2003b). I will return in a later section (see section 7.5.2) to the subject of 'developmental genes'.

7.3.5 Parent–offspring competition and the developmental origins of individuality

Whenever sexual reproduction occurs—that is, whenever we can speak of embryonic development beginning with the cleavage of an egg— there should be no problem tracing the origin and the temporal continuity of individuals. But things are not so simple. The zygote's genome, an obvious signature of individuality, is not expressed before the so-called mid-blastula transition. Therefore, during the first segment of embryonic development, all events, including for example the fixation of embryonic polarity, are still under strict maternal control, in the form of molecules, either mRNAs or proteins, obtained by transcription and eventual translation of maternal genes. This is perhaps the most subtle form of maternal control of the offspring's life, and one that challenges a rushed identification of fertilization, or the start of cleavage, with the origin of a new individual.

Parent–offspring competition is a classic in evolutionary biology, but the question is commonly reduced to a search for the best strategies a parent can adopt in splitting its resources between its own profit and the advantage of the offspring. Actually, competition is sometimes much more coarse, as in the case of those larvae of the beetle *Micromalthus debilis*, which literally devour their mother (Scott 1941), or the annelid *Polygordius*, where gametes eventually fill the whole body cavity, due in part to the lack, in this genus, of coelomoducts that could be used to discharge the sexual products into the water, as many other annelids do. This intergenerational competition is even worse in *Polygordius epitocus*, where the oogonia literally phagocytose the gut of the mother, then the nephridia and the musculature, until a thin cuticle is left, which breaks apart and sets the gametes free (Davydov 1905).

But let us return to the question of the origin of individuality. Besides the effects of parent–offspring competition, another phenomenon to be considered is polyembryony; that is, the development from one zygote of two or more genetically identical embryos. By definition, polyembryony is an instance of asexual reproduction, but instead it is widely perceived as being an atypical form of sexual reproduction. This attitude may have different explanations. One, and perhaps the viscerally strongest one, is the not-so-rare occurrence of this phenomenon in our own species. Speaking of asexual reproduction in humans may elicit reactions comparable to those raised by artificial human cloning today. But even if we disregard polyembryony in humans, the fact remains that asexual reproduction sounds odd, both in mammals (with the well-known example of the nine-banded armadillo, whose females generate sets of four identical twins) and in insects (where some tiny parasitic wasps like *Copidosoma* perform an 'explosive' polyembryony with up to a couple of thousand embryos from a single egg).

A more technical reason why polyembryony is often considered in the shadow of sexual reproduction is the fact that the organism eventually splitting into two or more embryos is, in turn, an embryo; that is—as a rule—the immediate product of sexual reproduction. Nevertheless, when the initially single embryo of a polyembryonic animal eventually splits into many units, any previously expressed element of body patterning is likely fated to destruction and the daughter embryos must start their patterning from scratch, or nearly so. This is probably what happens in a great many cases of asexual reproduction and regeneration.

The scattered occurrence of polyembryony across the metazoans is summarized in Table 7.3. Whether or not the events recorded under the name of polyembryony in the Digenea actually deserve this title, this is still a controversial question (e.g. James and Bowers 1967, Erasmus 1972, Galaktionov and Dobrovolskij 2003), but the concept seems at least to apply, for example, in the case of the so-called *Cercaria stagnicolae*, within which a rapid multiplication of germ

Table 7.3 Metazoan clades with examples of polyembryony. Splitting into two or more individuals may occur at either the embryonic or larval stage. Data mainly after Craig *et al.* (1997).

Taxon	Examples	Embryo	Larva
Cnidaria			
Hydrozoa Trachylina	*Pegantha* spp., *Cunina* spp., *Cunochtantha* spp.		+
Hydrozoa Hydroida	*Polypodium hydriforme*		+
Rhabditophora			
Gyrodactyloida	*Gyrodactylus elegans*	+	
Eucestoda	*Echinococcus* spp.		+
Digenea	*Schistosoma* spp.		+
Bryozoa	*Stenolaemata*	+	
Arthropoda			
Crustacea Rhizocephala	*Loxothylacus panopaei*		+
Insecta Hymenoptera	*Copidosoma* spp. (Encyrtidae), *Platygaster* spp. (Platygasteridae), *Macrocentrus* spp. (Braconidae), *Aphelopus theliae* (Dryinidae)	+	
Insecta Strepsiptera	*Halictoxenos simplicis*	+	
Echinodermata			
Asteroidea	*Luidia* sp.	+	
Ophiuroidea	*Ophiopluteus opulentus*		+
Mammalia	*Dasypus* spp.	+	

cells gives rise to germinal masses which eventually divide and develop into daughter sporocysts (Erasmus 1972).

Polyembryony is almost universal in the marine cyclostomatous bryozoans. This long-known fact has been definitively demonstrated by Hughes *et al.* (2005) by studying microsatellite genotypes of embryos of *Crisia denticulata*, a cyclostomate bryozoan encrusting rocky shores. Hughes *et al.* found genotypic differences among broods produced by the same colony, indicating outcrossing, but genetic identity within the same brood, a proof of polyembryony.

Summing up, polyembryony is a form of asexual reproduction (Boyden and Shelswell 1959) and its occurrence is a conspicuous example of the principle of developmental inertia I introduced above. Interestingly, polyembryony is sometimes coupled to phenomena of exactly opposite nature; that is, to the fusion of previously separate blastomeres, or even separate embryos.

A phenomenon known in the embryological literature as blastomere anarchy is known (Hallez 1887) from freshwater planarians (*Dendrocoelum*). Here, the products of the first cleavage are two seemingly identical cells that only barely have contact each other. Each of them divides again equally and the four-cell embryo then undergoes a further cleavage division in typical spiral fashion, giving rise to an apparently 'normal' eight-cell stage. But these eight cells then migrate a short distance away from each other within the yolk cell mass. A coherent embryo forms anew at a later stage.

In the small freshwater fishes of the genus *Cynolebias*, a zygote often gives rise to two separate blastoderms, which develop into two distinct advanced blastulae. At the time when these begin to gastrulate, the blastomeres of the two previously separate embryos reaggregate, giving rise to one embryo that soon loses any evidence of the original duplicity (Carter and Wourms 1993). In the embryos of stenolaematan bryozoans, early blastomeres commonly separate from each other, but eventually reaggregate to form a hollow primary embryo which increases in size and eventually buds off a few secondary embryos which mostly develop into larvae, but in some species split instead into a number of tertiary embryos (Zimmer 1997). That is, these animals combine three unusual phenomena: first, polyembryony; second, aggregation into one embryo of previously independent blastomeres; third, and the oddest, the development of a larva from something that is not an embryo formed by egg cleavage and gastrulation.

Comparable to the embryos formed by blastomere aggregation is the case of freshwater sponges (of the genera *Ephydatia* and *Spongilla*) derived from the fusion of two or more larvae (or of different gemmules) (Brien 1973). In the marine sponge *Esperella* (=*Mycale*), gemmules form by aggregation of archaeocytes, or by fusion of previously smaller groups of archaeocytes, and eventually turn into larvae identical to those produced through embryogenesis (Wilson 1894, 1902). Clearly, these animals do not to regularly pass through the one-cell bottleneck (see section 7.1.1).

7.3.6 Set-aside cells

Peterson *et al.* (1997) define set-aside cells as the cells from which the adult body develops in animals with a distinct larval stage. Clearly this is an adultocentric term for what would be better defined as 'temporarily marginalized cells', in terms of the internal competition by which other cell lineages successfully outcompete them for a while, only to succumb later by being more or less extensively discarded, or to be induced to apoptosis. My non-adultocentric perspective is in agreement with the view of Blackstone and Ellison (2000) and Michod and Roze (2001) who advocate a role of set-aside cells in the evolution of germline sequestration as a way by which the conflict between different cell lineages is reduced, a conflict that could otherwise result in a lowered fitness of the animal.

A very serious blow to the adultocentric interpretation of set-aside cells derives from the discovery of larval cloning in sea urchins and other echinoderms. Here, a juvenile (hence, prospectively, an adult) derives from cells other than those that would have been 'set-aside' with a view to building the future adult (Eaves and Palmer 2003).

A kind of set-aside cells are also those forming the imaginal discs of holometabolous insects. As in the case of the conventional set-aside cells of many marine invertebrates, a description in terms of competition among cell lineages helps understanding the relationship between the cells forming the bulk of the larval body and those confined to the imaginal discs. Across the holometabolans we find a clear trend towards an increasingly larger role of imaginal-disc-derived cells in forming the adult epidermis (Grimaldi and Engel 2005). In all holometabolan groups consistently regarded as basal, such as neuropterans and mecopterans, but also in the basal families of the other orders, including nematocerous flies and symphytan wasps, part of the larval epidermis is retained in the adult, whereas in the more advanced Diptera and Hymenoptera, as well as in the Lepidoptera, most or the whole of adult epidermis is formed anew from the imaginal discs. Therefore, in terms of cell lineage, the boundary between larval-cell-derived and imaginal-disc-derived cells is continuously shifting in evolution, as an effect of an increasing individualization of the holometabolous larva as a new independent target of adaptive selection. In agreement with this scenario, the imaginal discs of the more basal holometabolans (but also Lepidoptera; Nijhout and Wheeler 1996) are only active in the last larval instars, while those of insects with more dramatic metamorphosis like *Drosophila* begin developing in the embryo (Grimaldi and Engel 2005).

From the point of view of the dynamics of cell populations, the metamorphosis of a tadpole into a frog is probably less revolutionary than the metamorphosis of a typical holometabolous insect. In the amphibian we cannot speak of set-aside cells. For example, the progenitors of the functional intestinal epithelium of the adult frog are the cells of the functional intestinal epithelium of the larva (e.g. Marshall and Dixon 1978).

Worth exploring is the different perspective of Henry *et al.* (2007), according to whom all metazoan embryos set aside cells that act as 'organizers': inductive signalling centres that specify the fates of adjacent cells. In this sense, we will need to revise the traditional opinion that groups such as crinoids (Hyman 1955) do not have set-aside cells. But this is probably just one aspect of a pretty universal phenomenon of symmetry-breaking that begins with marking embryonic subpopulations of cells into those sets that we call embryonic layers and going on during the whole of development. In this vein, Held (2002) regards the establishment of separate groups of 'signaller' and 'receiver' cells as the *raison d'être* for the further subdivision of most imaginal discs into an anterior and a posterior compartment: cells in the latter compartment emit a signal (Hedgehog) to which cells in the anterior compartment can respond.

Strathman (2000) offers an adaptationist explanation for the evolution of set-aside cells. The availability of this 'dedicated' cell population, within which mitosis can be suddenly (re)activated and differentiation put into effect without major interference from the remaining (larval) cells, would be to the service of rapid metamorphosis, thus reducing exposure to danger over longer time. There is no need, however, to consider this advantage from a holistic (whole-organism) point of view. It is more parsimonious to rephrase the argument in terms of competition: the set-aside cell population will effectively produce a post-metamorphic juvenile because it is able to escape competition from the larval cells and suddenly acquire an advantage over it in the exploitation of both energetic and informational resources. A component of this competitive advantage over the other cells so suddenly obtained by previously resting cells is possibly related, in part, to a differential ability to effectively respond to a usually external metamorphosis-inducing signal (Hadfield 2000).

7.3.7 The germ layers

Germ layers have long been defined in terms of their prospective fate. This is what Hall (1998b) calls the germ layer theory (and aptly rejects it), that apparently enabled comparisons between the development of embryos where germ layers are actually present as distinct cell sheets and the development of embryos where this does not happen. With theory being taken more seriously than experimental evidence, germ layers have remained in the biological literature mainly as the sources of cells eventually fated to different roles, rather than as parts of the embryo recognizable after objective criteria. Things apparently changed with the identification of genes that are selectively expressed in one germ layer only. This would allow comparisons between 'classic' embryos with morphologically identifiable germ layers and those where germ layers cannot be discerned, as in the case of *C. elegans*, embryos of which contain too few cells to allow for the distinction of discernible germ 'layers', but have many endoderm-specific genes (Maduro and Rothman 2002).

Thus, mesoderm has emerged as the germ layer expressing *snail* and *twist*. However, although this molecular signature can be accepted as a good marker of mesodermal identity in bilaterians as different as insects and mammals, equating germ-layer identity with a specific gene's expression may lead to unexpected troubles. Interestingly, a *snail* homologue has been reported from diploblastic animals such as the the coral *Acropora millepora* (Hayward *et al.* 2004) and the sea anemone *Nematostella vectensis* (Martindale *et al.* 2004). Still more interesting is the fact that the expression pattern of *snail* in the two cnidarians shares much similarity with the mesoderm-restricted expression of its homologues in vertebrates and insects and arguably contributes to the specification of the endoderm with respect to the ectoderm (Ball *et al.* 2004, Martindale *et al.* 2004). It has been suggested that prior to the origin of mesoderm the so-called mesodermal genes were probably involved in cellular properties such as the regulation of cell proliferation, adhesion, and motility (Technau and Scholz 2003). I do not see how we could regard the nature of *snail* expression in cnidarians and bilaterians as so fundamentally distinct to justify our association of this gene's expression with the distinction between endoderm and mesoderm. I think that the answer is that this is not the case. Rather, I would argue that the localized expression of genes such a *snail* has the

effect of singling out an embryonic cell population, and that this, by becoming different from the remaining cells in terms of adhesion, mobility, and proliferation, becomes a developmental module in competition with other cell groups. What will eventually happen to these cells and, in particular, whether they will become epidermis, nerve, muscle, or gut lining, is something that history may have grafted on to one or other of the developmental units or modules (germ layers) identified by the initial differential gene expression. But that is another story and one of many evolutionary 'accidents'.

In hydrozoans, germ cells generally differentiate from ectodermal interstitial cells, but can proceed by developing within either germ layer. In the actinulids *Protohydra* and *Boreohydra* germ cells originate from endoderm (Van de Vyver 1993). In the hydrozoan *Phialidium gregarium*, the nervous cells seem to originate from the endoderm (Thomas *et al.* 1987), but the nervous system appears to be ectodermal in scyphozoans (Nielsen 2001). The excretory Malpighian tubules are endodermal in chelicerates, but ectodermal in insects. In most metazoans, the digestive tube (or, at least, the midgut) coincides with the endodermal derivatives, but even this widespread correspondence between a germ layer and a post-embryonic (or adult) structure is far from being a universal rule. In the Tardigrada, the midgut is of mesodermal origin (Kristensen 2003). In the ontogeny of many crustaceans (isopods and tanaids) ectodermal cells from the proctodeal region proliferate in an anterior direction, to eventually replace the original, endodermal anlage of the midgut; eventually, no endodermal derivative is left in the adult (summarized in Siewing 1969). Very different contributions of the individual germ layers in the formation of the same organ in related species are also known for the antennal gland, the excretory organ of crustaceans. Generally, bladder, labyrinth, and internal canals of this organ are mesodermal, with only the terminal efferent canal being of ectodermal origin, but in the lobster (*Homarus*) and other decapods, ectoderm contributes to most of the organ and the whole tube is ectodermal in ostracods (summarized by Beklemishev 1969). Shimkevich

(1908) proposed the term metorisis to denote a change in the boundaries between neighbouring anlagen, resulting in replacement of the anlage of a complex organ by another.

A further challenge to our categories is *Notentera ivanovi*, a parasitic flatworm living in the gut of the 'polychaete' *Nephthys ciliata*. *N. ivanovi* has no mouth or intestine, but its dorsal epidermis forms a thick pad very similar to a gut epithelium. At the ultrastructural level, this 'outer gastrodermis', arguably of ectodermal origin, closely resembles the transporting cells of specialized gut epithelia involved in the absorption of digestion products from the lumen of the gut (Joffe and Kornakova 1998).

Another conspicuous example of lack of germ-layer specificity in recruiting cells to form a given body structure is provided by segmentation. Segmentation may affect either ectodermal or mesodermal derivatives, or both, and tissues deriving from different germ layers are sometimes segmented independently. Transpatterning from segmented mesoderm to segmented ectoderm, or vice versa, may be suspected when the serial units within the epidermis and the nervous system are in register with muscular serial units, but this is not warranted (Holland 1988). There are few examples of segmented endoderm, but many leeches have intestinal caeca arranged in register with ectomesodermal segments. This segmental arrangement of endodermal derivatives depends in part on the local periodic pattern of expression of the *Lox3* gene, but this expression, in turn, is dependent on the mesoderm (Wedeen 1995, Wedeen and Shankland 1997). In the other segmented metazoans, both ectoderm and mesoderm, but not the endoderm, are generally segmented, often by different mechanisms. In arthropods, however, the process is first and foremost an ectodermal affair, whereas in vertebrates and annelids segmentation is mostly, but not exclusively, mesodermal (see section 8.1.5).

If we abandon the traditional germ-layer theory and opt for a less adultocentric view of embryogenesis, we can regard the traditional germ layers (ectoderm, mesoderm, endoderm) as members of a broader class of modular units

into which a developmental system becomes progressively articulated. Specific patterns of gene expression and, often, specific patterns of mitotic activity (or lack of it), together with other specific activities such as cell migration, identify the neural crest cells of vertebrates, the insect imaginal discs, and the clusters of set-aside cells of so many larvae of marine invertebrates as modular units somehow comparable with germ layers (Minelli 2003). Some kinds of modular units are very fleeting and do not leave behind any recognizably distinct trace in the late embryonic, larval, or adult organization. A good example is provided by the mitotic domains accurately described by Foe (1989) and Foe and Odell (1989) for pre-gastrulation embryos of *Drosophila* and *Calliphora* respectively. At a later stage, the fly embryos are subdivided into polyclonal compartments spanning the length of half a segment, whose boundaries cannot be crossed by the progeny of the restricted pool of the compartments' founder cells (García-Bellido *et al.* 1973). Similarly, the hindbrain of vertebrates is subdivided into units (rhombomeres) primarily identified by cell-adhesion properties, in that the cells from odd- and even-numbered rhombomeres do not mix together, whereas those of either odd- or even-numbered units would mix together among themselves (Guthrie and Lumsden 1991, Guthrie *et al.* 1993). Recently, developmental units separate by cell-adhesion properties have been described in the Scyphozoa. In this cnidarian group, there seems to be two types of endoderm, which form alternating longitudinal stripes spanning between the oral and the aboral end of the polyp. At the sites where the two types of endoderm meet, a septum is formed. As a consequence, the number of septa by which the gastral cavity of the polyp is subdivided is necessarily even. Indeed, this number is usually four, but in *Aurelia* polyps with zero, two, six, or eight septa have been also recorded (Berking and Herrmann 2007).

Usually, a given kind of developmental modularity—for example, the subdivision of the embryo into germ layers or the onset of mitotic domains—is manifested throughout the whole embryo in a short time span, but this is not necessarily always the case. In vertebrates, the caudal region is an area of continued growth and cell recruitment (Stern *et al.* 2006) where some cells are not yet committed to a specified germ layer and single cells can still contribute, with their progeny, to more than one layer (Selleck and Stern 1991, Catala *et al.* 1996, Brown and Storey 2000). In recent years, the ontogenetic origin and the phylogenetic significance of the vertebrate neural crest has been the focus of interesting discussion. On one hand, Hall (1998b, 1999) has strongly defended the argument that the neural crest should be regarded as a 'fourth germ layer' characteristic of vertebrates; that is, as a cell population with spatially and temporally specific origination and with a no less specific morphogenetic fate. The diversity and functional importance of the neural crest derivatives (Hall 1999, Butler 2000, Holland and Chen 2001) should be regarded as one of the keys to the structural complexification and functional success of vertebrates.

On the other hand, 'latent homologues' of neural crest cells have been identified in both cephalochordates and tunicates (Jeffery *et al.* 2004, Stone and Hall 2004). One of the most typical traits of vertebrate neural crest cells—that is, their migratory behaviour—has so far been reported only from tunicates, not from amphioxus (Delsuc *et al.* 2006). Ascidian neural crest-like and vertebrate neural crest cells may have had a common origin during early (stem?) chordate evolution. Otherwise, their presence in tunicates and vertebrates, but not in cephalochordates, would provide an argument other than gene or protein sequences in favour of the recently advocated sister-group relationship between vertebrates and tunicates rather than between vertebrates and cephalochordates. Jeffery (2006) suggests that their primitive function was in the generation of body pigmentation; this function has been retained in vertebrates, alongside the many other functions subsequently evolved.

7.3.8 Embryonic development

The phylogenetic signal present in the different patterns of cleavage is very variable. If the duet spiral cleavage sets the Acoela neatly apart from the core plathelminths (Catenulida and Rhabditophora), too much has been made in the past of other cleavage modes. A peculiarity of cleavage in the Hydrozoa and other cnidarians is its unstable nature (Beklemishev 1963), but instability is not an easy character to code in a data matrix. There is no reason to doubt that groups such as annelids and sipunculans, molluscs, entoprocts, nemerteans, and rhabditophorans inherited from a common ancestor their quartet spiral cleavage, with the typical orientation of the mitotic spindles during the first cell divisions, and a characteristic cell lineage with the identification of prototroch cells or equivalents along the border between first and second quartet of micromeres (Nielsen 2008). However, in other cases (e.g. Syndermata, Sørensen *et al.* 2000; and especially Hexactinellida, Boury-Esnault *et al.* 1999) it is still very doubtful how the putative similarities to a spiralian cleavage should be interpreted (Jenner 2004c). Clearly, a mechanistic reason why we should generally acknowledge high phylogenetic value to the occurrence of spiral cleavage is the fact that the transition from a radial to a spiral cleavage does not seem to be easy, whereas the opposite transition is possibly easier. In a spiralian group as are the Rhabditophora, two alternative cleavage modes have been described in one species (*Prorhynchus stagnalis*, a member of the Lecithoepitheliata), where the eight-cell stages can be represented either by eight blastomeres of equal size or by four macromeres and four micromeres, as in radial and spiral cleavage respectively (Steinböck and Ausserhofer 1950). It has also been suggested that different kinds of gastrulation may occur in different embryos of a popular medusa, *Aurelia aurita* (Franc 1993).

Wray (1994) studied the evolutionary history of echinoderm cell lineages against an independently obtained phylogenetic tree and found contrasting evolutionary patterns in different cell lineage characters, some of them being apparently conserved since the Cambrian, others having much more recent origin and being more variable. But the cleavage mechanisms of other groups are more uniformly conserved.

Guralnick and Lindberg (2001) used the timing of cell-lineage formation to produce phylogenetic trees in gastropods and found that cell lineage data reconstructed a phylogenetic hypothesis that was similar in several respects to patterns found in other more traditional analyses.

If comparative embryology invites to adopt a more flexible view of germ layers than we were accustomed to accept, moving away from a restricted range of model species also invites a revisitation of the concept of gastrulation. We have already seen in Chapter 4 that in sponges (and, to some extent, in some cnidarians) the differentiation of the two germ layers and the formation of an internal digestive cavity are not necessarily associated (e.g. Leys and Eerkes-Medrano 2005). This has two consequences. One is basically lexical (but with obvious consequences for the way we code the relevant data in a matrix to be used in a phylogenetic analysis); that is, what do we actually mean by gastrulation? The choice is between taking the term literally, to mean the formation of a digestive cavity, or to abandon fidelity to etymology to intend instead gastrulation as the differentiation of the two primary germ layers. The second consequence is that we must add gastrulation (in the traditional sense) to the growing list of biological terms which correspond to complex processes (or features) whose components are not necessarily associated, and must be treated separately if we want to make sensible statements of homology and, more interestingly, try to reconstruct the evolutionary history of those processes or features.

Overall, to have a term applying to the differentiation of the two primary germ layers seems to be more useful than to have a common term for the processes giving rise to a digestive cavity, as two germ layers are often identified even in animals without a gut, as is the case of many hydrozoan larvae with two cell layers separated by a mesohyl (e.g. Thomas *et al.* 1987). In animals such as echinoderms or chordates, there is an intimate connection between the process of gastrulation and the creation of mesoderm, to the extent that some authors (e.g. Nieto 2002) have even defined gastrulation as the process by which mesoderm is formed, but this implies stretching the concept too much away from its original meaning.

7.3.9 Growth

Although the yearly differences in maximum height or stem circumference cannot be appreciated without actually measuring them, old trees often give the impression of a truly indeterminate growth, only challenged by fungal or insect attack, or by other external agencies. On the contrary, the growth of animals is broadly perceived as intrinsically limited and often virtually ending with the onset of maturity. Indeed, this popular perception is supported by what we can observe in humans and many other vertebrates, and also by the fact that meter-long flies, or mountain-size dinosaurs are confined to the world of fiction. Nevertheless, there are animals with indeterminate growth and a recent review on trajectories and models of individual growth (Karkach 2006) lists examples from the following groups: annelids (the small freshwater clitellate *Pristina*), molluscs (long-lived bivalves), echinoderms (the sea urchin *Strongylocentrotus*), arthropods (cladoceran and decapod crustaceans), and vertebrates (many teleosts, various reptiles, and some big mammals like bison, giraffe, and elephant). Many invertebrates (e.g. sponges, cnidarians, and echinoderms) appear to lack true senescence, but this should not be considered a requisite for growth to be considered indeterminate (Sebens 1987).

As in the plant kingdom, among fishes there is a broad correlation between indeterminate growth, high longevity, and high fecundity (Reznick *et al.* 2002a). However, whereas in the plants the size of individual parts such as flowers or leaves is largely independent from the plant's total size, which correlates instead with their number, in animals it is organ size, rather than organ number, that correlates with body size (Desplan and Lecuit 2003). However, this is not true of small organs of fixed cell composition (one to a few cells) such as the epidermal sensilla and the ommatidia of insects, which do not vary in size with the animal's overall size.

Many animal species are regularly exposed to the risk of prolonged starvation, which can produce a sensible decrease in size. This phenomenon can be dramatic in some nemerteans and also in planarians: following months of starvation, a 10 mm-long planarian can be reduced to less than 1 mm (Abeloos 1930). Negative growth is also known for cnidarians, echinoderms, annelids, molluscs, and urochordates (Sebens 1987). Starvation not only causes planarians to decrease in size, but as they shrink they 'rejuvenate'; that is, they become similar to juveniles in both morphology and physiology (see Reuter and Kreshchenko 2004). It may seem very odd that a de-growing arthropod should waste additional resources by undergoing moults during starvation, but this is what actually occurs in some crustaceans and insects. Under normal conditions, the dermestid beetle *Trogoderma glabrum* develops through five or six larval stages; however, if water and food are not available, the larva may go on moulting at more or less regular intervals for several years, decreasing in weight and also in linear dimensions (Beck 1971). Producing new cuticles of size progressively reduced in proportion with the decreasing body size is perhaps a way by which these starving larvae can still keep control of their shape: this tentative interpretation leads us straight into the next topic.

7.3.10 Morphogenesis and morphostasis

Animal evolution is not simply a story of novel mechanisms through which unprecedented forms are produced, but also a story of innovative mechanisms by which form is frozen. I do not refer here to the effect of stabilizing selection, or to the putative evolutionary stasis occasionally punctuated by change, which form the core of the well-known mode of evolution according to Eldredge and Gould (1972). I am referring instead to the origination of morphostatic, as opposed to morphogenetic, developmental processes (Wagner and Misof 1993, Wagner 1994). It is probably legitimate to regard morphostasis as an effect of internal competition. The reason why morphostasis features so sparingly in the

literature on evolutionary developmental biology is probably that a lack of change is often uncritically taken as a default condition rather than as the result of dynamic confrontation. As a consequence, what is to be explained is only change, not the lack of it. However, I think that biology should learn a lesson from mechanics, where a state of rest is not necessarily an uninteresting given, but is often the effect of a complex and interesting balance of contrasting forces. Similarly, a lack of change in a developing organism is not necessarily due to a very low intake of molecules, just sufficient to sustain its minimum metabolic requirements, but can also be the result of a contrast, or competition, among different cell lineages within the developing organism: competition the escape from which will require increasing the autonomy of a developing part, that is, its transformation into a more or less distinct developmental module. Global morphostatic effects of competition must be expected, especially at early developmental stages, because the small size of the embryo may allow the involvement of the whole mass of blastomeres in the onset of a modularization such as with the origination of distinct germ layers. Later in development, morphostatic effects can be mainly expected at a local scale, as a result of interactions between neighbouring cells or cell groups.

An early suggestion that the morphostatic effects of competition have possibly played an important role in animal evolution was advanced by Buss (1987). His argument was based on the observation that in animal cells the centrioles cannot support the simultaneous presence of cilia and a mitotic spindle. As a consequence, a ciliated cell cannot divide, unless it loses its cilium and redirects its centriole to the assembly of spindle fibres. This may have been critically important in originating, and stabilizing, the bilaminated structure of the gastrula, with its typically ciliated ectoderm where mitotic activity is inhibited until the time the cilia are eventually lost, whereas the internal cells (endodermal and eventually also mesodermal) are not subject to this constraint. Did this major contrast between external and internal cells really originate as an effect of internal competition? Probably, we will be never able to provide a definitive answer to this question, but the way Buss framed it is one which sensibly addresses a developmental stage, in this case the gastrula, by asking what is going there at that very moment, rather than asking how it is working towards the production of later stages and, eventually, of the adult.

In recent times, an interpretation in terms of morphostatic development has been suggested for the ciliated peritoneum of the sipunculans (Nielsen 2001). In this case, as one may argue for the larvae of marine invertebrates, the advantages due to the mechanical role of the ciliature (circulation of coelomic fluid in the case of sipunculans; feeding and possibly locomotion in the case of larvae) would be explained as exaptation. A trade-off between cell division and ciliary locomotion is also clear in amphioxus. Beginning with the embryonic stage of neurula and down into early larval stages, all ectodermal cells stop dividing, while forming cilia used in locomotion. Later, an increasing number of epidermal cells will lose cilia and resume mitotic activity, but at that time the larva will already be provided with muscles, and therefore locomotion can easily switch from ciliary to muscular (Holland and Holland 2006). We could also speculate about a possible morphostatic role of cell junctions (Minelli 1975). If so, this way to reduce competition between neighbours within a superficial cell sheet may have been instrumental in transforming the latter into a true epithelium.

If we accept that morphostasis, as such, is something we must try to explain as a developmental event, and that good candidates to explaining it are, *prima facie*, the phenomena of internal competition, many other facts of major evolutionary importance can be added to the original example of Buss. In particular, we can consider different kinds of hard matrix, such as cuticles and mineralized skeletons. I have argued elsewhere (Minelli 2003) that competition between external and internal cell layers may explain the origin of the ecdysozoan cuticle. It would be difficult to explain why nematodes (e.g. Chin-Sang and Chisholm 2000) and arthropods (e.g. Konopová

and Zrzavý 2005) start producing cuticle at quite early embryonic stages, or why embryonic cuticles are shed at embryonic moults. Embryonic cuticles have recently been described for Cambrian embryos referable to the priapulid-like scalidophoran *Markuelia* (Donoghue *et al.* 2006b) (Fig. 7.6). In some *dumpy* mutants of *D. melanogaster*, the defective cuticle causes tracheae and mouthparts to grow out of proportion to the remainder of the body (Wilkin *et al.* 2000). The defects found in *D. melanogaster* embryos lacking chitin because of mutations in the *chitin synthase-1* (*CS-1*) gene demonstrate the role of chitin in maintaining epidermal morphology, as chitin is required to attach the cuticle to the epidermal cells (Moussian *et al.* 2005). Similarly, in nematodes, mutations in cuticular collagens cause gross morphological abnormalities (Kramer *et al.* 1990). On the other hand, retarded deposition of 'true' (post-embryonic) cuticle facilitates the morphogenesis of complex buccal structures in diplogastrid nematodes (Fürst von Lieven 2005).

In a broad phylogenetic perspective, morphostatic phenomena involving epithelia and cuticles have probably played a major role in the

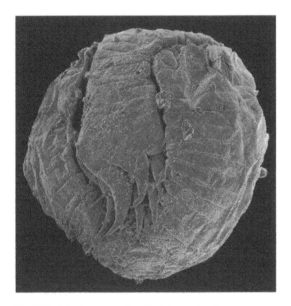

Fig. 7.6. A fossil embryo of the Cambrian scalidophoran *Markuelia*. Courtesy of Phil Donoghue.

evolution of metazoan lineages where the whole body organization has remained largely epithelial, as are nematodes and arthropods. One may even reverse the argument and argue that these animals have remained epithelial because of the difficulty to evolve an alternative to a body organization so strongly (but also so effectively) shaped by early epithelial morphostasis. In this context, Larsen (2003) has aptly remarked that a single gene change in an epithelial sheet can have a dramatic effect, not only on cell differentiation but also on coordinating changes in morphology. I would add that, if so, this might help to explain the wonderful species diversity and disparity of body forms in the nematodes and, especially, the arthropods. Another correlate of an epithelial organization subjected to epithelial and especially cuticular morphostasis is possibly the fact that cell movement is less frequent in arthropod than in vertebrate morphogenesis (Larsen 2003). A final speculation in this context is that a view of ecdysozoan development in non-adultocentric terms of competition among cells and cell groups along the animal's life may also help to explain moulting. If cuticle-dependent morphostasis is an expression of temporarily balanced competition, then a moult, which represents a break of an existing status quo, can be described in terms of reactivated competition, with conditions of agreement to be negotiated anew before the next mitosis.

The scenario suggested for the original role of cuticle in the ecdysozoans may apply to other animals too. Scyphozoan polyps such as *Aurelia* and *Rhizostoma* produce small cell masses (podocysts), some of which do not possess a chitinous cuticle and immediately develop into new polyps, whereas others have a chitinous cover and develop only months later (Franc 1993). Two interesting features support the comparison with the former examples involving either cilia or cuticles. First, the podocyst is the result of the migration of amoebocytes deriving from the transformation of flagellated cells formerly at the boundary between pharynx ectoderm and gastric endoderm (Chapman 1968). Second, removing the envelope stimulates the growth of the podocyst into polyp (Hérouard 1909). A

morphostatic role can tentatively be supposed also for the egg envelope of sipunculans, which is often retained as part of the cuticle of the pelagosphera larva (Jaeckle and Rice 2002). An original morphostatic role for mineralized sketetons and shells has been also suggested (Minelli 2003).

7.3.11 Fusion

In 1881 Wilhelm Roux, the founding father of the German school of developmental mechanics (*Entwicklungsmechanik*) published a little book with the title *Der Kampf der Teile im Organismus* (The Fight of the Parts in the Organism). I have sympathy with many aspects of his vision of the organism as a battlefield where smaller and larger parts are engaged in standing competition, as I repeatedly argue in this book. On two points at least, however, I think we must take distance from Roux's vision. The first point is that the material units involved in competition, such as molecules, cells, and organs, are not a given, but are historically determined and their boundaries are often far from clear, as I observed with respect to the cell and will argue again, in the next chapter, with respect to organs. The other point is that ontogeny and phylogeny are not always characterized by increasing degrees of splitting, segmentation, and individualization of parts, but also by the opposite of these. We must be careful, however, in distinguishing between ontogenetic and phylogenetic change. This has been clearly explained by Krell (1992) with respect to beetle antennae. The phylogenetic pattern is one thing, the nature of change in a developmental mechanism is another. In a group like the beetles, where the majority of species have retained the plesiomorphic 11 antennal articles, we can describe as fusion the phylogenetic transition to a smaller number of articles, as occurs in many genera in many families. In terms of development, however, there is no evidence that in any of these beetles has there been a developmental fusion of previously identifiable antennal articles. Probably, all these cases correspond to a lack of segmentation of units originally fated to divide (Minelli 2005a).

Different from a phylogenetic pattern of fusion, and probably less common, is fusion as a developmental process. Of course, the degree to which we may be able to identify events of fusion (or to regard this level of description as heuristically relevant) will depend on the degree to which we are able to identify growing and developing units as temporarily autonomous modules. In the vertebrate skeleton, several bones develop from multiple ossification centres whose products eventually merge together. This was already known to Geoffroy Saint-Hilaire (1807), who indicated ossification centres, rather than individual bones, as the most sensible units of comparison in what we call, in the modern meaning of the term, the identification of homologues. Scholtz (1995) illustrated a parallel event in the embryonic development of arthropods. Of all adult malacostracan crustaceans, only leptostracans have a pleon of seven segments; in all other lineages (the eumalacostracans) a seventh segment is formed transiently during embryogenesis, but eventually fuses with the preceding one. Much more impressive, however, are the case of the vertebrate heart and the dorsal closure of insect embryos. In zebrafish, the two bilaterally symmetric clusters of mesodermal cells are subdivided into future atrial and ventricular cells before the two halves eventually fuse (Thisse and Zon 2002). In insects, the dorsal aspect of an adult (and, generally, also that of a larva) would never suggest that in an earlier developmental stage the animal was a dorsally open embryo whose lateral margins converged dorsally until they eventually fused along the midline. It is difficult to ascertain how risky is this process for the developing insect, but rare specimens with left–right dorsal mismatch are sometimes found, even in nature.

7.4 The life cycle

Difficulties with the categories into which we daily try to shoehorn our growing knowledge of the biological world are often caused by the fact that some scientific concepts derive from pre-scientific notions that only adequately apply to organisms with properties comparable to those of the more familiar kinds of animals, such as mammals and other vertebrates. Less obvious is how the same problem can emerge also with respect to technical concepts directly originated in science, rather than derived from everyday language. Take, for example, the following sentence: 'Metagenesis, other than metamorphosis, as] a metamorphosis can only imply changes which occur in the same individual, but when from it other individuals originate, something more than a metamorphosis is concerned. [...] There is no transition from a metamorphosis to an alternation of generations' (Steenstrup 1845, p. 6). Japetus Steenstrup was one of the first zoologists who realized that a polyp and a medusa may belong to one and the same life cycle, rather than represent different species (Winsor 1976). And we owe him the concept of metagenesis, or alternation of generations, which has been included since then in the description (and thus, the interpretation) of the life cycle of those medusozoans that have retained both polyp and medusa. Problems with this model, or at least with its general applicability to all medusozoan polyp–medusa cycles, emerge, however, with the cubozoans, where the polyp does not give rise to the medusa by strobilation or budding; that is, by a mechanism that somehow preserves the identity of the polyp as parent individual, as distinct from the medusa to which it gives rise. In the cubozoans, the polyp metamorphoses into a medusa. In this case, why not to call the polyp a larva and admit that the life cycle includes one generation only? This point about cubozoans is not very controversial, but it leads to a flood of new questions. First, and most important, if we accept that the cubozoan polyp is a larva because the whole polyp is turned into a medusa, whereas a scyphozon polyp represents a generation distinct from the medusa as the latter is simply a part of the polyp that gives rise to it, where can we set a reasonable divide between metagenesis and metamorphosis? Is this simply matter of numbers? That is, in the case of cubozoans, the total number of individuals following the event we call a metamorphosis is one, as it was before, while in scyphozoans or hydrozoans the detachment of one or more medusae from the parent polyp involves an increase in the number of individuals; that is, reproduction. Hence, the medusa represents a new generation. What, however, if the polyp cannot produce more than one medusa? Would this case be different from the 'catastrophic' metamorphoses of many larvae in other animal groups, where most of the larval body is discarded, the juvenile deriving from an originally small cluster of set-aside cells? In the cnidarian example, it would remain to be checked how long the uniparous polyp would survive. Alternatively, we could say that an adult sea urchin is generated by the pluteus by asexual reproduction, as we currently claim for the medusae issued from cnidarian polyps. At least, this could be said in the case of the starfish *Luidia sarsi*, where the larva may continue swimming for 3 months after the juvenile it originated has dropped off (Tattersall and Sheppard 1934, Williamson 2006).

Clearly, we cannot blame Steenstrup for his original suggestion that no transition exists between metamorphosis and metagenesis. This view was in agreement with the knowledge available in year 1845. But today Steenstrup's distinction has lost its universality. The cnidarians are one of the groups that offer the best evidence of the many ways reproduction and development can be interlaced, to such an extent that the divide between the parent/offspring relationship and metamorphosis with the discarding of larval parts may be reduced to a matter of subjective, lexical choice. Perhaps, to go back to a real biological question we must turn our attention towards developmental inertia, and the diversity of constraints channelling its expression in the different metazoan species.

Indeed, if we need good examples of the principle that a new focus of development/reproduction can start virtually anywhere, provided that stem-like cells are available, no group provides more varied evidence than the Hydrozoa (Bouillon *et al.* 2006). First let us consider the production of buds. This is mainly the polyp's business, but there are also budding medusae that produce other medusae, or gonophores, or even polypoid structures, as in *Bougainvillia platygaster*, *Proboscidactyla ornata*, *Teissiera medusifera*, and *Zanclea medusopolypata*. As the medusa is more complex than the polyp, in the body of a budding medusa there is a large choice of sites where buds may form, including the manubrium, the radial canals, and the exumbrellar and subumbrellar rims.

Another dimension of diversity in the asexual reproduction of hydrozoans is the fate of the buds. These do not necessarily take, eventually, the form of a polyp or a medusa, but are sometimes similar to the planula larvae ordinarily produced by embryogenesis. Additionally, asexual reproduction in hydrozoans can be associated to progenesis, as several polyps can bud off from a single planula of, for example, *Oceania armata* and *Mitrocoma annae* (Bouillon *et al.* 2006).

Finally, there are alternatives to budding. Asexual reproduction does not necessarily occur through budding, but sometimes it takes the form of fragmentation or, most interestingly, of wholesale fission of either a polyp, as in *Protohydra* and *Hydra*, or a medusa, as in *Cladonema* and *Clytia*. But the most illuminating case is the life cycle of *Eirene hexanemalis*, where the planktonic polyp deriving from the metamorphosis of the planula buds off a single medusa which completely resorbs what remains of the polyp. In this case, the difference between the polyp and a sea urchin pluteus is no more one between an adult and a larva, but one between larvae with an earlier or later specification of the cell mass that will eventually give rise to the adult. Thus, there is one generation only in *Eirene hexanemalis*, exactly as in a sea urchin.

If in the case of this hydrozoan a life cycle conventionally described as metagenetic turns out not to differ from intragenerational development with metamorphosis, the following is the opposite case of larval metamorphosis deviating from the usual path in such a way as to suggest an intervening instance of asexual reproduction. To use the traditional categories, I shall discuss here a bivalve mollusc with alternation of generations. The lasidium larva of the freshwater mussel *Mutela bourguignati*, as soon as it is attached to a fish host, metamorphoses into a parasitic haustorial larva which in turn produces a bud that becomes the juvenile mussel (Fryer 1961).

How shall we classify this event: metamorphosis or asexual reproduction? In cases such as those of *E. hexanemalis* and *M. bourguignati*, two criteria can be used to decide between the two interpretations: the degree to which the body of an earlier stage is preserved in later stages and, perhaps more appealing, the number of individuals eventually issued from the initial stage. The issue of the continuity of the individual is at stake and we will hardly accept a one-to-many relation in development, whereas this is a defining property of reproduction. However, this is matter of semantics more than one of biology.

7.4.1 Plasticity

Within any major lineage, deviations from the ancestral life cycle do occur frequently. In several cases of change from indirect to direct development, we have evidence that this conspicuous transition has occurred recently, and we can guess that it was possibly completed within a very short time span. Particularly informative are the studies performed by Rudy Raff and associates (e.g. Kissinger and Raff 1998, Raff 1999, Nielsen *et al.* 2000) on a pair of closely related sea urchin species, *Heliocidaris erythrogramma* and *Heliocidaris tuberculata*. The former has a typical echinoderm development, with small eggs containing a limited amount of yolk from which a plankton-feeding larva is formed, whereas the latter produces larger, yolk-rich eggs which develop into larger larvae devoid of a mouth. Only at a later stage do the developmental trajectories of the two species converge and eventually result in adults of very similar habitus. The obvious differences between the larvae of the two species are largely explained as the effect of heterochronic change. This is clearly a good example of major structural and functional (and thus, arguably, adaptive) differences resulting from minor changes at the genetic level. This is further demonstrated by the fact that hybrids can be obtained between the two *Heliocidaris* species and the hybrid larva has traits intermediate between the typical pluteus of *H. erythrogramma* and the plump ovoid larva of *H. tuberculata*.

This example is indeed very informative, because of the detail to which it has been investigated. But there are groups of metazoans within which the diversity of life cycles is incomparably larger. The single best example is provided by digeneans. The study of reproduction and development of these parasitic flatworms is technically much less accessible than that of free-living animals, and especially of marine invertebrates with external fertilization, but there are excellent reasons to address the evolutionary developmental biology of parasitic flatworms, as the following summary, mostly inspired by the monograph of Galaktionov and Dobrovolskij (2003), will demonstrate. To the benefit of those readers who are less familiar with these worms, let us recall here the main features of a 'typical' digenean life cycle. A generally hermaphroditic adult (the so-called marita stage), living as a parasite in a vertebrate host, produces eggs from which an aquatic, free-living larva develops. This minuscule larva, the miracidium, moves thanks to the cilia of a small number of flat superficial cells which will be discarded at the moment the miracidium infects a molluscan host. In the mollusc's body, the miracidium will grow into a mother sporocyst, which will reproduce by means of a nature that is still contentious. Many researchers interpret it as parthenogenesis. The offspring is usually represented by one or more generations of rediae, but a second generation of sporocysts is often produced, in which case it is the daughter sporocyst that will produce the rediae. The rediae, in turn, by a process which is also (dubitatively) parthenogenetic, will produce a new kind of free-living larvae, the cercariae. These, often following an intercalary immobile stage of metacercaria, will finally metamorphose into a new vertebrate-parasitic marita. But this is simply a textbook generalization, with respect to which the life cycle of digeneans demonstrates an extraordinary degree of plasticity.

Similar to *Heliocidaris*, digeneans offer cases of major differences between the life cycles of species belonging to the same genus. For example, the life cycles of *Parvatrema borealis* and *Parvatrema borinquenae* follow the conventional sequence, but a strongly modified cycle is found instead in *Parvatrema homoeotecnum*, where conventional rediae and sporocysts are not recognizable, but a primary germinal sac (enclosed in an envelope regarded by James (1964) as equivalent to a miracidium/sporocyst) gives rise to daughter germinal sacs from which cercariae are issued that will turn into metacercariae and, finally, into maritae (Erasmus 1972).

But this is just a first level of deviation from the typical cycle. This is often subject to simplification, by loss of one or the other stage. In some Microphallidae and some Plagiorchiidae there is no cercaria, whereas in Azygiidae and Bivesiculidae the stage lost with respect to the standard life cycle is that of the metacercaria. The conventional, actively swimming miracidium is often replaced by passively infecting miracidia which remain protected inside the egg shell. This modification, which seems to have occurred several times and quite independently in different taxa, has allowed digeneans such as the fluke *Fasciola hepatica* to exploit terrestrial hosts (both vertebrate and mollusc) rather than aquatic ones. In other cases, a whole generation has been lost, as in some Cyathocotylidae, where sporocysts directly produce miracidia, which amounts to saying that in these worms the marita has disappeared. An opposite trend is seen in some Strigeida (*Strigea, Alaria*), with an increased life-cycle complexity, which runs through three different hosts, rather than two. In the intermediate host the cercaria turns into an additional stage of mesocercaria, to finally metamorphose to adult when the intermediate host is eventually eaten by the definitive host. In many digenean cycles, the distinction between larva and adult becomes blurred, because of phenomena we could classify as progenesis; that is, an animal which is still larval in its overall organization becomes sexually mature (and will never turn into the expected adult form). This happens, for example, in the miracidia of the Philophthalmidae, the Cyclocoelidae and some members of the Echinostomatidae and Allocreadiidae. These miracidia are so small that only one individual of the next generation is formed within each of them. Up to two embryos are found in the miracidia of Fasciolidae and Transversotrematidae. Progenesis is not limited to miracidia, as the other larvae (cercariae and metacercariae) may also become sexually mature. But in some cases the distinction between a progenetic metacercaria and a 'true' marita becomes a pure lexical question. Finally, the distinction between sporocyst and redia may disappear. In the Aporocotylidae, the mother redia simultaneously produces rediae and cercariae (Køie 1982), whereas in the Gymnophallidae, Schistosomatidae, Cyathocotylidae, Strigeidae, and Diplostomidae the daughter sporocysts are capable of producing both cercaria larvae and another generation of sporocysts at the same time (Galaktionov and Dobrovolskij 2003).

With all this diversity, the whole picture of the extraordinary plasticity of the digenean life cycle is not yet complete. The most extraordinary feature is probably seen in the cercariae and metacercariae of some gymnophalline species. These 'larvae' do not give rise to adult maritae, as expected, but revert to the function of germinal sacs similar to those from which they have been generated. This poorly known phenomenon deserves to be compared with the well-known example of reversal of developmental sequence known from the hydrozoan *Turritopsis nutricula*, where a medusa, under certain conditions, is able to revert to a polyp (Bavestrello *et al.* 1992, Piraino *et al.* 1996).

A final point, one well in agreement with the suggestion that developmental dynamics are a default condition in multicellular organisms, something that goes on everywhere unless repressed (see section 7.3.2), is the occurrence of multiple focal points of proliferation, as observed in the mother sporocysts of some Plagiorchiida. Their structure is described by Galaktionov and Dobrovolskij (2003) as modular. For example, the mother sporocyst of *Lechriorchis primus* grows into an arborescent structure with several large branches, each of them with its own centre of proliferation of germinal cells. Diffuse germinal masses are also found in the multi-branched sporocysts of several Bucephalidae.

A dramatic aspect of developmental plasticity is the occurrence of alternative types of larvae or juveniles in the same species, a phenomenon known as poecilogony (Giard 1905). Examples are known from insects (Greene 1999), nematodes, 'polychaetes', opisthobranch gastropods, and amphibians (Begun and Collins 1992). Within 'polychaetes', poecilogony is common in the Spionidae, with planktotrophic and lecithotrophic larvae occurring in a single species (Larink and Westheide 2006). In the cephalaspid gastropod *Haminaea callidegenita*, both lecithotrophic veligers and benthic juveniles are produced from the same egg mass (Gibson and Chia 1989, Chia *et al.* 1996), while in the sacoglossan opisthobranch *Alderia modesta* the offspring of a single clutch can be either uniformly planktotrophic or lecitotrophic, or mixed (Krug 1998). Two alternative developmental pathways are open to the parasitic nematode *Strongyloides ratti*: a direct (homogonic) pathway, where the developing larvae are infective to new hosts, and an indirect (heterogonic) pathway, where the larvae develop into free-living adults, a new generation of infective larvae showing up only after one or more intervening free-living generations (Fenton *et al.* 2004).

7.4.2 Maximum life-cycle complexity

With regard to the impressive complexity of the life cycle of many digeneans and of other parasitic metazoans, we may raise the question of whether there is an upper limit to the complexity a life cycle can achieve. The question cannot be expressed too formally, as the morphological and functional divide between subsequent stages is often arbitrary, as in some of the digenean examples cited above. More generally, the occurrence of phenomena such as progenesis, where mature gametes are released by individuals whose overall organization corresponds to the larval, rather than the adult, organization of their close relatives, demonstrates how subjective and, what more matters here, how easily overcome are the dividing lines between developmental stages which in other instances are separated by major morphological and functional differences. This is suggestive of developmental and evolutionary modularity, in the sense that individual body features and

functions are not necessarily associated in definitely integrated packages corresponding to the different stages in a life cycle, as we will discuss in the next two sections.

Thus, back to the question of life-cycle complexity, is there an upper limit to it? This is something we can expect to happen, indeed, as the progression from a given developmental stage to one with very different organization is likely to be accompanied by a dramatic shift in the profile of the genes that are expressed. This has been shown by Arbeitman *et al.* (2002) in their longitudinal analysis of the expression of 4028 genes during the embryonic and postembryonic development of *Drosophila*. For some 80% of these genes, the lowest level of expression during the whole span of the fly's development is at least four times lower than the highest level of expression of the same gene. The embryonic segment of the life cycle was the most eventful, in terms of transcription, as three-quarters of the studied genes were expressed then, and for two-thirds of these (half of the gross total) the level of expression changed significantly during embryonic life. But the beginning of post-embryonic life did not correspond to the beginning of a transcriptionally stable period, as the level of expression of a total of 445 genes changed during larval life, 646 during the pupal stage, and more than 100 during the first 5 days of adult life. Thus, a complex life cycle is based on a subtly orchestrated pattern of gene expression, in which dramatic switches occur. It would be profitable to enquire whether the number of different regimes of genomic expression corresponding to neatly differentiated developmental stages such as an insect larva, pupa, and adult is constrained to very low numbers because of the intrinsic networking properties among genes and their products, as suggested by Kauffman (1993): even accounting for the multiplication of different gene products obtained by processes like splicing, the size of an average genome would not be able to support more than four or five distinct, stable regimes. This covers the highest number of distinct phases we recognize in a life cycle. Interestingly, in the case of the digeneans, where the standard life cycle is complex, and grossly matching the series of environments the worm explores (aquatic vertebrate, water, aquatic mollusc, and water again), very limited increases in complexity are seen in those lineages where additional (paratenic) hosts are attacked during the worm's life cycle. Although up to two subsequent paratenic hosts can be added to the two basic hosts, the complexity of the life cycle increases, at most, by the addition of a mesocercaria stage, intercalated between the free-living cercaria and the adult marita, as mentioned above.

7.4.3 The life cycle as a unit of evolution

With some exceptions, animal taxonomy is virtually a science of adults. How many eggs, larvae, or pupae have ever featured in the original description of a beetle? How many families of mammals or reptiles have ever been diagnosed using characters of their embryos? There are obvious reasons for this strict focus on adults, but it would be unwise to immediately turn the page, having said that in most animals the adult morphology is much more informative than the morphology of earlier developmental stages. We must be aware that difficulties with the identification of non-adult stages do not justify the often unconscious but very strongly entrenched habit of regarding the adult as *the* animal, without further qualifications. We would never expect an atlas of bird species to contain only pictures of eggs or chicks; these may be a welcome extra, to complement an adequate illustration of adults. The same applies to an atlas of Lepidoptera, where we expect to find pictures of butterflies and moths while we will take any illustrated egg, caterpillar, or chrysalis as an added bonus.

To say that the adult is more important than the embryo, the larva, or the juvenile is either an extreme form of legalistic (rather than scientific) distinction between segments of a life cycle where an animal is, or is not yet, a 'full' representative of its species (something to be revisited, perhaps, in the light of current debates in bioethics), or a biased appreciation of the contribution to reproduction respectively provided by embryos, larvae, juveniles, and adults. The

seemingly obvious unique condition of the adult is actually challenged by several facts. First, the association between sexual maturity (or, at least, the ability to deliver mature gametes) and the animal's definitive (adult) shape is very common (and for good evolutionary reasons, selection being mostly ineffective against variation in a post-reproductive phase), but it is far from universal, as demonstrated by paedogenesis (see section 7.4.4). Second, many animals reproduce asexually from stages other than the (actual or conventional) adult: from juveniles, larvae, and even embryos (in the case of polyembryony). Third, and most obvious, a successfully reproducing adult depends on a successful series of earlier stages. Admittedly, in terms of economics one could object that an adult is more precious individually than a juvenile, because of the longer investment of resources from which it has benefited and also because of a simple demographic argument that the number of adults is smaller, often grossly, than the number of juveniles, larvae, or embryos from whose cohorts it has survived. Additionally, one may argue that adults are generally subject to selective regimes that are much more complex, and more species-specific (at least in the context of sexual selection) than earlier stages, but this should not obscure the fact that all developmental stages have their ecology and their adaptations, and face the challenges of peculiar selective regimes. It would be hard, however, to add further arguments in support to the current adultocentrism.

The next issue is whether as soon as we remove the adult from its fashionable condition of privilege, we are thus forced to embrace the idea that the fundamental unit of evolution is represented by the life cycle as an integrated whole. This position has been defended, for instance, by Sterelny (2001), who defines the life cycle (p. 335) as 'the set of developmental resources that are packaged together and interact in such a way that the cycle is reconstructed'. This is expressed from the perspective of the so-called developmental systems theory (Oyama 2000, Oyama *et al.* 2001), according to which organisms are dynamic systems integrating genetic, epigenetic, and environmental resources in such a way as to negate the possibility of any legitimate partition into internal and external factors. Other authors (e.g. Schmalhausen 1969) have defended opposite views, claiming for example independent evolution of larval and adult stages. To be sure, the major differences in larval type exhibited by closely related species as the two *Heliocidaris* sea urchin mentioned above would support the idea of a substantial evolutionary independence between larva and adult.

At a higher level of comparison, this is shown, for example, by the fact that trochophore larvae of similar structure give rise to very different adults, segmented and unsegmented, and broadly similar dipleurula larvae give rise to bilaterally organized hemichordates as well as to radially organized echinoderms (Peterson *et al.* 1997, 2000a). In sponges, different larval types may derive from eggs with identical cleavage patterns, while—more interestingly— different cleavage patterns may eventually result in identical larval types. This is the case of the haplosclerid *Reniera* and the halisarcid *Halisarca*, the former with chaotic cleavage followed by multipolar delamination, the latter with polyaxial cleavage and multipolar ingression, both of them further developing through a parenchymella larva (see the summary in Leys and Ereskovsky 2006).

The case of cnidarians is also interesting. In a first segment of their evolution, their adult was quite probably a polyp, as is still the case in the Anthozoa. With the origin of the medusozoan clade, the polyp was retained, but sexual maturity was displaced to a newly evolved stage, the medusa, but this is was not the end of the story, because some medusozoan clades have lost the medusa, thus recovering the polyp as the sexually reproducing stage, as in *Hydra* (which belongs to a clade Aplanulata, where also the ciliated planula larva has been lost; Collins *et al.* 2005), while other hydrozoan clades, for example the Trachylina, have completely abandoned the polyp (Martindale *et al.* 2002).

Scholtz (2004b, 2005, 2008) has strongly advocated the idea that development is a sequence of evolutionarily independent stages. As a consequence, later stages in a developmental sequence cannot be inferred from earlier ones. Clearly, the abundant available evidence that virtually any developmental stage can be modified without necessarily affecting the immediately following stage can be cited in support of this view. Scholtz summarized examples, mainly from insects and crustaceans, where novel gene-expression patterns (Abzhanov and Kaufman 1999, Jockusch *et al.* 2000, Davis and Patel 2002) or changed mechanisms of cell pattern formation (Dohle 1989, Scholtz and Dohle 1996) have distinctive effects only on one developmental stage and do not affect those upstream or downstream. To some extent, I think that Scholtz is right. If different developmental stages are subject to different selection regimes, it is no wonder that they will enjoy some degree of evolutionary independence. But I would argue that, as the developmental stages in a given life cycle are consecutive segments of the same thread of life, they will be hardly more independent than are genes on the same chromosome; that is, some 'linkage' effect is probably unavoidable, the only radical proof of independence being, in the end, their autonomous disappearance from the life cycle, as with the medusa of the ancestors of *Hydra*, and the polyp of the trachylines.

Rather than the evolutionary independence of the individual stages in a life cycle, I think we should pay more attention to the degrees of independence shown by individual aspects of body organization, and by individual developmental processes that can alternate among different degrees of coupling with other features of body organization, or other developmental process. This argument probably deserves be formulated in terms of modularity (see Wagner 1996, Schlosser 2002, Schlosser and Wagner 2004, Callebaut and Rasskin-Gutman 2005). Thus, I suggest that a developmental stage can be described as a temporarily stable association of morphological modules, which deserves recognition as a distinct segment of the life cycle only when it is robust enough with respect to evolutionary changes both upstream and downstream in development. But in practice, what 'robust enough' may mean will always remain matter of convenience and agreement. This brings us straight to the problem of the periodization of development.

7.4.4 Periodization

Periodization of events is a taxonomic problem faced by historians. How many major time slices can be recognized in Western history? How many geological eras are documented by the fossil record? How best to articulate our description of an insect's life, or our own? It seems difficult to identify a single set of guidelines that may help in addressing all problems of this kind, the only sensible suggestion apparently being that any two consecutive, non-terminal segments of any of these histories must be separated by a recognizable discontinuity, or marked by a uniquely important event. An increasingly better knowledge of the history to which we want to apply our criteria of periodization will eventually show whether the discontinuities we initially identified, or the singular events we selected as points of temporal divide, will retain their descriptive, if not euristic, value.

In developmental biology, periodization turns into a criterion according to which developmental stages, or events, can be compared. This is an important step towards a formulation of questions about the evolution (or the evolvability) of development. In a sense, this parallels the alignment of nucleotide or amino acid sequences, an obvious early step in a comparative analysis of molecular data. The problem concerns whether developmental (rather than molecular) sequences can actually be aligned. The answer to this question will be yes if the individual stages are strictly independent from an evolutionary point of view, as advocated by Scholtz (2004b, 2005, 2008). If so, any difference between the developmental schedule of two species would be either a point change, or an indel (insertion or deletion). But I think that in many cases any alignment would be arbitrary, because of a many-to-one, or many-to-many relationship between the developmental stages of the two species we are comparing. This is due to the fact that each developmental stage is not a monolithic unit, but an integrated complex of modules which are individually subject to changing patterns of association, disassociation and change, with respect to the other modules with which they are temporarily associated.

Let us start with the divide between embryonic and post-embryonic development. Where should we sensibly set this ideal line along the life of a kangaroo: at the time it is transferred from the uterus to the brood pouch, or later? Similarly, where does embryonic life end, and post-embryonic life begin, in a case where an otherwise embryonic animal has a pharynx and actively sucks its yolk? In planarians such as *Dendrocoelum lacteum*, the yolk is contained in yolk cells that eventually fuse into a yolk syncytium surrounding the egg. Following an initial phase of proliferation, several blastomeres differentiate into an embryonic pharynx and embryonic gut. Long before the remaining parts of the worm take form, the embryonic pharynx performs sucking movements by effect of which the yolk is progressively engulfed (Siewing 1969). An embryonic pharyngeal pump is also present, and active, in pseudoscorpions (Weygoldt 1969).

More generally, within the arthropods an equivalence of the divide between embryonic and post-embryonic condition is problematic because the juvenile, or the larva, leaves the egg shell with variable degrees of morphological differentiation and body segmentation (Minelli *et al.* 2006). Similarly, diplogastrid nematodes hatch at a stage equivalent to the second of the juvenile stages through which most nematodes develop before moulting to adult, because a moult equivalent to what is usually the moult from the first to the second juvenile stage takes place before hatching (Fürst von Lieven 2005).

Konopová and Zrzavý (2005) have argued that all winged insects hatch at equivalent stages because of the identical number (three) of embryonic cuticles they shed before hatching (an exception is provided, however, by the cyclorrhaphous flies). But the embryos of the basal wingless insects (jumping bristletails and silverfish) shed one embryonic cuticle less than the pterygotes, and this would be interpreted as the embryonization, in the pterygote lineage, of a previously post-embryonic instar.

A transition from anamorphic development (where several to many posterior segments of the trunk are formed post-embryonically) to epimorphic development (where all segments are present at the time of hatching) has happened in the evolution of the Chilopoda, but this is not the only reason to regard as non-equivalent the stages at which they begin their post-embryonic life, as the hatchlings of the anamorphic centipedes are active larvae with fully articulated appendages, whereas those of the epimorphic centipedes have an embryoid (or pupoid) aspect and require one more moult before developing well-jointed antennae, mouthparts, and legs.

Lack of a strict equivalence of their developmental conditions at the time of hatching are also manifest in other arthropods groups, and the association between the moulting schedule and the achievement of sexual maturity is also problematic. The condition shared by the vast majority of the winged insects, where the last moult turns the animal into an adult without any further possibility of change in external morphology, is a derived one. In a basal pterygote lineage, the mayflies, the first winged instar is a subimago that usually moults into the actually reproductive imago. In the basal wingless insects (jumping bristletails and silverfishes), as in the non-insect hexapods (proturans, collembolans, and diplurans), adult-to-adult moults are the rule. In the crustacean lineage, multiple adult instars are common in malacostracans, but are not known for copepods; a single case has been reported for branchiopods (Ferrari and Grygier 2003). In the chelicerate lineage, multiple adults are found in the horseshoe crabs and in the clade including the whip spiders and the true spiders, but not in other groups, for example in scorpions. No relationships are found between the occurrence of anamorphosis and the presence of multiple adult instars. For example, all millipedes are anamorphic, but some of them, for example the julids, have adult-to-adult moults, whereas others, like polydesmids and chordeumatids, have none.

A peculiar phenomenon occurring in some species of arthropods with multiple adult instars is periodomorphosis. Originally described for some julid millipedes (Verhoeff 1923; see also Sahli 1990), periodomorphosis is the alternance of sexually mature and intercalary instars, separated by moults, with an accompanying alternance of fully developed and partially regressed sexual appendages. A similar phenomenon has been reported for some crustaceans: the crayfish *Orconectes immunis*, the crab *Macropodia rostrata*, and some tanaidaceans (Pandian 1994).

These are just a few examples from an enormous mass of evidence demonstrating that even seemingly fixed points along the sequence of developmental events, as the hatching of an arthropod and the eventual achievement of its adult condition are not so uniquely equivalent as to represent non-contentious points around which to align and compare the developmental schedules of different species. As expected, even more questionable is the equivalence of individual post-embryonic instars separated by moults, as the nature and amount of morphological and physiological change associated with a moult is different in different arthropods, often also from moult to moult along the development of one animal. Many examples have been discussed by Minelli *et al.* (2006). Only two points will be mentioned here. The first is that equating moults with significant events in an ecdysozoan's schedule is very often unwarranted.

In ecdysozoans such as arthropods, nematodes, and kinorhynchs, whose external shape is largely dictated by a cuticle that mostly allows very limited change of form between successive moults, conventional descriptions of post-embryonic development are dominated by a 'cuticular point of view' (Minelli *et al.* 2006). Very limited attention is generally paid to the developmental changes occurring during intermoult periods, but a few clues suggest that these changes can be very substantial and largely uncoupled from the repeated moulting crises. For example, during the post-embryonic development of the kinorhynch *Pycnophyes kielensis* the development of longitudinal and dorsoventral muscles is not in register with moults (Schmidt-Rhaesa and Rothe 2006) (Table 7.4). Similarly, in the first post-embryonic ('larval') instars of the centipede *Lithobius forficatus*, neuromeres differentiate more or less continuously during intermoult time (Minelli *et al.* 2006). Perhaps more conspicuous, as it affects the epidermis and thus the cuticle, is a report about the pycnogonid *Achelia alaskensis*, whose appendages obtain

Table 7.4 Post-embryonic development of the kinorhynch *Pycnophyes kielensis*. There are six juvenile stages (J1–J6) before the adult. The progress of muscle development is not in register with the sequence of moults. Data after Schmidt-Rhaesa and Rothe (2006).

Developmental stage	Number of external segments	Number of dorsoventral muscles	Number of longitudinal muscles
J1	10	12	9
J2	10	12 (less frequently 13)	9
J3	10	13	9
J4	11	13	9 (less frequently 10)
J5	11	13	10 (less frequently 11)
J6	13	13	11
Adult	13	13	12

additional segments during the intermoult in the early, protonymphon phase of development (Okuda 1940, Gillespie and Bain 2006). A continuity of morphogenesis, irrespective of the apparent punctuation dictated by the moults, has been recorded in the regeneration of the eyestalk of the snapping shrimp *Alpheus heterochelis* (Gross and Knowlton 2002). Even if we focus on growth rather than morphogenesis, moults are often irrelevant as markers of discontinuous phases in the life of ecdysozoans. An impressive amount of intermoult and post-moult growth occurs in nematodes (e.g. Yeates and Boag 2002) and nematomorphs (Schmidt-Rhaesa 2002), not to mention the very conspicuous increase in size of some adult termite queens, or the adult females of several parasitic copepods, in both instances next to the final moult.

The last point to be discussed in this short account on periodization is the relationship between adulthood and sexual maturity. The two conditions are generally coupled, especially in those arthropods where the overall shape of the animal is 'frozen' since the last moult, or in animals where somatic mitoses are confined to embryonic life, but this association is loose and offers scope for evolutionary change. In many insects the freshly moulted imago is ready for reproduction, but in many others it is not. Diversity is sometimes found within one order, as in the case of the Diptera. Female mosquitoes require blood meals to complete the maturation of their eggs, but other dipterans anticipate sex-

ual maturity, to the extent that the parthenogenetic chironomid midge *Paratanytarsus grimmi* releases the eggs when still enclosed within the pupal skin; that is, as a non-feeding pharate adult (Langton *et al.* 1988). In still other dipterans, gametes are released by individuals that are, somatically, larvae.

This brings us straight to the phenomena of paedogenesis. The term is used to cover phenomena of reproduction from unfertilized oocytes produced by juveniles or larvae and must be kept distinct from the asexual reproduction of larvae, as in the budding of the metacestodes of caryophylloid tapeworms, and in several echinoderms (see section 7.3.6). Paedogenesis is uncommon, and its occurrence is phylogenetically scattered. Among the best investigated examples are several gall midges (Cecidomyiidae). Interestingly, the stage at which paedogenesis occurs varies in different groups of cecidomyiids: in *Henria psalliotae* paedogenesis occurs in the pupa (Wyatt 1961), in *Mycophila nikoleii* in the third larval instar (Nikolei 1961), and in *Mycophila speyersi* in the second larval instar (Wyatt 1964). Facultative paedogenesis has been reported in the beetle *Micromalthus debilis* (Pollock and Normark 2002), while the case of two groups of insects with parasitic larval habits and exceedingly complex (hypermetabolous) developmental schedule is more problematic. Some authors, for example Grimaldi and Engel (2005), do not hesitate to speak of paedogenesis in the case of some rhipiphorid beetles and the majority of the Strepsiptera. In both cases, indeed, first-instar larvae develop in the mother's body, which is larviform, but in the Rhipiphoridae the larviform female undergoes a conventional metamorphosis throughout a pupal stage before reproducing (Selander 1991). Thus—in my view—it does not qualify as an instance of paedomorphosis. Really questionable is the case of the Strepsiptera, where there is no female pupal stage (Kinbzelbach 1971), but one exists in the males, which fertilize the eggs, at variance with the 'typical' paedomorphosis of gall midges. A putative case of paedogenesis in the hoverfly *Eristalis* (Ibrahim and Gad 1975) is hard to believe. Recent reports of paedogenesis from Loricifera (Kristensen 2003, Gad 2005a, 2005b) suggest a widespread occurrence of this phenomenon in these minuscule invertebrates and, possibly, a diversity of mechanisms. Paedogenesis is arguably very rare, but its very existence demonstrates how myopic is the fashionable adultocentric perspective on animal development.

7.4.5 The larva

The lack of a satisfactory definition of larva is one of the main messages we can derive from the literature on this subject, including a multi-author, 12-chapter book on the origin and evolution of larval forms (Hall and Wake 1999). Take, for example, the following definition: 'A larva is a postembryonic stage that is free-living and is capable of developing into an adult' (Collier 1997, p. 207). I will not spend many words on the meaning of free-living, which is ambiguous at least, as the term is commonly used to denote a lifestyle opposite to that of parasites, rather than to denote autonomy from the parent organism, as it was probably meant here. This is anyway a minor lexical problem. More contentious is the term post-embryonic, for the reasons discussed in the two previous sections (remember, for example, the case of the kangaroo), but even this point is a minor one with respect to two further questions. First, Collier's definition does not include any hint at morphological or physiological differences between larva and adult, other than the implicitly considered difference

in terms of sexual maturity. This omission may have been inadvertent in the cited definition, but, as we will discuss shortly, determining the nature or the amount of differences with respect to the adult that may justify calling an earlier developmental stage a larva, rather than a juvenile, is often matter of purely arbitrary choice. Finally, and no less important, the above definition of larva is framed in terms of potentiality (to develop into an adult) rather than on actual properties or differences. This is a weak point, as is defining gametes as cells capable of fusing with other (complementary) gametes of the same species, or defining a biological species as the set of populations whose members would be able to engage in reproductive exchange, irrespective of the physical barriers than may currently exist between populations. Additionally, in the case of the larva defined as a developmental stage capable of developing into an adult, a further problem is adultocentrism: the larva, in a sense, is only explained as a scaffold in the construction of the adult.

I will not try to offer an alternative 'universal' definition of larva, because I am convinced that such an exercise would be futile. Larva is one of those concepts which rapidly lose their value whenever we want to apply them, without adequate critical analysis, in contexts quite remote from those where they originally proved to be useful. I do not object to calling both a caterpillar and a sea urchin pluteus a larva, but I do think that it is time to address larvae with a sharp analytical eye. The main target of this effort will be the dissection of this concept into a set of morphological, developmental, and phylogenetic aspects that are traditionally subsumed under this one term. Only following this exercise it will be sensible to revise questions about the origin and evolution of larvae and metamorphoses. The following account must be read as a step in this direction.

Let us start with an analysis of morphological features generally regarded as adequate (or even essential) to qualify an early developmental stage as a larva. A first feature is the main body axis. In many cases there are adequate morphological markers, such as trunk segments, or clear cellular phenomena, such an intensive mitotic activity localized in a generally posterior body part, which flag a contrast between early and late post-embryonic stages. A crustacean nauplius would be a good example of a 'head-only' larva to which more segments will be added later. Conspicuous as it may even be, this contrast is not always regarded as sufficient for granting the 'short-trunk' early stage the status of larva. A choice between 'larva' and 'juvenile' would be very subjective, for example, in the case of the early developmental stages of proturans and kinorhynchs, where a few small posterior segments or zonites, respectively, are not yet developed. In the case of young millipedes with an incomplete number of segments, the neuter term stadium is used, whereas the term larva is applied to the earliest post-embryonic instars of anamorphic centipedes, but this difference in usage is only the effect of a lack of standardization.

More trenchant differences between early and late developmental stages are observed when their main body axes are not the same. This happens at the metamorphosis of the phoronid actinotrocha, of the nemertean larvae (pilidium, Desor's larva, Iwata's larva), of the sea urchin pluteus, and of the ascidian tadpole larva. In the case of the pilidium or the Desor's larva, the anteriormost and dorsoventral axes are shifted by 90° relative to those of the larva, whereas in the case of the Iwata's larva the anteroposterior axis is reversed by 180° (Henry and Martindale 1997). Interestingly, a tilt by 180° with respect to the parent worm is found in the orientation of the bud which gives rise to a new individual in the asexual reproduction of the acoel *Convolutriloba retrogemma* (Hendelberg and Åkesson 1988, 1991).

Common sense suggests that an earlier developmental stage deserves be called a larva if there are substantial morphological differences between it and the following stage, so that the intervening change is so substantial as to deserve the name of metamorphosis. But such as vague indication cannot provide a firm criterion for distinguishing a larva (Fig. 7.7) from a 'generic' juvenile. For example, the use of 'larva' has been long abandoned in the description of the post-embryonic stages of non-holometabolous insects; nevertheless, the morphological differences between the nymphs and adults of mayflies and dragonflies are arguably comparable with those between the larvae and adults of some neuropterans.

Obviously, in the case of insects a divide is provided by the presence, or the lack, of a pupa, or by the largely associated presence, or lack, of imaginal discs. If so, why not to call larvae the earliest, active instars of the Thysanoptera,

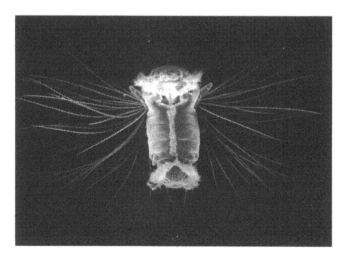

Fig. 7.7. A larva of a spionid 'polychaete'. Courtesy of Claus Nielsen.

where the adult is preceded by one or two immobile instars?

The case of holometabolous insects may invite restriction of the notion of metamorphosis (and the use of the term larva for the developmental stage involved in the process) to the developmental events where non-trivial parts of the earlier stage are discarded or resorbed. This is reasonable in principle but here, again, we find in nature a whole range of conditions, from those of many marine invertebrates where large parts of the larval body are discarded at metamorphosis, to those where the change into the adult is accompanied by vanishingly visible losses. Discarding larval parts can be viewed as one more instance of intra-individual competition. In the sea hare *Aplysia* the cells of the definitive cerebral ganglia are different from the cells that form the larval apical ganglion, which is lost at metamorphosis (Marois and Carew 1997). Similarly, in the limpet *Patella* the larval muscles are reduced or lost at metamorphosis and replaced by new adult muscles (Wanninger *et al.* 1999).

Major cytological markings are sometimes available to distinguish between two sets of cells: those that will eventually die at metamorphosis and those that will survive. In the *Drosophila* larva, the diploid cells forming the histoblasts and the imaginal discs are the survivors, whereas cell death will eventually affect a large mass of polyploid larval cells. Conditions are indeed quite diverse even within the holometabolous insects. Textbook examples such as the metamorphosis of a fruitfly cannot be generalized. As said before, in the basal holometabolans the role of imaginal disc-derived cells in forming the adult epidermis is much more limited than in flies or butterflies (Grimaldi and Engel 2005). But the trend towards increasingly sweeping restructuring of the insect body alongside the transition from preimaginal to imaginal condition is one that goes across the divide between hemimetabolans and holometabolans. A good example is provided by the kind of changes occurring in the skeletal muscles during post-embryonic development (Whitten 1968). In non-holometabolous insects, most muscles already present in the earlier instars also maintain their structure (but not necessarily their function) in the adult. In the holometabolans, most larval muscles are destroyed but sometimes replaced by new imaginal muscles with similar functions, but there are also new imaginal muscles which develop either through cleavage of larval fibres or following a novel accumulation and differentiation of myoblasts. However, the divide between the two main groups of insects is not simple and clear-cut. A behaviour of muscles similar to the one prevalent in holometabolans is also exhibited by the so-called neometabolous insects (the Thysanoptera and two groups of Hemiptera; that is, the scale-insects or Coccoidea and the white flies or Aleyrodoidea). Unfortunately, the term Holometabola has a long tradition as the name of a clade, to which the neometabolous insects clearly do not belong, but in view of their metamorphoses they would actually deserve to be called holometabolous (Heming 2003).

In addition to a substantial amount of morphological change accompanying the metamorphosis to adult, and often strictly correlated to it, are major changes in ecology or feeding habits, as observed in all marine invertebrates with planktonic larva and benthic adult. Other obvious examples are the changes from aquatic to terrestrial habitat in the life of frogs and toads, mayflies, and dragonflies. Why, then, are amphibian tadpoles classified as larvae, while the aquatic instars of these insects are not? It is clear that there are strongly different traditions in the description and interpretation of the post-embryonic stages of different animal groups.

I hope that the present short discussion will stimulate serious efforts in what might be called comparative larvology. Moving from the point that morphological and functional similarities and differences among all these developmental forms have evolved in a mosaic-like fashion, thus defeating any serious attempt to shoehorn all of them into the category of larva, it will be necessary to adopt a factorial approach to establish meaningful and diverse frameworks of comparison.

Sponge and cnidarian development even invites a revisitation of the criteria by which we divide the embryonic from the post-embryonic segment of a life cycle. This basic problem of periodization has been already mentioned with reference to the arthropods, but in that case it seemed sensible to frame the problem in terms of whether the hatchlings of two different arthropods can be regarded as comparable, while it seems convenient to align all their developmental schedules to the point of hatching. Things are different in the case of most marine invertebrates, where nothing happens that is comparable to hatching. Here, we may legitimately ask whether any 'larva' would be better called, under suitable criteria, a free-living embryo, rather than a post-embryonic stage. Van de Vyver (1993) admitted this with respect to the hydrozoan non-feeding planulae, which should be regarded as simple gastrula-like free-living stages.

The concept of metamorphosis is hardly better defined than the concept of larva. Change of form is clearly involved in metamorphosis, but this is not enough. Any form of allometric growth would otherwise qualify for inclusion in this class of events. Hardly better is to define metamorphosis as a set of genetically determined ontogenetic changes in whole-organism phenotype (Diggle 2002). Most of development and, specifically, any morphological (perhaps also physiological and behavioural) change associated with the onset of sexual maturity would thus be referred to as metamorphosis. Certainly, a somehow ill-defined 'large amount' of change is usually associated with current usage of the term, but the application of this criterion is subjective. The problem is also hardly solved by adding that more or less extensive body parts are discarded or resorbed at metamorphosis, while new tissues and organs are quickly deployed. In hydrozoans, the process everybody would describe as the metamorphosis of the larva into the polyp does not require differentiation of new tissue layers or cell types and can even proceed without DNA replication (Seipel and Schmid 2006). Changes, nevertheless, are massive, as fixation to substrate can involve a massive migration of cnidocytes, cnidoblasts, and interstitial cells from endoderm to ectoderm; a mouth will also open (Van de Vyver 1993).

Most discussions about the origin of larvae are founded on simple scenarios that do not derive from a rigorous use of comparative method. On the one hand, the widespread belief that 'early is old' betrays an unconscious survival of Haeckelian recapitulationism, on the other hand, discussions about primary versus secondary larvae are very seldom articulated with respect to precise nodes of the phylogenetic tree. In many instances, it seems to be implicitly accepted that the last common ancestor of all Recent metazoans had indirect development and, thus, that all animal groups where no larva is present must have lost it at some point of their evolutionary history, whereas the larvae of others are regarded as secondary because they must have evolved more recently, after their ancestor had lost the primary larva of their still more remote ancestors. In other cases, central to speculations about larval origin is the putative last common ancestor of the bilaterian animals (Urbilateria; see section 5.2) and this is again credited by many zoologists with having developed through larval stages and metamorphosis. In the recent past, Jägersten's (1972) very readable book on the evolution of the metazoan life cycle contributed much to the acceptance of indirect development as integral to the earliest segments of animal evolutionary history. Jägersten's model of a primitive alternation between a pelagic, planktotrophic larva and a benthic (either sessile or mobile) adult provided an idealized starting point from which is was easy to derive the whole diversity of life cycles of free-living marine invertebrates, sometimes with a prolongation of pelagic life until adulthood, sometimes with a reduction and eventual disappearance of the planktonic larva. It remained less easy to fit the arthropods and nematodes into this picture. On the one hand, the many crustaceans with a pelago-benthic existence, like the majority of crabs or the mantis shrimps, potentially suggested that the generalized occurrence of a pelago-benthic cycle in the other marine invertebrates (with ciliated larvae, like the annelid trochophore and the hemichordate tornaria) could have an ecological (adaptive) explanation, rather than a phylogenetic one. On the other hand, the problem remained of how to imagine development in the ancestors of what we now call the ecdysozoans, during the transition from their ciliated ancestors with a conventional pelago-benthic cycle to their new life as 'armoured' metazoans. At any rate, this scenario seemed an adequate foundation for a distinction of primary, ciliated larvae, and secondary larvae like the crustacean nauplius, but also the tadpole larva of the ascidians.

In the 35 years since the publication of Jägersten's book, our knowledge of both embryology and phylogeny has increased tremendously, but the question of primary versus secondary larvae is still with us. Unfortunately, these terms are used in a different and often loose sense. First, it is not always clear whether by primary larva an author means a characteristic developmental stage inherited from a corresponding stage in a very remote ancestor, or a recapitulative developmental stage conserving fundamental aspects of the adult organization of a Recent animal's putative ancestor. Second, there is seldom reference to a specific node in a phylogenetic tree, such as to justify discussing whether a larva or, perhaps better, which of a larva's traits are plesiomorphic or apomorphic for a given clade. In the absence of further qualification, it may be often assumed that evolution of larval types is discussed in the context of Bilateria, and thus a putative primary larva would be one of the attributes with which Urbilateria is credited. But this would apply to the complex Urbilateria whose attributes were inferred before the Acoela and the Nemertodermatida were hypothesized to occupy the most basal branches of the bilaterian tree. The direct development of these worms would indeed suggest that 'primary' larvae like the annelid trochophore evolved later; that is, within the Nephrozoa. At the same time, this phylogenetic scenario would rule out the hypothesis that trochophore-like larvae are recapitulative of the adult condition of their remote ancestors.

In recent years, the strongest defence of a phylogenetic scenario centred on these putative 'primary larvae' has been provided by Eric Davidson and his school, in a series of stimulating articles (e.g. Davidson 1991, Peterson et al.

1997, Cameron et al. 1998, Peterson and Davidson 2000). These authors suggest an evolutionary scenario that is recapitulationary as far as the overall bodily organization is concerned, but not so with respect to cell differentiation. Ancestral bilaterians would have lacked the genetic circuitry that is responsible for setting up the complex body structure of modern bilaterians; thus the modern bilaterian adult would be an evolutionary novelty added to the original body plan, which has been conserved in the larva. Associated with this prolongation of the developmental schedule would have been a progressive delay in terminal cell differentiation, which is supposed to have been largely conserved along the evolutionary history. Thus, trochophore-like and tornaria-like larvae would be primary both in the sense of being older (and recapitulative) with respect to the corresponding adults, and in the sense of being historically very old. An often implied corollary of this reconstruction is the supposed monophyletic origin of these larvae.

However, there are good arguments against this scenario (e.g. Knoll and Carroll 1999, Wolpert 1999, Jenner 2000, Sly et al. 2003). First, there are problems with phylogeny (Valentine et al. 1999, Jenner 2000, Sly et al. 2003). Within molluscs, for example, veligers are probably homoplastic (Ponder and Lindberg 1997, Waller 1998, Lindberg et al. 2004). In some groups there is frequent and reversible transition from a larval type to another; in particular, planktotrophic larvae that would correspond to the ideal type of the 'primary' larva are often lost and acquired anew (Haszprunar et al. 1995, McEdward and Janis 1997, McHugh and Rouse 1998, Willmer 2003). Independent transition from planktonic to non-planktonic larvae occurred many times within the genus *Conus* (Duda and Palumbi 1999). Many other examples were summarized by Hart (2000), including the repeated loss of tail in several molgulid and styelid tunicates, the parallel changes in larval form, habitat, and dispersal potential occurred, sometimes over a very short time span in *Patiriella* and *Asterina* sea stars, and two-way changes between feeding and non-feeding condition in the larvae of the marine gastropods of the family Lacunidae.

Larvae of marine invertebrates are traditionally divided into planktotrophic and lecithotrophic, but the two classes are not clearly divided. The existence of facultatively feeding larvae is well known (Thorson 1950, Miner *et al.* 2005, Allen and Pernet 2007). In the planktotrophic larvae of several subtropical echinoderms, the onset of the ability to feed does not coincide with the onset of the time they actually need to feed (Herrera *et al.* 1999). In several lineages, the transition from planktotrophy to lecithotrophy is marked by the persistence in the larva of feeding structures, as in the case of the sea slug *Tritonia hombergi*, whose larva retains a ciliary band, the activity of which results in the concentration of suspended phytoplankton, together with the ability to ingest the food collected that way, but these larvae have not retained the ability to digest phytopkankton and must be considered obligately lecithotrophic (Kempf and Todd 1989).

A second objection to the concept of very ancient and recapitulative 'primary' bilaterian larvae is that there seems to be no selective advantage for set-aside cells to have emerged before the evolution of the bilaterian body plan (Wolpert 1999). Third, larvae probably evolved by intercalation into early stages of what was a direct developmental schedule (Wolpert 1999, Valentine and Collins 2000). Thus, Sly *et al.* (2003) conclude that the bilaterian adult body plan is plesiomorphic, and that larval features evolved by interpolation.

The opposite idea, that all larvae are secondary, has been championed by many authors (e.g. Garstang 1922, Ivanov 1937, de Beer 1954, Hadži 1955, Steinböck 1963, Conway Morris 1998, Valentine and Collins 2000, Collins and Valentine 2001, Hadfield *et al.* 2001), but this view has hardly more merit than the opposite one, unless precise reference is made to a specific node in the metazoan tree. Even if it is only intended to mean that no larva is recapitulationary (something that very reasonably applies in a great many instances), this would be unwarranted as an unqualified generalization. It is clear that the phylogenetic significance of larvae has been considerably overestimated (Steinböck 1963). One of the pitfalls in according too much

phylogenetic significance to the 'primary' larvae of marine invertebrates is the likely convergence in the organization of pelagic larvae because of morphogenetic and functional constraints (Reisinger 1924, Steinböck 1963). In this vein, Raff and Byrne (2006) suggest that indirectly developing 'planktotrophic' larvae were likely to have arisen independently in lophotrochozoan and hemichordate+echinoderm clades as part of the Cambrian radiation.

When interpreting the phylogenetic signal of larval characters, caution is always required. Within a clade for which the presence of larvae is plesiotypic, as are the Annelida, some 'recapitulationary' trait may have been retained. This is what we must think of those trochophores of echiurans, where transient evidence of segmental organization is visible, between the prototroch and telotroch, in the form of rings of epidermal glands surrounding the trunk (Pilger 2002), while the adults have lost all externally visible evidence of the segmental organization of their putative ancestors.

The final, and possibly most serious objection to Davidson's scenario is that it implies a polyphyletic origin of the 'zootype'; that is, of the anteroposterior body patterning of the bilaterian body controlled by the patterned expression of the *Hox* genes. Nothing like this is found in the 'primary' larvae, although it would be perhaps improper to say that all of them lack *Hox* gene expression completely. Anyway, in the abalone (*Haliotis*) no *Hox* gene expression was found in pre-trochophore stages or trochophores (Hinman *et al.* 2003), whereas in the 'polychaete' *Chaetopterus*, *Hox* gene expression initiates when the adult body starts to develop and is confined to the growth zone (Peterson *et al.* 2000b). In the pluteus larva of the sea urchin *Strongylocentrotus purpuratus*, *Hox* gene expression is limited to the adult rudiment (and to two genes only, *SpHox7* and *SpHox11/13b*) (Arenas-Mena *et al.* 2000). Apparently, neither sea urchins nor annelids utilize the complete *Hox* complex in the development of the larva per se, although the *Hox* complex is expressed in the set-aside cells from which the adult body plan is formed.

The lack of *Hox* gene expression in these larvae is regarded by Nielsen (2003a) as an argument in favour of their ancestral nature. In particular, the lack of *Hox* gene expression in the head could indicate that this region is the remnant of oldest holopelagic, radial ancestor, while the bilateral body patterned through the 'zootype' expression of the *Hox* genes would represent the bilateral benthic stage 'added on' to the pelagic ancestor in the sense of Peterson and Davidson (2000). My interpretation is exactly the opposite: anteroposterior 'zootypic' patterning is a very old feature, probably evolved in stem-group Nephrozoa, and its absence in the larvae suggests that these have been intercalated into the developmental schedule, corresponding to a phase where *Hox* genes were still largely silent. If so, we must expect that in some instances a larval stage may have been intercalated at a somehow later stage, with respect to the expression of *Hox* genes in a directly developing metazoan. This is possibly the explanation for an expression of posterior group *Hox* genes in the posterior/lower half (the hyposphere) of the trochophore of the 'polychaete' *Nereis virens* (Kulakova *et al.* 2002). Convergent evolution of trochophore-like larvae has been defended by Haszprunar *et al.* (1995), based in particular on the lack of planktotrophic larvae in the 'basal' representatives of some metazoan lineages where these larvae occur in more 'advanced' subclades, and the strong similarity between the ciliated bands of these larvae and the ciliated bands independently evolved in rotifers; this is, outside the spiralians. A list of named post-embryonic juvenile and larval stages is given in Table 7.5.

7.5 Genes and animal development

For more than a century since Mendel's (1866) paper, genetics has essentially worked within the confines of individual populations, species, or, at most, interspecific hybrids. To a very large extent, this was transmission genetics only: the study of the mechanisms by which traits are inherited from one generation to the next. Progress has been very limited, however, in developmental genetics: the study of how hereditary information is eventually expressed in the offspring's phenotype. Things have changed dramatically in the last quarter of a century, due to the increasing availability of experimental methods and tools that have allowed us to address an impressive range of phenomena in developmental genetics. Most relevant to this book's subject, developmental genetics has turned comparative, thus offering unprecedented opportunities to address questions of the utmost relevance in phylogenetics. The single most important message has been that animals with very different general organization such as arthropods and vertebrates, nematodes and annelids, share an impressive number of genes, whose homologues in the different groups are somehow (often very specifically) involved in the animal's body patterning, or in the specification of organs traditionally regarded as analogous, rather than homologous.

To many researchers, this seemed to offer an unprecedented and unexpected opportunity to revive an old research target, or maybe dream of Étienne Geoffroy Saint-Hilaire, the French zoologist and comparative anatomist who at the beginning of the nineteenth century advocated a principle of fundamental anatomical equivalence among all kinds of animals. This was opposed to the concept of Geoffroy's rival, Georges Cuvier, who defended instead the idea of unbridgeable boundaries between the body plans of his Vertebrata, Articulata, Radiata, and Mollusca. This progress in comparative developmental genetics is generally understood as an advance that at last allows comparisons between distantly related phyla. I would argue that this appreciation holds only in the sense that comparative data on gene expression suggest that homologies can be found even between animals whose organization is so different that we have traditionally put them in distinct phyla, following Cuvier's example in dividing the animal kingdom into four *embranchements*. But there is no reason to stop here. Clearly, as soon as we discuss whether there are homologous components between the eyes of a fly, a human, and a squid, or between the brains of a snail and a leech, the very reason for granting special taxonomic

Table 7.5 Named post-embryonic juvenile and larval stages; and also resting stages and asexual propagules (marked with an asterisk).

Taxon	Stage
'Porifera'	*Gemmule
	*Sorite
	*Statoblast
Hexactinellida	Trichimella
Demospongiae	Dispherula[1]
Ceractinomorpha	Parenchymula=parenchymella[2]
	Petrosiid larva[3]
Demospongiae Clavaxinellida	Clavablastula
Demospongiae Tetractinellida	Hoplitomella
Homoscleromorpha	Cinctoblastula
Calcarea Calcaronea	Amphiblastula
Placozoa	None
Ctenophora	Cydippid larva
Lampea pancerina	Planuloid 'larva'[4]
Cnidaria	Planula
Anthozoa	Cerinula[5]
	Edwardsia
	Halcampoides
	Semper's larva[6]
	Zoanthella[7]
	Zoanthina[8]
Hydrozoa	Actinula[9]
	Calyconula[10]
	Conaria[11]
	Rataria[12]
	Siphonula[13]
Scyphozoa	Ephyra
	Metephyra
	Scyphistoma
Acoela	None
Nemertodermatida	None
Gastrotricha	None
Micrognathozoa	None
Syndermata	
'Rotifera'	None
Acanthocephala	Acanthor[14]
	Acanthella
	Cystacanth
Gnathostomulida	None
Catenulida	
Rhynchoscolex	Luther's larva
Rhabditophora	
Cestoda	Coracidium
	Lycophora
	Onchosphere

Table 7.5 (*Contd*)

Taxon	Stage
	Plerocercoid
	Procercoid
Digenea	Adolescaria
	Cercaria (amphistome, distome (cysticercous, echinostome, gasterostome, gymnocephalous, lophocercous, macrocercous, microcercous, monostome, rhopalocercous, trichocercous)
	Cercariaeum (gymnophallus, helicis, helveticum, leucochloridium, mutabile, squamosum)
	Cryptocercaria
	Cystocercaria
	Cystophora
	Diplostomulum[15]
	Macrocerca[16]
	Mesocercaria=agamodistomum
	Metacercaria
	Miracidium
	Neascus[17]
	Prohemistomulum[18]
	Redia
	Schistosomulum
	Tetracotyle[19]
	Xiphidiocercaria (armata, microcotyla, ornata, virgula)[20]
Digenea Aspidobothrea	Cotylocidium
Monogenea	Onchomiracidium
Polyclada[21]	Goette's larva[22]
	Müller's larva[23]
Cycliophora	Chordoid larva
	Male larva
	Pandora
	Prometheus
Bryozoa	Ancestrula[24]
	Coronate larva
	Cyphonautes
	Pseudocyphonautes
	*Statoblast[25]
Entoprocta	Trochophore
Dicyemida	Infusoriform larva
	Vermiform larva=Wagener's larva
Orthonectida	Orthonectid larva
Nemertea	Cephalothricid larva
	Desor's larva
	Hidden larva
	Iwata's larva

Table 7.5 (*Contd*)

Taxon	Stage
	Pilidium (gyrans, magnum, pyramidale, recurvatum)
	Planuliform larva
	Schmidt's larva
Phoronozoa	
Phoronida	Actinotroch
	Demersal lecithotrophic larva
'Brachiopoda'	Glottidia[26]
	Tripartite larva[27]
Mollusca	
Aplacophora	Pericalymma
	Trochophore
Bivalvia	Bivalved veliger
	Glochidium[28]
	Haustorial larva[29]
	Lasidium[30]
	Pediveliger
	Pericalymma
	Trochophore
Cephalopoda	Paralarva[31]
Gastropoda	Echinospire[32]
	Polytrochous larva[33]
	Trochophore
	Veliger
Polyplacophora	Trochophore
Scaphopoda	Stenocalymma
	Trochophore
	Veliger
Annelida	
Hirudinea	Cryptolarva
'Polychaeta'	Aulophora
	Endolarva
	Erpochaeta
	Exolarva
	Metatrochophore
	Mitraria
	Nectochaeta
	Nectosoma
	Trochophore[34]
Myzostomida	Trochophore
Sipuncula	Pelagosphaera
	Trochophore
Priapulida	Loricate larva
Loricifera	Higgins' larva
Kinorhyncha	Larva
Nematoda	Larva[35]
Nematomorpha	Larva
Tardigrada	None
Onychophora	None

Table 7.5 (*Contd*)

Taxon	Stage
Arthropoda[36]	
Merostomata	Prestwichianella larvoid of *Limulus*
	Trilobite larva
Pycnogonida	Protonymphon
Trilobita	Meraspis
	Phaselus
	Protaspis
Acari	Larva
	Proto-, deuto-, and tritonymph
Chilopoda	Larva[37]
Diplopoda	Juvenile[38]
Crustacea	Acanthosoma[39]
	Alima[40]
	Amphion[41]
	Antizoea[42]
	Calyptopis[43]
	Copepodid[44]
	Cyprid[45]
	Cyrtopia[46]
	Decapodid[47]
	Elaphocaris[48]
	Erichthus[49]
	Eryoneicus[50]
	Furcilia[51]
	Glaucothoe[52]
	Grimothea[53]
	Kentrogon[54]
	Manca[55]
	Mastigopus[56]
	Megalopa[57]
	Metanauplius
	Mysis[58]
	Nauplius[59]
	Nisto[60]
	Parva[61]
	Phyllosoma[62]
	Protozoea[63]
	Pseudibacus[64]
	Pseudozoea[65]
	Puerulus[66]
	Trichogon[67]
	Unnamed larva[68]
	Y larva[69]
	Zoea[70]
Insecta	Larva[71]
	Nymph[72]
	Pupa[73]
Chaetognatha	None
Xenoturbellida	None

Table 7.5 (*Contd*)

Taxon	Stage
Echinodermata	
Asteroidea	Barrel-shaped larva[74]
	Bipinnaria[75]
	Brachiolaria[76]
	Mesogen[77]
	Yolky brachiolaria[78]
Crinoidea	Ciliated larva
	Cystidean larva
	Doliolaria
	Pentacrinoid stage[79]
Echinoidea	Echinopluteus
	Lecithotrophic larva
	Vitellaria
Holothuroidea	Auricularia
	Doliolaria
	Pentactula
	Vitellaria
Ophiuroidea	Armless larva
	Benthic lecithotrophic larva
	Brooded larva
	Ophiopluteus
	Vitellaria
Hemichordata	
Enteropneusta	Planktosphaera
	Tornaria :
	Agassiz stage
	Heider stage
	Krohn stage
	Metschnikoff stage
	Müller stage
	Spengel stage
Pterobranchia	Lecitotrophic larva
Chordata	
Cephalochordata	Amphioxides
Urochordata	Tadpole larva[80]
Vertebrata[81]	Astrodermella[82]
	Histricinella[83]
	Leptocephalus[84]
	Luvarella[85]
	Stylophthalmella[86]
	Stylophthalmus[87]
	Tadpole larva[88]
	Tholichthys[89]
	Tilurella[90]
	Vexillifer[91]

[1] Halisarcida only.

[2] Most orders.

[3] Petrosiida.

Table 7.5 (*Contd*)

[4] A juvenile form, parasitic on salps, originally known under the name *Gastrodes parasiticum*. Not a primary larva, but secondarily evolved in relation to the parasitic lifestyle (Werner 1984).

[5] Ceriantharia only. Additionally, in Ceriantharia many 'larval genera' have been named: *Anactinia, Anthoactis, Apiactis, Calpanthula, Cerianthula, Dactylactis, Hensenanthula, Isapiactis, Isarachnactis, Isodactylactis, Isovactis, Ovactis, Ovanthula, Paradactylactis, Paranactinia, Peponactis, Plesiodactylactis, Solasteractis, Sphaeranthula, Synarachnachtis, Syndactylactis.*

[6] Zoanthidea only.

[7] Zoanthidea only.

[8] Zoanthidea only.

[9] Trachylinae only.

[10] Calycophoran Siphonophora only.

[11] Siphonophora only.

[12] Siphonophora only.

[13] Only in physonectan and and cystonectan Siphonophora.

[14] Matsuda (1987) regarded the acanthor as a juvenile rather than a larva.

[15] A kind of metacercaria.

[16] A kind of cercaria.

[17] A kind of metacercaria.

[18] A kind of metacercaria.

[19] A kind of metacercaria.

[20] A kind of cercaria.

[21] Two further names (the older lobophora and the more recent protrochula; Lacalli 1982) have been collectively applied to the two kinds of polyclad larvae.

[22] In the Acotylea only.

[23] The name was also used in the past for the larva of discinacean brachiopods.

[24] Actually, a post-metamorphic stage.

[25] A resting stage of freshwater bryozoans.

[26] In the 'Inarticulata' only

[27] In the 'Articulata' only.

[28] In the freshwater clade Unionacea only.

[29] In *Mutela bourguignati* only (Fryer 1961).

[30] In the freshwater clade Mutelacea only.

[31] According to Young and Harman (1988), this is not a larva in the ordinary sense of the term.

[32] In the Lamellariidae, Eratoidae, Capulidae, and a few other mesogastropod families.

[33] Gymnosomata only.

[34] Including Echiura.

[35] Actually, a juvenile.

Table 7.5 (*Contd*)

[36] In many arthropods belonging to different groups, from spiders to millipedes, from geophilomorph centipedes to grasshoppers, the first post-embryonic stage is embryoid and has poorly articulated, non-functional appendages. It is variously called pupoid, praelarva, praenymph, and others.

[37] The term is applied to the early post-embryonic stages of the Scutigeromorpha and Lithobiomorpha with a number of visibile trunk segments and complete leg pairs lesser than the definitive one. This use is quite disputable. Interestingly, equivalent post-embryonic stages of the Diplopoda are usually called juveniles or, simply, 'stadia', followed by a numeral.

[38] See previous footnote.

[39] Sergestidae (Decapoda) only: the equivalent of a mysis.

[40] Squilloidea (Stomatopoda) only.

[41] Comparable to a zoea.

[42] Lysiosquillidae (Stomatopoda) only.

[43] Dendrobranchiata (Decapoda) only.

[44] Copepoda. A juvenile rather than a larva; classified as protozoea by Williamson (1982).

[45] Several Cirripedia.

[46] Possibly corresponding to a furcilia (Harvey *et al.* 2002).

[47] Some Decapoda; the equivalent of a megalopa.

[48] Sergestidae (Decapoda) only: the equivalent of a protozoea.

[49] Lysiosquilloidea and Gonodactyloidea (both in Stomatopoda) only.

[50] Polychelidae (Decapoda) only.

[51] Dendrobranchiata (Decapoda) only.

[52] A few Decapoda only; a kind of megalopa.

[53] A few Decapoda only; a kind of megalopa.

[54] A stage in the female life cycle.

[55] Tanaidacea and Isopoda *partim*.

[56] Sergestidae (Decapoda) only: the equivalent of a megalopa.

[57] Regarded as a post-larva by Gurney 1942), but this has not been generally accepted.

[58] Dendrobranchiata only.

[59] In most crustacean lineages, but within the Malacostraca limited to the Euphausiacea and the dendrobranchiate Decapoda.

[60] Scyllaridae only: a kind of megalopa.

[61] A few Decapoda only; a kind of megalopa.

[62] Palinuridae and Scyllaridae only.

[63] Dendrobranchiata only.

[64] Scyllaridae only: a kind of megalopa.

[65] Stomatopoda excluding the Lysiosquillidae.

[66] Palinuridae only: the equivalent of a megalopa.

[67] A stage in the male life cycle.

[68] Pentastomida only.

[69] Facetotecta only.

Table 7.5 (*Contd*)

[70] Malacostraca only.

[71] Holometabola only.

[72] Heterometabola only.

[73] Holometabola only.

[74] In the Paxillosida *partim*.

[75] In the Forcipulatida *partim*, Notomyotida, Paxillosida *partim*, and Valvatida *partim*.

[76] In the Forcipulatida *partim* and Valvatida *partim*.

[77] Not a true larva.

[78] In the Forcipulatida *partim*, Spinulosida *partim*, Valvatida *partim*, and Velatida *partim*.

[79] Regarded as post-larval by Breimer (1978) and Holland (1991).

[80] Many ascidians.

[81] Some researchers would call larvae many newborn marsupials.

[82] Luvaridae (Osteichthyes).

[83] Luvaridae (Osteichthyes).

[84] Anguilliformes (Osteichthyes).

[85] Luvaridae (Osteichthyes).

[86] Myctophiidae (Osteichthyes).

[87] Myctophiidae (Osteichthyes).

[88] Many Anura and Caudata; possibly, independently originated in the two amphibian clades, perhaps even multiple times within the urodeles (Bruce 2005).

[89] Chaetodontidae (Osteichthyes).

[90] Anguilliformes (Osteichthyes).

[91] *Fierasfer* (Osteichthyes).

status to the Arthropoda, Chordata, Mollusca, and Annelida rapidly melts away. Phyla (traditional, as with the Mollusca, or newly established, as with the Xenoturbellida) have nothing special other than an arbitrarily high phenotypic gap with respect to other monophyletic groups that we can recognize in the tree of life.

Taxonomic quibbles aside, the important message deriving from developmental genetics is that we can study genes comparatively in a sense that goes far beyond the comparison of sequences between homologous genes in different animals. As we now have tools that allow us to study what a gene does, we may try to find out whether homologous genes do essentially the same things in two different animals. This means investigating when and where a gene is expressed, and also how its expression is turned on and off, and the effects of its expression, either on the expression of other (downstream) genes or, eventually, on the phenotype. If this is our programme, the focus will be on genes expressed during embryonic development, especially at quite early stages, when the main traits of body organization are established. Second, of particular interest will be those genes whose products are transcription factors; that is, proteins that interact in the nucleus with specific DNA sequences in such a way as to control the transcription of other genes. With this target, we may hope to find genes whose expression is very strictly associated with the origin of specific body features such as the segmentation of the trunk or the formation of the heart.

As often happens in science, exceedingly high expectations were raised initially with the discovery of comparable patterns of embryonic expression in animals as different as vertebrates and insects, of genes whose products are involved either in the general anteroposterior patterning of the trunk (as the members of the *Hox* family), or in the production of eyes (the *pax6* homologues) or the heart (the *tinman* homologues). Eventually it turned out that the early, rushing interpretations of these instances of shared expression of homologous genes in distantly related animals were not necessarily true or, at least, that many qualifications are necessary. The history of the changing reconstructions of the putative common ancestor of the Bilateria (Urbilateria; Chapter 5) summarizes many aspects of our changing appreciation of the phylogenetic signal to be read in this kind of comparative gene-expression data.

In the next subsection I will first deal with some general aspects of the genotype–phenotype relationship which are of consequence in extracting phylogenetic information from the results of comparative developmental genetics. I will then introduce some of the best known 'developmental genes', as well as a relatively new entry in the world of genetically encoded molecules, of possible relevance in the control of animal development (the miRNAs). The final section of this chapter will be devoted to a short discussion of some current models about the possible role of individual genes, or whole gene networks, in the origin of major novel aspects of animal organization.

7.5.1 Genotype–phenotype mapping

A one-to-one correspondence between a gene (or gene allele) and a phenotypic trait is a simplistic generalization that we can perhaps take for granted when introducing Mendel's laws in an elementary genetics course, but to know how poorly this represents reality biologists did not have to wait for the recent advances in developmental genetics. Nevertheless, a somewhat naïve attitude, in this respect, can be read in scientific literature of the recent past, when, for example, the involvement of the *Hox* genes in specifying positions along the anteroposterior axis in all bilaterians seemed to support the idea that the highly specific 'developmental role' of these genes might rule out pleiotropic effects such as those so often demonstrated for other genes. As a complement to this, the discovery that specific aspects of body organization are under the control of specific genes (better: under the control of the temporally and spatially highly specific expression of certain genes) seemed to rule out the possibility that those structural aspects could be obtained in more than one way. But all these views have rapidly vanished. Thus, before deriving phylogenetic implications from evidence of shared gene expression in development, we must be aware of this quite variegated nature of genotype–phenotype mapping.

As the elementary units in this relationship are genes and phenotypic characters, I should perhaps mention that neither of these concepts allows a simple definition to which all biologists (and all philosophers of biology) would subscribe. Matters are subtler with respect to a definition of gene (e.g. Beurton *et al.* 2000) and, perhaps counterintuitively, are becoming more and more difficult the more we know about the molecular structure of the genetic material and the processes (of which transcription is just one) in which this is involved. Further discussion would be far beyond our scope, as would be a critical discussion of the concept of character, which is, however, intimately related to a central concept in comparative and evolutionary biology: homology (Wagner 2001). To briefly introduce the complexity of genotype–phenotype mapping we can operationally take the gene as identical with a protein-coding sequence plus the regulatory DNA sequences on whose status transcription depends, while remarking that any procedure we may adopt in dissecting a phenotype into a number of unit characters will unavoidably bear the stamp of subjective choice.

This is the background against which we can appreciate the recurrent demonstration of major phenotypic change correlated with a point change in a single gene. For example, very large effects of a single protein have been demonstrated in the case of the organization of the insect compound eye. Each unit (ommatidium) in an insect eye includes an external dioptric apparatus (a cuticular corneole and the underlying crystalline cone made of cells with parietal nuclei and hyaline cytoplasm), and a retinula formed of elongate cells, each of which includes a series of packaged membranes (a rhabdomere) which act as a light guide leading towards the nerve endings. In brachycerous dipterans such as *Drosophila* or *Musca*, each ommatidium includes seven cells whose rhabdomes are separated from those of the remaining retinular cells and thus function as independent light guides. In other dipterans (mosquitoes) and insects of other orders, the rhabdomeres of all cells in an ommatidium are fused together into a single rhabdome; as a consequence, there is only one visual axis in each ommatidium. This major anatomical and functional difference depends on the expression, in insects like *Drosophila*, of Spacemaker, a protein which interacts with other molecules (which are also present in the insects with fused rhabdomeres) to keep the retinula cells spaced apart (Zelhof *et al.* 2006). Impressive as this example is, it is informative of a large, but anatomically circumscribed and, possibly, mechanistically simple, way by which a gene can affect phenotype. Many other examples demonstrate, instead, an extraordinary diversity of phenotypic effects of one point mutation. This will depend, in part, on

a cascade effect on the expression of other genes, more or less distantly downstream from it, but other aspects are a demonstration of the fact that the more immediate effect of a gene's expression is much more generic than we might have expected from the exquisitely specific nature of a patterning effect in which we had originally discovered its involvement.

Hox genes were initially shown to be involved in the anteroposterior patterning of the main body axis of bilaterians. Subsequently, a subset of these same *Hox* genes was found to pattern also the paired appendages (paired fins or limbs) of vertebrates. But their developmental function, or those of their relatives, does not end here. The cognate family of the *ParaHox* genes is involved, for example, in the patterning of the gut. Recently, up to 30 *Hox* genes are found to be expressed in vertebrates during kidney morphogenesis (di-Poï *et al.* 2007). As for less immediate consequences of *Hox* gene expression, in *Drosophila* hundreds of genes have been found to be more or less directly dependent on *Hox* genes for their own expression (Mastick *et al.* 1995, Botas and Auwers 1996). Among them there is a multiplicity of genes affecting elementary morphogenetic processes such as orientation and rate of cell division, cell–cell adhesion and communication, cell shape and migration, and cell death (Hueber *et al.* 2007). Therefore, the overall implications of *Hox* gene expression translate into a wonderful example of pleiotropy. This remains true even if we restrict our attention to the approximately 50 genes that are the direct targets of transcription factors encoded by *Hox* genes (Pearson *et al.* 2005). Among these genes controlled by *Hox* gene expression some are involved in apoptosis and others encode cell-adhesion effector proteins or are involved in the control of the cell cycle, in cell motility, in intercellular signalling, etc. (reviewed in Davidson 2006). In addition to these examples involving genes that have been the target of intensive and focused study, as are the *Hox* genes, we should perhaps not be surprised by other, numerous but less systematic accounts of gene pleiotropy affecting major aspects of animal development, as in the case of *BMP* gene involvement in both heart and limb development (Yang *et al.* 2006), in addition to gastrulation, mesoderm induction, and axial patterning (Reber-Müller *et al.* 2006). But we must also refrain from rushing to the currently fashionable conclusions about the homology of organs that share the expression of a particular gene. For example, in *D. melanogaster* diverse sensory organs might arise by segment-specific modification of a common developmental programme: a common proneural gene (*atonal*) is involved in the initial process of development of segment-specific organs, including compound eyes, auditory organs, and stretch receptors. This has been suggested to indicate that these organs share an evolutionary origin (Niwa *et al.* 2004), but this would only be justified in the very limited and uninteresting sense that these organs derive from the same, specifically marked set of embryonic cells involved in the production of nervous or sensory structures. This equates to saying that the wing of a bat and the wing of a bird are homologous, because both of them are vertebrate appendages supported by humerus, radius, ulna, etc., but this is only relevant to placing bats and birds in the lineage of tetrapod vertebrates, while it tells us nothing about the origin of their wings.

Eventually, problems with the phylogenetic interpretation of the shared expression of a gene with obvious pleiotropic effects can be reduced to the need to dissect the phenotype controlled by that gene into individual components, some of which may actually carry a phylogenetic signal at the level of analysis in which we are interested. In other terms, pleiotropy is one more argument in favour of the factorial notion of homology repeatedly advocated in this book.

Quite more misleading in terms of phylogenetic inference is the opposite circumstance provided by phenotypic features which can be obtained in different ways, but this must not be dismissed as evolutionarily uninteresting. I would even dare to say that, provided we avoid the pitfalls of suggesting unwarranted homologies, these many-to-one relationships between genotype and phenotype can instead open stimulating perspectives into the evolvability (see section 9.1.3) and the actual evolution of

characters. The most popular kind of many-to-one relationships between genotype and phenotype are those traditionally known as parallelism or convergence, and currently grouped, together with character reversal, under the broad concept of homoplasy. But I will not discuss these here, as the genetic underpinnings of parallelism and convergence are generally unknown; I dealt with them briefly in the opening chapter (section 1.2.2), in a different context. I will focus here instead on the circumstance that within one and the same animal there are often many ways to obtain the same result. This opposes the popular idea of a parsimonious nature which does things in the least expensive way and is thus unlikely to perpetuate useless duplications. In fact, nature is not always 'elegant' (in a sense akin to an 'elegant' mathematical demonstration). This has been noted in the case of the first serial elements forming at the blastoderm stage of *Drosophila* embryonic development, which are seven transversal stripes of expression of the products of the so-called pair-rule genes. Contrary to expectation, each stripe is under slightly different (and, in principle, separable) genetic control (Small and Levine 1991, Pick 1998). Similarly, genes such as *fringe*, which are expressed in each segment of the fly leg, are under different control in the individual segments of the appendage (Rauskolb 2001).

However, a multiplicity of mechanisms with similar outcome is not necessarily inelegant. Rather, it may suggest a selective advantage, in that a result that can be obtained in more than one way will probably be accomplished despite some degree of disturbance. In other words, the system is more robust than it would be in the absence of alternative routes to the same phenotype. From an evolutionary perspective, this circumstance invites a couple of remarks: one backwards, in the direction of the origin of the way a gene may become involved in the control of a phenotypic trait, and one forward, in the direction of the possible evolutionary consequences of a many-to-one relationship between genotype and phenotype.

From the point of view of the origin of the association between gene and phenotype, the many-to-one genotype–phenotype relationships we are considering gives strength to the argument discussed in the previous section, that the association between a new phenotype and a specific pattern of gene expression is more likely to follow the first manifestation of the phenotype, rather than to precede (or to 'prepare') it. The commonly accepted strict determinism of genes on phenotypes, so far as it actually occurs, is likely to be the result of a secondary fixation producing an increasingly narrow choice between alternative pathways to the same eventual phenotype.

As to the other aspect, that is, to the possible evolutionary consequences of a many-to-one mapping of genotypes on to phenotypes, it is necessary to introduce a distinction between two different scenarios. One is redundancy, or the presence of independent pathways that lead to an identical result. The other, probably more common scenario is the degeneracy of the network of relationships through which genes and their products, or more highly integrated units such as different cell types, are interconnected (Greenspan 2001). While redundancy is the effect of a one-to-one substitution of one or more elements in a network, while preserving its entire internal links, degeneracy refers to the functional equivalence of networks differing in some aspects of internal connectivity. By working on either redundant or degenerate systems, evolution can eventually produce either the fixation of one of the alternative pathways originally leading to the production of a given phenotype, or move away from the whole set of original alternatives. But here is where a difference between the two contexts can be expected. Moving away from a redundant system may prove more difficult than moving away from a degenerate system, as the internal connectivity of the latter has already proved to permit non-disruptive change. Changes in the genotype may anyway perhaps introduce acceptable changes in some aspects of the controlled phenotype, rather than disrupt it completely. As suggested elsewhere (Minelli 2003), a redundant system is likely to produce clearly identifiable homologues, while a degenerate system is likely to produce features with

complex and perhaps contradictory homologous components.

Whether the sometimes conspicuous differences by which identical results are obtained during the development of even strictly related species represent, or not, the fixation as alternative pathways of originally redundant or degenerate processes is something that must by determined case-by-case. In any case, this possibility should be kept in mind, as the suggested hypothetical scenario would represent the developmental equivalent of gene duplication: alternative developmental pathways deriving from the fixation of previously alternative forms of a redundant or, especially, degenerate network will be similar to paralogous, rather than orthologous, gene sequences: something obviously of consequence for our efforts to reconstruct phylogeny.

We can also tentatively argue that this sometimes loose degree of determination of phenotypes through specific gene expression has been involved in intercalary evolution. Intercalary evolution (Gehring and Ikeo 1999) is an anti-recapitulatory aspect of evolution, broadly comparable to the evolutionary trend once widely known as caenogenesis (e.g. de Beer 1951, Gould 1977), where novelties involving early developmental stages have no consequence on the later development of a largely conserved adult. Whereas authors in the past obviously considered morphological features (e.g. the evolution of the egg tooth used to open the egg shell, which cannot have derived from any imaginable adult feature, and therefore exemplifies caenogenesis), the focus today is on genes. In this context, a result of significant evolutionary interest has been obtained by a whole-genome comparison between *Drosophila* and the honeybee. Whereas the genes involved in brain function and adult behaviour, which appear to be markedly different between the two species, are largely the same, important differences have been found between the two insects as to the genes involved in early developmental pathways (Honeybee Genome Sequencing Consortium 2006). In nematodes, recent studies in developmental biology have revealed an unsuspected wealth of diversity in cellular mechanisms and genetic controls, which conflicts with the apparent morphological uniformity of most representatives of this group and with the simple and straightforward determinism originally expected from a taxon where eutely, either strict or not, is common. In particular, comparative analysis showed that the fate specification of early blastomeres in different nematodes varies considerably without influencing the resultant body structure (Schierenberg 2001, 2005, Schierenberg and Schulze 2008).

7.5.2 Developmental genes

Under the general term of developmental genes I will deal in the following pages with some classes of genes whose expression is known to have major effects on the establishment of gross body features, or to control important developmental events. This does not necessarily mean that their expression is limited to those early developmental phases during which the main traits of body architecture are actually laid down. For many of these genes, early embryonic expression has been studied in detail, in one or a few model species, so that we know exactly when their expression is first switched on, and which is the spatial and temporal pattern of their expression during the next few hours, or days, but longitudinal studies of gene expression throughout late embryonic and post-embryonic development are still very limited.

In the next few pages I will deal with genes involved in body patterning or in the control of developmental schedules. But a couple of words deserves here also a class of small molecules of non-coding RNA, called the microRNAs (or miRNAs), as these are revealing an unexpected and possibly diverse role in the control of developmental events. Moreover, their diversity in different animal groups is of phylogenetic significance per se. It remains to be seen how much this also directly mirrors differences in development and body organization between major metazoan lineages, but the little we already know is encouraging.

miRNAs control gene expression by negatively interfering with the stability of target mRNAs

(reviewed in Wienholds and Plasterk 2005). This behaviour is supposed to stabilize cell differentiation (Stark *et al.* 2005). The temporal pattern of expression of one of the best investigated among these molecules, *let-7*, is highly conserved in all organisms examined (Niwa and Slack 2007). For example, in *D. melanogaster* it is first expressed at the beginning of metamorphosis (Pasquinelli *et al.* 2000). In 'polychaetes' and molluscs, it is expressed in the adult but not in the larva, and in vertebrates (zebrafish, mouse) this molecule is not yet expressed in embryos at the time when the adult organization is already recognizable (Pasquinelli *et al.* 2000, Schulman *et al.* 2005, Thomson *et al.* 2006). The fact that the 22-nucleotide *let-7* RNA has been consistently found in a diversity of bilaterians but not in sponges, cnidarians, ctenophores, and acoels has been tentatively interpreted by Pasquinelli *et al.* (2003) as suggesting an association of *let-7* with the invention of feeding larvae but also, possibly, with an increase in lifespan, or with uncoupling of sexual maturity from a lengthening of total lifespan. As to the identity of its target mRNAs, it has been reported that several of the genes controlled by *let-7* are conserved (Grosshans *et al.* 2005, Lall *et al.* 2006), but other studies indicate that a conserved miRNA can have different targets and thus regulate different developmental processes in different organisms (Ambros and Chen 2007).

The expression of the miRNAs is often tissue- or organ-specific. This has lead to the probably unwarranted suggestion that these molecules may have had a fundamental role in the origin of organs such as the brain and the heart (Sempere *et al.* 2006). However, if a sensible evaluation of this hypothesis must be deferred to the time when we have a much better knowledge of the mechanisms and effects of miRNA expression, we probably already know enough about their diversity in several groups of metazoans to be entitled to use this information in phylogenetic inference. No miRNA has been found in the sponge *Reniera* and only two molecules of this type are known from the anthozoan *Nematostella vectensis* (Prochnik *et al.* 2007). Interestingly, only six miRNAs were found in the acoels *Childia* sp.

and *Symsagittifera roscoffensis* (Sempere *et al.* 2006, Baguñà *et al.* 2008), much fewer than reported for all other bilaterians investigated thus far. In particular, 'typical' plathelminths such as polyclads have a much larger set of miRNAs, broadly similar to that of other lophotrochozoans. This is a convincing piece of evidence in favour of a phylogenetic position of acoels as a basal branch of bilaterians, independent of rhabditophorans. Sets of miRNAs diagnostic for Epitheliozoa, Bilateria, Nephrozoa, Protostomia, Lophotrochozoa, Ecdysozoa, and Deuterostomia have been tentatively identified (Baguñà *et al.* 2008).

An open question is whether miRNAs existed in the earliest metazoans. Circumstantial evidence is contradictory, as the fact that miRNAs have not been found to date in sponges is counteracted by the presence of miRNA-mediated control of gene expression in the single-cell green alga *Chlamydomonas reinhardtii*. The latter evidence may suggest that these complex RNA-silencing systems evolved in a remote eukaryote stock before the advent of multicellularity (Molnár *et al.* 2007).

Let us now move to the *Hox* genes, the most famous class of 'developmental genes'. A key determinant of the increasing popularity of evolutionary developmental biology has been discoveries concerning the role of the *Hox* genes in patterning the main body axis of the Bilateria. In the early 1990s, the appreciation of the conserved relationship between the spatial patterns of expression of these genes and the orderly specification of positions along the anteroposterior axis was even pushed so far as to suggest a virtual identification of the origin of the Bilateria with the establishment of at least a core set of position-specifying genes, mainly belonging to the *Hox* family. This identification was embodied in the concept of the zootype (Slack *et al.* 1993), possibly an excessively typological notion (Schierwater and Kuhn 1998), but one that stimulated a burst of comparative research on the identity, sequence, and patterns of expression of the *Hox* genes in an increasing diversity of animals. Eventually, the generalized anteroposterior pattern originally implied in the concept of zootype became a feature of most (but not all) models of Urbilateria (see section 5.2), the putative last common ancestor of all living bilaterians. *Drosophila*, *Caenorhabditis*, and the mouse are the models around which many aspects of the nature, expression, and function of the *Hox* genes were originally studied, but many other metazoan species were soon added to the list.

A characteristic of the *Hox* genes is the presence of a highly conserved sequence, the 180-nucleotide homeobox, from which the gene family takes its name, but it has been subsequently discovered that these genes share this feature with other, less closely related genes (the *ParaHox*, *EHGbox*, and *NK-like* genes), which are also involved in key developmental processes. Thus, the search for the origin of the *Hox* gene family has been progressively extended to the *ParaHox* genes first, and to other homeobox genes more recently. This is the context within which we can understand the recent work of Larroux *et al.* (2007) on the demosponge *Amphimedon queenslandica* which settles, in a possibly definitive way, the question of the presence of *Hox* genes in sponges. In the mid-1990s there had been reports of the presence of perhaps one *Hox* gene in poriferans, but this proved eventually not to be correct. No *Hox*, *ParaHox*, or *EHGbox* gene is present in the completely sequenced genome of *Amphimedon*. Thus, it seems safe to say that these gene families originated subsequently to the split between the sponges (say, the Calcarea) and the Epitheliozoa. But the evidence obtained from the study of the *Amphimedon* genome is not just a negative one. This sponge possesses eight homeobox genes, all of them *NK-like*, of which six are clustered within a short chromosomal segment, like the *NK* genes in bilaterians. Larroux *et al.* (2007) regard this conserved organization of the *NK* gene cluster since the Late Precambrian as proof of functional advantages deriving specifically from this arrangement. We will deal soon with a similar relation between gene clustering and gene function in the case of the *Hox* genes.

Hox genes, and also members of the cognate *ParaHox* family, are present in the Cnidaria, but both families seem to have evolved in this lineage along lines other than those of the bilaterian *Hox* and *ParaHox* genes. The *Hox* gene complement of most bilaterians includes three sets called the anterior, the central, and the posterior cluster, the qualification referring to the patterning effect of each gene's expression being respectively visible in the anterior, central, and posterior part of the main body axis. The set of *Hox* genes found in *Hydra* and in the anthozoan *Nematostella* includes genes similar to those of the anterior *Hox* gene cluster of bilaterians, but the remaining cnidarian *Hox* sequences do not seem to correspond to the central or posterior ones of bilaterians, but rather represent specific evolution within the cnidarian lineage. Similar relationships hold between the cnidarian and bilaterian *ParaHox* genes. Based on this evidence, Chourrout *et al.* (2006) suggest that the original set of *ProtoHox* genes (ancestral to both the *Hox* and the *ParaHox* genes of modern metazoans) is likely to have only included two genes similar to those of the anterior cluster.

Within the bilaterians, the Acoela have the smallest complement of *Hox* genes and this is reflected in the limited and unstable anteroposterior patterning of their main body axis (see section 6.1). Nevertheless, in acoels all three main

groups of *Hox* genes are present, with one or two members each in the anterior, central, and posterior clusters. There are also two *ParaHox* genes. Thus, the acoels are in an intermediate position between the simpler set of *Hox*/*ParaHox* genes in cnidarians and the expanded set found in most bilaterians (Baguñà *et al.* 2008) (Fig. 7.8).

Across the bilaterians, the history of the *Hox* genes is one of gene duplications and gene losses. Duplications have produced, in the following order, the split of a putative *ProtoHox* gene into the ancestors of the *Hox* and *ParaHox* genes respectively, origination of the anterior and central-posterior *Hox* clusters, differentiation of the latter into a central and a posterior group, and also the increase in the number of members in each of the three clusters, with the eventual stabilization of the whole set in a range of up to 14 members. This maximum has been found in the amphioxus (Ferrier *et al.* 2000), the coelacanth, and a shark (Powers and Amemiya 2004) and it is unlikely that higher numbers will be found (Minguillón *et al.* 2005). However, in vertebrates many *Hox* genes are present in up to eight paralogous copies, quite likely the results of up to three events of whole-genome duplication, as mentioned in Chapter 2. Massive gene loss

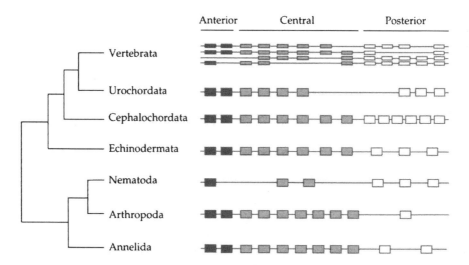

Fig. 7.8. The evolution of the *Hox* gene family within the Bilateria includes the origin of new family members by gene duplication (with up to 14 members, as in the cephalochordates), whole-genome duplications (as at the base of the vertebrate lineage), and also gene loss (as, quite extensively, in the nematodes).

has subsequently led, however, to a substantial reduction of the actual *Hox* gene complement. Thus, most vertebrates have four *Hox* gene clusters. Teleosts have had an additional genome duplication, with successive trimming of part of the potentially eight full sets of *Hox* genes; those eventually left have up to 48. It has been suggested, however, that in the absence of duplications the vertebrate *Hox* clusters are structurally less able to evolve than their invertebrate counterparts (Wagner *et al.* 2003).

The evolution of the *Hox* genes has not been simply one of duplication and loss, but also one of changing developmental effects. Most conspicuous have been the functional change of two arthropod genes which unequivocally belong to the *Hox* family, as shown by their nucleotide sequence, but are no more involved in the specification of positions along the animal's main body axis. One of these genes if *fushi tarazu* (*ftz*), which is involved in segmentation and neurogenesis; the other is *zerknüllt* (*zen*), which is involved in dorsoventral patterning. Comparative evidence suggests that *zen* may have acquired its current role at approximately the base of the insect radiation, whereas the segmentation role of *ftz* is probably older. But in the Diptera a much more recent duplication of *zen* has given rise to *bicoid*, whose functional role has continued to evolve within this insect order. In *Drosophila*, *bicoid* is required for the normal development of head and thorax, but in the phorid *Megaselia abdita* it is also required for the development of four abdominal segments (Stauber *et al.* 2000).

Besides their broadly conserved patterning functions across the Bilateria, one of the earliest unexpected discoveries about the *Hox* genes was the temporal and spatial collinearity of their expression. This means an orderly relationship between (1) the relative positions of individual *Hox* genes along the chromosome, (2) the relative positions along the main body axis of the animal of the structures specified by the gene, and (3) the time the gene is first expressed, compared with the expression of the other *Hox* genes. This means, for example, that the *Hox* genes specifying anterior positions are located at one end of the *Hox* gene cluster on the chromosome and are also expressed earlier than the *Hox* genes controlling more posterior positions. Admittedly, this collinearity was soon found to be far from perfect. Even more common are the examples of animals whose *Hox* gene complement is not tightly clustered in a short region of one chromosome, but is fragmented and more or less extensively interspersed with long non-*Hox* sequences.

In a very insightful analysis of comparative data about the clustering of *Hox* genes in the genomes of different animals, and of the consistency of the temporal and spatial collinearity of their expression in a subset of these, Duboule (2007) has pointed to several wrong messages conveyed by mainstream literature on the subject. First, the alignment of individual genes is often interpreted as suggesting a functional linkage, but this is an unwarranted inference. Second, the different *Hox* genes in a genome and the individual homologous *Hox* genes in different animals are far from being identical, as the usual representation of these genes as identical boxes (but for colour 'coding') would suggest. Finally, the non-*Hox* sequences separating the individual *Hox* genes along a chromosome are mostly overlooked, but this does not mean that these are unimportant. Duboule (2007) distinguishes four levels of clustering:

- organized clusters, as in the mouse and, generally, in vertebrates;
- disorganized clusters, as in sea urchins (Cameron *et al.* 2006), the malaria mosquito *Anopheles gambiae* (Holt *et al.* 2002), and the lesser flour beetle *Tribolium castaneum* (Brown *et al.* 2002);
- split clusters, as in *D. melanogaster*;
- atomized clusters, as in the larvacean tunicate *Oikopleura* (Seo *et al.* 2004), the ascidian tunicate *Ciona* (Ikuta *et al.* 2004), cnidarians (Chourrout *et al.* 2006, Kamm *et al.* 2006), the digenean *Schistosoma mansoni* (Pierce *et al.* 2005), and nematodes (Aboobaker and Blaxter 2003).

Sequence comparisons suggest that the bilaterian ancestor had a clustered *Hox* gene family. Duboule (2007) suggests that a functional integration within the cluster has been probably helped by the advantages found in phenotypes resulting by a more coordinated transcription of these genes. For example, the elimination of non-*Hox* sequences interspersed within the *Hox* cluster may have facilitated the integrated recruitment of several genes to a single function. An example is the *Hox* genes belonging to one of the four clusters originated in the vertebrate lineage by genome duplication (the *Hoxd* cluster), as these genes are globally involved in the morphogenesis of the digits (Kmita *et al.* 2002, Spitz *et al.* 2003). A more general result is that, contrary to an earlier widespread belief, no necessary relationship exists between the clustering of the *Hox* genes on the chromosome and the spatial and temporal collinearity of their expression (Holland 2001).

Point mutations in one *Hox* gene can have impressive consequences for an animal's organization. The best-known examples are probably still the oldest ones, such as the fruitflies with either four wings or with antennae replaced by an extra pair of legs. Thus, the discovery of the shared possession of a set of *Hox* genes among the metazoans and of their conserved role in specifying positions along the main body axis fuelled the hope of explaining major differences in body structure between different metazoan lineages as the direct result of changes in the expression of one or more *Hox* genes, or in the identity of the downstream genes whose transcription is under the control of the *Hox* genes. In the case of arthropod body regions and appendages, this research programme has been very successful (see the excellent review of Hughes and Kaufman 2002b). Changes in *Hox* gene expression are also correlated with the loss of limbs in snakes. In the python, hindlimb buds are formed during embryonic development, but no forelimb buds form at all. This is correlated with the peculiar expression pattern of *hoxc8* and *hoxc6* in the python embryo, which extends anteriorly to the head, while in the other tetrapods it mostly does not go beyond the level of the lateral plate mesoderm where the forelimb buds take form (Cohn and Tickle 1999).

Other regressive changes have been interpreted as being correlated to gene loss. This can be tentatively suggested for the relatively simple organization of the nematodes, compared to most other ecdysozoans, which possibly correlates with a secondarily reduced number of *Hox* genes. Within the arthropods, extreme reduction of the abdomen in the parasitic crustacean *Sacculina carcini* accompanies the loss of the *Hox* gene *abdominal-A* (Blin *et al.* 2003). In animals whose body structures depart more widely from that of a typical bilaterian, to the extent that even the identification of a main body axis is difficult (see section 8.1.1), the involvement of *Hox* gene expression in overall body patterning is less obvious. In the case of the Hawaiian bobtail squid *Euprymna scolopes*, which has nine *Hox* genes, these have been recruited in many ways in the developmental processes which generate cephalopod-specific structures (Lee *et al.* 2003).

In the meantime, we have learned that *Hox* genes are involved in many processes other than patterning the main body axis (see section 8.1.4). The involvement of the posterior members of the *Hoxd* cluster in patterning the limbs of vertebrates (but not in the appendages of arthropods or other animals) is briefly discussed in the next chapter.

Another gene which deserves be mentioned here, because of the phylogenetic interest of its involvement in many developmental systems, is *Brachyury/brachyenteron*. In hydra, *HyBra1*, a gene sharing high levels of sequence identity (75–80%) with vertebrate *Brachyury*, is expressed in a ring of endodermal epithelium cells encircling the mouth (Technau and Bode 1999). Comparisons with the expression of its homologues in other phyla seemed at first unreasonable, as in *Drosophila* the primary site of expression of *brachyenteron* is in the hindgut, while the expression of *Brachyury* in vertebrates involves the whole mesoderm at earlier stages and later becomes restricted to the notochord (McGhee 2000). However, the expression patterns of *Brachyury* are similar in the annelid *Platynereis dumerilii* and the hemichordate *Ptychodera flava*, suggesting homology of the foregut between protostome and deuterostome larvae (Arendt *et al.* 2001). Technau (2001) suggested that in the common ancestor of cnidarians and bilaterians *Brachyury* was involved in specifying, in the proximity of the blastopore, a region with organizer properties, responsible for the elongation of the main body axis. With the onset of a triploblastic condition, a set of blastoporal cells expressing *Brachyury* and other mesoderm-specific genes may have acquired its identity and eventually a spatial restriction as in vertebrates.

Another interesting gene is *engrailed* (*en*). As mentioned before (section 5.4.8), in arthropods *en* is expressed in transverse rows of cells immediately anterior to the future segmental margin. It is also expressed in a series of transverse stripes in the embryos of leeches, polyplacophorans, and onychophorans, as well as in amphioxus and the vertebrates (Jacobs *et al.* 2000). This does not mean, however, that in all these metazoans *en* is actually involved in segmentation. In *Drosophila*, *en* expression is limited to the ectoderm, where it marks compartment boundaries, besides being involved in patterning of the nervous system. In the leech, it is also expressed in the mesoderm and is not involved in the segmentation of the nervous system (Shankland 2003). Its vertebrate homologue is expressed in the somites only after these are formed (Holland and Holland 1998). Homologues of *en* have also been found in non-segmented animals such as the limpet *Patella* (Nederbragt *et al.* 2002). In several groups, including arthropods, molluscs, and echinoderms (Jacobs *et al.* 2000), the ectodermal expression of *en* is associated with skeletal development. For example, in molluscs a crown of *en*-expressing cells surrounds the ectodermal cells involved in the production of shell material (Moshel *et al.* 1998, Jacobs *et al.* 2000, Wanninger and Haszprunar 2001). In 'polychaetes', the chaetal sacks where the chaetae are produced also regularly express *en* (Seaver *et al.* 2001) and in the young brittle stars ectodermal cells expressing *en* encircle the spots where the ossicles are produced (Lowe and Wray 1997). Even in the case of arthropods it can be argued that ectodermal *en* expression is involved in establishing boundaries between skeletal units, rather than the segments themselves (Minelli 2004). Thus, the association of *engrailed* expression with segmentation is likely to be an indirect one, and

is no proof, per se, of common derivation from a segmented ancestor, as has been sometimes suggested in support of the Articulata hypothesis (see section 5.4.8). Even the involvement of this gene in specifying the boundaries of skeletal parts is probably an independent novelty of several metazoan lineages, if the ability to form skeletons was independently acquired by many animal lineages around the transition from the Precambrian to the Cambrian.

7.5.3 Gene duplication, co-option, and paramorphism

In the enormous literature on developmental genetics from the last quarter of a century, alternative alleles of a gene feature only as tools to investigate the function of a gene in a developmental context, but variation per se is virtually ignored. This amounts to saying that developmental genetics has progressed in isolation from population genetics. But a dialogue with evolutionary biology will not go very far as long as developmental biology looks at genes only as determinants of form, without taking variation into account (Gilbert 2003a). The enormous explanatory value of a combined approach where developmental genetics and population genetics are integrated in the study of the evolution of animal form has been demonstrated by the study of Abzhanov *et al.* (2004) of the relationship between the expression of the growth factor gene *BMP-4* in the mesenchyme of the upper beak and the eventual shape of the beak in Darwin's finches. A subsequent study by the same group demonstrated the role of expression levels of calmodulin, a molecule involved in mediating calcium signalling, in determining the difference between the long and pointed beaks of cactus finches and the more robust beak types of other species of Darwin's finches (Abzhanov *et al.* 2006). For the time being, however, our appreciation of the relationship between gene expression and phenotype is nearly universally based on much less specific evidence. Nevertheless, some general concepts are emerging about the ways genes may be involved in major evolutionary change.

A fashionable concept is gene co-option. For example, a co-option of the spatial patterning mechanism provided by the *Hox* genes would be at the origin of the bilaterian architecture (Lowe and Wray 1997). Co-option of additional networks complementing the ancestral set of genes already involved in producing a primitive pump has been suggested by Olson (2006) as responsible for the evolution of the vertebrate heart. Similarly, co-option is commonly advocated in the evolution of arthropod and vertebrate appendages (Tabin *et al.* 1999). In the latter case, as the genes involved in co-option are a subset of the trunk-patterning *Hox* genes, which are responsible for the body's anteroposterior polarity, the same polarity was translated from the main body axis to the axis of each limb (Tarchini *et al.* 2006). According to Wagner *et al.* (2003), co-option also explains the involvement of *Hox* genes in the cranial neural crest, the fins, and other organs of vertebrates. Pires-daSilva and Sommer (2003) do not hesitate to state that all developmental processes that are involved in the generation of new structures require co-option events.

In my opinion, the concept of gene co-option should be restricted to those instances where genes previously involved in a given context become subsequently involved in a different context, or body part, where until then they were not expressed, or were only expressed in a diffuse, aspecific way, but that body part, or context, already existed anyway. For example, the eyespot pattern on the wings of many butterfly species is specified by the localized expression of the gene *Distal-less* (Carroll *et al.* 1994): each wing eyespot is centred on a group of *Distal-less*-expressing cells. On the other hand, the same gene is known to mark the sites where the most diverse metazoan appendages will form, such as insect legs and vertebrate fins and limbs, but also sea urchin podia, 'polychaete' parapodia, and onychophoran lobopodia (Panganiban *et al.* 1997). This is clearly a much older role of *Distal-less* in animal development. Thus, it may be sensible (although, literally, a bit finalistic) to say that in butterflies this gene has been secondarily co-opted to mark the position of new 'virtual axes' represented by pigmentary circles rather than by appendages projecting outwards. But the presence of wings, independent of their patterning, is not dependent on this novel, co-opted expression of *Distal-less*. This point is probably non-contentious, but things are not so obvious in other contexts. In particular, what do we actually mean when we say that a group of *Hox* genes has been co-opted from its original role in patterning the trunk to their new role in patterning vertebrate limbs? Similar to the meaning of the term in the butterfly example, here co-option would mean that the paired appendages of the earliest gnathostomes (or of stem-group gnathostomes) developed without any involvement of *Hox* genes, and only at a later stage did some *Hox* genes eventually get expressed in the growing appendages too. This scenario would leave the actual origin of the vertebrate paired appendages completely unexplained, by advocating, for an initial stage of appendage evolution, the existence of an unknown mechanism for producing and patterning appendages independent from the one already available for the production and patterning of the main body axis, only to converge later on this one, as soon as events of co-option would have determined the expression on both primary and secondary body axes of an increasing number of identical genes. An alternative to this scenario (Minelli 2000b) suggests that a novel (additional) pattern of expression of a set of genes will essentially coincide with the origin of the additional body part where these genes are expressed, rather than follow it, perhaps by very long time spans. In the case of vertebrate limbs, this amounts to arguing that the novel appendicular expression of *Hox* and other patterning genes was not grafted (by co-option) on to long-existing appendages, but was directly inherited (by paramorphism; see section 8.1.4) from the trunk-patterning system when the first paired appendages appeared. Of course, during half a billion years of gnathostome evolution, *Hox* genes have continued to evolve, with possible consequences on the patterning of both the main body axis and the axes of their appendages. I will return to axis patterning in the next chapter (see section 8.1.1).

Overall, I am convinced that the history of the evolving phenotypic outcomes of gene expression is one of exaptation, rather than one of co-option. In a previous section of this chapter I mentioned the example of *nanos*, which may have turned from an earlier involvement in establishing axial polarity, as determinant of the posterior end of the trunk, to a specifier of germ-cell identity. Indeed, Rabinowitz *et al.* (2008) have recently demonstrated that *nanos* is required in somatic cell lineages in the posterior part of a mollusc embryo. Similarly, it is likely that the *pax6* gene, a gene whose importance in eye formation is so critical that it had been raised to the rank of 'master control gene' (Halder *et al.* 1995), was perhaps exapted from pigment specifier to specifier of an organ (the eye) where the visual pigment is consistently involved in a function of major adaptive significance: vision (Kozmik 2005).

It seems fitting to close this section on developmental genes with a necessarily short account of Davidson's (2003, 2006) masterly effort towards a comprehensive model of the regulatory control of development by the bilaterian genome, and of its evolution. According to Davidson (2006) the developmental regulatory genome is like a computer, the internal *cis*-regulatory nodes of which are hooked up through their functional *cis–trans* interactions. Evolution is thus reduced to changes in the DNA machinery for information processing, a sweeping generalization to which I am not ready to subscribe, but I will not stop here. Davidson (2006) distinguishes among four classes of network subcircuits: the batteries of genes involved in cell differentiation; a series of little invariant subcircuits repeatedly involved in different, less specific functions; switches; and complex, highly conserved networks with rigid and recursive wiring, which specify morphogenetic fields from which particular body parts arise. The latter are called 'kernels' by Davidson. An example would be a gene network responsible for endoderm specification. Of the four kinds of gene networks, kernels would be the most robust to change and are thus expected to be shared by large segments of the metazoan tree. Indeed, there will be bilaterian kernels, protostome kernels, ecdysozoans kernels, and so on. This should also explain the lack of new phyla evolving since the Cambrian. In theory, at least, it is not so clear how generally the developmental genes of Bilateria may actually deserve be regarded as kernels because, as Davidson himself admits, the criteria a kernel should obey are so strict. According to Davidson, the genes involved are expressed at the initial phase of regional specification and all their genes are required for their function. Once evolved, these conserved networks would be the less-evolvable components of the whole set of genes involved in development and their stability would be responsible for the long-term conservation since the Cambrian of the main traits of the organization of the individual phyla. At the opposite end of the range, genes responsible for cell differentiation would be the most evolvable; that is, those whose phylogenetic signal is more likely to fade away in the long term.

To know the extent to which these generalizations may hold true, we need to gather much more abundant evidence than we have today. In the meantime, a difficulty with maintaining the hypothesis of the scarce evolvability of chordate-specific kernels is provided by ascidians (Lemaire 2006), where studies by Imai *et al.* (2006) on *Ciona* demonstrate a low level of connectivity of the network in each embryonic territory. Whereas in echinoderms and vertebrates a small number of signalling molecules (fibroblast growth factor, Nodal) and transcription factors (ZicL, FoxD, FoxA-a, Otx) controls the expression of a very substantial number of regulatory genes, this is not true in *Ciona*, at least from the 16-cell stage to the gastrula.

Selected parts of a gene-regulatory network may show unequal rates of evolution. For example, within the gene-regulatory network controlling the specification of the endomesoderm in the nematodes, a preliminary analysis of genome sequences of *Haemonchus contortus* and *Brugia malayi* suggests that evolution is most rapid for the zygotic genes involved in the specification of blastomere identity (Maduro 2006). The evolvability of regulatory cascades is further shown by examples where the same molecule is regulated by different genes in different species or even within the same organism (Larsen 2003). For example, the gene *hedgehog*, which is involved in establishing the anteroposterior axis of the embryonic segments and in patterning the larval imaginal discs, is controlled by *bicoid* in flies and by *caudal* in the beetle *Tribolium* (Dearden and Akam 1999), whereas *engrailed*, whose expression is critically important in fixing segmental boundaries in *Drosophila*, is regulated by *paired* in some cells but by *fushi tarazu* in other cells, a few cell diameters away (Manoukian and Krause 1992).

Finally, we must consider that genes, including those which seem to be hardly dispensable for a metazoan, may eventually be lost. This is likely the reason why nematodes have a small number of *Hox* genes. A broader perspective is offered by Nam and Nei's (2005) study of 2031 homeodomain sequences from 11 species of bilaterian animals. This suggests that no fewer than 88 homeobox genes were present in Urbilateria. Of these, two-thirds are represented with at least one descendant gene in all the species in the sample, but about 30–40 genes were also lost in a lineage-specific manner.

CHAPTER 8

The evolution of animal body architecture

'It happens', he [Theophrastus] says, 'that when trees leaf very luxuriantly, they are more likely to be sterile, and whenever they bear copiously, they are more likely to leaf poorly, as though Nature could not satisfy both elements of the tree's growth but must spend her resources first on one and then on the other.'

Arber (1950), p. 20

8.1 The body

8.1.1 The main body axis

A lively issue in evolutionary developmental biology is the evolution of the genetic control of the anteroposterior and dorsoventral patterning of the main body axis. The discovery of a conserved set of body-patterning genes, of which the *Hox* genes are only the most popular and possibly the best investigated, suggests that the main body axis of animals as different as insects, nematodes, molluscs, annelids, and vertebrates can be meaningfully 'aligned', in a way somehow parallel to the alignment of homologous nucleotide or amino acid sequences. Certainly, there are problems with some metazoan groups, where an anteroposterior axis comparable to that of an earthworm or a millipede is not so obvious, and perhaps does not exist. Even ignoring sponges, which do not have *Hox* genes, we still have to deal with at least two large groups in whose body organization it is quite problematic to see an equivalent of the main body axis of most bilaterians. One group is the cnidarians, whose oral–aboral axis is not obviously equivalent to the anteroposterior axis of bilaterians. The

other is the echinoderms, whose disparate body forms may suggest a multiplicity of equivalent body axes, rather than the clear-cut dominance of a major one.

Both cases will be briefly addressed in the following, but it is perhaps more important to consider that both comparisons (cnidarians versus bilaterians and echinoderms versus 'typical' bilaterians) indirectly take the presence of a common main body axis for granted, albeit with many, sometimes deep modifications of an idealized model. This attitude is hardly justified. A main body axis is not a given of animal body organization that has been simply inherited, with modifications, since perhaps the origin of the Epitheliozoa. Rather, it is something that has evolved, and not necessarily just once, and also something that may eventually cease to exist, as in the irregularly branching sporocysts of some digeneans, or in the adults of rhizocephalan crustaceans. For example, what about the Acoela? Apparently, no positional interpretation is derived from the expression of one *Hox* gene in these little, basally branching bilaterians (Baguñà *et al.* 2008). Organs, in the acoels, do simply 'fill the bag' of the organism without occupying strictly fixed positions.

If the main body axis is not necessarily a given of animal organization, the same must be said of the Cartesian framework that *prima facie* seems so adequate to describe the spatial distribution of body features of many metazoans. Do animals have an 'internal description' of their spatial structure and, if so, is this broadly corresponding to a system of Cartesian coordinates?

Besides the anteroposterior body patterning largely controlled by the expression of the *Hox* genes, the specification of the dorsal and ventral side of the body is controlled, in both insects and vertebrates, by numerous, partly conserved genes (see section 7.5.2). Thus, it seems legitimate to speak of an internal description of the animal three-dimensional structure, even if many of the genes involved in its control are not clustered on the chromosomes in an order collinear with their expression, as the *Hox* genes often are. Genes other than the *Hox* genes are involved in specifying and patterning the head, which cannot thus be simply regarded as an extension of the trunk (Olsson *et al.* 2005). These head-specific regulatory genes are expressed early in development and form a set quite conserved during the evolution of the Bilateria (Bally-Cuif and Boncinelli 1997, Bouwnmeester and Leyns 1997, Beddington and Robertson 1998, Hartmann and Reichert 1998). Most of these genes belong to families of homeobox-containing genes (Paired-class genes: *Otx*, *Goosecoid*, and *Anf/Hesx1*; Antennapedia-class genes: *Emx* and *Hex/Prh*; Lim-class genes: *Lim-1*).

Anteroposterior and dorsoventral patterning are apparently independent in insects and perhaps also in the chordates (Yu *et al.* 2007) and the enteropneusts (Lowe *et al.* 2006). Stern *et al.* (2006) admit that one of the most important reasons why the problem of head–tail patterning has eluded a solution in vertebrates is that it is not even easy to define this axis precisely. This is still truer for the limbs: recent evidence demonstrates that to think in terms of signals that regulate the proximodistal or anteroposterior axes equates to trying to shoehorn developmental phenomena into artificial schemes (Niswander 2003).

Despite these shortcomings, treating the Cartesian axes as natural descriptors is justified, to some extent, by the presence of distinct gene cascades whose effects are primarily involved in the specification and/or in the patterning of the anteroposterior, or the dorsoventral axis. An implication is that a search for homologies between the body axes of different animals is in principle justified. Nobody thus objects to the implicitly accepted homology between the anteroposterior body axis of vertebrates, insects, or annelids. The only problems seems to be (1) the axial homologies between cnidarians and bilaterians, (2) the identification of the main body axis (or axes?) in echinoderms, and (3) the putative dorsoventral inversion of chordates with respect to the condition in the other bilaterians. Before dealing with these three 'popular' problems, I want to point to other, often overlooked problems in determining homologies between axes.

Let us first consider the nature of the criteria we currently adopt in deciding what is anterior and what is posterior in an animal. Three main criteria may concur to our appreciation of what can deserve the name of main body axis, and how to recognize its polarity. These three criteria often provide overlapping, non-contradictory results, but this is not true for the whole of the bilaterians, not to mention sponges, placozoans, ctenophores, and cnidarians. The three criteria are provided by the polarity of (1) locomotion, (2) food ingestion and processing, and (3) growth and development. Very frequently, the mouth opens at (or near) the body end, which is anterior in respect to the animal's preferred direction of displacement, and this body end differentiates first, whereas the opposite end is the place for most of the animal's embryonic or post-embryonic elongation. The criterion of locomotion is obviously useless in the case of sessile animals, not only those of polyp form, but also worm-like parasites as are the tapeworms. In the latter case, the total lack of a digestive tract adds to the uncertainty about their polarity. In non-bilaterians, locomotion is not necessarily decisive in our assessment of body polarity. In the case of ctenophores, it has been suggested that the oral pole may correspond to the anterior pole of bilaterians, not so much as comb-jellies swim with their mouth forward, as because their oral pole derives from the animal pole of the embryo (Martindale and Henry 1998). In cnidarians, evidence is ambiguous, as reported in Chapter 4 (see section 4.2.2): in the planula of the anthozoan *Nematostella*, the homologue of anterior *Hox* genes of bilaterians is expressed in the planula at the body end that locomotion identifies as posterior, but the opposite is seen in the planula of the hydrozoan *Podocoryne* (Yanze *et al.* 2001, Finnerty *et al.* 2004).

Based on expression patterns of genes such as *Nkx2.5*, *Otx*, *Gsc*, and *Brachyury*, Meinhardt (2002) suggested that the oral–aboral axis of hydra-like polyps does not correspond to the whole body axis of bilaterians, but only to the anterior segment comprising the brain, with the oral side being the most caudal part of this segment, which broadly corresponds to the part of the bilaterian body axis which is not patterned by the nested expression of the *Hox* genes. The remaining part (posterior head, when present, and the whole trunk) would be a bilaterian novelty with respect to the cnidarian organization. I would argue that this equivalence is reasonable with respect to the patterning system, but the whole-to-part correspondence of its cnidarian versus bilaterian domain of expression invites the adoption of a factorial approach to homology, in this case circumscribing the homology relationship to the patterning system, while not extending it to the patterned structures. This way of reasoning is similar to the interpretation by Wagner and Gauthier (1999) of the identity of digits in the bird wing: here, the positional identity of each digit (within the typical five-digit set of Recent tetrapod limbs) and the specification of their kind (from what in humans we call the thumb and the little finger, respectively) are not necessarily overlapping; digits II–IV in the bird wing take the identity of the three anterior fingers of most remaining tetrapods.

A mismatch between the position of the mouth and the anterior end of the body as identified by locomotion is also typical of the Acoela: taking the arrow of locomotion as the criterion to identify body polarity, the mouth is anterior in the Proporidae, but posterior in the Diopisthoporidae (Cannon 1986).

The relevance of a developmental criterion (polarized growth and differentiation) as a criterion to determine body polarity was given formal relevance by Huxley (1932), who formulated a 'law of anteroposterior development', according to which the anterior parts of an animal's body differentiate earlier and at a quicker pace

in respect to the posterior parts. The same is true of the proximal parts of appendages in respect to the distal ones, a point to which I will return later, when discussing the ontogenetic and phylogenetic relationships between the main body axis and the body's appendages. Rensch (1959) offered a simple mechanistic explanation of this principle, in terms of allometric growth, regulated by endocrine centres located close to one of the ends (probably the anterior one) of the main body axis. This is acceptable, provided that we refrain from taking for granted the existence of a strictly posterior growth zone of localized, or more intense, mitotic activity.

Mitosis is not necessarily required for elongation. Keller (2006) identifies the following classes of elongation mechanisms: (1) fibre-wound, hydraulic systems (elongation and strengthening of the vertebrate notochord, elongation of the nematode embryo), (2) convergent extension by cell intercalation, in both mesenchymal and epithelial tissues (intercalation of dorsal hypodermal cells in the nematode embryo, hindgut elongation in the *Drosophila* germ band, elongation of Malpighian tubules, development of *Drosophila* ovarioles), (3) elongation by oriented cell division (elongation of leech embryo, elongation of embryo in short and intermediate germ band insects), and (4) elongation by cell-substrate-mediated guidance (elongation of the pronephric duct in amphibians, fish lateral line).

Another important point is that the posterior end of the main body axis is perhaps indeterminate in some groups, such as the vertebrates, but in the arthropods and nematodes at least it is defined very early in development (Minelli 2005b). In nematodes, embryonic elongation is achieved by the alteration of cell shape (stretching), not through directional cell division (Priess and Hirsch 1986). The same is possibly true of molluscs such as chitons, where the whole body of the larva undergoes diffuse growth (Beklemishev 1969). According to Lartillot *et al.* (2002), axial development in Bilateria may primitively involve two organizers, one anterior and one posterior, both of them evolved from the single head organizer of a putative hydra-like ancestor. This brings us back to the topic of the eventual developmental fate of the blastopore with its frequent slit-like elongation and eventual mid-length closure giving rise to two residual openings (often, but not necessarily, coincident with mouth and anus respectively) (see section 8.2.5).

A set of two organizers plays a central role in the general model of positional specification formulated by Meinhardt (2006). In this model, the embryo is equated to a set of two or three tubes (ecto-, meso-, and endoderm), one inside the other. The positional specification of the three tubes is essentially a two-dimensional process requiring two organizing regions which will generate lines of reference for the anteroposterior and dorsoventral axes. Meinhardt observes that in the case where these tubes are quite long in relation to their diameter (a condition that often applies, but not always) one of these organizers has a stripe-like extension, defining the future midline. Crucial for axis formation in vertebrates are Spemann's organizer and its 'relatives' such as Hensen's node. It is amazing how much uncertainty surrounds the patterning mechanisms in which the organizers are involved. Contrary to the traditional opinion based on the usual model vertebrates, such as the frog, according to which the organizer determines the dorsal side of the future embryo while the anteroposterior body axis is parallel to the animal–vegetal axis, Gerhart (2002) suggests that the anteroposterior axis is parallel to the organizer–anti-organizer axis.

In the recent literature, elongation by a terminal or subterminal growth zone has been often called terminal addition, but this term was long used to indicate the addition of a novel developmental phase following a recapitulation of the whole ontogenetic schedule of the ancestors (Gould 1977). Whatever its name, elongation by a terminal or subterminal growth zone has been said to be most conspicuous where it is associated with segmentation (Jacobs *et al.* 2005), but this is also problematic, as segmentation does not necessarily happen by progressive addition of new elements at the posterior end of the series, in arthropods at least (see section 8.1.5).

Irrespective of segmentation, examples of intercalary growth and differentiation are common and taxonomically scattered. To the arthropod examples I provided elsewhere (Minelli 2003) I will add here the case of the trochophore larva of the aplacophoran mollusc *Neomenia carinata*, whose caudal bud elongates to form the definitive trunk, which spreads anteriorly over the larva's whole surface (Thompson 1960).

Tapeworms are possibly the metazoans whose body polarity is most controversial. This is true of both the dorsoventral and the anteroposterior polarity. The scolex is radially or biradially symmetrical and thus does not provide any criterion to fix the dorsal or ventral side of the worm's flat body. The position of the uterine pore on one of the two flat surfaces, or the proximity of the female organs to the same, has invited helminthologists to take that side as ventral (Hyman 1951), but this is not based on a comparison with other flatworms where dorsoventral polarity is less arbitrarily fixed. As for the anteroposterior polarity of the tapeworms, I have already discussed elsewhere (Minelli 2003, 2005b) a couple of comparative arguments suggesting that the scolex is perhaps the posterior, rather than the anterior body end. Comparative data on the expression of genes that usually mark the anterior body end may hopefully settle the question soon.

In addressing the issue of body axes in echinoderms, fossils play a very important role. On one hand, the phylum includes several extinct clades with overall shape and body symmetry quite different from the modern clades, on the other hand some ontogenetic sequences of extinct taxa have been preserved that help us to understand the organization of Recent forms. Five largely equivalent axes are, indeed, a character typical of the Eleutherozoa, the subphylum to which four of the five extant classes are referred: the Asteroidea, Ophiuroidea, Echinoidea, and Holothurioidea. In the other echinoderms, including the Recent Crinoidea, the ambulacral system is arranged in a 2–1–2 pattern (Sprinkle 1973), with three main branches radiating from the mouth, of which one (ambulacrum A) lies along the animal's plane of symmetry, while the other two (BC and DE) diverge from this plane and eventually split, one into B and C, the other into D and E (Sprinkle 1973, Bell 1976). Ontogeny of representatives of some extinct groups (edrioasteroids, glyptocystitoid rhombiferans, and gogiid eocrinoids; Bell 1975, 1976, Sumrall and Sprinkle 1999, Parsley and Zhao 2006) confirms that A, BC, and DE are the primary ambulacra. This body structure is still present in extant crinoids. Sumrall and Wray (2007) suggest that the pentaradiate symmetry of the other Recent echinoderms results from the simultaneous splitting of all five ambulacra, without a common root between B and C and between D and E. A remnant of the original 2–1–2 pattern is still evident, however, in the fact that only three ambulacra (A, C, and E) have the same morphology, whereas B is the mirror image of C and D is the mirror image of E. This is fixed in the peculiar arrangement of the ambulacral plates known as Lovén's law (Lovén 1874, David *et al.* 1995).

The last aspect of polarity of the main body axis to be discussed here is the putative dorsoventral inversion that has been hypothesized to oppose protostomes to deuterostomes (or, better, to chordates). In many protostomes, for example arthropods and annelids, the main neural axis is more or less definitely ventral, whereas the corresponding structure is dorsal in chordates. Based on this difference, Ulrich (1951) suggested the names Gastroneuralia and Notoneuralia for two major groupings roughly corresponding to Protostomia and Deuterostomia, respectively. Homologizing the dorsal side of a mouse or a frog with the ventral side of a fly or a crayfish may appear to be a futile exercise in theoretical morphology, and a similar suggestion by Geoffroy Saint-Hilaire (1822) was indeed ridiculed in his time. But comparative data on the spatial patterns of expression of genes involved in establishing dorsoventral polarity in vertebrates and insects have demonstrated the legitimacy of this comparison. Indeed, if we want to understand the origin and evolution of the dorsoventral patterning in metazoans, we must keep two aspects well distinct. The first aspect is how the two sides are differentiated from each other. The second is how they are functionally

interpreted as dorsal or ventral. Differential expression of a conserved set of genes seems to answer the first question. In vertebrates, the expression domain of *chordin* characterizes the dorsal side, the expression domain of *BMP-4* the ventral side of the embryo. In arthropods, the homologues of these genes (*short gastrulation* and *decapentaplegic*, respectively) are also involved in establishing the embryo's dorsoventral polarity, but the expression domain of *chordin*'s homologue corresponds here to the ventral, rather than to the dorsal, side of the embryo, and correspondingly inverted is the expression of *BMP-4*'s homologue in *Drosophila* (Arendt and Nübler-Jung 1994, Holley *et al.* 1995, Biehs *et al.* 1996, De Robertis and Sasai 1996). What is true of insects, with respect to vertebrates, is probably true of molluscs too. An orthologue of *hedgehog*, a gene involved in the dorsal midline patterning (neural tube) in vertebrates, is expressed along the ventral midline in the embryo of the mollusc *Patella vulgata* (Nederbragt *et al.* 2002).

Interestingly, a *BMP–chordin* developmental axis also underlies the anatomical dorsoventral axis of enteropneusts. In this group it is virtually impossible to say what is dorsal and what is ventral, as many hemichordates live vertically in burrows, uniformly surrounded by the substrate. But hemichordates develop the mouth on the non-*BMP* side, as in arthropods. In *Saccoglossus kowalevskii*, BMP and its antagonists are expressed on opposite sides of the embryo, inverted with respect to vertebrates and cephalochordates (Lowe *et al.* 2006), with BMP dorsal in hemichordates. The neural/anti-neural role of *BMP* and *chordin* respectively is conserved between protostomes and chordates, but not in hemichordates (possibly by loss). Homologues of *BMP* and *chordin* are expressed asymmetrically in cnidarian embryos (Matus *et al.* 2006a); however, even if cnidarians have dorsoventral patterning (Martindale 2005), the expression of these genes in cnidarians seems not to correlate with dorsoventral patterning but rather with the specification of the germ layers. Matus *et al.* (2006b) have suggested that the genetic control of dorsoventral patterning (and thus the origin of the anteroposterior axis; Ryan *et al.* 2007) by genes of the *BMP/chordin* system arose together with the evolution of mesoderm in the bilaterian lineage. Christiaen *et al.* (2007) suggest that mouth relocation in chordates is possibly dependent on a changed influence of the BMP/Nodal signal on the specification of oral ectoderm. Therefore, the relative position between the mouth and the *BMP–chordin* axis may have been rearranged in the chordate line (Gerhart *et al.* 2005). Lowe *et al.* (2006) suggest that the common ancestor of hemichordates and chordates did use its *BMP–chordin* axis to specify neural cell fates within a diffuse nervous system.

Lacalli (1996) contrasted two possible scenarios for the evolution of the dorsoventral axis in bilaterians. In the first scenario, the basal bilaterians would have possessed a protostome organization with ventral mouth and ventral nervous system, while deuterostomes would have evolved a dorsal mouth, either by migration of the original ventral mouth or by novel formation of a new anterior opening of the gut. In the alternative scenario, starting from a common ancestor with a blind gut, protostomes and deuterostomes would have evolved independently a mouth on opposite sides of the embryo with respect to the dorsoventral axis. This would not imply a dorsoventral inversion, but a functional change of polarity along the anteroposterior axis, where the mouth of the ancestor became the anus of deuterostomes. The hypothesis of a dorsoventral inversion of the main body axis at the origin of the Chordata has been also rejected by van den Biggelaar *et al.* (2002).

An aspect of anteroposterior patterning along the main body axis that has received inadequate attention is the relative position of male and female organs, or of male and female genital pores, whenever a comparison is possible, that is in hermaphrodites but also, to some extent, in gonochoric metazoans where a reasonably good alignment of organs is possible between the two sexes. In *Drosophila melanogaster* the genital imaginal disc is formed by cells from three abdominal segments (A8–A10), which express different proteins involved in the differentiation of the genital structures. *Abd-B m* is expressed in A8 and is needed for the development of this

segment's derivatives as female genitalia; *Abd-B r* is expressed in A9 and controls the differentiation in this segment, from the multisegmental genital disc, of the male genitalia (Foronda *et al.* 2006). One wonders how old, and how general, this mechanism actually is. Consistent reciprocal placement of female and male structures would be more sensible in hermaphrodites than in gonochoric animals, as largely are insects and arthropods (with very few exceptions, e.g. barnacles, all of them deeply nested within larger clades of gonochoric species).

I must admit that when I first mentioned this point of metazoan body patterning (Minelli 2003) I stressed unduly the condition of those groups where female organs are anterior with respect to male organs. Some clades respect, indeed, that generalization, but others have opposite arrangement and, characteristically, in the Acoela the relative position of male and female organs is unstable, as is the position of their mouth. Female organs are anyway anterior to the male organs in a majority of acoel families such as Anaperidae, Convolutidae, Mecynostomidae, Childiidae, Hallangidae, Haploposthiidae, and Antroposthiidae, whereas the opposite is found in the Otocelididae (Cannon 1986).

In annelids, male organs are anterior to the female organs. This is also true of less typical taxa such as vestimentiferans (Gardiner and Jones 1993), *Stygocapitella* (Westheide 1990), nerillids (Worsaae and Müller 2004), *Potamodrilus* (Bunke 1967), and also of myzostomids (Eeckhaut and Lanterbecq 2005). The only exceptions to this rule seem to be the spirorbine serpulids (which are simultaneous hermaphrodites) where oocytes are found in the first chaetigerous segments of the posterior ('abdominal') part of the trunk, while sperm cells develop in the posterior segments (Rouse and Pleijel 2001), and the bizarre *Aeolosoma*, where oocytes are produced in the mid-region of the body and sperm is produced in both anterior and posterior segments (Bunke 1967). In those members of the enchytraeid and naidid oligochaetes that alternate fission with sexual reproduction, the relative positions of testes and ovaries are preserved despite the fact that these organs collectively undergo a forward segmental shift when new segments are formed following fission. The ovaries are always found upon the new septum formed between the posterior new segment and the anterior old segment, while the testes are located in the segment in front of the ovarian segment (Christensen 1994).

Also in nemerteans the male organs are more or less distinctly anterior. This is very clear in *Dichonemertes*, where the anterior gonads are male, the posterior ones female (Coe 1938).

Male organs anterior to the female organs are also characteristic of a majority of the rhabditophoran lineages, for example the Haplopharyngida, Proseriata Monocelididae, Typhloplanoidea, Byrsophlebidae, and most of the Polyclada, but the opposite is true of the Macrostomida and of the Opisthogeniidae within the Polyclada (Cannon 1986), and also of the Monogenea.

Overall, male anterior seems to be a character of lophotrochozoans, whereas female anterior is the topographical arrangement we find in ecdysozoans, but also in Gnathostomulida, Chaetognatha and, possibly, Vertebrata. I dare say that the study of this aspect of body patterning, especially at the level of its genetic control, may turn out to be a rewarding novel avenue of research in evolutionary developmental biology.

8.1.2 Symmetry

With the specification of anteroposterior and dorsoventral axis, bilateral symmetry is implied.

Let us consider that bilateral symmetry is compatible with paired structures, such as our eyes and limbs, but also with single ones, such as our mouth or anus, provided that these structures are, in turn, bilaterally symmetrical. Thus, if two paired structures are close enough to the mirror plane of symmetry, they will eventually merge into a single one. This means that morphologically and functionally major differences as those between the paired genital pores and copulatory structures of mayflies and the single medial gonopore and the single penis of other insects are perhaps minor in developmental terms. In the Pterobranchia, gonads are paired in *Cephalodiscus* and *Atubaria*, but there is just

one gonad in *Rhabdopleura*. And the presence of a medial chitinous penis in male Histriobdellidae is more conspicuous for its structure and function than for the fact that it is not paired, like most of the structures of 'polychaetes', including the penises found in some other representatives of the group, for example in *Pisione*. But let us elaborate a little on the theme of body symmetry.

Next to the idea of recapitulation of phylogeny in ontogeny, one of the aspects of nineteenth-century tradition that most obviously survived in twentieth-century textbooks is the prominent place reserved for the geometry of animal architecture. Sometimes, the topic was presented under the title of 'promorphology', as if the properties of symmetry and repetition of parts would dictate a set of eternal archetypes which animal form would be obliged to obey. There was also an effort to combine this typological, ahistorical perspective with an evolutionary view of life, but the original sin proved to be impossible to cancel. Some 'principles' offered as a guide to understanding the general trends in the evolution of animal form were nothing other than formulations of morphoclines from putatively simpler to putatively more complex conditions; for example from animals without any evidence of symmetry to those with radial symmetry, to end up with those with bilateral symmetry, or from unsegmented to segmented forms. Exceptions to the general trends were only admitted in the case of 'degenerate' lineages, such as many parasites—for example, the bopyrid isopods with an asymmetric body with often incomplete segmentation—but the general 'progressive' perspective opposed strongly the acceptance of phylogenetic hypotheses that required substantial regressive transitions. For example, it seemed hard to follow, for example, Jägersten's (1955) idea that the anthozoan polyp, rather than a morphologically simpler and more symmetric type such as the hydra, could be the most primitive kind among the cnidarians. A corresponding bias has long troubled interpretations of evolutionary trends of serially repeated organs or body parts. In the last quarter of the past century it was fairly fashionable to refer to

THE EVOLUTION OF ANIMAL BODY ARCHITECTURE

so-called Williston's law, the principle according to which evolution would proceed from a series of numerous, identical parts (e.g. body segments, or vertebrae) to a more structured series formed by a smaller number of more differentiated parts. As a consequence, geophilomorph centipedes were taken to be archetypal arthropods, with a trunk organized in a long series of uniform segments directly comparable with that of the arthropods' putative closest relatives, the annelids. Happily, nowadays phylogenetic relationships are no longer derived from abstract promorphological principles. In our current versions of the tree of life, *Hydra* occupies a position deeply nested within the cnidarian branch and the anthozoans have been moved to a branching position that suggests their organization as closer to the ancestral cnidarian polyp than the *Hydra's* would be. In another branch of the tree, geophilomorphs are deeply nested within the Chilopoda, and it seems undisputable that both the number and the uniformity of trunk segments have increased in their clade, with respect to the condition in their ancestors, contrasting with expectations in accordance with Williston's principle (Minelli 2003).

But the evolution of 'promorphological' features of body architecture is not simply a question of phylogeny. It is also a question of development, of how body features like anteroposterior polarity, dorsoventrality, or a segmented body are produced. As a consequence, to understand the evolution of symmetry and segmentation we must address the problem of the evolution of the developmental mechanisms producing these general features of body architecture. These problems are quite popular among researchers in evolutionary developmental biology. Of the problems relevant in the present context that have been addressed through comparative studies of gene expression, one is the symmetry of the cnidarian polyps, another is segmentation in vertebrates, arthropods, and annelids. The latter question will be discussed later in this chapter (see section 8.1.5). Symmetry will be discussed briefly here.

It has been already reported in Chapter 4 (see section 4.2.2) that the expression patterns of multiple *Hox* genes and the transforming growth factor-β gene *decapentaplegic* (*dpp*) suggest that cnidarians are basically bilateral (Finnerty *et al.* 2004). Thus, contrary to the traditionally accepted contrast between a supposed radial condition in the diploblastic metazoans and bilateral symmetry in the triploblastic ones, bilateral symmetry can be assumed today to be the plesiomorphic condition in the Epitheliozoa (Ball *et al.* 2004). This does not mean, however, that the gross bilateral symmetry we see in the majority of animals (and their organs) is a simple heritage of such an ancient default condition. In several vertebrate model species it has been demonstrated that the bilateral symmetry of somites is conserved, despite the fact that the embryo is the target of left–right patterning cues, because a signal depending on retinoic acid shelters the somites from a laterally patterning influence (Kawakami *et al.* 2005, Vermot and Pourquié 2005).

Bilateral symmetry can be viewed as a default status deriving from balanced competition between two effectively equivalent developmental modules. Thus, fluctuating asymmetry is caused by minor, random deviations from this equilibrium condition, whereas directional asymmetry will require specific polarizing cues.

As in the other instances where I suggest an explanation in terms of competition between developmental modules, I do not reject an adaptational perspective. There is clearly no need to stress the nearly universal mechanical advantage of a symmetrical body over an asymmetrical one, especially in features involved in locomotion. Therefore, body symmetry is likely to be a very common target of selection. But this does not mean that there is anything like a gene or gene network *for* body symmetry. We shall rather expect a genetic control on the robustness and reliability of growth and development against internal or external disturbance, thus reducing the degree of asymmetry to be eventually expected. The hypothesis that fluctuating asymmetry represents, in fact, a visible expression of developmental noise—that is, the amount of stress experienced by the developing organism beyond the buffering power of its developmental system—has been championed by Møller in a long series of papers, largely summarized in a book (Møller and Swaddle 1997). Opposing views have been also voiced, for example by Palmer (2000).

Asymmetry is sometimes a transient feature of the embryo. In the amphipod crustacean *Orchestia cavimana* Scholtz and Wolff (2002) found, from the eight-cell stage onwards, two types of embryo that are mirror images of each other, the asymmetry eventually disappearing at later stages. Mirror images are reported in the development of several crustacean species, for example in another amphipod (*Tryphoselis*; Bregazzi 1973), in several species of euphausiacean (Taube 1909), in decapods (Hertzler and Clark 1992), in the cladocerans *Daphnia* and *Holopedium* (von Baldass 1937, 1941), and in the copepod *Cyclops* (Fuchs 1914). In all cases, the meaning is unclear and there is no obvious effect on adult morphology (Scholtz and Wolff 2002).

Lists of animal lineages with body parts exhibiting directional asymmetry have been provided by Palmer (1996). I (Minelli 2003) called attention to the fact that directional asymmetry is sometimes limited only to the male or female sex. Examples include the male terminalia of several insects, the copulatory organ of many beetles, and several genital and non-genital features of many copepods. Female-only asymmetries include the genital apparatus of many insects (examples are scattered among beetles, aphids, and dipterans) and birds, where only one of the two ovaries is developed.

In some groups, directional asymmetry has never been recorded, or is extremely rare, in striking contrast with its common occurrence in related groups. For example, among the arthropods, directional asymmetry is common in the Tetraconata (crustaceans and hexapods), although far from uniformly within the clade, but is virtually absent in myriapods and extremely rare in chelicerates. The few instances of asymmetry in genitalia of spiders, summarized by Huber (2006), include *Kaliana yuruani* and *Metagonia* spp. (Pholcidae) and the theridiid genus *Asygyna* (Agnarsson 2006), where asymmetric female

genitalia are present in both mirror forms within one species.

Symmetry is an interesting property of living beings, and one of obvious adaptive significance, which is probably obtained 'for free'; that is, there are no genes for symmetry, or distinct genes separately controlling the production of the left and the right half of the body, or the members of a symmetrical pair of body parts. It seems sensible to describe bilateral symmetry as the consequence of the establishment, in the egg or in the early embryo, of dorsoventral and anteroposterior polarity, followed by competition between the left and the right side of the growing embryo. But a bilaterally symmetric body will be the default status only if the whole organism is not too 'fluid', as probably were the first cell aggregates at the root of the metazoans (Forgacs and Newman 2005), where the default shape was quite likely a sphere, due to a steady redistribution of parts within a system devoid of axial polarity.

If symmetry is a default condition, directional asymmetry, as in the *situs viscerum* of vertebrates or the body of hermit crabs, must be genetically controlled. A growing body of literature discusses the role in determining the visceral asymmetry of vertebrates of the embryonic expression of two genes, *nodal* and *Pitx2*, the first of which is also known from ascidians and sea urchins, and thus probably qualifies as deuterostome-specific (Chea *et al.* 2005), whereas *Pitx2* is possibly chordate-specific, with asymmetric expression demonstrated also in amphioxus and in ascidians (*Ciona*; Boorman and Shimeld 2002).

8.1.3 The 'double animal'

Irrespective of the identity, number, and expression patterns of the genes eventually involved in determining the anteroposterior patterning of the main body axis, we can expect that many external and internal features of a bilaterian body will eventually align with this reference system. Especially in actively moving animals, the anteroposterior polarity of the body has its precise correspondence in the positioning of locomotory and sensory organs, of a brain, and possibly of a mouth; but consider the mouth of the triclads (planarians) which opens on the ventral side behind the animal's midlength. Certainly, even the fixing of a main body feature such as the mouth or the genital opening at a specific point along the animal's body axis is something that can be 'negotiated'. For example, comparative and especially palaeontological evidence suggests that the oral structures of Recent arthropods do not occupy the same position as in their stem-arthropod ancestors. The mouth was terminal in *Kerygmachela*, as it was in the Cambrian lobopods (e.g. in *Aysheaia*) and as it is also in tardigrades. This suggests that all the structures of the arthropod head which now appear to occupy a pre-oral position in the arthropods evolved from post-oral structures of stem-group arthropods (Budd 2003a).

Within the arthropods, it can be argued that the genital system originally opened close to trunk mid-length as, for example, in spiders and malacostracan crustaceans, but it moved to a more cephalic position in the millipedes (Progoneata) and to a more caudal position in other groups: more precisely, to the very end of the body in two distinct clades, the chilopods and the proturans, and to a different, subterminal position in the insects. Within the Prolecithophora clade of the Rhabditophora, the mouth coincides with the genital pore in the Combinata, but the two are distinct in the Separata (Cannon 1986). Among the tardigrades, the Eutardigrada have a cloaca, but the Heterotardigrada have distinct gonopore and anus (Kristensen 2003). In some clades, the same position along the anteroposterior body axis is 'interpreted' differently in the two sexes, as in the nematodes, where the females have the genital pore at mid-body and the anus in the proximity of the posterior end of the body, while the males have a cloaca; that is, coincident openings of genital and digestive system at a posterior, subterminal spot.

Other organs also would gain an advantage from longitudinal patterning, and there would be no danger if the resulting pattern was independent from the overall longitudinal pattern of the main body axis. This is the case, in principle, of externally discharging tubular structures with a large length-to-diameter ratio, as many nephridia are, and especially the digestive system, where longitudinal patterning would offer the opportunity for the evolution of an advantageous regionalization separating the mechanical processing of food, its digestion, the absorption of digestion products, and the elimination of faecal residues. We cannot take the functional polarity of the gut for granted, or as a necessary consequence of the general anteroposterior polarity of the body. Just the opposite: it is neither 'free' nor certain. For example, glyceriform 'polychaetes' have a complete gut, but the anus is not functional and faecal pellets are regurgitated (Klawe and Dickie 1957, Ockelmann and Vahl 1976). In some animals, external body patterning and the longitudinal patterning of the gut are in close register. A very clear example is provided by those leeches whose gut is provided with several pairs of blind lateral branches, precisely matching the longitudinal sequence of as many body segments. But things are clearly otherwise in many animals, which often have an intestine much longer than the main body axis, with which it remains coordinated only at the two ends, the mouth and the anus, commonly located at (or close to) the two ends of the animal. But even this link is optional, as demonstrated by the sipunculans, where the anus is closer to the anterior than to the posterior end of the roughly fusiform body, and by all kind of polypoid bilaterians (e.g. ectoprocts and endoprocts). Recognizing a body axis comparable to the longitudinal axis of other bilaterians is indeed difficult with most sessile forms, including barnacles and ascidians. In this respect, the oddest group is perhaps the Phoronida, where the entire adult trunk is arguably equivalent to a ventral structure, as the ventral midline of the average bilaterian corresponds in *Phoronis* to a line that goes from the mouth to the end of the trunk and back to the anus, whereas the dorsal midline is reduced to the short span between the mouth and the anus (Zimmer 1997).

It seems thus legitimate to say that with respect to body patterning the popular *Hox* gene-based system embodied by the zootype concept only concerns one set of body structures (basically, those of the epidermal and nervous systems or, more grossly, those of the ectodermally and mesodermally derived musculo-cutaneous system), while another set of structures (the endodermally derived digestive system) have their independent longitudinally patterning system, which is probably dependent, to some extent at least, on differential expression of *ParaHox* genes. Metaphorically, we can therefore speak of a 'double animal', with a somatic and a visceral component (Minelli 2003). The independent patterning of these two components is often masked by the fact that the two are mechanically tied to such an extent as to make their long axes coincident, with mouth and anus at the two ends of an elongate body, and a straight gut in between. This is true, for example, of most annelids, with their intersegmental septa and their longitudinal mesenteries 'obliging' the tubular gut to be simply a tube within another tube (the animal's musculo-cutaneous envelope). However, as soon as this constraint is removed, as in those annelids where segmentation has more or less completely vanished, the 'internal animal' becomes free from the 'external animal' and often becomes longer and coiled, thus showing its independence from the latter (Rensch 1959). Examples are *Poeobius* (Rouse and Pleijel 2001) and *Stygocapitella* (Westheide 1990). Gut loops are nevertheless present also in other polychaetes where segmentation is better conserved, for example the Terebellidae, Ampharetidae, Pectinariidae, and Sternaspididae (Rouse and Pleijel 2001).

8.1.4 From one to many axes: appendages and paramorphism

Many animals have appendages such as legs, fins, antennae, and tentacles. The morphological disparity of these structures and their scattered taxonomic distribution would suggest multiple independent origins. Anatomically and functionally there are enormous differences between the four limbs of tetrapod vertebrates, the multiarticulated antennae of a spiny lobster, the fleshy cephalic tentacles of a snail, and the hydraulic podia of a sea star. Therefore, the discovery that the most diverse kinds of animal appendages involve the expression of the same gene, *Distal-less*, to mark the sites where the appendages will eventually grow came largely as a surprise (Panganiban *et al.* 1997). This discovery was indeed one of those results of developmental genetics that contributed to imagining the Urbilateria (the putative common ancestor of the Bilateria; see section 5.2) as quite complex and provided, specifically, with some kind of appendages. The value given to shared gene-expression data as markers of common phylogenetic origin is even more directly shown by the hypothesis of Dong *et al.* (2001) that the common ancestor of arthropods and vertebrates possessed appendages with a proximal and a distal domain under separate genetic control. These novel reconstructions of ancestral states have found widespread acceptance among students of developmental genetics and evolutionary developmental biology, but are clearly based on a simple appreciation of developmental similarities independent of any application of phylogenetic inference methods. The question then becomes how these similarities in the genetic control of the production of appendages can be interpreted in light of our current understanding of animal phylogeny. In particular, if we accept that acoels and nemertodermatids are the most basally branching lineages of the bilaterian tree, it seems sensible to reconstruct the earliest bilaterians as acoeloid (or planuloid) worms without appendages. Appendages were also probably lacking in the most recent common ancestor of the insects and vertebrates. At the moment, however, we reject an incautious extrapolation from gene-expression data to the reconstruction of a complex urbilaterian with appendages, we are obliged to explain the origin of the aspects of genetic control shared by the appendages of metazoans belonging to distantly related clades. The simplest explanation is probably that the developmental parallels reflect conserved regulatory relationships between signalling pathways that are not specific to the development of the appendages as such, but are also involved, perhaps with more generic functions, in other developmental contexts (Mann and Casares 2002).

I have suggested (Minelli 2000b, 2003) that in order to understand the similarities between the appendages of different groups of animals we must start with considering what the main body axes of those animals have in common, especially in terms of mechanisms of elongation and patterning. I argued, indeed, that the secondary axes of the appendages arose by duplication of the primary axis, thus using since their first appearance much of the available machinery already involved in the construction of the main body axis. Thus, traits shared by two different animals A and B with respect to the production and patterning of the main body axis would be directly derived from the mechanisms responsible for the production and patterning of the main body axis in their common ancestor. This implies a kind of orthologous relationship, whereas the production and patterning of the appendages (the appendage *a* of A, the appendage *b* of B) would imply a kind of paralogy with respect to the mechanisms involved in the production and patterning of the main body axis of the same animal. Thus, any similarity between the appendages *a* and *b* of two animals A and B whose common ancestor probably lacked appendages would be explained transitively: *a* and A are 'paralogous', A and B are 'orthologous', and *b* and B are 'paralogous'. That is, *a* and *b* are not homologous in the conventional sense of being derived from the same feature in the last common ancestor of A and B. I do not see problems using the term homology to indicate the relation ('orthology' in the previous idealized example) between A and B, but I think that the relation between *a* and *b*, provisionally called 'paralogy', deserves a different name: for this concept, the term *paramorphism* (Minelli 2000b, 2003) is already available in the literature.

In a sense, paramorphism is just a special case of that widespread syndrome of autoreplication I alluded to in the previous chapter (see section 7.3.2). The patterns of gene expression shared by the main body axis of bilaterians and their appendages have a counterpart in the patterns of gene expression (e.g. *Cnox-2*) shared by individual polyps and their colonies in the Hydrozoa (Cartwright *et al.* 2006). Similarly, in the Bilateria

paramorphism is not necessarily restricted to a correspondence between the development and patterning of the main body axis and the axis of the appendages, but sometimes extends to other external or internal axes. In particular, this can be argued in the case of the genitourinary tract of the vertebrates, which can be considered a developmental axis longitudinally patterned, and subject to region-specific growth, in a manner similar to the limbs, with the involvement of genes such as *Hoxd13, Hoxa13, Hoxa10, Shh*, and *BMP-4* (Podlasek *et al.* 2002).

The idea that duplication followed by divergent evolution of the duplicates is a major pathway in morphological evolution is an old concept (e.g. Sewertzoff 1931), which has been occasionally revived in the recent literature on evolutionary developmental biology. Focus has generally been on serially repeated structures, such as teeth, vertebrae, and body segments (e.g. West Eberhard 2003), but we cannot take for granted that homonomous series—that is, series of identical elements—have always preceded the evolution of heteronomous series, with elements differentiated into two or more kinds. For example, a largely uniform dentition is possibly plesiomorphic in vertebrates, but this condition is certainly apomorphic in the toothed whales, which derive from terrestrial mammals with many kinds of teeth. In a sense, the origination of body appendages by paramorphism with respect to the main body axis is comparable with the origination of lateral buds which do not give rise, eventually, to separate individuals, but become anatomically and functionally integrated in a composite whole within which they will evolve divergently from the main axis.

Morphological evidence in support of the paramorphism hypothesis has been presented in detail elsewhere (e.g. Minelli 1996, 2000b). I will only mention here that the appendages of unsegmented animals are unsegmented, while those of many segmented animals are segmented; and also that the degree of regionalization of the main body axis in arthropods is mirrored by corresponding degrees of specialization of the segments in their appendages. I must admit that the appendages of annelids are often unsegmented, but segmented are the antennae and palps of *Nerilla attenuata*, despite the animal's very minute size (body length only 1.5–1.7 mm) (Westheide 1990), and the palps and/ or the caudal cirri of many Nephtyidae, Nereidae, Hesionidae, Dorvilleidae, and Syllidae.

As the main body axis is not necessarily elongated, and eventually segmented, in a strict anteroposterior sequence, the segmentation of the developing legs is not necessarily obtained in a strict proximodistal (or distoproximal) progression, but, at least in part, by intercalation. This has been found both in heterometabolous insects, where a completely articulated leg is produced during late embryonic development (Norbeck and Denburg 1991), and in holometabolous insects without larval legs, such as *Drosophila*, where the appendages are only formed from imaginal discs during the pupal stage (Rauskolb 2001). In *Drosophila*, again, the developing trunk is first subdivided into a few broad domains by the expression of the so-called gap genes, and only later subdivided into bisegmental and finally segmental stripes. Similarly, in the developing legs wide-zone genes with expression patterns comparable with those of the gap genes along the main body axis are expressed earlier than periodic-zone genes with more restricted 'striped' (or 'segmental') expression patterns (Held 2002).

I would regard as a good argument in favour of the paramorphism hypothesis the fact that the two levels of segmentation recognizable in the developing trunk (primary, early embryonic segments which are later split into secondary, or final segments; see below, section 8.1.5) have their equivalent in the developing appendages. In *Drosophila*, the expression of *spineless-aristapedia* (*ss*) is apparently responsible for the secondary subdivision into five tarsomeres of the originally undivided tarsus which originated, together with the femur, tibia, etc., from a primary process of segmentation of the leg anlage (Duncan *et al.* 1998). Similarly, Richardson *et al.* (2004) suggest that in vertebrates the proximodistal patterning of the limb is accomplished in two steps: first, the specification of broad domains in the early limb bud, followed by a process by which each

first-order domain is subdivided into the individual skeletal elements.

In vertebrates, the transcription factors coded by *Hox* genes of the *HoxA* and *HoxD* paralogous series are expressed in the developing limb in a nested pattern (collinear with their position on the chromosome) similar to the pattern with which these genes are also expressed in the trunk. For example, in the limb bud *Hoxd9* is expressed proximally in respect to the more 'caudal' *Hoxd11*, and the most distal parts of the growing appendage, fated to become hands and feet, express the last members in the series; that is, *Hoxd10–Hoxd13* (Kmita *et al.* 2002).

The paramorphism hypothesis accommodates also the recent evidence of shared genetic control on the development of median and paired fish fins, despite the different embryonic provenance of the cells from which they take origin. In fact, the paired fins develop from the mesoderm of the lateral plate, while the median fins are mainly derived from somatic (or paraxial) mesoderm. Nevertheless, the position of both paired and median fins is similarly specified by the expression of *Hoxd* and *Tbx18* genes, and the nested expression of *Hoxd* genes found in the developing buds of the paired fins is also found in the developing median fins (Freitas *et al.* 2006).

In the plant kingdom, the fundamental equivalence of main stem and lateral branches is often demonstrated by the ease with which a branch, once it has been cut off the stem, may produce roots and eventually turn into a new stem. Parallel phenomena are much rarer in animals, in natural conditions especially, but cnidarians can provide good examples. The most impressive are found in the boloceroidid sea anemones (Pearse 2002). In a couple of species belonging to this family, single tentacles are autotomized and shed into the gastral cavity, where they remain until each tentacle has given rise to a minute new polyp. In another species, the regenerating item which gives rise to the new polyp is not the single tentacle but a fanshaped cluster of short tentacles that are shed as a unit and remain together until they turn into a new polyp.

In addition to the main body axis and the secondary axes represented by the appendages, there may also be virtual, or 'honorary' axes, the best-known example of which are the eyespots on butterfly wings, whose position on the surface of the wing imaginal disc is marked by spots of expression of *Distal-less*, the gene we already know to be positional marker for the most diverse kinds of animal appendage (Carroll *et al.* 1994). Another example is provided by the horns of scarab beetles. These structures of epidermal origin grow very rapidly during the prepupal stage, and are localized and patterned by the same set of genes that are also expressed along the proximodistal axis of the insect appendage: *Distal-less* (*Dll*), *dachshund* (*dac*), *homothorax* (*hth*), and *extradenticle* (*exd*) (Moczek and Nagy 2005, Moczek 2006, Moczek *et al.* 2006).

Despite the theoretical attractiveness of the concept and the amount of evidence in its favour, I must concede that there are also difficulties with the notion of paramorphism. Some of these have been presented by Held (2002), who is nevertheless open to accepting paramorphism, pending future developments in this area. A problem is the different behaviour of the main body axis and the appendages of *Drosophila* with respect to evidence of genes specifying longitudinal spans corresponding to two future segments, despite the fact that three pair-rule genes (*hairy*, *odd-skipped*, and *odd oz*), which are expressed with bisegmental periodicity in the trunk, are also expressed in the leg (Cohen 1993, Godt *et al.* 1993, Kojima *et al.* 2000). I would argue, however, that bisegmentality, as such, is not a property of the expression pattern of the pair-rule genes themselves. The fact that their domains of expression are eventually split into two definitive segments depends on the expression of genes downstream of them, rather than on their own expression. It is thus in the downstream genes that we should look for a more sensible comparison. Another difficulty noted by Held (2002) is the lack of cell-lineage restriction in the leg (Bryant and Schneidermann 1969). In other terms, in the leg imaginal discs cell clones can freely cross the boundaries between the future femur, tibia, etc., whereas in the trunk of a developing *Drosophila*,

cell clones do not cross the boundaries between compartments, of which there are two (one anterior, one posterior) within the longitudinal span of each segment.

8.1.5 Segments

As we have already seen in Chapter 5, when discussing the Ecdysozoa/Articulata controversy, a segmented body has probably evolved independently more than once. In this section I will review morphological, developmental, and genetic evidence supporting this view. First it is advisable to start by browsing around some examples from the most disparate aspects of animal segmentation, as from this little bestiary will emerge the subjectiveness of current distinctions between what should and what should not be called a segment. To begin with, what is the relative importance of external versus internal (anatomical) features? In arthropods, segmentation is mainly an ectodermal affair (e.g. Minelli and Bortoletto 1988), so that the only internal feature expected to be segmentally arranged is, generally, the nervous system. In annelids (Fig. 8.1), however, segments are organized around internal structures of mesodermal origin, as are the coelomic pouches and the musculature, but in most species the nervous system and the body surface are also serially patterned. Borderline cases may thus be expected. Westheide (1990) has reviewed the problem with respect to the tiny annelids belonging to the Dinophilidae, where the external divisions of the body have been traditionally regarded as not indicative of the 'true' segment number. In these worms, even the internal anatomy suggests contentious interpretations. In *Dinophilus taeniatus*, Donworth (1986) recognized six segments based on the presence of six pairs of ganglia in the ventral nerve cords and of six nephridia in the female, but the most anterior pair of nephridia has a structure different from the other five pairs and has been tentatively interpreted as a remnant of larval excretory organs, which are traditionally regarded as presegmental.

Even in animals with more overtly expressed segmentation the identification of segments as

Fig. 8.1. Longitudinal section of the clitellate *Branchiobdella* sp., showing alternance of major and minor annuli. Courtesy of Marco Ferraguti.

unambiguous integrated units is not possible due to a gross mismatch between different series of structural elements. The most popular example is provided by millipedes (Diplopoda), where most of the trunk is articulated in rings, each of which corresponds to two pairs of legs as well as to two pairs of ganglia in the ventral nerve cord, but similar mismatches are widespread in arthropods generally (Minelli 2003, Minelli and Fusco 2004). Interestingly, this mismatch is also seen in stem-group arthropods such as *Fuxianhuia protensa* and *Chengjiangocaris longiformis* (Hou and Bergström 1997), suggesting that the individual series of repeated structures (e.g. dorsal sclerites and leg pairs) are primarily independent, while their integration into conventional segments is a secondarily acquired condition, rather than a 'promorphological' given of arthropod structure. That these features can be uncoupled again is shown by the frequent displacement of nervous ganglia out of the segment they innervate, especially in the case of extensive concentration of the nervous system, as in crabs and in many flies. In comparison to the segments of arthropods, annelid segments appear to be much more integrated; nevertheless, some examples of mismatch between different series of segmental structures have been recorded in this clade too. For example, in some Hermellidae and Serpulidae there are two pairs of ganglia per segment, and up to three per segment in *Pectinaria* (Beklemishev 1969). Individual

organs may also migrate from a segment to the other during development. In the Nephthyidae the eyes differentiate in the prostomium but are usually transferred backwards to one of the nearest segments during ontogeny (Rouse and Pleijel 2001).

How many segments are recognizable in an insect's head? The traditional interpretation in terms of six segments was challenged by Schmidt-Ott *et al.* (1994, 1995) and Urbach and Technau 2003), who reconized seven segments in the head of *Drosophila*, based on the number of neural structures under independent genetic control, but a study of the contribution of imaginal discs to produce the adult head of the fruitfly would suggest otherwise. Only cells from three of the six (or seven?) head segments actually contribute to the imaginal head, as cells from the intercalary, mandibular, and maxillary segments are not involved at all (Younossi-Hartenstein *et al.* 1993).

If in the case of some 'less typical' annelid or arthropod we accept that segmentation can be only present in one serially repeated feature, nothing other than tradition can explain why other animals are not described as segmented despite the presence of obviously repeated organs. A nice example is the nemertean *Neuronemertes* where the dorsal nerve shows regularly repeated ganglionic swellings corresponding to the peripheral branches of the lateral nerves (Coe 1933). The cryptically periodic organization of the nervous system of nematodes has also been noted (Johnson and Stretton 1980, Wood and Edgar 1994). For example, in the first larva of *Caenorhabditis elegans* blast cells with similar developmental potential are regularly spaced along the main body axis and give rise to homologous cell lineages (Salser and Kenyon 1994).

However, I think we must avoid lengthy discussion on purely lexical matters. Too much has been already said for or against the segmental nature of tapeworm proglottids and kynorhynch zonites, and of the serially repeated features in the internal anatomy of *Neopilina* as well as of the dorsal plates of chitons (see Minelli and Fusco 2004 for a summary). The problem virtually disappears if we accept that segmentation is probably a 'generic' property of bilaterians (Newman 1993); that is, a feature related to the general physicochemical properties of living tissue (see section 7.3.4) and thus likely to have evolved numerous times (e.g. Shankland 2003, Minelli and Fusco 2004). The problem then turns into a disentangling of this multiplicity of mechanisms likely involved in the production of serially repeated features in different metazoan lineages.

A good starting point is the appreciation of the fact that the notion of segmentation applies to organs rather than to organisms (Budd 2001, 2003a). That is, it is often much more objective to describe an animal in terms of serially repeated units of the central nervous system (neuromeres) or serially arranged bundles of muscles (myomeres) rather than in terms of integrated segmental units, unless several different structures present a repetitive pattern with the same period and phase (Holland 1990). The segment is probably the result of a continued integration of additional features onto an initial core of serially repeated units, probably represented, at first, by stereotyped sets of nervous cells or fibres (Scholtz 2003). Indirect proof of this evolutionary scenario is provided by the fact that evolution has also played it in reverse order, as individual features of segments can be lost or become uncoupled from the others, quite independent of all remaining segmental features (Scholtz 2003).

Evidence from developmental genetics supports this view. In the pill millipede *Glomeris marginata*, the genes involved in establishing the dorsal and the ventral pattern of segmentation are not the same, or have different roles in generating the two aspects of the animal's serial pattern (Janssen *et al.* 2004).

Neuromeres have been usually regarded as the most fundamental unit of segmented animals, but even this source may produce contradictory evidence. For example, Page (2004) demonstrated that two segmental commissures in the head of *Drosophila*, the tritocerebral and the mandibular ones, are not distinct at first appearance, but separate from an originally single commissure in the same way the anterior and posterior commissures of a typical trunk neuromere are formed. Thus, why do we regard the latter pair as belonging to one segment, whereas the tritocerebral and the mandibular commissures are regarded as belonging to separate neuromeres, and hence to separate body segments?

Segments are often produced by different processes within the same animal. In many annelids (Korn 1982), in amphioxus (Conklin 1932), and in arthropods (Minelli 2001), anterior segments have a different origin than the more posterior ones. On the other hand, serial units of different body structures do not necessarily follow the same periodicity along the body axis. This is seen, for example, in the sequence of legs versus exoskeletal rings in the tadpole shrimp *Triops cancriformis* (Linder 1952) or of muscles blocks versus gill slits in amphioxus (Conklin 1932).

During segmentation of the insect embryo, in the future anterior head (corresponding to the three anteriormost segments) there is no expression of the pair-rule genes which play instead a fundamental role in segmenting the posterior head and most of the trunk, but it is here that *orthodenticle, empty spiracles,* and *buttonhead* are expressed, as in the fore head of vertebrates. Thus, there is some reason to say that the boundary between the anterior head and the rest of the body in insects and vertebrates are homologous (Holland *et al.* 1992, Finkelstein and Boncinelli 1994). In many crustaceans, this boundary marks the division between naupliar and post-naupliar segments as the whole body of the nauplius is virtually limited to the most anterior segments bearing the antennules, the antennae and the mandibles, followed by a growth zone from which the remaining segments will later differentiate (Minelli 2001).

A different dimension along which several groups of metazoans have diversified their patterns of segmentation is the coexistence within one animal of primary and secondary serial units (Minelli 2000a). This feature is most conspicuous in the leeches, where most of the segments identified by internal anatomy, for example as neuromeres, are externally only little more distinct than the annuli (mostly three or five, but eventually up to 14) into which each primary segment is subdivided. However, annelid specialists have sometimes preferred features other than ganglia in determining the number of segments, as the 'polychaete' *Pectinaria* has been described as provided with two or three pairs of ganglia per segment (Nilsson 1912). Clearly, these examples show that the segment is not a given of annelid morphology, but a unit which possibly evolved by the regular alignment of different serially repeated units (e.g. coelomic pouches, ganglia, and nephridia) in a functionally integrated form which was eventually conserved because of its adaptive value.

Corresponding examples are found in other 'polychaetes', for example in the Glyceriformia, with two- or three-annulated segments and a prostomium divided into up to approximately 30 rings (Rouse and Pleijel 2001). Segments are biannulate in the tiny *Stygocapitella* and two ciliary rings per segment are found in another miniaturized form, *Dinophilus taeniatus* (Westheide 1990). Biannulate segments are also present in some tardigrades, suggesting that this phenomenon is 'generic' rather than clade-specific. This suspicion is strongly supported by the presence of secondary segmentation in a tapeworm, *Haplobothrium globuliforme*, whose primary proglottids divide into several secondary proglottids (Cooper 1919).

Sometimes, the segmental units recognizable in the adult are different from those formed at first during embryonic development. This process, known as resegmentation, is seen in vertebrates and insects alike. The concept was originally introduced to describe the vertebrae of the amniotes as deriving from the posterior half of a segmental sclerotome plus the anterior half of the next sclerotome. According to this interpretation, the definitive vertebral body would be intersegmental (Remak 1855). There is no general consensus today as to the validity of this interpretation, but there is recent morphological evidence in favour; for example, the dorsal vertebrae of turtles, whose neural arches (and ribs) are shifted forward by half a segment (Goette 1899, Rieppel 2001).

In *Drosophila*, each definitive segment corresponds to the posterior compartments of the original periodic units (parasegments) together with the anterior compartment of the immediately following parasegment. The parasegmental organization is maintained in the nerve cord, but not in the epidermis and in the musculature. It is possible that the primary parasegmental organization known from *Drosophila* is a plesiomorphic feature of the Arthropoda (Tautz 2004) or, at least, of the Tetraconata, as parasegments are formed in both insect and crustacean embryos (Dohle and Scholtz 1988), and also in spiders (Damen 2002). However, in the centipedes *Lithobius* (Hughes and Kaufman 2002b) and *Strigamia* (Chipman et al. 2004) the primary units of segmentation of the blastoderm are equivalent to the prospective definitive segments.

Segmentation is mostly sequential, often in association with the activity of a subterminal proliferative zone (annelids, malacostracan crustaceans, and vertebrates), but the process is virtually simultaneous in other cases, as in *Drosophila*. In the latter case, segmentation is not only uncoupled from cell proliferation, but the main steps in the gene-control cascade resulting in a segmented embryo are accomplished during a short interval during which all mitotic activity is temporarily stopped. A remnant of anteroposterior progression can be read, however, in a slight asynchrony in the formation of the strips of expression of pair-rule and segment polarity genes; this asynchrony is much more pronounced in dipterans other than *Drosophila*

(Rohr *et al.* 1999). Arguably, sequential and simultaneous segmentation are just different schedules whose outcomes can be identical. In insects, both conditions are known within one order, as in the beetles (Patel *et al.* 1994). Furthermore, both processes can occur in the same animal, and the products are morphologically a uniform series; for example, in amphioxus, the first five or six somite pairs form simultaneously, while the others form sequentially (Conklin 1932).

In clitellates, segments arise from five to seven bilateral pairs of longitudinal bandelets of primary blast cells which are generated by as many bilateral pairs of embryonic stem cells, called teloblasts (Zackson 1984, Bissen and Weisblat 1989, Shimizu *et al.* 2001). At each division, each teloblast gives off a cell which, together with those produced at the same time by the other teloblasts, will form the cell pool from which one segment will originate. Polychaetes are generally described as producing post-larval segments from a subterminal growth zone, but in *Hydroides elegans* four posterior thoracic segments arise from an area in the middle of the body (Seaver *et al.* 2005), similar to the temporal progression inferred by Donoghue *et al.* (2006a) for *Pseudooides prima*, a Lower Cambrian metazoan of uncertain affinities.

In arthropods, the segmental pattern is first specified in the ectoderm by the striped distribution of transcription factors coded by several classes of genes. In *Drosophila*, where the segmentation process is best known, the genes involved in the control of segmentation are classified into gap (e.g. *hunchback, giant, Krüppel,* and *knirps*), pair-rule (e.g. *fushi tarazu, even-skipped, hairy,* and *paired*) and segment-polarity (e.g. *wingless, hedgehog,* and *engrailed*) genes according to their spatial patterns of expression and their role in controlling increasingly restricted specified domains (e.g. Carroll *et al.* 2005). Homologues of most of these *Drosophila* segmentation genes have been discovered in bilaterians other than arthropods, but their developmental role is often different. Conversely, a role in segmentation is performed in arthropods other than *Drosophila* by genes that in this model insect are not involved in producing segments.

The segment-polarity gene *engrailed* (*en*) is expressed in a series of transversal stripes in leeches, polyplacophoran molluscs, onychophorans, arthropods, amphioxus, and vertebrates (Jacobs *et al.* 2000). This does not mean, however, that in all these metazoans it is actually involved in segmentation. In *Drosophila*, *en* expression is limited to the ectoderm, where it marks compartment boundaries, besides being involved in patterning the nervous system. In the leech, it is also expressed in the mesoderm but is not involved in the segmentation of the nervous system (Shankland 2003). In most arthropods where segmentation has been studied, *en* marks the posterior end of the developing segment, but its vertebrate homologue is expressed in the somites only after these are formed (Holland and Holland 1998). Homologues of *en* have also been found in non-segmented animals such as *Patella* (Nederbragt *et al.* 2002). In several groups, including arthropods, molluscs, and echinoderms (Jacobs *et al.* 2000), the ectodermal expression of *en* is associated with skeletal development. Expression patterns of *wg* and *en* similar to those in *Drosophila* have been found in all four of the major arthropod groups (Damen 2002, Davis and Patel 2002, 2003, Hughes and Kaufman 2002b), but a similar conservation has not been observed yet at the level of the pair-rule genes.

In vertebrates, the somites are sequentially produced by proliferation at the posterior extremity of the developing embryo under the control of a molecular oscillator called the segmentation clock, which results from the periodic activation of genes linked to the Notch pathway. This mechanism is first activated at the gastrula stage. Tightly linked to the production of somites is the control of the sequential activation of the different *Hox* genes, which thus specify with high precision the regional identity of each somite (Dubrulle and Pourquié 2002).

The discovery in the embryo of the spider *Cupiennius* of the involvement of the Delta/Notch signalling pathway in the segmentation process has been interpreted as a trait in common between vertebrate and arthropod segmentation (Stollewerk *et al.* 2003). In vertebrates, elongation of the embryo at the posterior tip is coupled with the activity of a 'somitogenetic clock'; that is, with the regularly cyclic pulses of expression of several genes, including *c-hairy1*, a vertebrate homologue of a pair-rule gene involved in the segmentation of the *Drosophila* embryo (Palmeirim *et al.* 1997), which is not involved in early segment specification in insect embryos (Tautz 2004). Peel (2004) hypothesizes that the activity of the pair-rule genes, which in *Drosophila* is not dependent on a segmentation clock involving the expression of *Notch* and *Delta* (Rusconi and Corbin 1999, Wesley 1999), may have been originally dependent on the latter; thus segmentation in spiders may still retain something of the original pathway (Rauskolb and Irvine 1999, Dearden and Akam 2000). However, the Delta/Notch system is far from being specific to segmentation. In different bilaterians, it is involved in the formation of boundaries in many tissues, independent of segmentation.

Another component of the cyclic gene expression in vertebrates is the Wnt pathway; *wingless* is also expressed at the tip of the growth zone in *Tribolium* (Nagy and Carroll 1994), the tadpole shrimp *Triops* (Nulsen and Nagy 1999), the spider *Cupiennius* (Damen 2002), and the centipede *Lithobius* (Hughes and Kaufman 2002a), but its involvement in the segmentation process has not been demonstrated.

By comparing segmentation in annelids, arthropods, and chordates, Tautz (2004) remarked that two genes are common to all three groups: these are the *hairy* homologues, involved in early specification processes, and *engrailed* homologues, which specify segment boundaries, although probably not uniformly in all members of the three groups. Convergent involvement of these gene classes in segmentation is likely: in the case of *hairy* there is evidence of independent recruitment to the segmentation process within the vertebrates (Leve *et al.* 2001). Correspondingly, the expression of orthologues of *engrailed*, *hedgehog*, and *wingless* studied by Seaver and Kamshige (2006) in larval and juvenile stages of two 'polychaetes' is consistent with distinct origins of segmentation between annelids and arthropods. The plasticity of these processes is additionally shown by the frequent changes of function of segmentation genes (Davis and Patel 2003). For example, *even-skipped* and *fushi-tarazu* are segmentation genes in *Drosophila*, but identify broader periodic domains in the grasshopper *Schistocerca*, while *fushi-tarazu* is expressed like a *Hox* gene in the mite *Archegozetes* (Telford 2000), and in the centipede *Lithobius* it has a Hox-like expression but also shows up in a posterior domain where transient segmental stripes are generated (Hughes and Kaufman 2002c).

8.2 Organs

'Do we know of a single organ that is essential to animal life in general, though it may be essential to the particular organization of which it forms a part? Or are we not justified in assuming that in certain organisms neither plant nor animal life may depend on any particular organ whatever?' This posthumously published sentence of Lamarck (1933, p. 185) suggests that no organ is indeed required to be an animal. It would require a stretching of the concept, probably, to say that animals with such simple organization as sponges or *Xenoturbella* have organs (Valentine 2004). Organs have also often disappeared, for example in interstitial, miniaturized animals, or in parasites. But organs are indeed recognizable in animals generally, and thus we shall spend a few pages on problems related to their origin and evolution. However, as an excellent book-size treatment has been recently devoted to the topic (Schmidt-Rhaesa 2007), I will deal here with selected cases only, which are nevertheless exemplary of more general phylogenetic biases and evolutionary trends.

8.2.1 Organogenesis

Next to reproduction and development, the origin of organs is an aspect of biology where a sensible framing of questions strongly depends on our overall concept of life phenomena. Within the premises of developmental biology, searching for the origin of organs means to investigate patterns of gene expression and cellular behaviours eventually resulting in the appearance of, for instance, a lung, a stomach, or a nephridium. From the point of view of evolutionary biology, searching for the origin of organs means reconstructing the phylogenetic sequence of changes that these organs have undergone, beginning with what is generally described as an evolutionary innovation. Such a historical picture looks particularly attractive if we are able to place it in a plausible scenario of adaptation to changing environmental conditions.

To be sure, the functional performance of lungs, stomachs, and nephridia is under constant selection pressure, but this circumstance does not tell us anything about the way these organs can actually change over the generations. The problem is the usual one in approaching the evolution of phenotypes more complex, and less mechanically determined, than the proximate products of gene expression. In this respect, we can learn an important lesson from the work of Paul Brakefield and his associates on the evolvability of the colour patterns on butterfly wings (e.g. Monteiro *et al.* 1997, Beldade and Brakefield 2002, Brakefield 2008) (Fig. 8.2). In their accurately planned experiments of artificial selection, what to an external observer, including a potential predator, appears to be an eyespot of concentric rings of different colours turns out to be an intricate melting pot of cellular processes, each with a very different evolvability. For example,

the colour of a black ring is somehow independent from the yellow colour of a concentric ring, but the positions of the rings are not independent; and the overall shape and position of an eyespot can change quite easily in some direction in the plane of the wing's surface, while demonstrating strong resilience in another. Back to our initial question: body parts that in their wholeness seem to be the targets of natural selection turn out, in fact, to be products of entangled networks of unit processes in development.

If such a dramatic lack of matching between selective units and morphogenetic modules underlies the evolvability and the actual evolution of a seemingly simple phenotypic trait such a butterfly wingspot, how could we expect complex anatomical and functional units (organs), each of which can be singled out, to some extent, as a unit of adaptation, to also behave as units in terms of development?

Our descriptions of animals and plants and their anatomical parts are often biased by excessive, although often unconscious, attention to functional aspects. Hand, finger, arm, and heart are obvious functional units. Their adaptive importance *may* have led to the selection of genes whose expression helps a secure development of these parts, but this has nothing to do with these parts' origin and may still have little to do with the specifics of their development. What is probably more important is micro-function (local growth, local interactions, whether competitive and/or cooperative, and local morphostasis and the breaking thereof).

As with many other key questions in biology, it seems advisable to distance ourselves from an obsessive adultocentric interest in the 'final product'. If we want to understand what has happened during the last two centuries in the history of human transport, it is not enough to stress the selective pressure of a desire for greater speed and improved fuel economy, or in the satisfaction of the ever-changing requirements of fashion. In addition to these questions about adaptation, we must also learn a lot about development; that is, what kinds of materials were available at different times, and how these materials were shaped, not simply into

Fig. 8.2. Eyespots on the ventral side of the wings of *Bicyclus anynana*. Courtesy of Marco Uliana.

sumptuous limousines or racing cars, but also, and perhaps most importantly, into metal sheets, pipes, and the like. Shifting one's focus from the car as a whole to its individual parts shows that there is nothing like automobilogenesis, other than in a very rough sense. But this dissection of the production of a car into individual steps, each of which represents a choice among a set of alternative processes, is not that different from the dissection into individual and individually evolvable developmental steps of what in biology is currently called organogenesis.

Thus, to study the evolution of organs we should be able to combine two different kinds of information. On the one hand, there is the distribution of the 'final products' (mainly adult organs, but also larval ones if distinct) on to the phylogenetic tree; on the other there is the phylogenetic history of change in the developmental processes out of which these organs are built. The traditional, widespread concern for homoplasy points to the deep gap existing between these two aspects of the problem. It may be interesting to see whether current progress in evolutionary developmental biology offers promise of rapid progress. Eventually, it may turn out that there is an even greater pervasivity of homoplasy than was suspected in the past.

To build a vertebrate heart requires contributions by several progenitor fields and the involvement of many different developmental processes. In addition to the mass of the myocardium and endocardium deriving from the lateral plate mesoderm, and to the components (epicardium, cardiac interstitium, and coronary blood vessels) originating from the proepicardium, a primarily extracardiac progenitor cell population, Wilting *et al.* (2007) demonstrated a further contribution of immigrating lymphangioblasts to the cardiac lymphatic system, thus showing that coronary blood and lymph vessels, although growing in close association with each other, are derived from different sources. Things are possibly even more complicated, as recent tissue ablation, genetic ablation, and lineage labelling experiments in chick and mouse embryos have demonstrated the possible existence of two distinct heart fields. On the other hand, the signalling pathways involved in heart development are not specific to cardiac morphogenesis but are widely used in different embryonic processes (Dunwoodie 2007).

Similarly, the insect dorsal vessel is the final outcome of different processes, beginning with the alignment of cardioblasts as continuous rows of cells on both sides of the embryo, followed by their transition from mesenchymal to epithelial status. Subsequently, the two cell rows migrate towards each other, as an effect of the dorsal closure of the embryo. Only at this time do the assembly of cardioblasts into a dorsal vessel and the shaping of its lumen take place. Each process eventually resulting in the formation of this heart is under separate genetic control (Tao and Schulz 2007). On the other hand, completely different organ systems can be patterned by identical mechanisms. For example, in vertebrates there is much in common between nerve and blood-vessel wiring (Carmeliet 2003, Carmeliet and Tessier-Lavigne 2005)

8.2.2 Epithelia and epidermis

We can argue that epithelia serve two main functions: protection from the environment and the separation of two different chemical milieus (Schöck and Perrimon 2002). But this does not help much in our understanding of the origin of these cell sheets, as I suggested in the previous chapter, when discussing cell–cell competition and morphostasis. An epithelium is a sheet of cells with distinct apical–basal polarity, lying on a basal lamina and laterally bound by junctions (zonulae adhaerentes) (Bartolomaeus 1993).

Contrasting epithelia with mesenchyme, Tyler (2003) was inclined to regard the first as the default type in the Eumetazoa, because epithelial cells are the first cell type to be specified in embryos and mesenchyme originates from epithelium-like cells, but this hypothesis may deserve second thoughts, especially when we realize that modern animals most closely approaching the ideal of pure epitheliality, like *Hydra*, are somewhat simplified with respect to their ancestors. Rieger (1984) sketched a series of hypothetical steps in the evolution of metazoan integument, a pathway explored independently by multiple lineages, convergently, by first depositing a glycocalyx on the apical membrane of epidermal cells, or between microvilli of the same. Later on, a true exoskeleton composed of multilayered fibre lattices developed by incorporating collagenous fibres into the extracellular matrix. Today, the only metazoans without extracellular matrix are the placozoans, the ctenophores and, probably, the acoels (Rieger 1984; but Tyler and Rieger 1999 believe that some extracellular matrix is present in the Acoela). As an extracellular matrix with collagen, fibronectin, and integrin is present in sponges (Labat-Robert *et al.* 1981, Akiyama and Johnson 1983, Brower *et al.* 1997, Pancer *et al.* 1997, Garrone 1998), the condition found in *Trichoplax* and in acoels might be derived. The glycocalyx is the only outer extracellular structure in flatworms, gnathostomulids, micrognathozoans, nemerteans, phoronids, crinoids, asteroids, echinoids (Fig. 8.3), enteropneusts, pterobranchs, and cephalochordates (summarized in Schmidt-Rhaesa 2007).

In phylogenetic reconstructions (e.g. Nielsen 2001) much value has been accorded to the monociliated versus multiciliated conditions, but this contrast is arguably of limited significance. Within the Bilateria, exclusively monociliated cells are found in both protostomes (Gnathostomulida, some Gastrotricha, two annelid genera, Phoronozoa, and Chaetognatha) and deuterostomes (almost all Echinodermata, Pterobranchia, and Cephalochordata) (Rieger 1976, Lammert 1991, Ruppert 1991b, Gardiner 1992, Westheide and Rieger 2007). Multiciliated cells have evolved multiple times (Rieger 2003), but some instances of reversal of multiciliated into monociliated epithelia are also implied by phylogeny (Meyer and Bartolomaeus 1996).

8.2.3 The cuticle

Epithelial cilia and cuticle are, in principle, incompatible, but their joint occurrence in gastrotrichs, lobatocerebrids, and orthonectids demonstrates that this is not strictly true. The cuticle of the Gastrotricha (Fig. 8.4), which is devoid of chitin, is possibly homologous to the epicuticle of the ecdysozoans. A contrast can be tentatively made between cuticles containing α-chitin, such as those of the Ecdysozoa, and cuticles containing β-chitin, such as those of the Annelida and other Lophotrochozoa. However, α-chitin has been reported from the operculum of the tube of the polychaete *Pomatoceros triqueter* (Bubel *et al.* 1983) and β-chitin is said to occur in the pentastomid *Raillietiella gowrii* (Karuppaswamy 1997).

8.2.4 The nervous system

In a stimulating elaboration on the concept of zootype, originally introduced by Slack *et al.* (1993) to embody the notion of a bilaterian conserved body-patterning system due to conserved

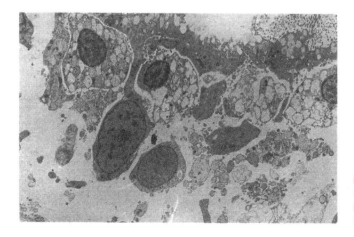

Fig. 8.3. Peristomial 'epithelium' of the sea urchin *Paracentrotus lividus* showing the lack of a distinct basal membrane. Courtesy of Daniela Candia and Francesco Bonasoro.

Fig. 8.4. The gastrotrich *Thaumastoderma ramuliferum*. Gastrotricha are the major example of coexistence of cuticle and epidermal cilia. Courtesy of Antonio Todaro.

patterns of expression of a conserved set of genes (mostly, the *Hox* genes), Deutsch and Le Guyader (1998) introduced the notion of the neural zootype. Under this term they formulated the hypothesis that the original target of what eventually became the conserved patterning of the anteroposterior axis of the bilaterian body was the animal's nervous system. In a non-finalistic view of evolution, the differential expression of genes coding for different transcription factors is possibly a means by which distinct differentiation and metabolic states could be stably maintained in different parts of the body. If so, the neural zootype would imply an ancestral role of the nervous system in maintaining regional differentiation along the animal's development.

The presence in the cnidarians of both ectodermal and endodermal nerve nets suggests that these neural patterning genes (*Hox* genes and others) may have been originally involved in patterning an endodermal nervous system. This would have been the source from which these genes were later recruited to generate the central nervous system of ectodermal origin found in bilaterians (Matus *et al.* 2006b).

At the molecular level, the roots of the nervous system are very deep. By comparing a fraction of the planarian genome with completely sequenced genomes of other metazoans and non-animal eukaryotes, Mineta *et al.* (2003) found in the yeast and in flowering plants

(*Arabidopsis*) the homologues of one-third of the 116 nervous system-related genes that they studied. This means that these genes predate the origin of the Metazoa and thus their relatively recent recruitment into the specification or patterning of the nervous system. Similarly, the *Pax2/5/8* genes, which specify the midbrain of many bilaterians, have their homologues in the sponges, which lack nerve cells, not to mention a brain (Hoshiyama *et al.* 1998). The continuous recruitment in neural roles of number of existing genes is further shown by the coral *Acropora millepora*, whose genome contains many of the genes whose expression is required to specify and pattern the nervous system of insects and vertebrates, which are vastly more complex than the simple nervous system of cnidarians (Kortschak *et al.* 2003).

A brain or cerebral ganglion (Haszprunar 1996) is arguably a apomorphy of the Nephrozoa (Jenner 2004c), but other elements of the central nervous system can be possibly traced back to more basal segments of the metazoan tree. Nielsen (2005a) argues that the nerve ring encircling the mouth of cnidarians has its equivalent in the nerve ring which differentiates around the blastopore of the bilaterians, which can give rise to part of the adult brain and to a pair of ventral nerve cords. This would be supported by the expression of *Hox* genes around the cnidarian mouth paralleled by their expression along the ventral nervous system of the protostomes. As to the Acoela and the Nemertodermata, a concentration of four to eight nerves radiates from the statocyst, and this was regarded by Beklemishev (1969) as a rudiment of brain. We may question the terminology, but the role of the statocyst in fostering the development of a centralized component of the nervous system in these basal bilaterians is probably warranted.

Within the Nephrozoa, a distinctly recognizable brain is lacking is some groups, for example the Echiura (Pilger 1993, Purschke 1996, Purschke *et al.* 2000), where immunohistochemical studies of the ontogeny of the central nervous system have nevertheless revealed the presence of slight swellings along the nerve loop in the prostomium, possibly corresponding to the supra-oesophageal ganglia of typical annelids (Hessling 2002, Hessling and Westheide 2002). In this case, the lack of a brain is obviously a secondary feature.

I doubt whether a well-defined brain was ever present in the earliest ecdysozoans (no brain is found, for example, in the nematodes) or in the earliest deuterostomes, as no brain is recognizable in echinoderms or enteropneusts (Bullock 1965b). Lowe *et al.* (2003) have suggested that the common ancestor of Recent deuterostomes had a diffuse rather than centralized nervous system. This is suggested by the peculiar pattern of expression in the hemichordate *Saccoglossus kowalevskii* of the orthologues of 22 genes involved in the patterning of the central nervous system. The expression domains of all these genes are arranged in the hemichordate in an antero-posterior sequence identical to the sequence in which they are expressed in the chordates, despite the fact that the latter have a centralized nervous system, whereas the hemichordates have a diffuse net. We are therefore facing less than obvious relationships of equivalence (see Davidson 2006). The epidermis of the hemichordate prosoma and mesosoma would in fact correspond to the parts of the vertebrate neural tube respectively fated to become the forebrain and the anterior midbrain, while the epidermis of the hemichordate metasoma would correspond to the remains of the vertebrate main neural axis, from the posterior midbrain to the hindbrain and the spinal chord.

Clearly, the hypothesis that the earliest deuterostomes lacked a brain amounts to doubting that all bilaterian brains are homologous, something that other gene-expression data would indeed suggest. In both insects and vertebrates, the regionalization of the brain is controlled by homologous genes. Of these, *orthodenticle* (*otd/Otx*) and *empty spiracles* (*ems/Emx*) identify the most anterior part of the brain, an intermediate segment is characterized by the expression of *Pax2/5/8* and *unpg/gbx2* orthologues, whereas the posterior part of the brain corresponds to the most cephalic expression of *Hox* genes (*lab/hox1*). Some authors (e.g. Arendt and Nübler-Jung 1996, Sprecher and Reichert 2003, Lichtneckert and Reichert 2005, Reichert 2005) argue that these conserved patterns of gene expression would demonstrate that a similarly controlled brain is a feature evolved at the base of the bilaterian radiation. However, as in the case of the role of *tinman* and *Pax6* homologues in the control of heart and eye formation respectively, multiple parallel recruitment of orthologous genes in the production of functionally equivalent organs cannot be ruled out. According to Davidson (2006), the joint involvement of these (and other) genes in brain patterning reveals the phylogenetic conservation of a patterning regulatory network which is nevertheless activated, in different bilaterian groups, downstream of morphogenetic events that are different in the different groups.

It is quite likely that the plesiomorphic state for the nervous system of the Bilateria occupies a basiepithelial (Bullock and Horridge 1965) or perhaps intraepithelial (Fig. 8.5) position. The latter is found among representatives of distantly related phyla, both protostome and deuterostome, including nemerteans, annelids, phoronids, echinoderms, and hemichordates. However, a sinking towards a subepithelial position has occurred many times. In this respect (as in most aspects of body organization), a great disparity of conditions is present within the annelidan lineage. The plesiomorphic nature of the basiepithelial condition of the nervous system in this group is supported by the fact that

Fig. 8.5. Two subepithelial nerves appear in cross-section between epithelium and musculature in the body wall of the clitellate annelid *Enchytraeus* sp. Courtesy of Roberto Valvassori.

this position is independent from body size. Derived is thus the largely intradermal nervous system of the Oweniidae (Rouse and Pleijel 2001) and especially the subepidermal position found in *Nephthys* and *Nereis*, and also in *Myzostoma* (Orrhage and Müller 2005). Similarly, Hyman (1951a) described a gradual sinking of the nervous system in the nemerteans, but the trend must be checked again against the modern views on the internal phylogeny of the group (e.g. Thollesson and Norenburg 2003).

Harzsch and Müller (2007) list as at least putative apomorphies of the nervous system of the Protostomia the presence of a circumoral brain ring, of serial longitudinal fibre tracts embedded within the peripheral plexus and centralizations of the neural plexus linked to the brain by longitudinal tracts.

In a phylogenetic perspective, major questions about the bilaterian nervous system are:

• the relationships between larval and adult nervous system,
• the possible independent origin of the brain with respect to the longitudinal nerve cords,
• the homology between the brains (and individual brain areas) of distantly related clades,
• the homology between the longitudinal nerve cords of distantly related clades.

The relationships between larval and adult nervous systems have been particularly addressed in the case of the spiralians. Here, 'larval' and 'adult' components are generally recognized (Croll and Dickinson 2004, Nielsen 2008). Two distinct elements are present in the larva: the apical ganglion and the cerebral ganglion. The apical ganglion is connected to the apical tuft of cilia found in a majority of trochophores and trochophore-like larvae, and degenerates before metamorphosis, or at metamorphosis, together with the loss of the apical cilia. Therefore, it is not retained as a part of the post-larval nervous system. Different is the fate of the paired cerebral ganglia, linked together by a commissure, which are often closely apposed to the apical ganglion, but do not share the fate of the latter. Instead, the cerebral ganglia survive metamorphosis as the anterior part of the adult brain, except for the endoprocts, where the larval frontal (or 'cerebral') ganglion is lost at metamorphosis (Nielsen 2008). In the larvae of echinoderms and enteropneusts but also in those of the phoronozoans (phoronids and brachiopods), the larva does not differentiate the equivalent of a trochophore's episphere (the pre-oral body half where both apical cilia and ganglion and cerebral ganglia are formed) or larval neural centres comparable to the cerebral ganglia. Eventually, the adult will lack a well-defined brain (Nielsen 2005a).

The post-larval component is represented by the posterior part of the brain and by the longitudinal nerve cords. However, as to the common or independent origin of brain and longitudinal neural axes, development offers contrasting evidence. Independent origin of the (anterior) brain and longitudinal nerve cord is assumed in the case of annelids (Golding 1992, Dorresteijn *et al.* 1993) and a majority of molluscs (Cephalopoda, Gastropoda, Bivalvia, and Scaphopoda; Haszprunar 2000, Raineri 2000). This double origin of the central nervous system is not limited to bilaterians developing from trochophore-like larvae, as comparably distinct components are recognizable in the developing nervous system of the onychophorans (Eriksson *et al.* 2003). On the other hand, in the plathelminths and nemerteans, the longitudinal nerve cords develop from the cerebral ganglia as posteriorly directed outgrowths (Hartenstein and Ehlers 2000, Younossi-Hartenstein and Hartenstein 2000). The same is apparently true of the pedal and lateral (pleurovisceral) nerve cords of the polyplacophoran molluscs (Hammersten and Runnström 1925, Hyman 1967, Ponder and Lindberg 1997) and at least the lateral (pleural) nerve cords of the Solenogastres (Thompson 1960, Hyman 1967). The origin of the ventral nerve cords is thus variable, even within the same phylum (Mollusca). This also opens the question of the homology of the longitudinal nerve cords in different bilaterian phyla, which is uncertain (Jenner 2004c), not to mention the homology of the tranversal commissures uniting them, which have likely evolved independently many times within groups such as the Rhabditophora (Reuter and Gustafsson 1995, Reuter *et al.* 1998, Reuter and Halton 2001).

Studies of gene expression suggest that in the case where two main components can be recognized in the central nervous system of many bilaterians, these are not necessarily the brain and the longitudinal nerve cords running along the trunk, but rather the anterior brain and posterior brain plus main longitudinal nerves. The differentiating anterior brain is characterized by the expression of *orthodenticle* homologues (*otd*, *Otx*) and the surrounding anterior region of the body is not interested by the expression of *Hox* genes. Overall, evidence of *Hox* gene expression in trochophore (or equivalent) larvae is very limited, but the most accurate study performed to date (Kulakova *et al.* 2007) shows that these are limited to the inferior/posterior part of the larva and do not extend to the episphere, and

are therefore lacking in the region of the larval cerebral ganglia, the precursors of the future anterior brain.

Studies on cell-level mechanisms of neurogenesis and the identity and expression patterns of genes involved in this process have revealed interesting phylogenetic signatures, despite the fact that the great effort necessary to gather the corresponding data does not offer reasonable hopes that evidence on a great number of representative data will be available shortly. Within arthropods, similarities in neurogenesis support the Tetraconata hypothesis; that is, a close phylogenetic relationship between insects and crustaceans (Whitington 1996, Whitington and Bacon 1998, Harzsch 2003), where individual neural stem cells (neuroblasts) can be recognized that will adopt a neural fate during the development of the nervous system. This condition is presumably apomorphic with respect to the condition found in spiders (Stollewerk *et al.* 2001, Stollewerk 2002), millipedes (Dove and Stollewerk 2003), and centipedes (Kadner and Stollewerk 2004, Chipman and Stollewerk 2006), where the neural fate is adopted by previously identified groups of cells, with characteristic position and behaviour, rather than by independent neuroblasts. Of course, if the latter condition is plesiomorphic for arthropods, it will be of no use in assessing the phylogenetic position of myriapods, either with chelicerates (the Myriochelata hypothesis) or with crustaceans plus insects (the Mandibulata hypothesis) (Stollewerk and Chipman 2006, Stollewerk 2008) (Fig. 8.6).

Let us now move to the sensory organs. Several authors (most recently, Jacobs *et al.* 2007) distinguish three kinds of sensory structure: 'naked' sensory neurons, sensilla, and sensory organs. Naked sensory neurons, as found in cnidarians and acoels but also in true flatworms and nemerteans, include an afferent projection (dendrite) formed by a modified cilium (Chia and Koss 1979, Wright 1992). In sensilla, non-neural cells are integrated structurally and functionally with individual neurons or groups of a few neurons. When numbers of cells, both neural and non-neural, are associated in larger and often

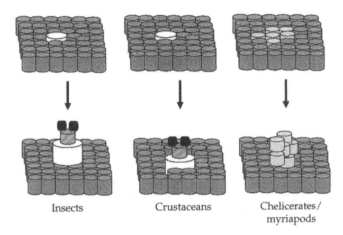

Fig. 8.6. Different modes of neurogenesis in arthropods. In insects and crustaceans (malacostracans) individual neuroblasts are specified in the ventral neuroectoderm. Insect neuroblasts delaminate from the neuroectoderm, while crustacean neuroblasts remain in the outer cell layer. In both arthropod groups, neuroblasts divide in a stem-cell-like mode, producing smaller ganglion mother cells which divide once to generate two neural cells. In contrast, in chelicerates and myriapods, groups of neural precursors are recruited from the neuroectoderm which invaginate and directly develop into neural cells. Grey, ventral neuroectoderm; white, neuroblasts; light grey, neural precursors; hatched, ganglion mother cells; black, neural cells. Courtesy of Angelika Stollewerk.

more complex structures, these are generally called sensory organs. By applying this definition Jacobs *et al.* recognize the presence of sense organs, for example statocysts and eyes, in cnidarians, but no complex sense organ is present in the Enteropneusta (Beklemishev 1969).

Statocysts are a major, or *the* major, component of the sensory endowment of several basal branches of the metazoan tree and have probably evolved several times. Their presence is often connected with a pelagic life style. Statocysts are thus a generalized feature of ctenophores, but among the cnidarians these sensory organs are an apomorphy of the medusae of medusozoans (Bullock 1965a, Bridge *et al.* 1995, Grimmelikhuijzen and Westfall 1995, Schäfer 1996). But statocysts are also widespread among benthic, interstitial animals, such as *Ototyphlonemertes*, which is unique among the nemerteans in possessing these sensory structures (Bruggeman and Ehlers 1981). Similarly, in annelids statocysts are only known from a few families of burrowing (or tubicolous!) 'polychaetes' (Jamieson 1992, Verger-Bocquet 1992, Rouse and Fauchald 1997) and also from *Grania*, a genus of marine enchytraeid clitellates (Locke 2000).

Morphologically and functionally, different kinds of sensory organs would hardly suggest a common origin, but their specification by the same or similar genes has suggested otherwise (e.g. Treisman 2004). In *Drosophila* the external sensory organs are specified by transcription factors encoded by genes belonging to the *achaete-scute* complex (Campuzano and Modolell 1992), while the precursors of the chordotonal organs (the insect's proprioceptors) are identified by another factor, Atonal (Ato) (Jarman *et al.* 1993). However, a single transcription factor (Cut) is sufficient to transform embryonic chordotonal organs into external sensory organs (Bodmer *et al.* 1987, Blochinger *et al.* 1991). Additionally, differentiation of all these sensory organs depends on the transcription factor Senseless (Nolo *et al.* 2000). It is very arduous to also reconcile within this scheme the most complex sensory organs of the adult insect (the auditory organs and the eye). Treisman (2004) suggests that a common origin of all sensory organs might also be argued for the vertebrates, where many sensory organs form from placodes, a set of ectodermal thickenings that arise, more or less simultaneously, anteriorly and laterally to the cranial neural plate. Clearly, this is not exactly the same as to say that all organs in an animal are homologous because they all derive from the same zygote, or to predicate homology of muscle and bone in vertebrates because they both share mesodermal origin, but it would be hard to fix a divide between these fictitious examples and the less hyperbolic examples above. Once more, I would argue that the question must be addressed by adopting a factorial notion of homology. Cranial placodes are responsible for many evolutionary novelties of the vertebrate head, including the specialized paired sense organs. Developmental origin of the placodes is traced back to an area of the anterior neural plate defined by the expression of transcription factors belonging to the *Six1/2*, *Six4/5*, and *Eya* families, an expression that is maintained in the differentiating placodes and seems to control their activities, from cell proliferation to eventual neurogenesis. Schlosser (2007) argues that these genes were recruited in this function in the common ancestor of vertebrates and urochordates. Some of the differentiation pathways in which they are involved did nevertheless develop only within the vertebrate lineage.

Moving next to the eye, it is fair to remark that photoreceptors are not exclusive of animals, as they have evolved in several genera of protists; neither are they universal among the metazoans.

Unique, but more closely comparable to protist structures than to the visual systems of the other animals, are the 10–15 ocelli of the planula larva of the cubozoan *Tripedalia cystophora* which are independent of any connection to the nervous system. Each ocellus is formed by a single cell containing a cup of screening pigment, a surface covered by photosensory microvilli and a well-developed cilium, whose motility is apparently controlled by the intensity of the surrounding light registered by the cell (Nordström *et al.* 2003).

All other metazoan photoreceptors are integrated with the animal's nervous system, but only the more complex organs may deserve be called eyes. Within a classification of the animal kingdom accepting a partitioning into 33 phyla,

about one-third of these major groups include animals without specialized light-detecting organs, one-third have simple light-sensitive structures, and only one-third have organs of such complexity as to be called eyes. Of these, only six phyla (Cnidaria, Mollusca, Annelida, Onychophora, Arthropoda, and Chordata) include species with image-forming eyes (Fernald 2006). Their phylogenetic distribution is hardly suggestive of a common origin, contrary to the hypothesis recently defended, based on the phylogenetically widespread involvement in eye formation of control genes such as *Pax6* and *Six*. But the origin of *Pax* genes antedates the origin of eyes and even the origin of the nervous system (Scott 1994, Hoshiyama *et al.* 1998, Kozmik 2005). Expression of *Pax* genes has been demonstrated even in some cnidarians, the most basal animals with lens-containing eyes (Kozmik *et al.* 2003, Nordstrom *et al.* 2003, Piatigorski and Kozmik 2004).

Pax transcription factors are complex regulatory proteins with two distinct functional elements, the paired domain and the homeodomain, which selectively bind to different DNA sequences. Kozmik (2005) has suggested that these two domains eventually evolved as an integrated tool involved in the production of eyes because each of them is responsible for a function which is essential for building an eye: the paired domain controls the production of the dark pigment required to screen the photoreceptor from the light coming from unwanted directions, while the homeodomain controls the production of opsin; that is, the photosensitive pigment. *Pax* genes, however, are not exclusively colocalized with the eye, but are involved, for example, in the production of melanocytes within the progeny of the neural crest cells (*Pax3*) and of the sensory pigment cells of ascidians (*Pax6*, *Pax3/7*) (Martinez-Morales *et al.* 2004). *Pax6* expression is common to the two main types of metazoan photoreceptor: the ciliary type (e.g. found in vertebrates) and the rhabdomeric type (e.g. found in arthropods) (Arendt 2003).

Arendt and Wittbrodt (2001) provided a reconstruction of the urbilaterian eye, moving from the circumstance that the most widespread eye type in bilaterians is represented by a pair of larval pigment-cup eyes at both sides of the apical organ in both protostome and deuterostome larvae. In their simplest condition, these eyes consist of one pigment cell and one photoreceptor cell in inverse orientation. In the adults of several protostomians there are more complex pigment-cup eyes with photoreceptor cells in everse arrangement; that is, directly facing the incoming light. Both kinds of eye possess rhabdomeric photoreceptors, whereas the chordate cerebral eyes have ciliary photoreceptors. The difference between the two kinds of photoreceptor is not simply morphological, but also molecular, as ciliary photoreceptor cells do not employ for phototransduction the same rhodopsins and non-orthologous G-proteins, rhodopsin kinases, and arrestins as their rhabdomeric counterparts. This suggests homology of cerebral eyes within the Protostomia, but challenges the homology of chordate and non-chordate cerebral eyes. However, Arendt *et al.* (2004) have recently demonstrated that the polychaete *Platynereis dumerilii* possesses both kinds of photoreceptor. Rhabdomeric photoreceptors, traditionally regarded as a protostomian character, are present in the eyes of *Platynereis*, while ciliary photoreceptor cells, once regarded as typical of the vertebrates, are found in its brain. The similarity of the ciliary photoreceptors of *Platynereis* to those of the vertebrates is stressed by their use of a photopigment similar to the opsins of vertebrate rods and cones.

Common to protostomes and deuterostomes the opsins are involved in the transduction of photons into cellular signals across the cell membrane. There are several members in this protein family that control sensitivity to light of different wavelengths (Fernald 2006). Regulation of the expression of the opsin gene is under different control in the vertebrates, where *Crx* genes are involved, and in *Drosophila*, where this role is taken by another paired-domain gene, *Otd* (Punzo *et al.* 2001). The gene encoding opsin was recruited repeatedly during eye evolution. But the most obvious example of multiple independent co-option of functionally equivalent genes in eye formation is provided by the eye

lens proteins (crystallins), which are structurally very heterogeneous and were often co-opted from other functions, including as enzymes (Fernald 2006).

The *Pax6* gene has undoubtedly a major role in controlling the production of eyes, as most conspicuously demonstrated by its ability to induce ectopic eyes (for example, along a fly's legs) if experimentally expressed outside its natural domain (Halder *et al.* 1995) but it seems not to be justified to call it straight a master control gene for eye development (Gehring 1998, 2002, Gehring and Ikeo 1999). Ectopic eyes are also induced by genes for other transcription factors (*sine oculis, optix, eyes absent, dachshund, eye gone,* and *teashirt*; Silver and Rebay 2005), some of which are also involved in the development of photoreceptors in metazoan lineages other than the Arthropoda (Arendt *et al.* 2002). More important, *Pax6* and many other genes in the eye developmental cascade also have a function in the development of other tissues, including in animal clades were no eyes exist. For example, *Pax6* homologues are expressed in the peripheral sensory organs of nematodes (Harris 1997) and in the tube feet of sea urchins (Czerny and Busslinger 1995).

It seems, thus, that the entire regulatory circuit, more than single genes, has been repeatedly co-opted in the developmental control of different cell types or organs (Kozmik 2005). The *Pax6* gene seems to have evolved at the base of the Bilateria subsequent to their separation from the Cnidaria, as in the cubozoan *Tripedalia cystophora* there is a *PaxB* gene, which unites functional domains that are present in the bilaterian *Pax6* gene and another domain that is typical of a distinct *Pax* subfamily (*Pax2*) (Kozmik *et al.* 2003).

The origin of a third kind of photoreceptor found in some annelids, the phaosomous sensory cells, is uncertain, as these possess both cilia and microvilli, as otherwise typical of the ciliary and rhabdomeric type, respectively (Purschke *et al.* 2006). Besides the two major kinds of metazoan eyes, the ciliary and the rhabdomeric type, there are isolated examples of the independent evolution of unique types of photoreceptor, as in the catenulid *Stenostomum virginianum* with eyes consisting of single nerve cells containing a large, light-reflecting mitochondrion (Ruppert and Schreiner 1980).

While the case for a derivation of all bilaterian eyes from a complex photoreceptor already present in Urbilateria seems unlikely, it is also clear that many components of the developmental machinery producing the diverse types of metazoan eyes are indeed homologous. Thus, as in the case of most, if not all, of the complex features of animal organization, the best way to deal with this problem is a factorial approach, inviting the dissection of eye production into individual components (such as those involving the topographical localization of the eye field, the production of opsin, etc.), each of which has its own, not necessarily linear, history of involvement in eye development. A nice example of independent recruitment of gene-regulatory networks has been shown by Ogura *et al.* (2004), who compared expressed gene sequences from octopus and human eye tissue. They found that not less than two-thirds of the expressed genes are homologous between the mollusc and the mammal; 97% of these shared genes probably being of urbilaterian or prebilaterian origin. Among the about 875 genes conserved between humans and octopuses, a subset is supposed to form a regulatory network responsible for the construction of chambered eyes, that has been recruited independently at least twice, in the molluscs and the vertebrates (Fernald 2006).

8.2.5 The digestive system

Until now, the origin of the anus has been considered from the point of view of its relationships with the blastopore, but the main problem in reconstructing its history lies probably elsewhere. For an opening of the digestive cavity to function as either a mouth or an anus, the digestive tract must possess a distinct polarity, which is a mouth-to-anus patterning. In principle at least, this longitudinal patterning is independent from the anteroposterior patterning of what we usually read as the animal's main body axis: this is the meaning of the 'double animal' concept suggested above (see section 8.1.3). Evolving

a longitudinal patterning of the digestive system is also independent from the ability to open new pores (something required to make a through-gut, save for the case of amphistomy, when the blastoporal slit is sealed at midlength and eventually gives rise to both mouth and anus). For example, in ctenophores there are small anal pores, but the majority of faecal matter is nevertheless expelled through the mouth, thus showing that in these animals the longitudinal patterning identifies a real digestive 'tube' (Martindale *et al.* 2002).

It is possibly too early to evaluate the real significance of the recent demonstration by Shimizu *et al.* (2007) that the digestive tract of *Hydra* has a second, extremely small opening, in addition to the mouth. This very narrow aboral pore is coincident with the site where the synthesis of an extracellular matrix-degrading enzyme is particularly high; transfer of material through this pore, both inward and outward, has been documented.

Is a through-gut plesiomorphic within the Bilateria? In the past—that is, before the advent of molecular phylogenetics—the answer to this question was uniformly negative. Flatworms materialized the concept of primitive bilaterians still lacking an anus, something that was supposed to have been 'invented' by nemerteans, together with the circulatory system. The modest cephalization of the flatworms supported this *scala naturae* scenario, which was only contradicted by the presence, in these animals, of a genital apparatus of unusual complexity. Current phylogenetic views, with the acoels and nemertodermatids close to the bilaterian root, may still support the idea of a bilaterian common ancestor with a blind gut, if not with a virtual one, as typical of the acoels. Therefore, I would not argue that the anus is an apomorphy of the Bilateria (e.g. Zrzavý *et al.* 1998b, Giribet *et al.* 2000, Nielsen 2001, Peterson and Eernisse 2001, Jenner 2004c) but, rather, that it is apomorphic for the Nephrozoa.

But the problem now arises of whether the Catenulida and the Rhabditophora are also primitively provided with a blind gut, or have secondarily reverted to this condition. Among the clades grouped with Catenulida and Rhabditophora in the Platyzoa, an anus is present in the Gastrotricha and in the basal Syndermata (i.e. in the former 'Rotifera'). It is obviously lacking in the Acanthocephala, but these members of the Syndermata have lost mouth, gut, and anus by becoming parasites, and the condition in the Gnathostomulida and the Micrognathozoa suggests that an anus is quite probably plesiomorphic for the whole of the Gnathifera. Indeed, the gnathostomulids were initially described as lacking an anus (one of the reasons why Ax (e.g. 1995) united them with the flatworms in a taxon Plathelminthomorpha), but ultrastructural evidence suggests the presence of a functional opening of the gut, opposite to the mouth (Knauss 1979, Sterrer *et al.* 1985, Lammert 1991) and the same seems to hold for the micrognathozoans (Kristensen and Funch 2000). Whether this anus is homologous or not to the anus of most protostomes is uncertain (Jenner 2004c), but I am not aware of a specific reason to doubt this. Thus, Catenulida and Rhabditophora are the only members of the Platyzoa for which there is no evidence of a derivation from animals provided with anus, but I would wait for positive, rather than negative evidence before regarding the point as settled. Marginally, it may be interesting to note that a temporary anal pore was found in a rhabditophoran, the kalyptorhynchid *Tabaota curiosa* (Steinböck 1963).

Davidson (2006) suggests broad homologies between the main sections of the digestive system (mouth, pharynx, midgut, hindgut, and anus) of distantly related phyla such as Arthropoda and Chordata, based on similarities of regional patterning based, for example, on Gata transcription factors. Midgut formation in *Drosophila* depends on the expression of two genes belonging to the *gata* class, *serpent* and *gatae* (Rehorn *et al.* 1996, Okumura *et al.* 2005), and related genera are also required for midgut formation in *C. elegans* (Maduro and Rothman 2002), echinoderms, and vertebrates. Similarly, the same set of genes (*brachyury*, *caudal*, and *foxa/hnf3β/fkh*) is jointly involved in the invagination of the hindgut in insects and vertebrates (Lengyel and Iwaki 2002). In detail, however, the genetic control cascades in which these and other genes are involved are not identical, and partly clade-specific (Davidson 2006). However, these molecular markers do not tell the whole story. Novelties must be expected even within one phylum. For example, the terminal gut of ascidians is formed late in ontogeny and Dohle (2004) has suggested that it represents a neoformation in respect to the terminal gut of the chordate common ancestor. Heterochronies may cause the problem to be more complex, unless we adopt a relaxed, factorial approach to homology (see Minelli 1998, and this book, section 4.2.3). A good example is provided by those insects (e.g. the honey bee and the ant-lions) where the midgut is blind until the end of the larval development and only connects to the hindgut at metamorphosis. In this case, the main pieces of the puzzle are phylogenetically old, but the times and ways they are linked together are evolutionarily novel (and independently developed within the Planipennia and the Hymenoptera).

Gastrodermal cells are monociliate in brachiopods and cephalochordates, multiciliate in flatworms, nemerteans, and rotifers, and non-ciliate in other bilaterians (Ax 1996). The phylogenetic signal derived from their distribution in the main animal clades is weak. In particular, intestinal cilia have been lost many times independently, not just in animals that lack cilia generally, as arthropods and other ecdysozoans, but also in animals provided with a ciliated epidermis, as gnathostomulids and micrognathozoans.

Among the metazoans, intracellular digestion is possibly more widespread than generally assumed. It appears to have a major role in a few groups such as molluscs and several crustaceans and arachnids (Mathieson and Lehane 2002, Hamilton *et al.* 2003), but it has been also recorded in parasitic flatworms (e.g. Dalton *et al.* 2004), chaetognaths (Perez *et al.* 2001), and fishes such as the goldline, *Sarpa salpa* (Kozaric *et al.* 2006). Intracellular digestion is sometimes a property of embryos, as in ophiuroids (Gliznutsa and Dautov 2005) or newborn animals (e.g. the hatchlings of the sea bream, Calzada *et al.* 1998; or the suckling rat, where this property is restricted to the middle segment of the small intestine, Baba *et al.* 2002). The phylogenetic distribution of intracellular digestion suggests that this property has probably been gained and lost independently many times. Interestingly, within the appendicularians intracellular digestion has been demonstrated in *Oikopleura* (Burighel *et al.* 2001, Cima *et al.* 2002) but seems not to occur in *Fritillaria* (Brena *et al.* 2003). The divide between intracellular digestion and the phagocytic activity of cells not related to nutrition is often arbitrary, as in the case of those insects where substantial amounts of sperm cells not involved in the fertilization of eggs are phagocyted, either in a broadly conventional spermatheca, as in the bee *Melipona bicolor* (da Cruz Landim 2002), or in specialized organs as in the bed bugs and related groups of heteropterans (Carayon 1966).

8.2.6 Body cavities

Homology of body cavities has traditionally featured as one of the main criteria in assessing the phylogenetic relationships of the main bilaterian lineages, but it is increasingly clear that several well-entrenched assumptions are unwarranted. The main points at issue are:

• the distinction between primary and secondary body cavities,
• the probability that a secondary body cavity can be lost,
• the phylogenetic value to be acknowledged to the distinction between secondary body cavities formed by schizocoely and secondary body cavities formed by enterocoely.

Primary cavities corresponding to blastocoelic spaces surviving in the post-embryonic stages of development are assumed to be necessarily older, phylogenetically, than secondary cavities forming in the embryo at a later stage, either by cavitation of a solid mesodermic mass (schizocoely) or by outpouching of the wall of the archenteron (enterocoely). This expectation reflects a recapitulationist view of animal evolution to which very few zoologists would subscribe today, but the consequences of rejecting this view for an understanding of the evolution of body cavities is not appreciated to the same extent. Similarly,

scepticism with respect to the likelihood of a loss of coelomic cavities is largely rooted in the belief that 'coelomate is better' and, thus, losing the coelom would represent a regressive trend to be only expected in exceptional situations. This amounts to identifying evolution with progress, where progress, additionally, is measured against some arbitrary scale of morphological 'perfection' or of functional performance at least. Finally, the putatively deep phylogenetic divide between schizocoelic and enterocoelic coelomates was rooted, in part, in an inadequate appreciation of how easily the mechanics of cavitation of a solid tissue mass and the outpouching of an epithelial sheet can be set in. Indeed, we should probably regard these processes as generic properties of animal tissues.

Jenner (2004c) lists a long series of developmental processes mechanically identical to enterocoely, in that they involve outpouching, folding, or invagination of epithelial sheets, including the formation of the archenteron by emboly, the growing of the neural fold into a neural tube in chordates, the opening of the pharyngeal pouches in vertebrates, and the production of the stomodeum and proctodeum in many metazoan clades. Schizocoely is also probably generic, and this invites caution with respect to declaring as homologous the coelom of polychaetes, the gonocoels of nematomorphs (Lanzavecchia *et al.* 1995), and the entocodon of hydrozoans (see section 4.2.3).

Interpreting the relationships between the body cavities of different bilaterian groups is complicated by difficulties in classifying these as either primary or secondary. Ruppert (1991a) remarked that the traditional definition of the primary body cavities as persisting blastocoel is too restrictive, as the blastocoel often disappears completely and primary body cavities are formed anew. It is therefore at the histological level that the distinction between primary and secondary body cavities must be maintained. But this is also problematic. As remarked by Jenner (2004c), in many invertebrates scattered among different phyla the body cavities (currently regarded as secondary body cavities; that is, as coeloms) are lined by myoepithelial cells rather than by a non-contractile peritoneum. This is true of some polychaetes (Fransen 1988, Bartolomaeus 1994), hemichordates (Benito and Pardos 1997), and echinoderms (Rieger and Lombardi 1987). On the other hand, non-muscular epithelial cells line the rhynchocoel of the nemerteans (Turbeville and Ruppert 1985, Turbeville 1991) and the internal spaces in the podia of the echinoderms (Rieger and Lombardi 1987), both of which should therefore be described as coelomic cavities. If we take the presence of a peritoneal lining as a distinguishing feature of the coelomic cavities, these have evolved many times (Rieger 1986, Rieger and Lombardi 1987, Bartolomaeus 1994, Jenner 2004c). It seems reasonable to follow Nielsen (1995, 2001) in adopting a definition of peritoneum that includes both non-contractile coelomic linings and myoepithelia. Nevertheless, a polyphyletic origin of the coelom is increasingly accepted (e.g. Kristensen 1995, Minelli 1995, Valentine 2004, Schmidt-Rhaesa 2007). Consensus is not universal even about the homology of the coeloms of annelids and molluscs, but this is due in part to reluctance to accept as secondary the lack of a coelom in groups such as the Myzostomida and Entoprocta (see Jenner's 2004c comments on Haszprunar's 1996 phylogeny). There is no convincing argument, however, to support the idea that two other groups lacking a body cavity (the gnathostomulids and the flatworms) derive from coelomate ancestors (Jenner 2004c).

A widely disputed question is the presence of a coelom in nemerteans. Difficulty in accepting the presence of a secondary body cavity in these animals was probably due to their traditional association with flatworms, in old zoological classification. In practice, the ribbon worms were often described as a kind of advanced flatworms that had 'invented' the anus and the circulatory system. But the histology of the nemertean blood vessels suggests a coelomic nature, and the same is true of the rhynchocoel (the fluid-filled cavity into which the proboscis of ribbon worms is retracted at rest). Thus, histology (that is, the presence in the vessels and the rhynchocoel of the nemerteans of a lining epithelium comparable to the peritoneum of a polychaete coelomic

pouch) suggests homology between body parts involved in very different functions.

Another group where the traditional interpretation of the origin and nature of body cavities was strongly biased by preconceived views of phylogenetic affinities is the Arthropoda, or the Panarthropoda. Accepting close affinities of the Arthropoda with the Annelida, arthropod body cavities have been generally regarded as secondarily reduced, an adaptive explanation for this trend being obviously offered by the presence of new locomotory solutions offered by the articulated appendages. But there is no evidence that the transient coelomic cavities seen in the embryonic stages of these animals have any role in the production of the musculature of the body wall, as is the case, for example, with annelids (Bartolomaeus and Ruhberg 1999). Within the Ecdysozoa hypothesis, annelid and arthropod body cavities are convergent rather than homologous.

A different problem is presented by tardigrades, which possess a body cavity lined by a basal lamina rather than by a peritoneal epithelium, and thus should be defined as a primary rather than secondary body cavity (Dewel *et al.* 1993). It would not be too difficult to imagine a loss of coeloms in very small animals, as are tardigrades, especially if the starting point was comparable to the condition of secondary body cavities in the onychophorans or the arthropods, but novel phylogenetic results may suggest an alternative scenario. As we have seen in Chapter 6, tardigrades might in fact be more closely related to the nematodes than to the onychophorans and arthropods. If so, the lack of a secondary body cavity in tardigrades would be a primitive rather than a secondarily acquired condition.

The embryonic origin of the body cavity in the Chaetognatha is unique, as are many other traits of this phylogenetically untractable clade. Here, rather than having an archenteron (the 'primitive' gut) giving rise to the coelomic pouches as in typical enterocoely, the endoderm first develops into a provisional body cavity whose wall will eventually give rise to both the remaining (anterior) part of the body cavity system and also to the animal's gut (Kapp 2000). This ontogenetic sequence is unique and thus of no help in fixing the affinities of the arrow-worms.

The phylogenetic position of the Priapulida has long been questioned, due to contrasting interpretations of the nature of their body cavities, but this small group has eventually turned into an exemplary case of the limited phylogenetic significance of this trait of animal architecture. In fact, histological evidence suggests that the body cavities in Priapulida are mostly primary, but a coelom is nevertheless present in the neck region of a single species (*Meiopriapulus fijiensis*; Storch 1991).

Finally let us comment briefly on three groups where the lack of a secondary body cavity is possibly a derived feature. I have already mentioned myzostomids. The presence of coeloms and coelomoducts in these animals is accepted by some authors (Rouse and Fauchald 1995, 1997, Fauchald and Rouse 1997, Rouse 1999) but not by others (Haszprunar 1996, Zrzavý *et al.* 1998b, Eeckhaut *et al.* 2000, Jenner 2004c). Next, the acanthocephalans. Their lacunar system distributed within the epidermis and the musculature seems to be neither a primary nor a secondary body cavity, as usually defined, but a network of clefts opening within the syncytial epidermis (Schmidt-Rhaesa 2007, citing Herlyn's unpublished observations). Dunagan and Miller (1991) found that this lacunar system opens through pores through which

nutrients probably pass from the outside (i.e. from the host's digestive cavity) into the acanthocephalan's body. Third, the urochordates. These deuterostomes are generally described as lacking typical body cavities formed by enterocoely, as found in the other deuterostomes (Nielsen 1995a, Welsch 1995, Presley *et al.* 1996). However, if the pericardium of ascidians and thaliaceans is to be regarded as a coelom, as suggested by Godeaux (1990) and Burighel and Cloney (1997), this would be the only surviving evidence of a secondary reduced coelom.

8.2.7 Respiratory organs

Respiratory organs are often very plastic and thus of limited phylogenetic interest. For example, there is little doubt that tracheae evolved many times independently in terrestrial arthropods (Ripper 1931, Hilken 1998) (Fig. 8.7). However, recent developments in comparative developmental genetics suggest that there might be a common ground to the respiratory structures of arthropods, gills as well as tracheae, starting from specific areas of the body surface specialized for gas exchange and osmoregulation. A strict developmental and, by inference, phylogenetic association between gills and appendages has been demonstrated in crustaceans (Averof and Cohen 1997) and a corresponding association

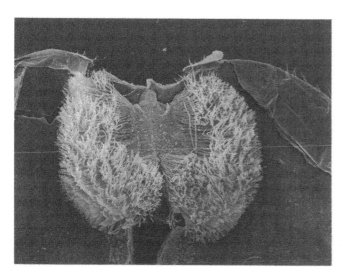

Fig. 8.7. Unique among the terrestrial arthropods is the respiratory apparatus of the scutigeromorph centipedes. This consists of eight dorsal units, each with a longitudinal slitlike opening giving entrance to a dorsal chamber. From the bottom of the latter, a huge number of short tracheae reach the dorsal vessel. The photo shows an internal view of one of these units in *Parascutigera festiva*. Courtesy of Gero Hilken.

has been found between leg primordial and tracheal placodes in *Drosophila* (Franch-Marro *et al.* 2006). Moreover, homologues of genes known to induce the formation of tracheae in insects are also expressed in crustacean gills. Additionally, the insect wing is likely to represent a modified kind of crustacean gill (Averof and Cohen 1997). We can therefore assume that within the arthropodan radiation respiratory organs re-evolved multiple times, but always exploiting a common pool of tissue types and control genes.

8.2.8 The circulatory system

As noted by Schmidt-Rhaesa (2007), most vessels (blood vascular systems *sensu* Ruppert and Carle 1983) are bordered by the extracellular matrix and must therefore be regarded as primary body cavities, or parts of these. Vessels provided with their own epithelium and thus corresponding to coelomic cavities (coelomic circulatory systems *sensu* Schmidt-Rhaesa 2007) are typical of nemerteans, cephalopods, and vertebrates (Nielsen 2001), but vessels lined by an endothelium have been also found in the proximity of the ganglia in the snail *Helix pomatia* (Pentreath and Cottrell 1970). Several coelomate animals (e.g. Priapulida, Sipunculida, Ectoprocta, Chaetognatha, and some 'polychaete' families like the Psammodrilidae and the Capitellidae) lack a circulatory apparatus (Beklemishev 1969, Rouse and Pleijel 2001).

Schmidt-Rhaesa (2007) restricts the use of the term 'blood' to the fluid present in coelomic circulatory systems, while the fluid in 'blood' vascular systems should be better called haemolymph. However, this distinction cannot be used uniformly throughout the metazoans, as there are mixed conditions between the two main classes of circulatory system (mixocoels): this is one more example of the strictures deriving from our efforts to describe nature through simple, rigid classifications.

8.2.9 The muscles

As in the case of many other cell types of metazoans and the peculiar molecules they express,

molecules characteristic of muscle, such as actin and myosin, are also present in organisms where muscles do not exist, like sponges (Harrison and De Vos 1991). Sponges, however, can perform coordinated pumping and interrupt this pumping, despite the lack of muscular and neural elements (Jacobs *et al.* 2007). Different adaptive scenarios can be suggested for the earliest steps in the evolution of a muscular system. Rieger and Ladurner (2003) favoured an interpretation based on the functional significance of the myocytes in a basal bilaterian clade—the small, ciliated nemertodermatids—where these cells control swimming and gliding motion. Muscular and ciliary activity do indeed also cooperate in locomotion in more 'modern' metazoans, and ones of much larger size than nemertodermatids, for example planarians. Even in the most improbable context for ciliary gliding—out of water—locomotion may still retain a ciliary component, as in the case of terrestrial planarians (Minelli 1981).

As myocytes are the only kind of mesodermal cells in nemertodermatids, it has been suggested that they may represent the original type of mesodermal cell (Rieger and Ladurner 2003). But the condition found in nemertodermatids is perhaps due to secondary simplification, because in the acoels, putatively more basal in the bilaterial tree than the nemertodermatids, there are also other mesodermal cells in the form of a peripheral parenchyma and the tunica cells of the gonads. All these mesodermal cells share an endodermal origin. A further area of uncertainty is the nature of the earliest muscular cells. Acoels and nemertodermatids have 'true' myocytes, but many coelomates (both protostomes and deuterostomes) have myoepithelial cells. Whether these are actually more primitive than typical myocytes, as suggested by their similarity to cell types found in cnidarians, is far from certain. We must avoid thinking in terms of evolution necessarily following a morphocline from epithelial through myoepithelial to fully muscular cells, which is rooted in typological thinking more than in sound phylogenetic comparisons.

A different scenario for the early evolution of the muscular system was suggested by

Beklemishev (1969), who envisaged a primary function of contractile cells in the seizure and swallowing of food. This hypothesis could make sense if related to the idea that early metazoans may have been sessile, but it was written many years before the discovery of carnivorous sponges. Following the first description by Vacelet and Boury-Esnault (1995) of a novel crustacean-crunching poriferan found in the Mediterranean Sea, many more sponges with a similar habit have been described (e.g. Vacelet 2006, Ereskovsky and Willenz 2007). In these poriferans, the filtering chambers are strongly reduced or completely lacking, but the adaptation to a carnivorous diet has not required the development of muscular cells or provided the opportunity for that. The tiny prey is trapped with thread-like cellular projections that eventually engulf them.

Since the 1970s, a time of rapidly expanding ultrastructural studies and of technical progress in transmission electron microscopy, a lot of information about the fine structure of muscles in the most diverse metazoan groups has been obtained and used, to some extent, in phylogenetic analysis. Caution is required, however. It is highly unlikely, for example, that a useful phylogenetic signal will be recovered from the presence of an intermediate epithelial cell, between the muscle and the structure to which it attaches, as this condition, peculiar as it may be, has been reported for the most diverse kind of taxa and muscles, including the stylet muscles of tardigrades (Dewel *et al.* 1993), the introvert retractor muscles of cephalorhynchs (Kristensen and Higgins 1991), the muscles at the service of the cephalopod beak (Budelmann *et al.* 1997), and other muscles of gastrotrichs (Ruppert 1991b), ectoprocts (Mukai *et al.* 1997), cyclophorans (Funch and Kristensen 1997), arthropods (Mellon 1992), and chaetognaths (Shinn 1997). There are conspicuous developmental aspects whose potential for phylogenetic inference has not been exploited yet, however. An example is the relative timing of the development of circular and longitudinal muscle fibres, which may differ between phyla, with the acoel *Convoluta pulchra* first forming circular fibres, followed by the development of longitudinal fibres, while the 'polychaete' *Capitella* sp. forms longitudinal fibres first (Ladurner and Rieger 2000, Hill and Boyer 2001). Another example is the high degree of independence between larval and adult muscle systems in molluscs (Wanninger *et al.* 1999, Haszprunar and Wanninger 2000), which contrasts with the close ontogenetic association of these systems in the recently investigated acoels and 'polychaetes', where the larval muscle system appears to forms a scaffold for the development of the adult muscles (Ladurner and Rieger 2000, Hill and Boyer 2001).

8.2.10 Skeletons and biomineralization

The origin of metazoan skeletons features is a central topic in many discussions about the reality of the Cambrian explosion. The sudden appearance in the fossil record of armoured animals belonging to many different clades is indeed undisputable, and this argues definitely in favour of a multiple origin of skeletons, whereas it may not be decisive (see section 3.5) in providing an answer to the question of the timing of metazoan origins. Changing from a soft body to one provided with mineralized parts, or vice versa, is something that does not require major structural changes, such as those corresponding to the differences between two phyla in the traditional classifications. A more sensible evolutionary question is thus to discuss the adaptive significance of the earliest biomineralization. Three main different scenarios have been suggested.

First is the seemingly obvious competitive advantages provided by the presence of a hard, protecting armour, as well as by internal and external rigid structures to which muscle can attach, thus providing opportunities for new or better locomotory solutions, especially when coupled with the development of articulated appendages. A second, different perspective was presented by Cohen (2005), who proposed that biomineralization caused a major ecological shift in the life of small pelagic bilaterians of the Late Proterozoic, because it allowed them to grow to much larger size, while at the same time becoming too heavy to continue a pelagic life. This would have caused them to settle in the benthic zone, where they could exploit new ecological opportunities. Cohen's scenario suggests an interesting explanation for the eventual success of biomineralization, rather than accounting for its origin, but the same objection could be raised in respect to the current adaptationist hypothesis framed in terms of selective advantage provided by biomineralization by improving mechanical efficiency and defence from predators. Although his model does not elaborate further on this point, Cohen's views have the merit of suggesting that mineralization is likely to have been mechanically and energetically advantageous *per se*, independent of any advantage it might eventually provide by generating armour against predation. Whatever the ecological context within which these rigid structures first appeared, their protective effect is indeed something that could hardly have been selectively relevant since the beginning. The defensive advantage eventually provided by a shell or a thick exoskeleton is likely to have evolved by exaptation of structures originally selected for something else. One of the possible scenarios was sketched in the previous chapter, in terms of the role of rigid structures, either cuticular or skeletal, in reducing intraorganismal competition and stabilizing form. A third aspect is the role of biomineralization as a way to store valuable cations of uncertain availability (Minelli 2003), or to remove them if they are in excess of safe metabolic levels. No doubt, the ability to control mineralization is older than the use of minerals in animal skeletons. Mollusc shells and vertebrate bones may still today act as stores of mobilizable calcium ions.

Within the individual lineages, the existing mineralized structures have probably followed diverging evolutionary paths where different selective advantages played the leading role, but it might be difficult to decide. What was most important to the earliest molluscs, between feeding with the help of a radula and exploiting the protective advantage provided by dorsal shell? Due to the persisting uncertainties about

the affinities and the internal phylogeny of the group (section 6.15), we should probably refrain from risking an answer. But a plausible scenario can be suggested in the case of the vertebrates, if we accept conodonts as vertebrates (Donoghue *et al.* 1998, 2000). There is no reason to dispute the major role the characteristic 'teeth' of these Palaeozoic animals must have had in their life as predators or scavengers. This has suggested that the vertebrate skeleton did not evolve first as a way to reduce damage from predators, but as an advantageous tool to be used in predation or scavenging (Purnell 1995, Sweet and Donoghue 2001).

8.2.11 The excretory system and osmoregulatory devices

Most bilaterians are provided with excretory organs, exceptions being acoels, nemertodermatids, and xenoturbellids. Two kinds of nephridium are commonly distinguished, the protonephridia and the metanephridia. Protonephridia are ciliated tubes that open through the epidermis with a nephropore and are proximally closed by one or more terminal cells (cyrtocytes), typically perforated in the distal part. Metanephridia also open proximally into an internal cavity filled with body fluids. Ruppert and Smith (1988) found a correlation between the presence or absence of a circulatory system and the type of nephridium present in the different bilaterian groups. As remarked by Schmidt-Rhaesa (2007), a good example of this correlation is provided by annelids. Here, those adult 'polychaetes' which lack a circulatory system retain the larval protonephridia rather than developing metanephridia, as do those provided with a circulatory system.

Homology of protonephridia throughout the protostomes has been widely disputed: Wilson and Webster (1974), Brandenburg (1975), and Ruppert and Smith (1988) have regarded them as independently evolved within several lineages, their basic similarities being explained by specific functional constraints coupled with the nature and distribution of fluid-filled cavities within the animal. Bartolomaeus and Ax (1992),

Bartolomaeus (1999), and Schmidt-Rhaesa (2007) favour instead the hypothesis of a single origin of the protonephridia, due to their structural uniformity in comprising a terminal cell covered by extracellular matrix, a canal cell, and a nephroporus cell. The simplest organization of the protonephridia, with just these three cells, is found in gastrotrichs, gnathostomulids, and some annelid larvae.

There are also problems with the homology of the metanephridia. Schmidt-Rhaesa (2007) stresses the different origin of the ciliated funnel, which develops from the duct cells in annelids, but is composed of coelothel cells in phoronids, and suggests that metanephridia may have evolved three times independently: in molluscs, in the ancestor of annelids inclusive of

sipunculans and echiurans, and in the deuterostome ancestor.

In the leech *Erpobdella octoculata* Quast and Bartolomaeus (2001) described two pairs of transitory nephridia which degenerate in a late embryonic stage (Fig. 8.8). The authors assumed that these organs are homologous to the larval protonephridia of 'polychaetes', but found instead ultrastructural similarities between these transitory nephridia and the segmental metanephridia of other leeches. As some annelid protonephridia resemble initial developmental stages of metanephridia, Bartolomaeus and Quast (2005) hypothesized that annelid protonephridia may have derived from truncated metanephridial development. This anti-recapitulationist argument deserves attention.

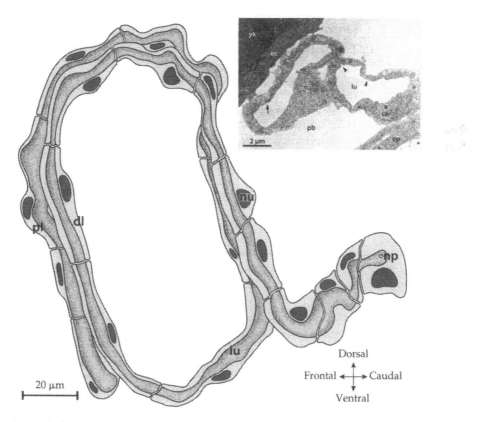

Fig. 8.8. A transitory nephridium in the leech *Erpobdella octoculata*. This organ is functionally comparable to a protonephridium, but morphologically it is closer to an immature metanephridium. ce, centriole; dl, distal loop; ec, entoderm cell; ep, epidermis; lu, lumen; np, nephridiopore; nu, nucleus; pb, primary body cavity; pl, proximal loop; yk, yolk; arrow, adhaerens junction; arrowheads, membrane pit. Courtesy of Thomas Bartolomaeus and Björn Quast.

8.2.12 The immune system

Immune molecules are not a privilege of vertebrates or, indeed, of 'higher' animals. On the contrary, their presence has been ascertained for groups as different as sponges, cnidarians, arthropods, molluscs, echinoderms, and nonvertebrate chordates (Loker *et al.* 2004). Vistas on the immune-system evolution have been recently widened by taking into account other classes of biological phenomena in which are involved some of the molecules hitherto regarded as specific to immune reactions. This is of major consequence both for understanding the origin of immune phenomena and for throwing light on to other processes which, in turn, can be regarded as derived from them, by exaptation.

An unexpected link has been found between some developmental events and immune defence. In vertebrates and insects alike, cellular response to microbial attack is mediated by homologues of two genes, *toll* and *dorsal*, which are involved in embryonic patterning (Cooper 2003).

A connection between the evolution of the adaptive immune system of vertebrates and the evolution of the nervous system was suggested long ago (e.g. Hildemann *et al.* 1979), based on the common properties of the two systems of performing specific recognition, retaining memory of events and giving rise to specifically directed aggressive responses. Moreover, the origin of the adaptive immune system may be temporally coincident with the origin of the neural crest. At the molecular level, several chemicals originally described as neurotrasmitters have been subsequently found to be active in the immune system too (Schauenstein *et al.* 2000, Bayne 2003), whereas some immunoglobulins play a role in controlling the growth path of neuronal axons in the developing nervous system of *C. elegans* (Aurelio 2002).

If the immune system is borne as a by-product of the activity of molecules previously involved in other functions, it is instead arguably at the origin of a very different function; that is, feeding an animal's offspring. Vorbach *et al.* (2006) postulate that the mammary gland of mammals evolved into a source of nutrition starting with a function in immune defence. Lactation may have evolved as an inflammatory response to tissue damage and infection. This should have been the original function of anti-microbial enzymes, xanthine oxidoreductase or lysozyme, which are remained involved in the production of milk.

8.2.13 Somatic structures involved in reproduction

Of all the main features of animal organization, gonads are possibly those that evolved, so to say, from scratch the highest number of times (Extavour 2008). In many sponges, cnidarians, acoels, and in some members of other groups, including *Xenoturbella*, gametes simply develop in the gastral cavity or in the coelom. Species with gonads and species without gonads are found even within restricted groups, such as the acoels and the annelids. Of the acoels, some lack gonads but others have well-defined paired ovaries. Within annelids, in most 'polychaetes' gametes form on the walls of the coelomic pouches, but in the gonochoric *Pherusa* and *Flabelligera* gonads separate from the coelom are present in the anterior region of the body (Amor 1994, Spies 1997).

In several instances, to decide whether the cavity hosting the germ cells is a gonad or a coelom is simply a matter of terminology; this is the case of nemerteans and clitellate annelids. In still other cases, germ cells are found in cavities lacking any kind of lining epithelium. This happens, for example, in crinoids, in male *Caenorhabditis* (Wright 1991), and in the female of the gastrotrich *Lepidodermella squamata* (Ruppert 1991b). In the female of the nematomorph *Nectonema*, ovogenesis does apparently take place in the gonoparenchyma that occupies most of the worm's body but eventually regresses, leaving the mature eggs free in the body cavity, without any epithelial gonad to contain them (Schmidt-Rhaesa 2002).

CHAPTER 9

The overall picture

In this golden age of biology, a book faces the danger of becoming obsolete before its publication. It is my belief that in order to avoid early obsolescence, the author, judging on the basis of the scant evidence available, is obliged to anticipate future developments and paint a picture with broad strokes of his brush.

Ohno (1970), p. 2

Thirty years ago, in April 1978, 125 participants attended a meeting held by the Systematics Association at the University of Hull, UK, to discuss *the origin of major invertebrate groups* (see House 1979). Results were clearly far from definitive, as 5 years later, at another Systematics Association meeting, held in London at what was then called the British Museum (Natural History), 150 participants discussed *the origins and relationships of lower invertebrates* (see Conway Morris *et al.* 1985). In the quarter of century elapsed since the latter event, our appreciation of phylogenetic relationships among the main animal lineages has completely changed. An overview of the recent discussion and a provisional synthesis of its still evolving results has been given in the first half of this book. In June 2007, while I was writing these chapters, another international conference attracted to the Royal Society of London several hundred researchers to discuss *the evolution of the animals* (see Telford and Littlewood 2008). Basically, the theme of this more recent conference was the same as those that took place 25 and 30 years ago, but the scenario had completely changed. This is not simply due to recent discoveries of remarkable Late Precambrian and Early Cambrian fossils, nor to the unexpected patterns of metazoan interrelationships emerging from a previously unavailable and perhaps unexpected wealth of molecular evidence. What is also changing is type of question being asked.

In the past, focus was on origins, whereas now the key word is relationships. Cladistic methods and tree-thinking are now firmly established in the community of zoologists working in this field, and the residues of *scala naturae* models and concepts such as the 'lower invertebrates' have largely disappeared. But this does not mean that we can content ourselves with the current awareness of metazoan phylogenetic relationships.

The critical point is not so much the uncertainty that still surrounds many nodes in the tree, but the widespread attitude of regarding the reconstruction of phylogeny as the most pressing problem in evolutionary studies. I think that we must be more ambitious and go on using phylogeny as a background against which to investigate a potentially unlimited number of interesting evolutionary problems. The topics briefly treated in Chapters 7 and 8 show that there is much more to evolution than reconstruction of the phylogenetic tree. One might even argue that a substantial amount of phylogenetic reconstruction is nothing more than playing with numbers in a matrix. This is indeed an objection raised many times against so-called pattern cladistics, the discipline within which were developed most of the algorithms currently implemented in phylogenetic packages. But evolution is process rather than pattern. A few years ago, Jenner (2004b) lamented that recent phylogenetic studies make no attempt to evaluate their cladograms in terms of character-state transformations. There are laudable exceptions to this generalization, but still not enough.

If we are interested in evolution, the tree is not the final target of our investigations, but the branching topology against which we can study a long and not necessarily progressive history of change.

Ironically, whereas the recent major advances in phylogenetic reconstruction have to date failed to stimulate a lively revisitation of the evolutionary history of morphological and developmental features, many bold stances about this history have been pronounced without being based on a critical use of the comparative method, but simply based on comparisons of evidence from a couple of model species.

This is particularly true of comparative developmental genetics, the discipline which in the last two decades has had a leading role in launching a new entry in life sciences; that is, evolutionary developmental biology, or evo-devo. The complex Urbilateria discussed in Chapter 4 is a typical example of this effort.

I hope that in the near future we will be able to combine an increasingly accurate reconstruction of phylogenetic relationships with an increasingly detailed knowledge of developmental mechanisms, and especially of the identity and patterns of expression of the genes involved in these, and of the evolution of the genes themselves and of the way they operate. The picture of animal evolution thus emerging will be arguably much more sensible and accurate than those produced to date. But this picture will not be necessarily framed by the same categories we have used to until now. As noted earlier in this book, evolution is not simply a history of the emergence of new products, but also a history of changes in the rules by which products are obtained. This means that we cannot expect to be able to describe the whole history of the metazoans by using concepts and terms that are adequate to describe many of the forms around us, though not all of them, and arguably still less adequate to describe animal forms of the most remote past. This final chapter is devoted to a critical overview of some evolutionary problems we can address and of some conceptual pitfalls we must avoid, perhaps at the price of disposing of many cherished concepts and terms.

9.1 Evolutionary novelties

9.1.1 Apomorphies versus innovations

As soon as a reasonably robust phylogenetic tree is available, one of the first questions we are probably going to ask is where along the tree did some major features originate; for example, the bird wing, the amniotic egg, or the insect pupa. Nobody will imagine these features to have originated in one day, but there must be a segment in the tree where the transition from arm to wing happened, or any other change of similar tremendous consequence in terms of adaptation. However, any hope of eventually fixing the point where these major transitions happened is strictly bound to the use of very coarse, poorly resolved trees. The more numerous the taxa included in the tree, the more vanishing is the chance that what we regard as a lineage's novelty or key feature will be found as an apomorphy defining a node in the tree. This is indeed a seldom appreciated point of clash between research traditions. At variance with evolutionary biologists interested in the origin of adaptations, phylogeneticists look at robust apomorphies, irrespective of their nature and of their functional significance and adaptive value. Deeply branching clades corresponding to high-ranking taxa in the traditional classification turn out to be diagnosed by a few molecular signatures or by seemingly minor ultrastructural details such as the fine structure of the ciliary roots in epithelial cells, whereas gross morphological characters are less and less likely to appear in a list of their apomorphies, the better we resolve the tree.

As the scale of what is traditionally called a family, a recent study of the alpheid shrimps (Anker *et al.* 2006) has provided an exceptionally clear example of the need to revise our approach to the evolution of the so-called key adaptations, or major changes. Many members of this group of decapod crustaceans are provided with a modified 'snapping' claw (Fig. 9.1) and are thus called snapping shrimps. A functional snapping claw differs in many respects from a conventional appendage and a phylogenetic analysis based on a rich taxon sample suggests that this adaptation has evolved only once in the group. However, several individual features which, combined with others, would eventually produce a snapping claw have evolved independently many times within this family. Asymmetry, folding, inverted orientation, adhesive plaques that enhance claw cocking, and tooth-cavity systems on opposing claw fingers are found in many alpheids which have not evolved into 'true' snapping shrimps. What eventually turned these features into 'preadaptations' to the evolution of a snapping appendage was a relatively minor additional change, like cutting the edges of the pages of a book, which may require this small action to become readable from cover to cover. But, as we would hardly say that the book is created at the exact time when we snip off its folded leaves (something young readers may never have had a chance to do), so why should we say that the snapping appendages of a lineage of alpheids originated exactly at the time the last piece was added to what eventually turned to be a snapping device?

Most if not all of the features in which we are most interested are no less complex than a snapping shrimp's claw. As a consequence, to search for their origins is likely to produce increasingly disappointing results, as long as our knowledge of phylogeny becomes more detailed. This is happening, indeed, with the reconstruction of the water-to-land transition in vertebrates. It is by now clear that the Palaeozoic forms somehow

intermediate between a perfectly aquatic 'fish' and a largely terrestrial tetrapod do not form a linear sequence of animals provided with an increasingly larger and uniformly progressive series of land adaptations, but a branched (at least, comb-shaped) lineage, whose members variously evolved anatomical features compatible with terrestrial life. Many useful features first or uniquely appeared in this or that animal were not necessarily inherited by the eventually surviving lineage of terrestrial vertebrates (e.g. Ahlberg 2003, Shubin *et al.* 2004, Daeschler *et al.* 2006).

9.1.2 Factoriality

Mayr (1960) suggested that major changes are often accomplished by progressively adding new components to a central, pivotal structure. Splitting the origin of a complex organ into a number of individual components is clearly a perspective that helps us to accept the idea that structures as complex as a vertebrate eye could evolve within a short time (on the scale of geological time) (e.g. Nilsson and Pelger 1994). An attractive feature of Mayr's approach to the problem of the origin of evolutionary innovations is its factorial nature. It invites the splitting of complex features into the smallest components we can individually trace in their evolutionary history. A factorial approach to the study of evolutionary innovation has helped us to frame the

Fig. 9.1. The snapping shrimp *Alpheus* cf. *sulcatus.* Courtesy of Arthur Anker.

question about the origin of arthropod append-ages in a way that is compatible with what we know about body patterning along the main body axis of bilaterians in general. Traditional scenarios (Snodgrass 1935, Raff and Kaufman 1983) suggested that the lineage leading to arthro-pods originated from segmented ancestors (in agreement with the Articulata hypothesis), first developing a set of identical pairs of append-ages, one pair per segment, and subsequently acquiring an increasingly complex anteroposter-ior patterning expressed in the differentiation of different kinds of appendages (sensory, feeding, locomotory, copulatory). Besides the question of the likely independent origin of segmentation in annelids and arthropods, which is not at issue here, the weakest point in this scenario is that it presupposes a lack of anteroposterior body pat-terning in the immediate ancestors of the arthro-pods, or at least the irrelevance of any patterning of the main body axis to the kind of appendages eventually developed on each segment. But the most proximate ancestors of the arthropods, either segmented or not, were certainly pat-terned anteroposteriorly and it is highly unlikely that this would not have affected the structure of the appendages they developed at different posi-tions. In other terms, we must assume that some degree of regional differentiation among the segmentally repeated pairs of appendages must have been present in the earliest members of the arthropodan lineage. Thus, if a novelty can be singled out at the origin of the Arthropoda, this was possibly a new way to 'read' the axial infor-mation already available, thus giving rise to a series of appendages that were segmentally pat-terned since their first appearance (Minelli and Fusco 2005).

Back to the principle of factoriality impli-citly present in Mayr's suggestion, there is no doubt that this is a valuable point, but there is a problem with Mayr's original formulation. The problem is whether we can accept as a rule the existence of a 'pivotal structure' around which further elements are progressively added. In current terminology, this corresponds to accept-ing a very long persistence of an evolutionary module that progressively expanded with time

without losing its identity. Intuitively appealing as the idea may be, I think that we cannot accept it as a rule. It is only through a case-by-case investigation that we will be able to see whether a 'pivotal structure' has actually existed, as an element upon which selection has continued to act in basically the same direction irrespective of the additional elements progressively accrued to the feature. In other cases, the long-term sta-bility of what we recognize as a conserved core element of a complex organ could be a develop-mentally or structurally stable feature, which is conserved as an 'attractor of morphological design' (Striedter 1998, Müller and Newman 1999, G.B. Müller 2003) irrespective of any pos-sible change of its adaptive value. This perspec-tive can be regarded as an elaboration of the current scenario of evolution by exaptation, by which a feature is regarded as having origin-ally performed a function other that the one it performs today. However, the new perspective differs basically from the change of function *à la* Gould and Vrba (1982), as it points to the fact that features can first appear as developmentally stable elements of structure without necessarily being involved in a recognizable function since the first day they are present.

Speculative aspects apart, a factorial approach to the study of the origination of evolutionary novelties helps us to trace their steps across our well-entrenched taxonomic boundaries. For example, the neural crest is a vertebrate apomor-phy, but the cellular basis of this feature is likely a chordate, rather than a vertebrate apomorphy. Indeed, neural crest-like cells have been also found in ascidians (Jeffery 2007), where these cells migrate from the dorsal midline, express a few typical neural markers and eventually dif-ferentiate into pigment cells. Indeed, pigment cell development and dispersal was possibly their original role, to which many other roles have been added in the vertebrate lineage.

9.1.3 Evolvability and constraints

Several biologists and philosophers of biol-ogy regard the emerging field of evolutionary developmental biology as the source of a view of evolution fundamentally different from the traditional neo-Darwinian view. Others think, instead, that evolutionary developmental biology is not an alternative to neo-Darwinism, but integrates it by focusing on a hitherto neglected aspect of the evolutionary scenario; that is, on evolvability.

What is evolvability? In somewhat loose terms, we can define the evolvability of a population as the range of new phenotypes that are immediately within reach, starting from its current characteristics. Most animals, for example, will have some degree of evolvability in body size, and all snakes have arguably some degree of evolvability in the number of vertebrae, but in the mammals the evolvability in the number of cervical vertebrae is virtually zero. Indeed, in the long neck of the giraffe there are only seven cervical vertebrae, exactly the same number found in the neck of a human, or in virtually any other mammal. Selection managed to elongate the neck of the giraffe's ancestor by acting on the size and shape of the individual vertebrae, but not on their number, because the latter offered no variation. That is, zero evolvability.

Evolvability thus represents a constraint on evolution (see Maynard Smith *et al.*1985, Fusco 2001). Dworkin *et al.* (2001) introduced a distinction between qualitative (in a sense, absolute) and quantitative constraints. Quite often, anyway,

constraints are of contingent nature, as demonstrated in the case of mammals by the existence of a couple of genera where the number of cervical vertebrae is slightly different from seven (actually, between six and nine) and in one case at least (the dugong) with some degree of intraspecific variation, as obviously expected from a population genetics-based perspective on evolution. (From a strictly neo-Darwinian perspective, Hansen (2006) defined evolvability as the ability of the genetic system to produce and maintain potentially adaptive genetic variants, but this is somehow reductive in a wider perspective.) Eventually, it seems sensible to follow Richardson and Chipman (2003) in remarking that all we say about constraints are in fact statements about the relative frequency of a particular transformation; that is, within limits, never say never.

A few years ago, Eberhard (2002) discovered the existence of movable appendages on the fourth abdominal sternum of the males of several sepsid Diptera. These appendages (Fig. 9.2) are not homologous to the 'true' segmental appendages of arthropods, but are independently (and recently) derived from originally fixed projections which eventually became articulated with respect to the supporting sclerite. It has been recently shown that these novel abdominal appendages develop from ventral histoblasts of the fourth abdominal segment. Similar to the

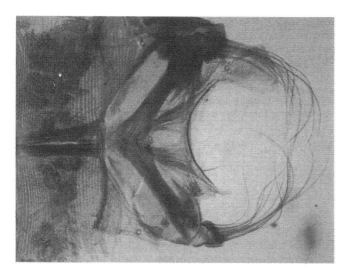

Fig. 9.2. Fourth abdominal sternite of the male of the sepsid dipteran *Themira putris*, showing the newly evolved articulated appendage on the right side. Courtesy of William Eberhard.

imaginal discs, histoblasts are clusters of larval cells surviving the extensive histolysis occurring during the pupal stage. However, whereas adult thoracic legs, antennae, etc. derive from proliferating and differentiating imaginal discs, histoblasts are usually involved in the production of adult epidermis and musculature. Thus, their involvement in the production of the abdominal appendages of the male sepsids implies a degree of evolvability we would never have suspected before knowing about this case (Bowsher and Nijhout 2007)

Whether all this is of consequence for our appreciation of Dollo's law (the principle according to which evolution of any non-trivial trait is irreversible) is another question. Recently it has been claimed that a gastropod family, the Calyptraeidae, first evolved uncoiled shells and then, after conserving this condition for not less than 20 million years, re-evolved shell coiling simply by prolonging the developmental time during which genes relevant for shell coiling are expressed in the larva (Collin and Cipriani 2003, Pagel 2004); but this analysis has been rejected on methodological grounds by Urdy and Chirat (2006).

9.2 Evolutionary trends

9.2.1 Evolutionary trends: facts or fiction?

A fashionable topic in the early literature on biological evolution was the putative orthogenetic series within which morphological and functional change seemed to proceed steadily in the same direction over long time spans. Most famous among these series was the sequence of equids from the small Eocene *Hyracotherium* (generally cited as *Eohippus* in the works dealing with orthogenesis) to the modern *Equus* species. A convenient selection of fossils and a more or less conscious silence about other branches of the horse family tree amounted to shaping a seemingly linear series spanning some 60 million years. Along this sequence, body size increased steadily, accompanied by a reduction in the number of toes and by a progressive specialization of the teeth towards the kind

found in modern horses, asses and zebras. This manifold morphological trend was presented against the background of a plausible history of major climatic change along the Cenozoic, with an increasing reduction of the forests (where *Eohippus* was supposed to live) to the benefit of increasingly wider grassland areas, the typical habitat of modern equids.

Orthogenesis has been long relegated in the archives of old and increasingly forgotten theories. If interpreted as the result of an internal drive pushing an animal lineage along a preferred and eventually unavoidable fate, it smelled too much of the Lamarckian views of change. Alternatively, it might be construed as the proof of a Divine design, something that has no place in a scientific enquiry. Eventually, the horse family series and the like have disappeared from the professional literature on evolution (but not necessarily from all school textbooks), not simply because of their methodological inadequacy but also because the very evidence on which this series was supposedly based failed to sustain the challenge of more accurate phylogenetic studies. Nevertheless, a 'domesticated' form of the frame of mind that supported orthogenesis is still alive. In the literature on biological evolution there is still a place for so-called evolutionary trends. Think of that group of green algae which pushed a simple stem above the water surface and eventually became less and less dependent on water, for example disposing of flagellate ('male') gametes, which are 'still' present in mosses, but which were notoriously lost in the 'more advanced' clade of the flowering plants. Think of vertebrates moving up along the ladder of terrestriality, progressively changing fins into limbs, evolving a couple of lungs, and eventually 'inventing' both the amniotic egg and a body cover much more protective against to desiccation than the 'poor' naked, moist skin of frogs and salamanders. Are these trends not real?

To be sure, to reconstruct the steps of this transition from water to land, either in the plant or the vertebrate lineage, we do not need to invoke either internal drive or providential design. We can easily frame the individual changes in terms of selective advantage of specific morphological and functional traits with respect to the physical conditions in the terrestrial environment, or with respect to the biological context within which the plants and the animals on their way from water to land could find competitors, or enemies, or resources to be exploited. However, it is difficult to avoid the feeling that most of our reasoning about evolutionary trends is based on a selective reading of the phylogenetic tree, from which we prune off the branches that do not fit with our largely preconceived scenario of directional transitions. It is all too easy to dismiss as secondary (in the sense of uninteresting, rather than in the sense of more recent) the land-to-water transitions so frequently observed among the tetrapods. Looking to Recent forms, we can easily argue that aquatic lineages of mammals or birds with terrestrial ancestors are more numerous than the other way around. Moreover, we cannot dismiss these geologically recent transitions from land to water as evolutionary events that have not changed much of the structural and functional organization of the animals involved. Whichever way we estimate change, it would be hardly possible to say that whales and dolphins are minimally modified with respect to what we can imagine was the last common ancestor of all Cetartiodactyla.

On the other hand, segments of evolutionary history characterized by broadly progressive changes in a given morphological and functional direction are something we must expect to find, for two different but complementary reasons. One is evolvability, the other is selection. As we have seen, evolvability refers to the range of variation that is more immediately accessible to an organism. In this respect, a vertebrate lineage where the ancestral five-digit pattern has been destabilized, for whatever reason, is likely to vary not just from a five-digit to a four-digit condition, but also to produce, sooner or later, further degrees of reduction. This is the material upon which selection will operate and in a broadly stable environmental context we can expect directional selection to continue as long as variation is available. But these directional trends are nothing other than chapters of local evolutionary history, and different if not opposite

trends could well be at work at the same time in other parts of the big crown of the evolutionary tree, or could be started as soon as changed patterns of variation or new selective regimes, or both, dictate a new evolutionary direction.

Thus we have plasticity on one side and constraints on the other. Plasticity is probably much more conspicuous than we generally assume, because evolutionary change is not simply matter of available variation and differential fitness. It is also change in the rules of the game. In the leeches, segmentation is a strictly constrained process, such that the number of body segments is fixed at 32. It is the same number for the flat, leaf-like glossiphoniids with a body length of just three or four times their maximum width, and for the longest cylindrical erpobdellids, with a diameter less than one-tenth of the body length at rest. Nevertheless, leeches have evolved annulation—that is, a regular subdivision of segments into secondary units—which is not reflected in the internal anatomy, but is of some importance with respect to the manoeuvrability of the body. In this way the developmental constraint fixing the segment number at 32 units is partly overcome, not by modifying a probably intractable mechanism of segmentation, but by exploiting an independent, multiplicative mechanism of annulation. Changing the rules, or adding new rules, is arguably a very powerful way by which living beings evolve.

Thus a sensible question is how common are evolutionary dead ends. I do not mean here narrow endemics, or animals with a strictly specialized diet, frequently fated to disappear because of unsustainable fluctuations in populations size, nor obligate parthenogens, whose long-term survival is possibly higher than often suspected in the past. I mean, instead, no-return morphological specializations accompanied by steadily decreasing evolvability. For example, is there any chance of reversing a trend towards a miniaturized organization?

9.2.2 Miniaturization

Body size is an obvious target of natural selection, but it correlates very loosely with both phylogeny and the general traits of body architecture, including complexity. In several lineages, including mammals, insects, crustaceans, and bony fishes, the largest living representatives of the group are about 1000 times as long (that is, roughly, 1 billion times as heavy) as the smallest members of the same group. Compare, for example, the Etruscan pygmy shrew (*Suncus etruscus*) or the Thailand bumblebee bat (*Craseonycteris thonglongyai*) with the huge blue whale (*Balaenoptera physalus*) or, within crustaceans, the smallest copepods and tantulocarids with the largest of lobsters and crabs. Nematodes are even more impressive, the 1 mm size of *Caenorhabditis elegans* is barely one ten-thousandth of the length of *Placentonema gigantissimum*, a parasitic roundworm living in the genital apparatus of the cachalot.

I will discuss here briefly a recurrent trend in body-size variation that often correlates with the occurrence of novel morphologies (Hanken and Wake 1993); that is, miniaturization.

Miniaturization is not simply reduction in size, it is an adjustment of cells, organs, and functions to an adult size dramatically smaller than in an animal's ancestor. Lineages with a very long history as diminutive animals are often easily distinguishable from species very recently derived from much larger ancestors. Mites, for example, are a very old lineage of small-sized

arthropods (Fig. 9.3), whereas the tiny spider *Comaroma bertkaui*, just 1.6 mm long (Kropf 1998), is an example of much younger miniaturization. This contrast is reflected in their anatomy. Whereas in mites all integumentary structures are rigorously symmetrical and precisely patterned, in *Comaroma* the distribution of the slit sensory organs, which in the larger spiders are distributed in a regularly symmetrical pattern, are irregularly asymmetric, as if the mechanisms involved in body patterning had not yet got a chance to be fine-tuned to the dramatically reduced size of the miniaturized animal.

Similar to the mites, the minuscule loriciferans, mostly one-quarter of a millimetre in total length when adults, arguably also have a long history as a lineage of miniaturized animals. Their very complex armour (lorica) and the numerous set of scalids regularly arranged around the anterior section of their body are the outwards visible aspects of a compact but anatomically complex animal composed of many thousands of extremely small cells. On the contrary, in recently miniaturized animals the average size of cells in any given organ is the same as in their bigger relatives. This has important consequences for their overall organization, because the number of cells in certain organs cannot be arbitrarily reduced without impairing function. This is particularly true of the central nervous

Fig. 9.3. The eriophyid mite *Ditrymacus* (≈0.1 mm). Mites are an example of a large metazoan lineage with a very long evolutionary history as miniaturized organisms. Courtesy of Giorgio Nuzzaci.

system, which is much less prone to a reduction in size than are, for example, the skeleton or the musculature.

Other size (or, better, volume) constraints are at work in the reproductive organs. A change to small body size may result in the loss of the ability to produce large quantities of gametes that can be freely discharged into the environment. Small-bodied taxa may thus shift towards specialized modes of sperm transfer, internal fertilization, and the consequent evolution of modified sperm morphology to ensure fertilization success. This can be argued for the rotifers (Melone and Ferraguti 1999) and for small-sized gastropod lineages (Ponder and Lindberg 1997) which frequently evolved direct sperm transfer through copulation, internal fertilization, and modified filiform sperm morphology.

Thus, recently miniaturized species are likely not to be a simple diminutive equivalent of their bigger relatives, but to show conspicuous differential changes in different body parts (Hanken and Wake 1993). This is observed in many groups of miniaturized animals, from the tiny *Thorius* salamanders (Hanken and Wake 1998) to the smallest geophilomorph centipedes such as *Nannarrup hoffmani* (Foddai *et al.* 2003). In his excellent book on *Evolution Above the Species Level*, Rensch (1959) had already remarked that, in contrast to what is observed in their larger relatives, the brain of very small insects (e.g. gall midges and plant lice) occupies all the space of the relatively large head capsule, whereas the midgut of such small species is distinctly reduced. In the minute marine opisthobranch *Caecum glabrum*, just 1 mm long, Götze (1938) found that the cells of the head epithelia and those lining the intestine are of about the same size as in large gastropods like *Littorina littorea*; this is accompanied by a simplification of the internal anatomy, from which the nervous system largely escapes.

Thus, is there any limit to miniaturization? Perhaps more interesting, is this trend reversible?

To some extent, no organ seems to be so vital as not to be dispensable when size reduction is pushed to the extreme. In some tiny parasitoid wasps, a larval type has evolved which not only

lacks appendages completely, but also a respiratory, circulatory, and even nervous system (Heming 2003). Eventually the animal can be reduced to a very small number of cells. Within the Cnidaria, the Microhydrulidae are reduced to a spherical body, 20–300 μm in diameter, with no mouth, gastral cavity, or tentacles. In the absence of a recognizable main body axis, asexual reproduction is not by paramorphic origination of a new body axis, as in the budding of *Hydra*, but by simple mobile frustules that split off from the parent animal (Bouillon 1993). Extremely small forms have evolved several times among the digeneans, with hermaphroditic progenetic Microphallidae and Lecithodendrioidea approximately 0.5 mm long, and only 150–200 μm in the case of sexually mature *Parvatrema* (Galaktionov and Dobrovolskij 2003). In the same group, active miracidium larvae are usually 80–150 μm long, but can be reduced to 20–60 μm; a further reduction (to just 4–10 μm in Microphallidae, Lecithodendrioidea, and Brachylaimoidea) is accompanied by the loss of locomotion, these smallest miracidia infecting their mollusc host by ingestion (Galaktionov and Dobrovolskij 2003). Very small larvae have also evolved in the cestodes, with oncospheres containing some 50–100 cells only (Hartenstein and Jones 2003). Small larval size, however, is the rule among marine invertebrates; the larvae of many sponges, hydrozoans, bryozoans, and 'polychaetes' metamorphosing with body lengths of 100 μm or less (Hadfield 2000). Much more interesting is the extremely reduced body size of many adult interstitial annelids, the male of *Dinophilus gyrociliatus* being just 50 μm and unsegmented, while the sensibly larger female is segmented (Scharnofske 1984, cited by Westheide 1990). Of its 330 cells, 68 are nervous; in addition, this tiny animal is provided with a copulatory stylet: as expected, the organs less affected by miniaturization are the nervous system and the reproductive organs.

The plasticity of animal architecture with respect to dramatic changes in overall body size is further demonstrated by the following examples. One is again from the 'polychaetes', where the Sabellidae include big animals like

Sabella spallanzanii, up to 45 cm in total length, and also the tiny *Fabriciola minuta*, just 0.85 mm long (Rouse and Pleijel 2001). The second example is from the vertebrates, with the Cyprinidae ranging in size between 300 cm total length and 300 kg maximum weight of the giant barb *Catlocarpio siamensis* (Roberts and Warren 1994) and the 7–8 mm of *Paedocypris progenetica*, the smallest vertebrate described to date (Kottelat *et al.* 2006). But let us go back to the 'polychaetes', to discover a fine example of a reverse transition in body size, accompanied by an increase in the number of segments, contrary to predictions in accordance to the so-called Williston's law (see section 1.3.2). The hesionid *Microphthalmus hamosus* is a non-interstitial member of a genus otherwise including diminutive interstitial species (Westheide 1982). The increase in size and segment number observed in this species accompanies an ecological shift from interstitial meiofauna to non-interstitial macrofauna (Westheide 1987). The previous history of miniaturization has left its mark, as the typical 'polychaete' chaetae lost by this lineage along its history of adaptation to the interstitial environment have not re-evolved in *M. hamosus* (Worsaae and Kristensen 2005).

9.2.3 Ecological trends

Reconstructing phylogeny is the only way we can hope to determine whether a major ecological shift happened only once or multiple times, and also whether there have been reversals. This will be the background against which to look for morphological and functional correlates of the ecological shift, but the more we know about this, the more cautious we must become in the face of even seemingly obvious correlations. Take, for example, Bergmann's rule, the principle according to which larger size would be favoured in cooler environments (Bergmann 1847). How generally does this rule apply? As for ectotherms, a recent study (Adams and Church 2008) has shown that amphibians do not follow Bergmann's rule. This rule makes more sense, indeed, for endothermic animals like mammals, as larger size (and compact shape) will obviously reduce heat loss. Many studies report

size-to-temperature relationships following Bergmann's rule in mammals and birds, but with important caveats. For example, Ashton *et al.* (2000) confirmed its validity for mammals, but their study failed to confirm heat conservation as the driving force. Another study (Rodríguez *et al.* 2008), based on a large sample of mammal species from the Western Hemisphere, reported that the rule applies in cold areas only, while Guillaumet *et al.* (2008) reported that two species of North African larks of the genus *Galerida* follow Bergmann's rule, but a third congeneric species does not.

Moreover, extrapolating from intraspecific to trans-specific patterns requires additional care, a methodological question we cannot discuss here in depth. We need to develop an effective link between the traditional evolutionary biology rooted in population genetics and the emerging evolutionary developmental biology interested in major changes in animal organization (Gilbert 2003a), despite the conceptual and technical difficulties it may imply.

It may thus be useful to revisit Anderson's (1948) concept of 'hybrid habitat', introduced (in botany) to describe the environmental conditions where a population is likely to maintain an unusually elevated variation such as deriving from recent hybridization. Similarly, I think that of the environmental scenarios proposed as the theatre of singularly active evolutionary change, those pointing to the 'creative' nature of transitional or ecotonal environments are the most plausible, in principle at least. Those environments are more likely to support the existence of organisms that simultaneously present alternative ways to survive, such as terrestrial as well as aquatic respiration, a peculiarity which can be exploited sequentially during an animal's life, as in dragonflies and frogs, as long as these animals remain in the transitional environments, but can also eventually be fixed in the condition opposite to the original one. Besides the familiar scenario of Devonian and Carboniferous swamps imagined to have been the theatre of important steps in water-to-land transitions, among both the vertebrates and the insects, a less familiar but no less suggestive scenario has been proposed by Jablonski (2005) for the evolution of many groups of marine invertebrates, within which innovations, as a rule, arise onshore, or preferentially survive onshore, while the more stable offshore environments are mostly colonized by largely stabilized forms.

To be sure, we will never know how many lineages have independently crossed the water-to-land divide. The best we can do is to try to identify the minimum number of such transitions in the ancestry of the terrestrial animals living today, something we can only base on sound phylogenetic hypotheses. Let us see, for example, what current phylogenetic awareness suggests in the case of the terrestrial arthropods. Within the Chelicerata, Scholtz and Kamenz (2006) have provided evidence supporting the homology of the book lungs of Scorpiones and Tetrapulmonata (Araneae+Pedipalpi): this suggests a single terrestrialization event in the arachnid lineage. Combining together data from millipede palaeontology and molecular phylogeny as well as the plant fossil record, Wilson (2006) hypothesizes that the Diplopoda colonized land no later than the Ordovician along with the origin of the lineage leading to Penicillata (the pin-cushion millipedes) and the Palaeozoic Arthropleurida. The oldest palaeontological evidence of spiracles is provided by the mid-Silurian millipede *Pneumodesmus newmani*, more recent indeed than the earliest fossil evidence of terrestrial animal activity, which dates from the Ordovician, approximately 450 Ma (Shear *et al.* 1984, Jeram *et al.* 1990, Pisani *et al.* 2004). Whether a single event of terrestrialization occurred along the history of myriapods or not is a different question. In particular, if the two main lineages of extant myriapods (Chilopoda and Diplopoda) are not strictly related, as tentatively suggested by Negrisolo *et al.* (2004), it is quite possible that their last common ancestor was aquatic rather than terrestrial.

Terrestrial lifestyle is probably the basal state for the Collembola and possibly for the Hexapoda as a whole. The oldest known insect, *Rhyniognatha hirsti*, seems to be quite derived morphologically, indicating that insects originated in the Silurian period and were members

of some of the earliest terrestrial communities (Engel and Grimaldi 2004). Aquatic life is inferred to have evolved several times independently in Collembola and in Pterygota but it remains ambiguous whether the first Pterygota had aquatic juveniles and turned back towards fully terrestrial life in Neoptera, or aquatic nymphs evolved in Odonata and Ephemeroptera independently. Aquatic nymphs or larvae were secondarily acquired by Plecoptera and Trichoptera and by many lineages of Coleoptera, Diptera, etc. (D'Haese 2004).

It is possible that evolutionary events of major ecological consequence such as the origination of the main groups of terrestrial animals (and plants) may have been triggered by important changes in the physical environment (although I feel totally incompetent to judge this). The mechanism is far from obvious that might have causally linked the conquest of land to the availability of high oxygen concentration in the atmosphere (Ward *et al.* 2006), something that happened twice in the Palaeozoic: the first approximately 410 Ma, at a time which corresponds to a first burst of water-to-land transitions among the arthropods, the second time after the Devonian mass extinction, approximately at the time of the terrestrialization of the vertebrates and of other arthropods (Berker *et al.* 2007).

But asking what may have favoured a major environmental transition is probably an inadequate way to address the question. In so doing, we are inclined to regard these transitions as exceptional events and this opens the way to suggesting the widest, or wildest, array of *ad hoc* hypotheses, similar to what happened with dinosaur extinction. Let us then try to reverse the question. In the absence of contrary evidence, it is more parsimonious to suppose that events such as moving from water to land, or vice versa, are not special, except for the fact that these events did not happen more often. In other terms, rather than asking what may explain these environmental transitions, let us ask what can explain their rare occurrence. This is the logic behind the macroevolutionary scenario of Vermeij and Dudley (2000) who, in particular, addressed the puzzling question of why two otherwise enormously successful groups such as insects and the flowering plants have been so ineffective in colonizing the sea. Vermeij and Dudley suggested that an explanation can be found in the competitive disadvantage an animal usually faces when starting to invade a physically different habitat, such as water from the land, or even freshwater from seawater. Facing novel physical conditions, the invader is forced to shut down its metabolic machinery and this sets it at disadvantage with respect to the inhabitants with a long history of adaptation of that environment. This adds strongly to other problems such as risk of attack from predators previously unexperienced, and represents a limitation from which only endothermic vertebrates can easily escape. This would explain the high frequency at which major environmental transitions have occurred among mammals and birds.

In describing evolutionary patterns of change, it is often remarked that an initial burst of rapid and intense change is followed by a long history of changes going on at a much reduced pace. This has been observed, for example, in early tetrapods (Ruta *et al.* 2006). This may be construed as evidence for an eventual stabilization of an adaptively preferred organization, but this explanation cannot be taken for granted, at least as long as we do not know enough about the further evolvability of the characters we are considering. Take, for instance, the case of the parasitic flatworms, flukes, and tapeworms alike. Brooks and McLennan (1993) analysed morphological changes in larval and adult features against a phylogeny of these worms and showed that only a limited fraction of those changes could be classified as regressive. Apparently, this seemed to contradict the common opinion that the evolution of internal and intestinal parasites is strongly dominated by regressive trends. However, this analysis failed to consider that regressive trends are unavoidably fated to confront the principle of decreasing returns. The more you erase, the less remains to be erased. Thus, if regressive changes are abundantly manifested at the level of the deepest branches of the tree of the Neodermata, what we must expect is exactly the picture found by Brooks and

McLennan; that is, an abundance of changes, in the more recent branchings, that do not affect the complexity of an already pruned bodily organization and occasionally even increase it. If a temporal pattern of decreasing evolvability is so manifest in the case of a trend towards simplified forms, evolvability may not be so obvious in other cases. But it is in this direction, rather than towards explanations in terms of adaptation, that we must look in many cases, for example with the extraordinary stability of the number of cervical vertebrae in mammals, or in the number of body segments, which is literally frozen in many large and old arthropodan lineages.

9.3 Intercalary evolution

In Chapter 7 (section 7.5.1) and also in the previous paragraph I mentioned intercalary evolution, a pathway of change opposite to the once-fashionable Haeckelian scenario that emphasized the terminal addition of evolutionary novelties on the top of a steadily abbreviated recapitulation of the ontogenetic schedule of the ancestors. Intercalary evolution deserves a few words, first to provide evidence that it is all but rare, and also to discuss how much its occurrence may depend on a higher evolvability of earlier rather than later developmental stages, and how much it may depend instead on differently strong selection.

During the last decade, major advances in nematode embryology have demonstrated that the largely conservative architecture of the adult is obtained along very different but eventually convergent paths (Schierenberg and Schulze 2008), including different mechanisms of fate specification of early blastomeres (Schierenberg 2001), different mechanisms of axis specification (Goldstein *et al.* 1998), and different patterns of cell migration, apoptosis, etc. at later developmental stages (Sulston *et al.* 1983, Houthoofd *et al.* 2006).

When discussing larvae in Chapter 7, we have seen that simple does not necessarily mean old. But taking the simpler organization of a larva as an indication that it recapitulates the adult condition of a remote ancestor is only the most obvious pitfall. Much more deceptive than the lack of 'more advanced' traits is the presence in the larva of specific structures that are eventually lost in the adult, but are in some way reminiscent of features that were probably present in the adult of some remote ancestor. This is the case of the articulated abdominal appendages found in the larvae of some holometabolous insects such as *Sialis* (Megaloptera). In former but not too distant times, when myriapods (or a part of them) were regarded as the sister group of the hexapods, any evidence of polypody in the latter was likely to attract attention and to suggest atavism. But this conclusion would be no safer than interpreting the similarity between the two pairs of wings in a four-winged *Drosophila* mutant as evidence that the Diptera derive from ancestors with two pairs of identical wings. No phylogenetic analysis has ever put the Megaloptera, or any other holometabolan lineage, in a basal position within the hexapods (or the insects); thus there is no doubt as to the derived nature of the holometabolan clade and, specifically, of the vaguely polypodous organization of the larva of *Sialis* (Fig. 9.4). But also, more interestingly, there is no doubt as to the fact that holometabolan adults are, on a whole, more conservative than the corresponding preimaginal stages (see Klingenberg 1998).

In fairly recent times, Truman and Riddiford (1999) have revived an evolutionary scenario originally proposed nearly 100 years ago by Berlese (1913). These authors have suggested that the larva of the holometabolans corresponds to an usually fleeting phase observed in the development of other insects (and other arthropods), variously called the prolarva, or the pronymph, or still otherwise. Crickets, for example, hatch as prolarvae, with a body already complete in all its main features, but with incompletely articulated appendages, so that it requires one more moult for them to transform into the first mobile and feeding nymphal stage. According to Berlese (and Truman and Riddiford), the larval stages of the holometabolans do not correspond to the active nymphal stages of the heterometabolans, but to their prolarva (or pronymph), which became active despite retaining some more 'embryoid' traits than a typical nymph. A trace of the original sequence of nymphal stages would have been retained in the form of a nonfeeding and increasingly less mobile pupa. But these far-reaching changes affecting the preimaginal stages did not produce significant changes in the adult. A nice piece of evidence recently produced in support of this hypothesis is the development of the larval legs in the tobacco hawkmoth *Manduca sexta*, which are

Fig. 9.4. Jointed abdominal appendages of the larva of *Sialis lutaria* (Insecta Megaloptera). Courtesy of Paolo Grilli.

produced by a temporary arrest in the otherwise conserved adult leg-patterning process in insects (Tanaka and Trueman 2007).

As a consequence of this evolutionary trend, the more important became the novelties in the organization of the larva, the more dramatic became the metamorphosis throughout the pupal stage, by which the conserved adult form is produced (Heming 2003). To read this sentence in the opposite and perhaps a more informative way, a 'catastrophic' metamorphosis has not evolved to allow for the production of a significantly new adult, but to compensate for the production of a significantly divergent larva, while the adult did not change too much. This fact is a further blow to the adultocentric view of development I have already denounced in this book (see section 1.3.3), although one might naively claim that the conservative character of the adult condition is exactly the evidence supporting the view that nothing in a life cycle is important if it is not to the service of the adult. But I think that we should examine a different interpretation. Let us consider the evolution of insect post-embryonic development by taking into account both the evolvability of the different stages and the nature of the selection acting on them.

One of reasons why early developmental stages are more prone to evolutionary change than later stages is arguably because their phenotypes, or those of their component cells, are less definitely committed, or determined, than at later stages. This is also why they can be extensively manipulated by maternal genes, especially when the first expression of the zygotic genome is significantly delayed by a very rapid sequence of cleavage mitoses that does not offer interphase intervals of a length sufficient for transcription. Multipotency of embryonic cells is not necessarily lost very early. Arenas-Mena *et al.* (2007) have demonstrated, in the 'polychaete' *Hydroides elegans* and in the sea urchin *Strongylocentrotus purpuratus*, that in the embryonic cells expressing the histone 2A.Z the achievement of an irreversible state of differentiation is delayed because this protein prevents the burying of transcription factor-binding sites into heterochromatin, thus

maintaining their sensitivity to transcriptional regulators. Nevertheless, cell differentiation and body patterning eventually go on inexorably as a one-way process and the resulting phenotypes become more and more entrenched and thus less and less evolvable.

Let us now move to the adult. In comparing holometabolous to heterometabolous insects, I have remarked on the evolutionarily stability of the adult (with exceptions!). Is this mainly due to stabilizing stability or to a limited amount of available variation? I think that in the case of insects and some other arthropods, a particular circumstance limits very strongly the selectable variation. Individuals in reproductive condition do not moult any more, but remain—in all their external traits at least—literally frozen in the shape they obtained with the last moult. This is not necessarily true of other arthropods, such as crabs, or julid millipedes, or collembolans, which continue to moult after they have reached adulthood. Thus, before the end of the reproductive age—that is, before selection ceases to be effective—an insect does not offer a range of different phenotypes on which selection may operate. The likely relevance of this fact is demonstrated by the Helminthomorpha, the largest and more diversified clade of millipedes. In these arthropods, the males are provided with gonopods, one or two pairs of modified legs turned into sexual appendages.

Interestingly, during the early stages of post-embryonic development these appendages still appear as normal, locomotory legs, but subsequently, following a moult, they reduce to minimal scale-like projections and are finally substituted by gonopods, when the millipede moults to become an adult (Fig. 9.5). In most helminthomorph millipedes, gonopods are not divided into segments and their shape is often extraordinarily complex and does not preserve any evidence of their origin from normal walking legs. The most derived kinds of gonopods are arguably those of the juloid millipedes and this circumstance provides evidence of the scope for morphological evolution that can be offered by the existence of multiple reproductive stages. In fact, juloid millipedes are euanamorphic; that

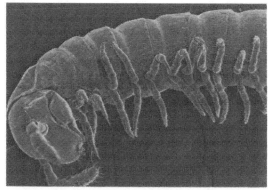

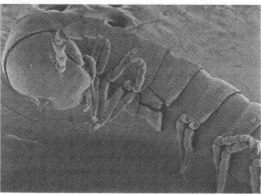

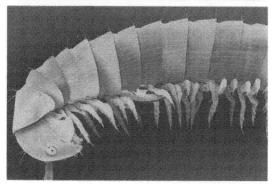

Fig. 9.5. Three steps in the post-embryonic development of the millipede *Nopoiulus kochii*, showing the changing fate of the eight and ninth pairs of legs. In stadium IV (upper panel), these leg pairs are normal walking legs, but with the moult to stadium V (middle panel) they turn into squamiform appendages. Finally, in stadium VI (lower panel) the appendages are developed anew as functional gonopods. Courtesy of Leandro Drago.

is, they continue to moult and to add body segments after they have reached the adult condition, whereas other millipedes such as the polydesmids, with simpler and less diversified

gonopods, are teloanamorphic, meaning that they do not add segments and do not moult once they become sexually mature (Enghoff *et al.* 1993). Thus, in terms of gonopod morphology, polydesmids, with 'frozen' adults (like insects), are more conservative and less variable than julids, which have more than one adult instar.

9.4 Patterned polyphenism

The effects on the adult phenotype of the nutritional conditions to which are exposed the female larvae of many social hymenopterans are well known. It does not depend on genes whether a honey bee, or an ant, eventually develops into a reproductive queen or into a sterile worker. Conspicuous as they are, these instances of phenotypic plasticity are simply the most familiar examples of a pervasive phenomenon (West Eberhard 2003) which can arguably open dramatically important vistas into the working of evolution. In this section I will briefly introduce some major aspects of morphological evolution where polyphenism is likely to represent the starting point of evolutionary changes of major consequence, generally involving a stabilization of spatial and temporal patterns under increasingly effective genetic control.

At the first level, or kind, of morphological and functional complexity which may derive from a sort of frozen simultaneous polyphenism is the presence, in animals and multicellular organisms at large, of multiple differentiated cell types.

There is little doubt that several differentiated cell states have remote origins, much older than the highly structured spatial patterns according to which the different cell types are distributed in the body of most Recent animals. This is, indeed, one of the basic points in Davidson's (1991, 2003, 2006) model of evolution of animal organization. However, this is a somewhat simplified view of history. On the one hand, a number of highly specialized cell types have clearly evolved during the history of the metazoans and, in particular, *after* the origin of the bilaterian body plan. On the other hand, even the most conservative kinds of differentiated cells cannot be strictly assumed to have been invariant. Nevertheless, the first kind of patterned heterogeneity within a multicellular assemblage was very likely some form of cell differentiation rather than an overall body patterning. The fact that in modern metazoans polarities, seriality, and positions are specified before the final differentiation of the different kind of cells is a subsequently acquired feature, and also a good example of evolution acting on early developmental stages (in a non-recapitulatory, intercalary way) rather than adding novelties (in recapitulatory form) at the end of the original developmental schedule.

Nowadays, cell differentiation is a typical example of a biological process dependent on strictly controlled and spatiotemporally restricted gene expression, but this is arguably an acquired condition, probably preceded by a stage when alternative phenotypes within a multicellular organism were just variants within an environmentally inducible polyphenism. Increasingly repeatable forms of control of the internal dynamics of specifically positioned cells by their immediate neighbours, or even by slightly more remote cells within the reach of their diffusible products, may have eventually lead to predictable differential patterns of gene expression; that is, to the stabilization of alternative differentiated cell phenotypes. In this interpretation, differential gene expression would have allowed a stable coexistence of a set of alternative phenotypes originally expressed at unpredictable times and in different locations within the multicellular organism (e.g. Valentine 2004, Arenas-Mena 2007). In this scenario, what is new in the multicellular organisms is the evolution of cell–cell interactions that allow the coexistence of alternative phenotypes, while the production of the latter, as such, was already manifested by their unicellular precursors.

According to Davidson, this deepest but eventually conserved degree of bodily complexity was dramatically improved, at the origin of the Bilateria, with the evolution of a robust body patterning system, by effect of which the conserved batteries of genes involved in tissue specification were spatially distributed according to increasingly complex and eventually conserved rules embodied in newly emerging genetic networks

(Davidson 2003, 2006). Summing up: the first step was the fixation of alternative phenotypes that came to coexist stably within the same cluster of cells; the second step was the fixation of the positions at which those phenotypes (by now turned into alternative states of cell differentiation) were eventually expressed.

What next? I would argue that a further dimension in this evolutionary scenario is the temporal distribution of both cell differentiation and body patterning along the arrow of individual development. With this, I do not refer simply to heterochrony; that is, to the temporal shifting of one developmental event in respect to another, or to a change in the rate at which a developmental process occurs. I suggest, instead, that complex developmental schedules such as those punctuated by dramatic larva to adult metamorphoses resulted from the temporal control of differentiation patterns in the same way as complex morphologies resulted from a control of their spatial occurrence. Sly *et al.* (2003) regarded the nematode dauer larva as an example of how 'true' larvae might have originated, as alternative phenotypes eventually triggered by particular environmental conditions and subsequently becoming constitutive.

I have already mentioned (see section 7.4.2) the dramatic changes in gene expression patterns that accompany the not less dramatic metamorphosis of *Drosophila* (Arbeitman *et al.* 2002). I only need to add that the spatial and the temporal regulation of gene expression are not necessarily separate, to the extent that at a given instant different parts of the same animal can have a very different 'physiological age'. This has been observed, for example, in some larvae of the blue crab *Callinectes sapidus*, which combine traits otherwise characteristic of two different zoeal stages (Costlow 1968), as the anterior part of the body is morphologically more advanced than the posterior part. Similar shifts can be expected in isopods, where moulting is completed in two steps, the anterior and posterior parts of the exuvia been often released some days apart. In this group of crustaceans, the most dramatic example of 'age mismatch' between the anterior and posterior segments is offered by *Hemioniscus balani*, a parasite of barnacles (Goudeau 1977). This is a protandric hermaphrodite whose anterior half does not proceed further than the male phase and does not undergo further moults, whereas the posterior part of the body continues to moult, increases conspicuously in size, and eventually reaches maturity as a female. This is probably an unparalleled example of sex-related age mismatch, but one could argue that this has something in common with a phenomenon found in several arthropods, where the number of moults required to reach maturity is higher in females than in males. This happens in many spiders, most conspicuously in the strongly sexually dimorphic *Nephila*, whose females are giants, with respect to their closest relatives, whereas the males have evolved in the exactly opposite direction (Coddington *et al.* 1997) (Fig. 9.6). Again, in a very small group of malacostracan crustaceans, the Amphionidacea, males mature as megalopae whereas females reach maturity at the next moult (Williamson 1973).

Thus, an original polyphenism can eventually evolve into a structured array of parts 'frozen' in a predictable spatial pattern or in a sequence of phenotypes corresponding to different developmental stages in a life cycle. But the eventual fate of a polyphenism is not necessarily found in its patterned maintenance. I have already suggested (see section 7.1.5) a 'chorus effect' as one of the possible origins of homogeneity between neighbouring cells in a tissue. Similar forms of assimilation of neighbouring elements are arguably frequent in evolution. One example is provided by the tail of vertebrates (Minelli 2003), which in developmental terms is much closer to an appendage than to a portion of the main body axis, but it has been extensively assimilated morphologically to the latter, in some lineages at least and most conspicuously in the limbless forms such as the snakes. The limited patterning of the vertebral column of the latter can be interestingly compared with the extensively regionalized axial skeleton of the 'ancestral tetrapod' *Ichthyostega* (Ahlberg *et al.* 2005).

Another example is provided by geophilomorph centipedes (especially the mecistocephalids) whose worm-like, multisegmented body is

Fig. 9.6. Male (above) and female (below) of *Nephila* sp.; Madagascar. Courtesy of Francesco Barbieri.

formed by strictly uniform segments, in a series which is much less overtly patterned than the trunk of their putative ancestors (Fig. 9.7). Other forms of assimilation involve the disappearance of the morphological differences originally distinguishing different developmental stages in a life cycle, as in the case of the bivesiculid digeneans, where the rediae have became similar to the cercariae.

9.5 Hierarchies, or not

Most descriptions of the living world are obsessively dominated by the notion of inclusive hierarchy. For example, organelles are parts of cells, and cells are parts of organs, the sum of which eventually represents the individual organism. The latter, in turn, is part of a population, which interacts with other populations

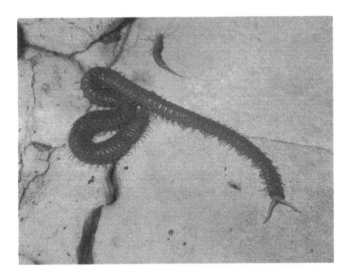

Fig. 9.7. A specimen of *Himantarium gabrielis* (Chilopoda Geophilomorpha) with 131 pairs of legs; Puglia, Italy. Courtesy of Lucio Bonato.

in a local ecosystem. Inclusive hierarchies of this sort have many interesting properties (e.g. Ahl and Allen 1996), one of which is that we can deconstruct a system into its intermediate level components without forcing it to split into its ultimate parts. Read the opposite way, this means that we can assemble a complex object by using readily available building blocks that have an interesting degree of complexity themselves. In other terms, with hierarchical objects there is no need to start every time from scratch. Back to living organisms, it is indeed impossible to deny that mitochondria, chloroplasts, and cells represent integrated units that are not built again by assembling their constituent molecules every time a new individual animal or plant is born. The problem is, instead, whether anything of relevance is lost when we analyse an organism in terms of a linear hierarchy of levels of inclusiveness.

Thinking in terms of hierarchies has provided the background to apply the concept of heterochrony in describing evolutionary change. This classic topic (e.g. de Beer 1958, Gould 1977, McKinney 1988, McKinney and McNamara 1991, McNamara 1995) has attracted much favour in the last two decades, first with the discovery of genes whose expression specifically modifies the timing of developmental events (the so-called heterochronic genes; e.g. Ambros 1989, Ruvkun and Giusto 1989, Ambros and Moss 1994), then with the appreciation of how heterochrony interferes with phylogenetic reconstruction (Wiens *et al.* 2005, Minelli *et al.* 2007) or represents a challenge to a strictly typological concept of the phylotypic stage, requiring it to be softened in terms of phylotypic period (Richardson 1995). A reconceptualization of this area went through extensively revising and rationalizing nomenclature (McNamara 1986, Gould 2000), an effort largely swamped by the development of new comparative protocols to analyse heterochrony, generally in terms of changes in the relative sequences of many pairs of developmental events between two broadly comparable sequences (Smith 1997, 2001, Velhagen 1997, Bininda-Emonds *et al.* 2002, Jeffery *et al.* 2002a, 2002b, Schulmeister and Wheeler 2004).

Explicitly or not, heterochrony implies a large degree of independence of developmental events; that is, a large degree of developmental modularity. The same condition is presupposed by the parallel concepts of heterotopy, heterotypy, and heterometry (Arthur 2002, Webster and Zelditch 2005), which refer to evolutionary changes in the relative position of body parts, in their kind, or in their relative size. All these terms refer to patterns rather than to actual processes, and this is the reason why they have become fashionable in palaeontology, whereas the attitude of comparative developmental biologists has been more critical (e.g. Raff and Wray 1989).

The problem with heterochrony and, more generally, with all the analyses based on a strictly hierarchical view of the modularity of living beings is, however, much deeper than the risk of being so satisfied with a nice description of patterns that the underlying mechanisms are overlooked. My objection is that all descriptions of living beings in terms of a single, all-inclusive hierarchy are, in principle, selective and incomplete. Evolution does not proceed by changing body patterning while keeping cell differentiation unchanged, nor by changing cell differentiation while keeping body patterning constant. A change in average cell size or in the rate of cell division, not to mention a change in a cell membrane-bound molecule affecting cell-adhesion properties, can have dramatic transversal consequences that no description in terms of structural hierarchy would adequately encapsulate. And this is exactly what we can expect to happen in evolution, with changes cutting obliquely across the hierarchies we depict on paper.

McShea (2001) has argued that a major trend in evolution is the increasing loss of internal complexity of lower-level entities as a consequence of the increasing performance of a higher-level system within which those entities are integrated. For example, as soon as the multicellular organisms became able to feed, to reproduce, and to defend themselves, their component cells became individually less complex than were their unicellular ancestors. Of course, the anucleate red cells of mammalian blood would provide a good example in support of this thesis, but the sting cells of the cnidarians or the colloblasts of the ctenophores would probably suggest otherwise. But still more important is the fact that *all* cells of multicellular organisms continue to evolve, and several evolving properties are shared by all or nearly all cells in a multicellular organism, especially those depending on housekeeping genes. Undeniably, many protist cells are more complex than many animal or plant cells (McShea 2002), but the opposite is

also true. Comparisons, anyway, should involve suitable pairs of taxa (e.g. choanoflagellates and 'sponges') rather than, indiscriminately, unicellular and multicellular organisms.

On the whole, the large degree of independence with which individual features can evolve with respect to other aspects of the phenotype (see Gass and Bolker 2003 for a review) cannot be taken as proof of a stable hierarchy of organized sub-elements, because the next burst of evolutionary change will possibly exploit modules obtained by a completely different partition of the organism. The evolutionary change that turned a pair of walking legs of an arthropod into a pair of maxillipedes involved in the manipulation of food may well be followed by another change resulting in the intimate coaptation of a limited part of the maxillipedes (not necessarily a single, 'modular' segment of these appendages) together with a part of the head capsule, or a localized element in another pair of appendages (Fig. 9.8).

9.6 Summing up

Summing up, where are we, in our appreciation of the evolution of metazoans? Recent progress in phylogenetics has not yet produced a single best tree of the metazoan relationships with a satisfactory degree of robustness and of branching resolution. However, many old cherished hypotheses have been convincingly demonstrated to be wrong and many unexpected but increasingly corroborated relationships have been emerging. Of these, the Ecdysozoa hypothesis is one of the most convincing while the basal position of the Acoela within the Bilateria is perhaps less definitively demonstrated but of no less importance for our appreciation of animal phylogeny. Less robust or even contradictory results of other analyses has increasingly singled out a few taxa, such as the Chaetognatha, the Gastrotricha, the Gnathostomulida, the Cycliophora, and the Myzostomida as especially worthy of renewed research effort. At the same time, from the dark depths of the evolutionary past emerges from time to time the spectre of star phylogenies: the possibility that the nodes in the phylogenetic

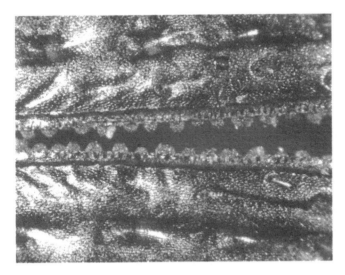

Fig. 9.8. Complementary teeth and notches between the two elytra of the weevil *Ocladius amharicus* (Coleoptera Curculionidae). Courtesy of Massimo Meregalli.

tree at the level of which many large metazoan clades diverged are chronologically too close to each other to give us serious hope of eventually resolving their topology and temporal order. Additional efforts to better understand the resolution power of our data (and of our analytical tools) are obviously welcome.

But I hope that the many aspects of animal evolution very briefly presented in the second part of this book will have convincingly shown that to study character transitions along the branches of the tree is more important, even if not easier, than reconstructing the tree itself. This analysis requires many conceptual tools additional to those required by a phylogenetic analysis but also, as I have argued repeatedly, a fresh revisitation of many traditional concepts. Let me outline in this final section a series of possible guidelines for future investigations into animal evolution.

1. *Let us adopt a factorial but not uniquely hierarchical view of living organisms.* There is a virtually unlimited number of ways according to which we can partition an organism, most of which are indeed hierarchical. One is the usual description of an animal's structure in terms of organ systems formed by organs which are made of tissues, and so on. Another is cell lineage, tracing—hierarchically again—the origin

of each cell in the animal back in time to the zygote. The study of ontogenetic processes reveals the existence of more or less largely independent developmental modules (which can be defined either in terms of processes or in terms of distinct embryonic territories), and morphological evolution demonstrates that single body features can extensively change with limited consequences for the remaining of the organism. Thus, resolving the complexity of even the simplest of the organisms into a multiplicity of constituent parts can be critically important in singling out an evolutionary or developmental problem to be investigated. But there is no single way, or even a single best way, do to that. Additionally, and not less important, there is no reason to expect that independent evolutionary modules will necessarily correspond to developmental modules. And there is no reason to expect that the existing boundaries between two evolutionary (or developmental) modules or their position in a module hierarchy will necessarily last for long. Summing up, every bit of structure in an organism belongs, at one and the same time, to a plurality of modules of different origin, nature, and robustness.

2. *Let us acknowledge a principle of developmental inertia and describe development in terms of competition between local dynamic modules.* A way to characterize a living being in a language common to sciences other than biology is to describe it as the site of a complex set of self-perpetuating dynamics of which metabolism, cell division, and development are the most conspicuous aspects. Factors limiting these dynamics are obviously found in the environment, in a limited availability of material and energetic resources, but other limits are set by internal forms of control. The expression of some genes, or the lacking expression of others, may limit or forbid mitosis, or even induce a cell to undergo apoptosis. Read the other way around, we can take as the (admittedly, somewhat idealized) default state of living matter a condition of everlasting dynamics which, in multicellular organisms, easily translates into unlimited growth and fractal-like iteration of developmental patterning. The finely dissected leaves of ferns provide a familiar example. Locally, developmental dynamics are arguably controlled by limiting resources, so that competition eventually results between units such as cells, or other integrated compartments, for the access to them. Under specific circumstances, competition can be kept alive indefinitely, although we may predict that local interactions between competing units may eventually generate results other than a prolonged *status quo*. We can expect, for example, the splitting of a local integrated module into two or more distinct modules between which competition will soon begin, or the merging of previously distinct modules into a larger unit involved in unitary dynamics. I would argue that the evolution of development is, to a large extent, the evolution of the patterns of competition among local units within a multicellular organism.

3. *Let us abandon the current adultocentric view of development.* There would be no continuity in life across the generations, where it not for reproduction. And the adult is, by definition, the developmental stage that is responsible for reproduction. But we should not feel authorized to take this undisputable fact as a reason to grant the adult a unique position in a life cycle, as if all other stages would be only relevant because they represent obligate steps towards the eventual production of the adult. To the contrary, all developmental steps should be interpreted for what they are, with the immediately previous stages setting limits to their origination and evolution, and with their own problems of sustainability and robustness. What will eventually derive from them is not their business. Refreshing our view of development by disposing of the current adultocentrism will help us to formulate new, interesting questions and avoid several dramatic blind alleys. For example, we will not look any more at the egg as at the archetypal starting point of development, depositary of the programme for originating the future adult. Much more realistically, we will identify the egg as one of the most specialized kinds of cells, and will eventually ask in non-finalistic terms about the origin of its unique peculiarities. Can we explain the origin of the storage of yolk in the egg in terms of competition between the oocyte and the surrounding cells, rather than as a means to provide nourishment to future generations of blastomeres? Correspondingly, what about the male gamete? Why not to suggest that its eventual union with a female gamete is a way to escape from the otherwise unavoidable lack of future, because the commitment to its centriole to control the cilium does not allow a chance of building a mitotic spindle? Again, and perhaps in less controversial terms, I strongly argue that we cannot single out within a developmental schedule specific processes of organogenesis such as making a lung, a heart, or a stomach, because these organs are unitary, integrated modules of structure and function only when finished, but are built by processes like folding epithelia, sealing cell sheet margins, and making epithelial tubes, which have their own origin, history, and mechanics which clearly do not care much for the functional organs eventually emerging from them.

4. *Acknowledging the pervasivity of convergence.* Moore and Willmer (1997) presented an impressive overview of convergent features in metazoans, but this did not elicit many interested reactions. Judging from conversations with several students of animal phylogenetics, I got the impression that most of them would regard

Fig. 9.9. An ant-mimicking salticid spider; Malaysia. Courtesy of Francesco Barbieri.

most putative instances of convergence as nothing more than uninteresting gross similarities, too coarse to be considered as a threat to the reconstruction of phylogeny (Fig. 9.9). Declaring two features convergent involves an unavoidable amount of subjectivity, but this can be easily reduced by adopting in this context too the same factorial analysis I have repeatedly advocated with respect to homology. But I think that the average reaction to Moore and Willmer's paper was inappropriate for a much deeper reason. Unfortunately, the wonderful efforts towards a reconstruction of phylogeny set in motion by the publication of Hennig's seminal works (1965, 1966) caused convergence (or homoplasy, at large) to be commonly regarded as a source of potential error in recognizing relationships, rather than as an evolutionary phenomenon worth investigation. Opposite to the trend prevailing among evolutionary biologists, Conway Morris (2003a) has developed a worldview largely based on the appreciation of the prevalence of convergence in evolution. I would suggest that we can take an intermediate path, by acknowledging that convergence is there, and frequent, and that this deserves investigation.

Arguably, a leap ahead in understanding convergence may derive from taking into account evolvability, in addition to selection. That is, I expect that a better knowledge of the landscape of forms within which selection can operate (Arthur 2004b) may offer a way to understand whether, or when, convergence reveals more robust adaptive solutions or, as probably happens more frequently, nothing other than more easily reachable solutions.

5. *Finally, let us accept that the 'rules' of evolution are also evolving.* This means that the very categories we use in describing the living beings correspond to kinds of organization that have been generated by evolution, perhaps multiple times, and will not necessarily last for ever (that is, as long as the thread of life continues on our planet). There is nothing wrong, in principle, with describing animals in terms of individuals, segments, larvae, germ layers, or biological species, but one of the major tasks for evolutionary biology is not to search for the origin of this or that species, or this or that larval type, but to investigate the circumstances where things can originate, to which the concept of species, larva, individual, or segment can be eventually applied. These are not Kantian *a priori* categories, but concepts to be applied, whenever it is sensible to do so, to some of the products of evolution.

References

Abascal, F., Posada, D., Knight, R.D., and Zardoya, R. 2006. Parallel evolution of the genetic code in arthropod mithochondrial genomes. *PLOS Biology* 4: 711–718.

Abeloos, M. 1930. Recherches expérimentales sur la croissance et la régénération chez les planaires. *Bulletin biologique de la France et de la Belgique* 64: 1–140 [in French].

Aboobaker, A. and Blaxter, M. 2003. Hox gene evolution in nematodes: novelty conserved. *Current Opinion in Genetics and Development* 6: 593–598.

Abzhanov, A. and Kaufman, T.C. 1999. Homeotic genes and the arthropod head: expression patterns of the *labial, proboscipedia*, and *Deformed* genes in crustaceans and insects. *Proceedings of the National Academy of Sciences USA* 96: 10224–10229.

Abzhanov, A., Protas, M., Grant, B.R., Grant, P.R., and Tabin, C.J. 2004. *Bmp4* and morphological variation of beaks in Darwin's finches. *Science* 305: 1462–1465.

Abzhanov, A., Kuo, W.P., Hartmann, C., Grant, B.R., Grant, P.R., and Tabin, C.J. 2006. The calmodulin pathway and evolution of elongated beak morphology in Darwin's finches. *Nature* 442: 563–567.

Adams, C.L., McInerney, J.O., and Kelly, M. 1999. Indications of relationships between poriferan classes using full-length 18S rRNA gene sequences. *Memoirs of the Queensland Museum* 44: 33–43.

Adams, D.C. and Church, J.O. 2008. Amphibians do not follow Bergmann's Rule. *Evolution* 62: 413–420.

Adanson, M. 1763. *Familles des Plantes*. Vincent, Paris.

Adl, S.M., Simpson, A.G.B., Farmer, M.A. *et al.* 2005. The new higher level classification of eukaryotes with emphasis on the taxonomy of protists. *Journal of Eukaryotic Microbiology* 52: 399–451.

Agnarsson, I. 2006. Asymmetric female genitalia and other remarkable morphology in a new genus of cobweb spiders (Theridiidae, Araneae) from Madagascar. *Biological Journal of the Linnean Society* 87: 211–232.

Aguinaldo, A.M.A., Turbeville, J.M., Linford, L.S. *et al.* 1997. Evidence for a clade of nematodes, arthropods and other moulting animals. *Nature* 387: 489–493.

Ahl, V. and Allen, T.F.H. 1996. *Hierarchy Theory: a Vision, Vocabulary, and Epistemology*. Columbia University Press, New York.

Ahlberg, P.E. 2003. Fossils, developmental patterning and the origin of tetrapods, pp. 45–54 in A. Legakis, S. Sfenthourakis, R. Polymeni, and M. Thessalou-Legaki (eds), *The New Panorama of Animal Evolution*. Proceedings of the 18th International Congress of Zoology. Pensoft, Sofia/Moscow.

Ahlberg, P.E., Clack, J.A., and Blom, H. 2005. The axial skeleton of the Devonian tetrapod *Ichthyostega*. *Nature* 437: 137–140.

Ahlrichs, W. 1993. Ultrastructure of the protonephridia of *Seison annulatus* (Rotifera). *Zoomorphology* 113: 245–251.

Ahlrichs, W.H. 1995. *Ultrastruktur und Phylogenie von Seison nebaliae (Grube 1859) und Seison annulatus (Claus 1876)*. Cuvillier, Göttingen [in German].

Ahlrichs, W.H. 1997. Epidermal ultrastructure of *Seison nebaliae* and *Seison annulatus*, and a comparison of epidermal structures within the Gnathifera. *Zoomorphology* 117: 41–48.

Ahlrichs, W.H. 1998. Spermatogenesis and ultrastructure of the spermatozoa of *Seison nebaliae* (Syndermata). *Zoomorphology* 118: 255–261.

Akam, M. 2000. Arthropods: developmental diversity within a (super) phylum. *Proceedings of the National Academy of Sciences USA* 97: 4438–4441.

Akimenko, M.A., Marí-Beffa, M., Becerra, J., and Géraudie, J. 2003. Old questions, new tools, and some answers to the mystery of fin regeneration. *Developmental Dynamics* 226: 190–201.

Akiyama, S.K. and Johnson, M.D. 1983. Fibronectin in evolution: presence in invertebrates and isolation from *Microciona prolifera*. *Comparative Biochemistry and Physiology B* 76: 687–694.

Alberti, G., Gegner, A. and Witalinski, W. 1999. Fine structure of the genital system in the females of *Pergamasus mitis* (Acari: Gamasida: Pergamasidae). *Journal of Morphology* 240: 195–223.

Albrecht, C., Wilke, T., Kuhn, K., and Streit, B. 2004. Convergent evolution of shell shape in freshwater limpets: the African genus *Burnupia*. *Zoological Journal of the Linnean Society* 140: 577–586.

Albrecht, H., Ehlers, U., and Taraschewski, H. 1997. Syncytial organization of acanthors of *Polymorphus minutus* (Palaeacanthocephala), *Neoechinorhynchus rutili* (Eoacanthocephala), and *Moniliformis moniliformis* (Archiacanthocephala) (Acanthocephala). *Parasitology Research* 83: 326–338.

Aldridge, R.J. and Purnell, M.A. 1996. The conodont controversies. *Trends in Ecology & Evolution* 11: 463–468.

Aldridge, R.J., Briggs, D.E.G. Clarkson, E.N.K., and Smith, M.P. 1986. The affinities of conodonts—new evidence from the Carboniferous of Edinburgh, Scotland. *Lethaia* 19: 279–291.

Aldridge, R.J., Briggs, D.E.G. Smith, M.P., Clarkson, E.N.K., and Clark, N.D.L. 1993. The anatomy of conodonts. *Philosophical Transactions of the Royal Society of London Series B, Biological Sciences* 340: 405–421.

Allen, J.D. and Pernet, B. 2007. Intermediate modes of larval development: bridging the gap between planktotrophy and lecithotrophy. *Evolution & Development* 9: 643–653.

Alvarez, B.B., Delfino, G., Nosi, D., and Terreni, A. 2005. Ultrastructure of poison glands of South American frogs: a comparison between *Physalaemus albonotatus* and *Leptodactylus chaquensis* (Anura: Leptodactylidae). *Journal of Morphology* 263: 247–258.

Amaral Zettler, L.A., Nerad, T.A., O'Kelly, C.J., and Sogin, M.L. 2001. The nucleariid amoebae: more protists at the animal-fungal boundary. *Journal of Eukaryotic Microbiology* 48: 293–297.

Ambros, V. 1989. A hierarchy of regulatory genes controls a larva-to-adult development switch in *C. elegans*. *Cell* 57: 49–57.

Ambros, V. and Moss, E.G. 1994. Heterochronic genes and the temporal control of *C. elegans* development. *Trends in Genetics* 10: 123–127.

Ambros, V. and Chen, X. 2007. The regulation of genes and genomes by small RNAs. *Development* 134: 1635–1641.

Amor, A. 1994. Gametes, fertilization and development of *Pherusa* sp., an endolithic worm (Polychaeta, Flabelligeridae). *Mémoires du Muséum d'Histoire naturelle, Paris* 162: 612.

Anderson, C.L., Canning, E.U. and Okamura, B. 1998. A triploblast origin for Myxozoa? *Nature* 392: 346–347.

Anderson, D.T. 1973. *Embryology and Phylogeny in Annelids and Arthropods*. Pergamon Press, Oxford.

Anderson, E. 1948. Hybridization of the habitat. *Evolution* 2: 1–9.

Andreone, F., Vences, M., Vieites, D.R., Glaw, F., and Meyer, A. 2005. Recurrent ecological adaptations revealed through a molecular analysis of the secretive cophyline frogs of Madagascar. *Molecular Phylogenetics and Evolution* 34: 315–322.

Anker, A., Ahyong, S.T., Noel, P.Y., and Palmer, A.R. 2006. Morphological phylogeny of alpheid shrimps: parallel preadaptation and the origin of a key morphological innovation, the snapping claw. *Evolution* 60: 2507–2528.

Arango, C.P. 2002. Morphological phylogenetics of the sea spiders (Arthropoda: Pycnogonida). *Organisms, Diversity and Evolution* 2: 107–125.

Arbeitman, M.N., Furlong, E.E.M., Imam, F. *et al.* 2002. Gene expression during the life cycle of *Drosophila melanogaster*. *Science* 297: 2270–2275.

Arber, A. 1950. *The Natural Philosophy of the Plant Form*. Cambridge University Press, Cambridge.

Arduini, P., Pinna, G., and Teruzzi, G. 1983. *Eophasma jurasicum* n.g. n.sp.: a new fossil nematode of the Sinemurian of Osteno in Lombardy. *Atti della Società Italiana di Scienze Naturali e del Museo Civico di Storia Naturale di Milano* 124: 61–64.

Arenas-Mena, C. 2007. Developmental transcriptional-competence model for a histone variant and a unicellular origin scenario for transcriptional-multipotency mechanisms. *Evolution & Development* 9: 209–211.

Arenas-Mena, C., Cameron, A.R., and Davidson, E.H. 2000. Spatial expression of Hox cluster genes in the ontogeny of a sea urchin. *Development* 127: 4631–4643.

Arenas-Mena, C., Wong, K.S.-Y., and Arandi-Foroshani, N.R. 2007. Histone 2A.Z expression in two indirectly developing marine invertebrates correlates with undifferentiated and multipotent cells. *Evolution & Development* 9: 231–243.

Arendt, D. 2003. Evolution of eyes and photoreceptor cell types. *International Journal of Developmental Biology* 47: 563–571.

Arendt, D. and Nübler-Jung, K. 1994. Inversion of dorsoventral axis? *Nature* 371: 26.

Arendt, D. and Nübler-Jung, K. 1996. Common ground plans in early brain development in mice and flies. *BioEssays* 18: 255–259.

Arendt, D. and Nübler-Jung, K. 1997. Dorsal or ventral: similarities in fate maps and gastrulation patterns in annelids, arthropods and chordates. *Mechanisms of Development* 61: 7–21.

Arendt, D. and Wittbrodt, J. 2001. Reconstructing the eyes of Urbilateria. *Philosophical Transactions of the Royal Society of London Series B, Biological Sciences* 356: 1545–1563.

Arendt, D., Technau, U. and Wittbrodt, J. 2001. Evolution of the bilaterian larval foregut. *Nature* 409: 81–85.

Arendt, D., Tessmar, K., de Campos-Baptista, M.I., Dorresteijn, A., and Wittbrodt, J. 2002. Development of pigment-cup eyes in the polychaete *Platynereis dumerilii* and evolutionary conservation of larval eyes in Bilateria. *Development* 129: 1143–1154.

Arendt, D., Tessmar-Raible, K., Snyman, H., Dorresteijn, A.W., and Wittbrodt, J. 2004. Ciliary photoreceptors with a vertebrate-type opsin in an invertebrate brain. *Science* 306: 869–871.

Arnold, J.M. 1974. Intercellular bridges in somatic cells of *Loligo pealei*. *Differentiation* 2: 335–341.

Arthur, W. 2002. The emerging conceptual framework of evolutionary developmental biology. *Nature* 415: 757–764.

Arthur, W. 2004a. The effect of development on the direction of evolution: toward a twenty-first century consensus. *Evolution & Development* 6: 282–288.

Arthur, W. 2004b. *Biased Embryos and Evolution.* Cambridge University Press, Cambridge.

Asher, R.J. 1999. A morphological basis for assessing the phylogeny of the "Tenrecoidea" (Mammalia, Lipotyphla). *Cladistics* 15: 231–252.

Asher, R.J., Novacek, M.J., and Geisler, J.G. 2003. Relationships of endemic African mammals and their fossil relatives based on morphological and molecular evidence. *Journal of Mammalian Evolution* 10: 131–194.

Ashton, K.G., Tracy, M.C., and de Queiroz, A. 2000. Is Bergmann's Rule valid for mammals? *American Naturalist* 156: 390–415.

Atkin, N.B. and Ohno, S. 1967. DNA values of four primitive chordates. *Chromosoma* 23: 10–13.

Aurelio, O., Hall, D.H., and Hobert, O. 2002. Immunoglobulin-domain proteins required for maintenance of ventral nerve cord organization. *Science* 295: 686–690.

Aury, J.-M., Jaillon, O., Duret, L. *et al.* 2006. Global trends of whole-genome duplications revealed by the ciliate *Paramecium tetraurelia*. *Nature* 444: 171–178.

Averof, M. and Cohen, S.M. 1997. Evolutionary origin of insect wings from ancestral gills. *Nature* 385: 627–630.

Ax, P. 1956. Die Gnathostomulida, eine rätselhafte Wurmgruppe aus dem Meeressand. *Akademie der Wissenschaften und der Literatur Mainz. Abhandlungen der Mathematisch-Naturwissenschaftlichen Klasse* 8: 1–32 [in German].

Ax, P. 1984. *Das phylogenetische System. Systematisierung der lebenden Natur aufgrund ihrer Phylogenese.* Gustav Fischer Verlag, Stuttgart [in German].

Ax, P. 1987. *The Phylogenetic System: the Systematization of Organisms on the Basis of their Phylogenesis.* Wiley, Chichester.

Ax, P. 1995. *Das System der Metazoa—ein Lehrbuch der phylogenetischen Systematik,* vol. 1. Springer, Berlin [in German].

Ax, P. 1996. *Multicellular animals. Volume I—a new approach to the phylogenetic order in nature.* Vol. 1. Springer, Berlin.

Ax, P. 1999. *Das System der Metazoa—ein Lehrbuch der phylogenetischen Systematik,* vol. 2. Springer Verlag, Berlin [in German].

Ax, P. 2000. *Multicellular Animals. Volume II—The Phylogenetic System of the Metazoa.* Springer Verlag, Berlin.

Ax, P. 2001. *Das System der Metazoa—ein Lehrbuch der phylogenetischen Systematik,* vol. 3. Spektrum Akademischer Verlag Gustav Fischer, Heidelberg [in German].

Ax, P. 2003. *Multicellular Animals. Volume III: Order in Nature—System Made by Man.* Spektrum Akademischer Verlag Gustav Fischer, Heidelberg.

Baba, R., Fujita, M., Tein, C.E., and Miyoshi, M. 2002. Endocytosis by absorptive cells in the middle segment of the suckling rat small intestine. *Anatomical Science International* 77: 117–123.

Baccetti, B. and Dallai, R. 1978. The evolution of myriapod spermatozoa. *Abhandlungen und Verhandlungen des naturwissenschaftlichen Vereins in Hamburg, Neue Folge* 21–22: 203–217.

Baccetti, B., Gaino, E., and Sarà, M. 1986. A sponge with acrosome: *Oscarella lobularis*. *Journal of Ultrastructure and Molecular Structure Research* 94: 195–198.

Baguñà, J. and Riutort, M. 2004a. Molecular phylogeny of the Platyhelminthes. *Canadian Journal of Zoology* 82: 168–193.

Baguñà, J. and Riutort, M. 2004b. The dawn of bilaterian animals: the case of acoelomorph flatworms. *Bioessays* 26: 1046–1057.

Baguñà, J., Ruiz-Trillo, I., Paps, J. *et al.* 2001. The first bilaterian organisms: simple or complex? New molecular evidence. *International Journal of Developmental Biology* 48: S133–S134.

Baguñà, J., Martinez, P., Paps, J., and Riutort, M. 2008. Unravelling body-plan and axial evolution in the Bilateria with molecular phylogenetic markers, pp. 217–228 in A. Minelli and G. Fusco (eds), *Evolving Pathways: Key Themes in Evolutionary Developmental Biology.* Cambridge University Press, Cambridge.

Bailey, J.V., Joye, S.B., Kalanetra, K.M., Flood, B.E., and Corsetti, F.A. 2007. Evidence of giant sulphur bacteria in Neoproterozoic phosphorites. *Nature* 445: 198–201.

Balavoine, G. 1997. The early emergence of platyhelminths is contradicted by the agreement between 18S rRNA and *Hox* genes data. *Comptes Rendus de l'Académie des Sciences, Paris* 320: 83–94.

Balavoine, G. 1998. Are Platyhelminthes coelomates without a coelom? An argument based on the evolution of *Hox* genes. *American Zoologist* 38: 843–858.

Balavoine, G. and Adoutte, A. 2003. The segmented *Urbilateria*: a testable scenario. *Integrative and Comparative Biology* 43: 137–147.

Ball, E.E., Hayward, D.C., Saint, R., and Miller, D.J. 2004. A simple plan—cnidarians and the origins of developmental mechanisms. *Nature Reviews Genetics* 5: 567–577.

Ballard, J.W., Olsen, G.J., Faith, D.P., Odgers, W.A., Rowell, D.M., and Atkinson, P.W. 1992. Evidence from 12S ribosomal RNA sequences that onychophorans are modified arthropods *Science* 258: 1345–1348.

Bally-Cuif, L. and Boncinelli, E. 1997. Transcription factors and head formation in vertebrates *BioEssays* 19: 127–135.

Bartolomaeus, T. 1993. Die Leibeshöhlenverhältnisse und Verwandtschaftsbeziehungen der Spiralia. *Verhandlungen der Deutschen Zoologischen Gesellschaft* 86: 42 [in German].

Bartolomaeus, T. 1994. On the ultrastructure of the coelomic lining in the Annelida, Sipuncula and Echiura. *Microfauna Marina* 9: 171–220.

Bartolomaeus, T. 1999. Structure, function and development of segmental organs in Annelida. *Hydrobiologia* 402: 21–37.

Bartolomaeus, T. and Ax, P. 1992. Protonephridia and metanephridia—their relation within the Bilateria. *Zeitschrift für Zoologische Systematik und Evolutionsforschung* 30: 21–45.

Bartolomaeus, T. and Ruhberg, H. 1999. Ultrastructure of the body cavity lining in embryos of *Epiperipatus biolleyi* (Onychophora, Peripatidae)—a comparison with annelid larvae. *Invertebrate Biology* 118: 165–174.

Bartolomaeus, T. and Quast, B. 2005. Structure and development of nephridia in Annelida and related taxa. *Hydrobiologia* 535/536: 139–165.

Bartolomaeus, T., Purschke, G., and Hausen, H. 2005. Polychaete phylogeny based on morphological data—a comparison of current attempts. *Hydrobiologia* 535/536: 341–356.

Bateson, W. 1885. The later stages in the development of *Balanoglossus kowalevskii* with a suggestion as to the affinities of the Enteropneusta. *Quarterly Journal of Microscopic Science* 25: 81–122.

Bavestrello, G., Sommer, C., and Sarà, M. 1992. Bi-directional conversion in *Turritopsis nutricula*, pp. 137–140 in J. Bouillon, F. Boero, F. Cicogna, J.M. Gili and R.G. Hughes (eds), *Aspects of Hydrozoan Biology*. Scientia Marina 56. Institut de Ciències del Mar, Barcelona.

Bayne, C.J. 2003. Origin and evolutionary relationships between the innate and adaptive arms of immune systems. *Integrative and Comparative Biology* 43: 293–299.

Beaton, M.J. 1988. *Genome Size Variation in the Cladocera*. MSc Thesis, Department of Biological Sciences, University of Windsor, Windsor, ON.

Beck, S.D. 1971. Growth and retrogression in larvae of *Trogoderma glabrum* (Coleoptera: Dermestidae). 1. Characteristics under feeding and starvation conditions. *Annals of the Entomological Society of America* 64: 149–155.

Beddington, R.S.P. and Robertson, E.J. 1998. Anterior patterning in mouse. *Trends in Genetics* 14: 277–284.

Begun, D.J. and Collins, J.P. 1992. Biochemical plasticity in the Arizona tiger salamander (*Ambystoma tigrinum nebulosum*). *Journal of Heredity* 83: 224–227.

Beklemishev, V.N. 1944. *Principles of Comparative Anatomy of the Invertebrates*. Akademia Nauk, Moscow [in Russian].

Beklemishev, V.N. 1963. On the relationships of the Turbellaria to other groups of the animal kingdom, pp. 234–244 in E.C. Dougherty (ed.), *The Lower Metazoa. Comparative Biology and Phylogeny*. University of California Press, Berkeley, CA.

Beklemishev, W.N. 1969. *Principles of Comparative Anatomy of Invertebrates*. Oliver & Boyd, Edinburgh.

Beldade, P. and Brakefield, P.M. 2002. The genetics and evo-devo of butterfly wing patterns. *Nature Reviews Genetics* 4: 442–452.

Bell, B.M. 1975. Ontogeny and systematics of *Timeischytes casteri*, n.sp.: an enigmatic Devonian edrioasteroid. *Bulletins of American Paleontology* 67: 33–56.

Bell, B.M. 1976. Phylogenetic implicatiuons of ontogenetic development in the class Edrioasteroidea (Echinodermata). *Journal of Paleontology* 50: 1001–1019.

Bely, A.E. and Wray, G.A. 2001. Evolution of regeneration and fission in annelids: insights from engrailed- and orthodenticle-class gene expression. *Development* 128: 2781–2791.

Bengtson, S. 1976. The structure of some Middle Cambrian conodonts and the early evolution of conodont structure and function. *Lethaia* 9: 185–206.

Bengtson, S. 2003. Tracing metazoan roots in the fossil record, pp. 289–300 in A. Legakis, S. Sfenthourakis, R. Polymeni and M. Thessalou-Legaki (eds), *The New Panorama of Animal Evolution* . Proceedings of the 18th International Congress of Zoology. Pensoft, Sofia/ Moscow.

Bengtson, S. and Budd, G. 2004. Comment on "Small bilaterian fossils from 40 to 55 million years before the Cambrian". *Science* 306: 1291.

Bengtson, S. and Hou, X.-G. 2001. The integument of Cambrian chancelloriids. *Acta Palaeontologica Polonica* 46: 1–22.

Benito, J. and Pardos, F. 1997. Hemichordata, pp. 15–101 in F.W. Harrison and E.E. Ruppert (eds), *Microscopic*

Anatomy of Invertebrates. Vol. 15. Hemichordata, Chaetognatha, and the Invertebrate Chordates. Wiley-Liss, New York.

Bergmann, C. 1847. Über die Verhältnisse der Wärmeökonomie der Thiere zu ihrer Grösse. *Göttinger Studien* 1: 595–708 [in German].

Bergquist, P.R. 1985. Poriferan relationships, pp. 14–27 in S. Conway Morris, J.D. George, R. Gibson and H.M. Platt (eds), *The Origins and Relationships of Lower Invertebrates.* Clarendon Press, Oxford.

Bergström, J. 1979. Morphology of fossil arthropods as a guide to phylogenetic relationships, pp. 3–56 in A. Gupta (ed.), *Arthropod Phylogeny.* Van Nostrand Reinhold Company, New York.

Bergström, J. 1980. Morphology and systematics of early arthropods. *Abhandlungen des naturwissenschaftlichen Vereins in Hamburg, Neue Folge* 23: 7–42.

Bergström, J. 1991. Metazoan evolution around the Precambrian-Cambrian transition, pp. 25–34 in A.M. Simonetta and S. Conway Morris (eds), *The Early Evolution of Metazoa and the Significance of Problematic Taxa.* Cambridge University Press, Cambridge.

Bergström, J. 1992. The oldest Arthropoda and the origin of the Crustacea. *Acta Zoologica* 73: 287–291.

Bergström, J. and Hou, X.-G. 2001. Cambrian Onychophora or Xenusians. *Zoologischer Anzeiger* 240: 237–245.

Bergström, J. and Hou, X. 2003. Cambrian arthropods: a lesson in convergent evolution, pp. 89–96 in A. Legakis, S. Sfenthourakis, R. Polymeni and M. Thessalou-Legaki (eds), *The New Panorama of Animal Evolution.* Proceedings of the 18th International Congress of Zoology. Pensoft, Sofia/Moscow.

Berker, R.A., VandenBrooks, J.M., and Ward, P.D. 2007. Oxygen and evolution. *Science* 316: 557–558.

Berking, S. and Herrmann, K. 2007. Compartments in Scyphozoa. *International Journal of Developmental Biology* 51: 221–228.

Berlese, A. 1913. Intorno alle metamorfosi degli insetti. *Redia* 9: 121–136 [in Italian].

Berto, D., Fusco, G., and Minelli, A. 1997. Segmental units and shape control in Chilopoda. *Entomologica Scandinavica* Supplement 51: 61–70.

Beurton, P., Falk, R., and Rheinberger, H.-J. (eds), 2000. *The Concept of the Gene in Development and Evolution: Historical and Epistemological Perspectives.* Cambridge University Press, Cambridge.

Beutel, R.G. and Weide, D. 2005. Cephalic anatomy of *Zorotypus hubbardi* (Hexapoda: Zoraptera): new evidence for a relationship with Acercaria. *Zoomorphology* 124: 121–136.

Beutel, R.G. and Pohl, H. 2006. Endopterygote systematics—where do we stand and what is the goal?

(Arthropoda, Hexapoda). *Systematic Entomology* 31: 202–219.

Biehs, B., François, V. and Bier, E. 1996. The *Drosophila short gastrulation* gene prevents Dpp from autoactivating and suppressing neurogenesis in the neuroectoderm. *Genes and Development* 10: 2922–2934.

Biella, S., Smith, M.L., Aist, J.R., Cortesi, P., and Milgroom, M.G. 2002. Programmed cell death correlates with virus transmission in a filamentous fungus. *Proceedings of the Royal Society of London Series B, Biological Sciences* 269: 2269–2276.

Bininda-Emonds, O.R.P., Jeffery, J.E., Coates, M.I., and Richardson, M.K. 2002. From Haeckel to event-pairing: the evolution of developmental sequences, *Theory in Biosciences* 121: 297–320.

Bischoff, M. and Schnabel, R. 2006. Global cell sorting is mediated by local cell-cell interactions in the C. *elegans* embryo. *Developmental Biology* 294: 432–444.

Bissell, M.J., Mian, I.S., Radisky, D., and Turley, E. 2003. Tissue specificity: structural cues allow diverse phenotypes from a constant genotype, pp. 103–117 in G.B. Müller and S.A. Newman (eds), *Origination of Organismal Form. Beyond the Gene in Developmental and Evolutionary Biology.* MIT Press, Cambridge, MA.

Bissen, S.T. and Weisblat, D.A. 1989. The durations and compositions of cell cycles in embryos of the leech, *Helobdella triserialis. Development* 106: 105–118.

Bitsch, J., Bitsch, C., Bourgoin, T., and D'Haese, C. 2004. The phylogenetic position of early hexapod lineages: morphological data contradict molecular data. *Systematic Entomology* 29: 433–440.

Blackstone, N.W. 1998. Individuality in early eukaryotes and the consequences for metazoan development. *Progress in Molecular and Subcellular Biology* 19: 23–44.

Blackstone, N.W. and Ellison, A.M. 2000. Maximal indirect development, set-aside cells, and levels of selection. *Journal of Experimental Zoology (Molecular and Developmental Evolution)* 288: 99–104.

Blackstone, N.W. and Jasker, B.D. 2003. Phylogenetic considerations of clonality, coloniality, and mode of germline development in animals. *Journal of Experimental Zoology Part B (Molecular and Developmental Evolution)* 297B: 35–47.

Blackwelder, R.E. 1963. *Classification of the Animal Kingdom.* Southern Illinois University Press, Carbondale, IL.

Blair, J.E., and Hedges, S.B. 2005. Molecular phylogeny and divergence times of deuterostome animals. *Molecular Biology and Evolution* 22: 2275–2284.

Blair, J.E., Ikeo, K., Gojobori, T., and Hedges, S.B. 2002. The evolutionary position of nematodes. *BMC Evolutionary Biology* 2: 7.

Blanco, M.J. and Alberch, P. 1992. Caenogenesis, developmental variability, and evolution in the carpus and tarsus of the marbled newt *Triturus marmoratus*. *Evolution* 46: 677–687.

Bleeker, P. 1859. *Enumeratio specierum piscium hucusque in Archipelago Indico observatarum, adjectis habitationibus citationibusque, ubi descriptiones earum recentiores reperiuntur, nec non speciebus Musei Bleekeriani Bengalensibus, Japonicis, Capensibus Tasmanicisque*. Lange, Batavia [in Latin].

Bleidorn, C. 2007. The role of character loss in phylogenetic reconstruction as exemplified for the Annelida. *Journal of Zoological Systematics and Evolutionary Research* 45: 299–307.

Bleidorn, C., Vogt, L., and Bartolomaeus, T. 2003a. A contribution to sedentary polychaete phylogeny using 18S rRNA sequence data. *Journal of Zoological Systematics and Evolutionary Research* 41: 186–195.

Bleidorn, C., Vogt, L., and Bartolomaeus, T. 2003b. New insights into polychaete phylogeny (Annelida) inferred from 18S rDNA sequences. *Molecular Phylogenetics and Evolution* 29: 279–288.

Bleidorn, C., Podsiadlowski, L., and Bartolomaeus, T. 2006. The complete mitochondrial genome of the orbiniid polychaete *Orbinia latreillii* (Annelida, Orbiniidae)—a novel gene order for Annelida and implications for annelid phylogeny. *Gene* 370: 96–103.

Bleidorn, C., Eeckhaut, I., Podsliadlowski, L. *et al*. 2007. Mitochondrial genome and nuclear sequence data support Myzostomida as part of the annelid radiation. *Molecular Biology and Evolution* 24: 1690–1701.

Blin, M., Rabet, N., Deutsch, J.S., and Mouchel-Vielh, E. 2003. Possible implication of Hox genes *Abdominal-B* and *abdominal-A* in the specification of genital and abdominal segments in cirripedes. *Development, Genes and Evolution* 213: 90–96.

Blochlinger, K., Jan, L.Y., and Jan, Y.N. 1991. Transformation of sensory organ identity by ectopic expression of *Cut* in *Drosophila*. *Genes & Development* 5: 1124–1135.

Bock, W. 2004. Species: the concept, category and taxon. *Journal of Zoological Systematics and Evolutionary Research* 42: 178–190.

Bode, H.R. 1996 The interstitial cell lineage of hydra: a stem cell system that arose early in evolution. *Journal of Cell Science* 109: 1155–1164.

Bodmer, R., Barbel, S., Sheperd, S., Jack, J.W., Jan, L.Y., and Jan, Y.N. 1987. Transformation of sensory organs by mutations of the *cut* locus of *D. melanogaster*. *Cell* 51: 293–307.

Boero, F., Gravili, C., Pagliara, P., Piraino, S., Bouillon, J., and Schmid, V. 1998. The cnidarian premises of metazoan evolution: from triploblasty, to coelom formation, to metamery. *Italian Journal of Zoology* 65: 5–9.

Boero, F., Bouillon, J., and Piraino, S. 2005. The role of Cnidaria in evolution and ecology. *Italian Journal of Zoology* 72: 65–71.

Böger, H. 1988. Versuch über das phylogenetische System der Porifera. *Meyniana* 40: 143–154.

Bolker, J. and Raff, R.A. 1996 Developmental genetics and traditional homology. *BioEssays* 18: 489–494.

Bonato, L. and Minelli, A. 2002. Parental care in *Dicellophilus carniolensis* (C. L. Koch, 1847): new behavioural evidence with implications for the higher phylogeny of centipedes (Chilopoda). *Zoologischer Anzeiger* 241: 193–198.

Bonner, J.T. 1988. *The Evolution of Complexity by Means of Natural Selection*. Princeton University Press, Princeton, NJ.

Bonner, J.T. 2003. Evolution of development in the cellular slime molds. *Evolution & Development* 5: 305–313.

Boore, J.L. 2006. The use of genome-level characters for phylogenetic reconstruction. *Trends in Ecology & Evolution* 21: 439–446.

Boore, J.L. and Brown, W.M. 2000. Mitochondrial genomes of *Galathealinum*, *Helobdella*, and *Platynereis*: sequence and gene arrangement comparisons indicate that Pogonophora is not a phylum and Annelida and Arthropoda are not sister taxa. *Molecular Biology and Evolution* 17: 87–106.

Boore, J.L. and Straton, J. 2002. The mitochondrial genome of the sipunculid *Phascolopsis gouldii* supports its association with Annelida rather than Mollusca. *Molecular Biology and Evolution* 19: 127–137.

Boore, J.L. Collins, T.M., Stanton, D., Daehler, L.L., and Brown, W.M. 1995. Deducing arthropod phylogeny from mitochondrial DNA rearrangements. *Nature* 376: 163–165.

Boore, J.L., Lavrow, D.V., and Brown, W.M. 1998. Gene translocation links insects and crustaceans. *Nature* 392: 667–668.

Boorman, C.J. and Shimeld, S.M. 2002. Pitx homeobox genes in *Ciona* and amphioxus show left–right asymmetry is a conserved chordate character and define the ascidian adenohypophysis. *Evolution & Development* 4: 354–365.

Borchiellini, C., Boury-Esnault, N., Vacelet, J., and Le Parco, Y. 1998. Phylogenetic analysis of the Hsp70 sequences reveals the monophyly of Metazoa and specific phylogenetic relationships between Animals and Fungi. *Molecular Biology and Evolution* 15: 647–655.

Borchiellini, C., Manuel, M., Alivon, E., Boury-Esnault, N., Vacelet, J., and Le Parco, Y. 2001. Sponge paraphyly and the origin of Metazoa. *Journal of Evolutionary Biology* 14: 171–179.

Borchiellini, C., Chombard, C., Manuel, M., Alivon, E., Vacelet, J., and Boury-Esnault, N. 2004. Molecular phylogeny of Demospongiae: implications for classification and scenarios of character evolution. *Molecular Phylogenetics and Evolution* 32: 823–837.

Borda, E. and Siddall, M.E. 2004a. Arhynchobdellida (Annelida: Oligochaeta: Hirudinida): phylogenetic relationships and evolution. *Molecular Phylogenetics and Evolution* 30: 213–225.

Borda, E. and Siddall, M.E. 2004b. Review of the evolution of life history strategies and phylogeny of the Hirudinida (Annelida: Oligochaeta). *Lauterbornia* 52: 5–25.

Börner, C. 1910. Die phylogenetische Bedeutung der Protura. *Biologisches Zentralblatt* 30: 633–641.

Bornstein, O., Twig, G., Benda, J., Weiler, R., and Perlman, I. 2002. Dynamic changes in the receptive fields of L1-type horizontal cells in the retina of the turtle *Mauremys caspica*. *Visual Neuroscience* 19: 621–632.

Borucki, H. 1996. Evolution und Phylogenetisches System der Chilopoda (Mandibulata, Tracheata). *Verhandlungen des naturwissenschaftlichen Vereins in Hamburg* 35: 95–226 [in German].

Botas, J. and Auwers, L. 1996. Chromosomal binding sites of Ultrabithorax homeotic proteins. *Mechanisms of Development* 56: 129–138.

Boudreaux, H.B. 1979a. Significance of intersegmental tendon system in arthropod phylogeny and a monophyletic classification of Arthropoda, pp. 551–586 in A.P. Gupta (ed.), *Arthropod Phylogeny*. Van Nostrand Reinhold, New York.

Boudreaux, H.B. 1979b. *Arthropod Phylogeny with Special Reference to Insects*. John Wiley & Sons, New York.

Bouillon, J. 1993. Classe des Hydrozoaires, pp. 29–416 in P.P. Grassé (ed.), *Traité de Zoologie*, 3(2). Masson, Paris.

Bouillon, J., Gravili, C., Pagès, F., Gili, J.-M., and Boero, F. 2006. *An Introduction to Hydrozoa*. (*Mémoires du Muséum national d'Histoire naturelle, Tome 194*). Muséum national d'Histoire naturelle, Paris.

Bourgoin, T., Steffan-Campbell, J.D., and Campbell, B.C. 1997. Molecular phylogeny of Fulgoromorpha (Insecta, Hemiptera, Auchenorrhyncha). The enigmatic Tettigometridae: evolutionary affiliation and historical biogeography. *Cladistics* 13: 207–224.

Bourlat, S.J., Nielsen, C., Lockyer, A.E., Littlewood, D.T.J., and Telford, M.J. 2003. *Xenoturbella* is a deuterostome that eats molluscs. *Nature* 424: 925–928.

Bourlat, S.J., Juliusdottir, T., Lowe, C.J. *et al.* 2006. Deuterostome phylogeny reveals monophyletic chordates and the new phylum Xenoturbellida. *Nature* 444: 85–88.

Boury-Esnault, N. and Vacelet, J. 1994. Preliminary studies on the organization and development of a hexactinellid sponge from a Mediterranean cave, *Oopsacas minuta*, pp. 407–415 in R.W.M. Van Soest, T.M.G. Van Kempen and J.C. Braekman (eds), *Sponges in Time and Space*. A.A. Balkema, Rotterdam.

Boury-Esnault, N. and Jamieson, B.G.M. 1999. Porifera, pp. 1–20 in B.G.M. Jamieson. (ed.), *Progress in Male Gamete Ultrastructure and Phylogeny*. Wiley, Chichester.

Boury-Esnault, N., Muricy, G., Gallissian, M.F., and Vacelet, J. 1995. Sponges without skeleton: a new Mediterranean genus of Homoscleromorpha (Porifera, Demospongiae). *Ophelia* 43: 25–43.

Boury-Esanult, N., Efremova, S., Bezac, C., and Vacelet, J. 1999. Reproduction of a hexactinellid sponge: first description of gastrulation by cellular delamination in the Porifera. *Invertebrate Reproduction and Development* 35: 187–201.

Boury-Esnault, N., Ereskovsky, A., Bezac, C., and Tokina, D. 2003. Larval development in the Homoscleromorpha (Porifera, Demospongiae). *Invertebrate Biology* 122: 187–202.

Boute, N., Exposito, J.Y., Boury-Esnault, N. *et al.* 1996. Type IV collagen in sponges, the missing link in basement membrane ubiquity. *Biology of the Cell* 88: 37–44.

Bouwmeester, T. and Leyns, L. 1997. Vertebrate head induction by anterior primitive endoderm. *Bioessays* 19: 855–863.

Bowerbank, J.S. 1864. *A Monograph of the British Spongiadae*, vol. 1. Ray Society, London.

Bowsher, J.H. and Nijhout, H.F. 2007. Evolution of novel abdominal appendages in a sepsid fly from histoblasts, not imaginal discs. *Evolution & Development* 9: 347–354.

Boyan, G.S., Williams, J.L.D., Posser, S., and Bräunig, P. 2002. Morphological and molecular data argue for the labrum being non-apical, articulated, and the appendage of the intercalary segment in locust. *Arthropod Structure and Development* 31: 65–76.

Boyden, A. and Shelswell, E.M. 1959. Prophylogeny: some considerations regarding primitive evolution in lower Metazoa. *Acta Biotheoretica* 13: 115–130.

Brakefield, P.M. 2008. Prospects of evo-devo for linking pattern and process in the evolution of morphospace, pp. 62–79 in A. Minelli and G. Fusco (eds), *Evolving Pathways: Key Themes in Evolutionary Developmental Biology*. Cambridge University Press, Cambridge.

Brandenburg, J. 1975. The morphology of the protonephridia. *Fortschritte der Zoologie/Progress in Zoology* 23: 1–17.

Bregazzi, P.K. 1973. Embryological development in *Tryphosella kergueleni* (Miers) and *Cheirimedon femoratus*

(Pfeffer) (Crustacea: Amphipoda). *British Antarctic Survey Bulletin* 32: 63–74.

Breimer, A. 1978. General morphology, recent crinoids, pp. 9–58 in R.C. Moore and C. Teichert (eds), *Treatise on Invertebrate Paleontology. Part T. Echinodermata 2*, vol. 1. The Geological Society of America, Boulder, CO.

Brena, C., Cima, F. and Burighel, P. 2003. The highly specialised gut of Fritillariidae (Appendicularia: Tunicata). *Marine Biology* 143: 57–71.

Bresslau, E. and Reisinger, E. 1928. Allgemeine Einleitung zur Naturgeschichte der Plathelminthes, pp. 34–51 in W. Kükenthal and Th. Krumbach (eds), *Handbuch der Zoologie*, volume 2, part 1. De Gruyter, Berlin-Leipzig [in German].

Bridge, D., Cunningham, C.W., DeSalle, R., and Buss, L.W. 1995. Class-level relationships in the phylum Cnidaria: molecular and morphological evidence. *Molecular Biology and Evolution* 12: 679–689.

Brien, P. 1967. Les éponges. Leur nature metazoaire, leur gastrulation, leur état colonial. *Annales de la Société Royale zoologique de Belgique* 97: 197–235 [in French].

Brien, P. 1973. Les Démosponges. Morphologie et reproduction, pp. 133–461 in P.P. Grassé (ed.), *Traité de Zoologie*, 3(1). Masson, Paris [in French].

Briggs, D.E.G. and Fortey, R.A. 2005. Wonderful strife: systematics, stem groups, and the phylogenetic signal in the Cambrian radiation. *Paleobiology* 31: 94–112.

Briggs, D.E.G., Erwin, D.H., and Collier, F.J. 1994. *The Fossils of the Burgess Shale*. Smithsonian Institution Press, Washington.

Brooks, D.R. 1996. Explanations of homoplasy at different levels of biological organization, pp. 3–36 in M.J. Sanderson and L. Hufford (eds), *Homoplasy: the Recurrence of Similarity in Evolution*. Academic Press, San Diego, CA.

Brooks, D.R. and McLennan, D.A. 1993. Macroevolutionary patterns of morphological diversification among parasitic flatworms (Platyhelminthes: Cercomeria). *Evolution* 47: 495–509.

Brower, D.L., Brower, S.M., Hayward, D.C., and Ball, E.E. 1997. Molecular evolution of integrins: genes encoding integrin β subunits from a coral and a sponge. *Proceedings of the National Academy of Sciences USA* 94: 9182–9187.

Brown, J.H., Marquet, P.A., and Taper, M.L. 1993. Evolution of body size: consequences of an energetic definition of fitness. *American Naturalist* 142: 573–584.

Brown, J.M. and Storey, K.G. 2000. A region of the vertebrate neural plate in which neighbouring cells can adopt neural or epidermal fates. *Current Biology* 10: 869–872.

Brown, S.J., Shippy, T.D., Beeman, R.W., and Denell, R.E. 2002. *Tribolium Hox* genes repress antennal development in the gnathos and trunk. *Molecular Phylogenetics and Evolution* 24: 384–387.

Bruce, R.C. 2005. Did *Desmognathus* salamanders reinvent the larval stage? *Herpetological Review* 36: 107–112.

Bruggeman, J. and Ehlers, U. 1981. Ultrastruktur der Statocyst von *Ototyphlonemertes pallida* (Keferstein, 1862) (Nemertini). *Zoomorphology* 97: 75–87 [in German].

Bryant, P.J. and Schneidermann, H.A. 1969. Cell lineage, growth, and determination in the imaginal leg discs of *Drosophila melanogaster*. *Developmental Biology* 20: 263–290.

Bubel, A., Stephens, R.M., Fenn, R.H., and Fieth, P. 1983. An electron microscope, X-ray diffraction and amino acid analysis study of the opercular filament cuticle, calcareous opercular plate and habitation tube of *Potamoceros lamarckii* Quatrefages (Polychaeta: Serpulidae). *Comparative Biochemistry and Physiology B* 74: 837–850.

Buchholz, K. and Ruthmann, A. 1995. The mesenchyme-like layer of the fiber cells of *Trichoplax adhaerens* (Placozoa), a syncytium. *Zeitschrift für Naturforschung Section C, Biosciences* 50: 282–285.

Budd, G.E. 2001. Why are arthropods segmented? *Evolution & Development* 3: 332–342.

Budd, G.E. 2003a. Arthropods as ecdysozoans: the fossil evidence, pp. 479–487 in A. Legakis, S. Sfenthourakis, R. Polymeni, and M. Thessalou-Legaki (eds), *The New Panorama of Animal Evolution*. Proceedings of the 18th International Congress of Zoology. Pensoft, Sofia/Moscow.

Budd, G.E. 2003b. The Cambrian fossil record and the origin of the phyla. *Integrative and Comparative Biology* 43: 157–165.

Budd, G.E. and Jensen, S. 2000. A critical reappraisal of the fossil record of the bilaterian phyla. *Biological Reviews* 75: 253–295.

Budelmann, B.U., Schipp, R., and von Boletzky, S. 1997. Cephalopoda, pp. 119–414 in F.W. Harrison and A.J. Kohn (eds), *Microscopic Anatomy of Invertebrates. Vol. 6A. Mollusca II*. Wiley-Liss, New York.

Bullock, T.H. 1965a. Coelenterata and Ctenophora, pp. 459–534 in T.H. Bullock and G.A. Horridge (eds), *Structure and Function in the Nervous Systems of Invertebrates*. W.H. Freeman and Company, San Francisco, CA.

Bullock, T.H. 1965b. Chaetognatha, Pogonophora, Hemichordata, and Tunicata, pp. 1559–1592 in T.H. Bullock and G.A. Horridge (eds), *Structure and Function in the Nervous Systems of Invertebrates*. W.H. Freeman and Company, San Francisco, CA.

Bullock, T.H. and Horridge, G.A. 1965. *Structure and Function in the Nervous System of Invertebrates*. Freeman, San Francisco, CA.

Büning, J. 2005. The telotrophic ovary known from Neuropterida exists also in the myxophagan beetle *Hydroscapha natans*. *Development, Genes and Evolution* 215: 597–607.

Bunke, D. 1967. Zur Morphologie und Systematik der Aeolosomatidae Beddard 1895 und Potamodrilidae nov. fam. (Oligochaeta). *Zoologische Jahrbücher, Abteilung für Systematik Ökologie und Geographie der Tiere* 94: 187–368 [in German].

Burighel, P. and Cloney, R.A. 1997. Urochordata: Ascidiacea, pp. 221–347 in F.W. Harrison and E.E. Ruppert (eds), *Microscopic Anatomy of Invertebrates. Vol. 15. Hemichordata, Chaetognatha, and the Invertebrate Chordates*. Wiley-Liss, New York.

Burighel, P., Brena, C., Martinucci, G.B., and Cima, F. 2001. Gut ultrastructure of the appendicularian *Oikopleura dioica* (Tunicata). *Invertebrate Biology* 120: 278–293.

Burke, W.D., Malik, H.S., Lato, III, W.C., and Eickbush, T.H. 1998. Are retrotransposons long-term hitchhikers? *Nature* 392: 141–142.

Burlando, B. 1990. The fractal dimension of taxonomic systems. *Journal of Theoretical Biology* 146: 99–114.

Buss, L.W. 1987. *The Evolution of Individuality*. Princeton University Press, Princeton, NJ.

Butler, A.B. 2000. Chordate evolution and the origin of craniates: an old brain in a new head. *Anatomical Record* 261: 111–125.

Bütschli, O. 1884. Bemerkungen zur Gastraea-Theorie. *Morphologische Jahrbücher* 9: 415–427 [in German].

Bütschli, O. 1910. *Vorlesungen über vergleichende Anatomie, 1. Einleitung. Vergleichende Anatomie der Protozoen. Integument und Skelett der Metazoen*. W. Engelmann, Leipzig [in German].

Butterfield, N.J. 1990. A reassessment of the enigmatic Burgess Shale (British Columbia, Canada) fossil *Wiwaxia corrugata* (Matthew) and its relationship to the polychaete *Canadia spinosa* Walcott. *Paleobiology* 16: 287–303.

Butterfield, N.J. 2003. Exceptional fossil preservation and the Cambrian explosion. *Integrative and Comparative Biology* 43: 166–177.

Butterfield, N.J. 2006. Hooking some stem-group "worms": fossil lophotrochozoans in the Burgess Shales. *BioEssays* 28: 1161–1166.

Byrum, C.A. and Martindale, M.Q. 2004. Gastrulation in the Cnidaria and Ctenophora, pp. 33–50 in C. Stern (ed.), *Gastrulation. From Cells to Embryos*. Cold Spring Harbor Press, Cold Spring Harbor, NY.

Cadigan, K.M. and Nusse, R. 1997. Wnt signaling: a common theme in animal development. *Genes & Development* 11: 3286–3305.

Callebaut, W. and Rasskin-Gutman, D. 2005. *Modularity: Understanding the Development and Evolution of Natural Complex Systems*. MIT Press, Cambridge, MA.

Calzada, A., Medina, A., and Gonzalez De Canales, M.L. 1998. Fine structure of the intestine development in cultured sea bream larvae. *Journal of Fish Biology* 53: 340–365.

Cameron, C.B. 2002. Particle retention and flow in the pharynx of the enteropneusts worm *Harrimania planktophilus* [sic]: the filter-feeding pharynx may have evolved before the chordates. *Biological Bulletin* 202: 192–200.

Cameron, C.B. 2005. A phylogeny of the hemichordates based on morphological characters. *Canadian Journal of Zoology* 83: 196–215.

Cameron, C.B., Garey, J.R., and Swalla, B.J. 2000. Evolution of the chordate body plan: new insights from phylogenetic analyses of deuterostome phyla. *Proceedings of the National Academy of Sciences USA* 97: 4469–4474.

Cameron, R.A., Peterson, K.J., and Davidson, E.H. 1998. Developmental gene regulation and the evolution of large animal body plans. *American Zoologist* 38: 609–620.

Cameron, R.A., Rowen, L., Nesbitt, R. *et al.* 2006. Unusual gene order and organization of the sea urchin Hox cluster. *Journal of Experimental Zoology (Molecular and Developmental Evolution)* 306B: 45–58.

Cameron, S.L., Miller, K.B., D'Haese, C.A., Whiting, M.F., and Barker, S.C. 2004. Mitochondrial genome data alone are not enough to unambiguously resolve the relationships of Entognatha, Insecta and Crustacea sensu lato (Arthropoda). *Cladistics* 20: 534–557.

Campuzano, S. and Modolell, J. 1992. Patterning of the *Drosophila* nervous system: the achaete-scute gene complex. *Trends in Genetics* 8: 202–208.

Canning, E.U., Curry, A., Feist, S.W., Longshaw, M., and Okamura, B. 2000. A new class and order of myxozoans to accommodate parasites of bryozoans with ultrastructural observations on *Tetracapsula bryosalmonae* (PKX organism). *Journal of Eukaryotic Microbiology* 47: 456–468.

Cannon, L.R.G. 1986. *Turbellaria of the World. A Guide to Families & Genera*. Queensland Museum, Brisbane.

Caprette, C., Lee, M.S.Y., Shine, R., Mokany, A., and Downhower, J.F. 2004. The origin of snakes (Serpentes) as seen through eye anatomy. *Biological Journal of the Linnean Society* 81: 469–482.

Carapelli, A., Nardi, F., Dallai, R., and Frati, F. 2006. A review of molecular data for the phylogeny of basal hexapods. *Pedobiologia* 50: 191–204.

Carayon, J. 1966. Traumatic insemination and the paragenital system, pp. 81–166 in R.L. Usinger (ed.), *Monograph* of Cimicidae (Hemiptera-Heteroptera). Entomological Society of America, College Park, MD.

Cardona, A., Hartenstein, V., and Romero, R. 2005. The embryonic development of the triclad *Schmidtea polychroa*. *Development, Genes and Evolution* 215: 109–131.

Carmeliet, P. 2003. Blood vessels and nerves: common signals, pathways and diseases. *Nature Reviews Genetics* 4: 710–720.

Carmeliet, P. and Tessier-Lavigne, M. 2005. Common mechanisms of nerve and blood vessel wiring. *Nature* 436: 193–200.

Caron, J.-B. 2005. *Banffia constricta*, a putative vetulicolid from the Middle Cambrian Burgess Shale. *Transactions of the Royal Society of Edinburgh: Earth Sciences* 96: 95–111.

Caron, J.-B., Scheltema, A., Schander, C., and Rudkin, D. 2006. A soft-bodied mollusc with radula from the Middle Cambrian Burgess Shale. *Nature* 442: 159–163.

Caron, J.-B., Scheltema, A., Schander, C., and Rudkin, D. 2007. Reply to Butterfield on stem-group "worms": fossil lophotrochozoans in the Burgess Shale. *BioEssays* 29: 200–202.

Carré, D. 1984. Existence d'un complexe acrosomal chez les spermatozoids du cnidaire *Muggiaea kochi* (Siphonophore, Calycophore): differentiation et reaction acrosomale. *International Journal of Invertebrate Reproduction & Development* 7: 95–103 [in French].

Carroll, S.B. 2005. *Endless Forms Most Beautiful: the New Science of Evo Devo and the Making of the Animal Kingdom*. W.W. Norton, New York.

Carroll, S.B., Gates, J., Keys, D. *et al.* 1994. Pattern formation and eyespot determination in butterfly wings. *Science* 265: 109–114.

Carroll, S.B., Grenier, J.K., and Weatherbee, S.D. 2005. *From DNA to Diversity: Molecular Genetics and the Evolution of Animal Design*, 2nd edn. Blackwell Scientific, Malden, MA.

Carter, C.A. and Wourms, J.P. 1993. Naturally occuring diblastodermic eggs in the annual fish *Cynolebias*: implications for developmental regulation and determination. *Journal of Morphology* 215: 301–312.

Cartwright, P. and Collins, A. 2007. Fossils and phylogenies: integrating multiple lines of evidence to investigate the origin of early major metazoan lineages. *Integrative and Comparative Biology* 47: 744–751.

Cartwright, P., Schierwater, B., and Buss, L.W. 2006. Expression of a *Gsx* parahox gene, *Cnox-2*, in colony ontogeny in *Hydractinia* (Cnidaria: Hydrozoa). *Journal of Experimental Zoology (Molecular and Developmental Evolution)* 306B: 460–469.

Carvalho, A.B. and Clark, A.G. 2005. Y chromosome of *D. pseudoobscura* is not homologous to the ancestral *Drosophila* Y. *Science* 307: 108–110.

Castelli-Gair Hombría, J., and Brown, S. 2002. The fertile field of *Drosophila* JAK/STAT signalling. *Current Biology* 12: R569–R575.

Castro, L.R. and Dowton, M. 2007. Mitochondrial genomes in the Hymenoptera and their utility as phylogenetic markers. *Systematic Entomology* 32: 60–69.

Catala, M., Teillet, M.A., De Robertis, E.M., and Le Douarin, M.L. 1996. A spinal cord fate map in the avian embryo: while regressing, Hensen's node lays down the notochord and floor plate thus joining the spinal cord lateral walls. *Development* 122: 2599–2610.

Cavalier-Smith, T. 1981. Eukaryotic kingdoms, seven or nine? *BioSystems* 14: 461–481.

Cavalier-Smith, T. 1991. Cell diversification in heterotrophic flagellates, pp. 113–131 in D.J. Patterson and J. Larson (eds), *The Biology of Free-Living Heterotrophic Flagellates*. Clarendon Press, Oxford.

Cavalier-Smith, T. 1998. A revised six-kingdom system of life. *Biological Reviews* 73: 203–266.

Cavalier-Smith, T. and Chao, E.E.-Y. 2003. Phylogeny of Choanozoa, Apusozoa, and other Protozoa and early eukaryote megaevolution. *Journal of Molecular Evolution* 56: 540–563.

Cavalier-Smith, T., Alsopp, M.T.E.P., Chao, E.E., Boury-Esnault, N., and Vacelet, J. 1996. Sponge phylogeny, animal monophyly, and the origin of the nervous system: 18S rRNA evidence. *Canadian Journal of Zoology* 74: 2031–2045.

Cerra, A. and Byrne, M. 2004. Evolution of development in the sea star genus *Patiriella*: clade-specific alterations in cleavage. *Evolution & Development* 6: 105–113.

Chapman, D.M. 1968. Structure, histochemistry and formation of the podocyst and cuticle of *Aurelia aurita* (L.). *Journal of the Marine Biological Association of the United Kingdom* 48: 187–208.

Chapman, E.G., Foote, B.A., Malukiewicz, J., and Hoeth, W.R. 2006. Parallel evolution of larval morphology and habitat in the snail-killing fly genus *Tetanocera*. *Journal of Evolutionary Biology* 19: 1459–1474.

Chea, H.K., Wright, C.V., and Swalla, B.J. 2005. Nodal signalling and the evolution of deuterostome gastrulation. *Developmental Dynamics* 234: 269–278.

Chen, E.H. and Olson, E.N. 2005. Unveiling the mechanisms of cell-cell fusion. *Science* 308: 369–373.

Chen, J.-Y. and Li, C.-W. 1997. Early Cambrian chordate from Chengjiang, China. *Bulletin of the National Museum of Natural Science* 10: 257–273.

Chen, J.-Y. and Zhou, G.-Q. 1997. Biology of the Chengjiang fauna. *Bulletin of the National Museum of Natural Science Taiwan* 10: 11–106.

Chen, J.-Y., Dzik, J., Edgecombe, G.D., Ramsköld, L., and Zhou, G.Q. 1995. A possible Early Cambrian chordate. *Nature* 377: 720–722.

Chen, J.-Y., Huang, D.-Y., and Li, C.-W. 1999. An early Cambrian craniate-like chordate. *Nature* 402: 518–522.

Chen, J.-Y., Oliveri, P., Li, C.-W. *et al.* 2000. Precambrian animal diversity: putative phosphatized embryos from the Doushantuo Formation of China. *Proceedings of the National Academy of Sciences USA* 97: 4457–4462.

Chen, J.-Y., Bottjer, D.J., Oliveri, P. *et al.* 2004a. Small bilaterian fossils from 40 to 55 million years before the Cambrian. *Science* 305: 218–222.

Chen, J.-Y., Oliveri, P., Davidson, E., and Bottjer, D.J. 2004b. Response to comment on "Small bilaterian fossils from 40 to 55 million years before the Cambrian". *Science* 306: 1291.

Chen, J., Waloszek, D., and Maas, A. 2004c. A new 'great-appendage' arthropod from the Lower Cambrian of China and homology of chelicerate chelicerae and raptorial antero-ventral appendages. *Lethaia* 37: 3–20.

Chen, J.-Y., Bottjer, D.J., Davidson, E.H. *et al.* 2006. Phosphatized polar lobe-forming embryos from the Precambrian of Southwest China. *Science* 312: 1644–1646.

Chia, F.S. and Koss, R. 1979. Fine structural studies of the nervous system and the apical organ in the planula larva of the sea anemone *Anthopleura elegantissima*. *Journal of Morphology* 160: 275–298.

Chia, F.-S., Gibson, G., and Qian, P.-Y. 1996. Poecilogony as a reproductive strategy of marine invertebrates. *Oceanologica Acta* 19: 203–208.

Chin-Sang, I.D. and Chisholm, A.D. 2000. Form of the worm: genetics of epidermal morphogenesis in *C. elegans*. *Trends in Genetics* 16: 544–551.

Chipman, A.D. and Stollewerk, A. 2006. Specification of neural precursor identity in the geophilomorph centipede *Strigamia maritima*. *Developmental Biology* 290: 337–350.

Chipman, A.D., Arthur, W., and Akam, M. 2004. Early development and segment formation in the centipede *Strigamia maritima* (Geophilomorpha). *Evolution & Development* 6: 78–89.

Chourrout, D., Delsuc, F., Chourrout, P. *et al.* 2006. Minimal protoHox cluster inferred from bilaterian and cnidarian *Hox* complements. *Nature* 442: 684–687.

Christ, B., Schmidt, C., Huang, R., and Wilting, J. 1998. Segmentation of the vertebrate body. *Anatomy and Embryology* 197: 1–8.

Christensen, B. 1994. Annelida-Clitellata, pp. 1–23 in K.G. Adiyodi and R.G. Adiyodi (eds), *Reproductive Biology of Invertebrates, vol. VI, Part B, Asexual Propagation and Reproductive Strategies*, Oxford and IBH Publications, New Delhi.

Christiaen, L., Jaszczyszyn, Y., Kerfant, M., Kano, S., Thermes, V., and Joly, J.-S. 2007. Evolutionary modification of mouth position in deuterostomes. *Seminars in Cell & Developmental Biology* 18: 502–511.

Chuang, P. and McMahon, A.P. 2003. Branching morphogenesis of the lung: new molecular insights into an old problem. *Trends in Cell Biology* 13: 86–91.

Ciccarelli, F.D., Doerks, T., v. Mering, C., Creevey, C.J., Snel, B., and Bork, P. 2006. Toward automatic reconstruction of a highly resolved tree of life. *Science* 311: 1283–1287.

Cima, F., Brena, C., and Burighel, P. 2002. Multifarious activities of gut epithelium in an appendicularian (*Oikopleura dioica*: Tunicata). *Marine Biology* 141: 479–490.

Cisne, J.L. 1974. Trilobites and the origin of arthropods. *Science* 186: 13–18.

Clark, R.B. 1964. *Dynamics in Metazoan Evolution; the Origin of the Coelom and Segments*. Clarendon Press, Oxford.

Coddington, J. 1994. The roles of homology and convergence in studies of adaptation, pp. 53–78 in P. Eggleton and R. Vane-Wright (eds), *Phylogenetics and Ecology*. The Linnean Society of London, London.

Coddington, J.A., Hormiga, G., and Scharff, N. 1997. Giant female or dwarf male spiders? *Nature* 385: 687–688.

Coe, W. R. 1933. Metameric ganglia connected with the dorsal nerve in a nemertean. *Zoologischer Anzeiger* 102: 237–240.

Coe, W.R. 1938. A new genus and species of Hoplonemertea having differential bipolar sexuality. *Zoologischer Anzeiger* 124: 220–224.

Cohen, B.L. 2000. Monophyly of brachiopods and phoronids: reconciliation of molecular evidence with Linnaean classification (the subphylum Phoroniformea nov.). *Proceedings of the Royal Society of London Series B, Biological Sciences* 267: 225–231.

Cohen, B.L. 2005. Not armour, but biomechanics, ecological opportunity and increased fecundity as keys to the origin and expansion of the mineralized benthic metazoan fauna. *Biological Journal of the Linnean Society* 85: 483–490.

Cohen, B.L. and Weydmann, A. 2005. Molecular evidence that phoronids are a subtaxon of brachiopods (Brachiopoda: Phoronata) and that genetic divergence of metazoan phyla began long before the early Cambrian. *Organisms Diversity and Evolution* 5: 253–273.

Cohen, S.M. 1993. Imaginal disc development, pp. 747–841 in M. Bate and A. Martinez-Arias (eds), *The Development of Drosophila melanogaster*. Cold Spring Harbor Press, Cold Spring Harbor, NY.

Cohn, M.J. and Tickle, C. 1999 Developmental basis of limblessness and axial regionalization in snakes. *Nature* 399: 474–479.

Colgan, D.J., Ponder, W.F., Beacham, E., and Macaranas, J. 2007. Molecular phylogenetics of Caenogastropoda (Gastropoda: Mollusca). *Molecular Phylogenetics and Evolution* 42: 717–737.

Colitti, M., Allen, S.P., and Price, J.S. 2005. Programmed cell death in the regenerating deer antler. *Journal of Anatomy* 207: 339–351.

Collier, J.R. 1997. Gastropods, the snails, pp. 189–217 in S.F. Gilbert and A.M. Raunio (eds), *Embryology: Constructing the Organism*, Sinauer Associates, Sunderland, MA.

Collin, R. and Cipriani, R. 2003. Dollo's law and the re-evolution of shell coiling. *Proceedings of the Royal Society of London Series B, Biological Sciences* 270: 2551–2555.

Collins, A.G. 1998. Evaluating multiple alternative hypotheses for the origin of Bilateria: an analysis of 18S rRNA molecular evidence. *Proceedings of the National Academy of Sciences USA* 95: 15458–15463.

Collins, A.G. 2002. Phylogeny of Medusozoa and the evolution of the cnidarian life cycles. *Journal of Evolutionary Biology* 14: 418–432.

Collins, A.G. and Valentine, J.W. 2001. Defining phyla: evolutionary pathways to metazoan body plans. *Evolution & Development* 3: 432–442.

Collins, A.G., Winkelman, H., Hadrys, H. and Schierwater, B. 2005. Phylogeny of Capitata and Corynidae (Cnidaria, Hydrozoa) in light of mitochondrial 16S rDNA data. *Zoologica Scripta* 34: 91–99.

Collins, A.G., Schuchert, P., Marques, A.C., Jankowski, T., Medina, M., and Schierwater, B. 2006. Medusozoan phylogeny and character evolution clarified by new large and small subunit rDNA data and an assessment of the utility of phylogenetic mixture models. *Systematic Biology* 55: 97–115.

Colombo, J.A., Fuchs, E., Hartig, W., Marotte, L.E., and Puissant, V. 2000. "Rodent-like" and "primate-like" types of astroglial architecture in the adult cerebral cortex of mammals: a comparative study. *Anatomy and Embryology* 201: 111–120.

Colosi, G. 1967. *Zoologia e Biologia generale*. UTET, Torino [in Italian].

Coluzzi, M., Sabatini, A., della Torre, A., Di Deco, M.A., and Petrarca, V. 2002. A polytene chromosome analysis of the *Anopheles gambiae* species complex. *Science* 298: 1415–1418.

Conklin, E.G. 1932. The embryology of the *Amphioxus*. *Journal of Morphology* 54: 69–118.

Conway Morris, S. 1976. A new Cambrian lophophorate from the Burgess Shale of British Columbia. *Palaeontology* 19: 199–222.

Conway Morris, S. 1977. A new entoproct-like organism from the Burgess Shale of British Columbia. *Palaeontology* 20: 833–845.

Conway Morris, S. 1993. Ediacaran-like fossils from the Cambrian Burgess Shale type faunas of North America. *Palaeontology* 36: 593–635.

Conway Morris, S. 1994. Early metazoan evolution: first steps to an integration of molecular and morphological data, pp. 450–459 in S. Bengtson (ed.), *Early Life on Earth*. Nobel Symposium no. 84. Columbia University Press, New York.

Conway Morris, S. 1997. The cuticular structure of the 495-Myr-old type species of the fossil worm *Palaeoscolex, P. piscatorum* (?Priapulida). *Zoological Journal of the Linnean Society* 119: 69–82.

Conway Morris, S. 1998. Metazoan phylogenies: falling into place or falling to pieces? A palaeontological perspective. *Current Opinion in Genetics & Development* 8: 662–666.

Conway Morris, S. 2003a. *Life's Solution: Inevitable Humans in a Lonely Universe*. Cambridge University Press, Cambridge.

Conway Morris, S. 2003b. The Cambrian "explosion" of metazoans, pp. 13–32 in G.B. Müller and S.A. Newman (eds), *Origination of Organismal Form. Beyond the Gene in Developmental and Evolutionary Biology*. MIT Press, Cambridge, MA.

Conway Morris, S. 2006. Darwin's dilemma: the realities of the Cambrian 'explosion'. *Philosophical Transactions of the Royal Society of London Series B, Biological Sciences* 361: 1069–1083.

Conway Morris, S. and Peel, J.S. 1995. Articulated halkieriids from the Lower Cambrian of North Greenland and their role in early protostome evolution. *Philosophical Transactions of the Royal Society of London Series B, Biological Sciences* 347: 305–358.

Conway Morris, S. and Caron, J.-B. 2007. Wiwaxiids and the early evolution of the lophotrochozoans. *Science* 315: 1255–1258.

Conway Morris, S., George, J.D., Gibson, R., and Platt, H.M. (eds) 1985. *The Origins and Relationships of Lower Invertebrates*. The Systematics Association Special Volume no. 28. Clarendon Press, Oxford.

Cook, C.E., Smith, M.L., Telford, M.J., Bastianello, A., and Akam, M. 2001. *Hox* genes and the phylogeny of the arthropods. *Current Biology* 11: 759–763.

Cook, C.E., Jiménez, E., Akam, M., and Saló, E. 2004. The *Hox* gene complement of acoel flatworms, a basal bilaterian clade. *Evolution & Development* 6: 154–163.

Cook, C.E., Yue, Q., and Akam, M. 2005. Mitochondrial genomes suggest that hexapods and crustaceans are mutually paraphyletic. *Proceedings of the Royal Society of London Series B, Biological Sciences* 272: 1295–1304.

Cooper, A. and Fortey, R. 1998. Evolutionary explosions and the phylogenetic fuse. *Trends in Ecology and Evolution* 13: 151–156.

Cooper, A.R. 1919. North American pseudophyllidean cestodes from fishes. *Illinois Biological Monographs* 4: 1–243.

Cooper, E.L. 2003. Comparative immunology of the animal kingdom, pp. 117–125 in A. Legakis, S. Sfenthourakis, R. Polymeni, and M. Thessalou-Legaki (eds), *The New Panorama of Animal Evolution*. Proceedings of the 18th International Congress of Zoology. Pensoft, Sofia/Moscow.

Copley, R.R., Aloy, P., Russell, R.B., and Telford, M.J. 2004. Systematic searches for molecular synapomorphies in model metazoan genomes give some support for Ecdysozoa after accounting for the idiosyncrasies of *Caenorhabditis elegans*. *Evolution & Development* 6: 164–169.

Cori, C.J. 1929. Kamptozoa, pp. 1–64 in W. Kükenthal and T. Krumbach (eds), *Handbuch der Zoologie*, vol. 2. W. de Gruyter, Berlin [in German].

Corley, L.S. and Lavine, M.D. 2006. A review of insect stem cell types. *Seminars in Cell & Developmental Biology* 17: 510–517.

Costlow, Jr, J.D. 1968. Metamorphosis in crustaceans, pp. 3–41 in W. Etkin and L.I. Gilbert (eds), *Metamorphosis: a Problem in Developmental Biology*. North-Holland, Amsterdam and Appleton-Century-Crofts, New York.

Cotton, T.J. and Braddy, S.J. 2004. The phylogeny of arachnomorph arthropods and the origin of the Chelicerata *Transactions of the Royal Society of Edinburgh Earth Sciences* 94: 169–193.

Coutinho, C.C., Fonseca, R.N., Mansure, J.J.C., and Borojevic, R. 2003. Early steps in the evolution of multicellularity: deep structural and functional homologies among homeobox genes in sponges and higher metazoans. *Mechanisms of Development* 120: 429–440.

Craig, S.F., Slobodkin, L.B., Wray, G.A., and Biermann, C.H. 1997. The 'paradox' of polyembryony: a review of the cases and a hypothesis for its evolution. *Evolutionary Ecology* 11: 127–143.

Croll, R.P. and Dickinson, A.J.G. 2004. Form and function of the larval nervous system in molluscs. *Invertebrate Reproduction and Development* 46: 173–187.

Crow, K.D., Stadler, P.F., Lynch, V.J., Amemiya, C., and Wagner, G.P. 2006. The "fish-specifics" hox cluster duplication is coincident with the origin of teleosts. *Molecular Biology and Evolution* 23: 121–136.

Cuénot, L. 1940. Remarques sur un essai d'arbre généalogique du règne animal. *Comptes Rendus de l'Académie des Sciences de Paris* 210: 23–27 [in French].

Cunha, A., Azevedo, R.B.R., Emmons, S., and Leroi, A.M. 1999. Variable cell numbers in nematodes. *Nature* 402: 253.

Cunningham, C.W., Blackstone, N.W., and Buss, L.W. 1992. Evolution of king crabs from hermit crab ancestors. *Nature* 355: 539–542.

Cuvier, G. 1795. Second mémoire sur l'organisation et les rapports des animaux à sang blanc, dans lequel on traite de la structure des Mollusques et de leurs division en ordres. *Magasin encyclopédique* 2: 433–449 [in French].

Cuvier, G. 1798. *Tableau elementaire de l'histoire naturelle des animaux*. Baudoin, Paris [in French].

Cuvier, G. 1802. Mémoire sur l'animal de la lingula (*Ligula anatina* Lam.). *Annales du Muséum national d'Histoire Naturelle, Paris* 1: 69–80 [in French].

Cuvier, G. 1812. Sur un nouveau rapprochement à établir entre les classes qui composent le Règne Animal. *Annales du Muséum d'Histoire naturelle, Paris* 19: 73–84 [in French].

Cuvier, G. 1816. *Le règne animal distribué d'apres son organisation, pour servir de base à l'histoire naturelle des animaux et d'introduction à l'anatomie comparée*. Déterville, Paris [in French].

Czerny, T. and Busslinger, M. 1995. DNA-binding and transactivation properties of *Pax-6*: three amino acids in the paired domain are responsbile for the different sequence recognition of Pax-6 and BSAP (Pax-5). *Molecular and Cellular Biology* 15: 2858–2871.

da Cruz Landim, C. 2002. Spermiophagy in the spermatheca of *Melipona bicolor* Lepeletier, 1836 (Hymenoptera, Apidae, Meliponini). *Anatomia Histologia Embryologia* 31: 339–343.

Daeschler, E.B., Shubin, N.H., and Jenkins, Jr, F.A. 2006. A Devonian tetrapod-like fish and the evolution of the tetrapod body plan. *Nature* 440: 757–763.

Dalton, J.P., Skelly, P., and Halton, D.W. 2004. Role of the tegument and gut in nutrient uptake by parasitic platyhelminths. *Canadian Journal of Zoology* 82: 211–232.

Damen, W.G.M. 2002. Parasegmental organization of the spider embryo implies that the parasegment is an evolutionary conserved entity in arthropod embryogenesis. *Development* 129: 1239–1250.

David, B., Mooi, R., and Telford, M. 1995. The ontogenetic basis of Lovén's Rule clarifies homologies of the echinoid peristome, pp. 155–164 in R.H. Emson, A.B. Smith and A.C. Campbell (eds), *Echinoderm Research 1995*. Balkema, Rotterdam.

Davidson, E.H. 1991. Spatial mechanisms of gene regulation in metazoan embryos. *Development* 113: 1–26.

Davidson, E.H. 2003. *Genomic Regulatory Systems: Development and Evolution*. Academic Press, San Diego, CA.

Davidson, E.H. 2006. *The Regulatory Genome*. Academic Press-Elsevier, Amsterdam.

Davidson, E.H., Peterson, K.J. and Cameron, R.A. 1995. Origin of bilaterian body plans: evolution of developmental regulatory mechanisms. *Science* 270: 1319–1325.

Davies, J., Lyon, M., Gallagher, J., and Garrod, D. 1995. Sulphated proteoglycan is required for collecting duct growth and branching but not nephron formation during kidney development. *Development* 121: 1507–1517.

Davis, G.K. and Patel, N.H. 2002. Short, long, and beyond: molecular and embryological approaches to insect segmentation. *Annual Review of Entomology* 47: 669–699.

Davis, G.K. and Patel, N.H. 2003. Playing by pair-rules? *BioEssays* 25: 425–429.

Davydov, K.N. 1905. Scientific results of travels in Java and other islands of the Malay Archipelago. III. Morphology of Archiannelida. Biological observations on an epitokous form of *Polygordius*. *Izvestiya imperatorskoi Akademii nauk, ser.* 5, 13 [in Russian].

Dearden, P.K. and Akam, M. 1999. Developmental evolution: axial patterning in insects. *Current Biology* 9: R591–R594.

Dearden, P. and Akam, M. 2000. A role for *Fringe* in segment morphogenesis but not segment formation in the grasshopper, *Schistocerca gregaria*. *Development, Genes and Evolution* 210: 329–336.

de Beer, G.R. 1951. *Embryos and Ancestors*. Clarendon Press, Oxford.

de Beer, G.R. 1954. The evolution of the Metazoa, pp. 24–33 in J. Huxley *et al.* (eds), *Evolution as a Process*. Allen and Unwin, London.

de Beer, G.R. 1958. *Embryos and Ancestors*, 3rd edn. Clarendon Press, Oxford.

De Bodt, S., Maere, S., and Van de Peer, Y. 2005. Genome duplication and the origin of angiosperms. *Trends in Ecology and Evolution* 20: 591–597.

Degnan, B.M., Leys, S.P., and Larroux, C. 2005. Sponge development and antiquity of animal pattern formation. *Integrative and Comparative Biology* 45: 335–341.

De Jong, D.M., Hislop, N.R., Hayward, D.C. *et al.* 2006. Components of both major axial patterning systems of the Bilateria are differentially expressed along the primary axis of a 'radiate' animal, the anthozoan cnidarian *Acropora millepora*. *Developmental Biology* 298: 632–643.

Delage, Y. and Hérouard, H. 1897. *Traité de zoologie concrète. Vol. 5. Les Vermidiens*. Schleicher, Paris.

De Ley, P. and Blaxter, M. 2004. A new system for Nematoda: combining morphological characters with molecular trees, and translating clades into ranks and taxa, pp. 633–653 in R. Cook and D.J. Hunt (eds), *Nematology Monographs and Perspectives*. E.J. Brill, Leiden.

Delfino, G., Nosi, D., and Giachi, F. 2001a. Secretory granule-cytoplasm relationships in serous glands of anurans: ultrastructural evidence and possible functional role. *Toxicon* 39: 1161–1171.

Delfino, G., Nosi, D., Brizzi, R., and Alvarez, B.B. 2001b. Serous cutaneous glands in the paludiculine frog *Physalaemus biligonigerus* (Anura, Leptodactylidae): patterns of cytodifferentiation and secretory activity in premetamorphic specimens. *Acta Zoologica* 82: 149–158.

Dellaporta, S.L., Xu, A., Sagasser, S. *et al.* 2006. Mitochondrial genome of *Trichoplax adhaerens* supports Placozoa as the basal lower metazoan phylum. *Proceedings of the National Academy of Sciences USA* 103: 8751–8756.

Delle Cave, L. and Simonetta, A.M. 1991. Early Paleozoic arthropods and problems of arthropod phylogeny; with notes on taxa of doubtful affinities, pp. 189–244 in A.M. Simonetta and S. Conway Morris (eds), *The Early Evolution of Metazoa and the Significance of Problematic Taxa*. Cambridge University Press, Cambridge.

Delle Cave, L., Insom, E., and Simonetta, A.M. 1998. Advances, divisions, possible relapses and additional problems in understanding the early evolution of the Articulata. *Italian Journal of Zoology* 65: 19–38.

Delsuc, F., Brinkmann, H., Chourrout, D., and Philippe, H. 2006. Tunicates and not cephalochordates are the closest living relatives of vertebrates. *Nature* 439: 965–968.

de Quatrefages, M.A. 1846. Études sur les types inférieurs de l'embranchement des annelés. Mémoire sur la famille des nemertiens (Nemertea). *Annales de Sciences naturelles* (3)6: 173–303 [in French].

Derelle, R., Lopez, P., Le Guyader, H., and Manuel, M. 2007. Homeodomain proteins belong to the ancestral molecular toolkit of eukaryotes. *Evolution & Development* 9: 212–219.

De Robertis, E.M. 1997. The ancestry of segmentation. *Nature* 387: 25–26.

De Robertis, E.M. and Sasai, Y. 1996. A common plan for dorsoventral patterning in Bilateria. *Nature* 380: 37–40.

de Rosa, R., Grenier, J.K., Andreeva, T. *et al.* 1999. Hox genes in brachiopods and priapulids and protostome evolution. *Nature* 399: 772–776.

de Rosa, R., Prud'homme, B., and Balavoine, G. 2005. *caudal* and *even-skipped* in the annelid *Platynereis dumerilii* and the ancestry of posterior growth. *Evolution & Development* 7: 574–587.

Desplan, C. and Lecuit, T. 2003. Flowers' wings and flies' petals. *Nature* 422: 123–124.

Desutter-Grandcolas, L., Legendre, F., Grandcolas, P., Robillard, T., and Murienne, J. 2005. Convergence and parallelism: is a new life ahead of old concepts? *Cladistics* 21: 51–62.

Desutter-Grandcolas, L., Legendre, F., Grandcolas, P., Robillard, T., and Murienne, J. 2007. Distinguishing between convergence and parallelism is central to comparative biology: a reply to Williams and Ebach. *Cladistics* 23: 90–94.

Deutsch, J. and Le Guyader, H. 1998. The neuronal zootype. A hypothesis. *Comptes Rendues de l'Académie des Sciences, Paris, Sciences de la Vie* 321: 713–719.

Dewel, R.A. 2000. Colonial origin for Eumetazoa: major morphological transitions and the origin of bilaterian complexity. *Journal of Morphology* 243: 35–74.

Dewel, R.A., Nelson, D.R., and Dewel, W.C. 1993. Tardigrada, pp. 143–183 in F.W. Harrison and M.E. Rice (eds), *Microscopic Anatomy of Invertebrates. Vol. 12. Onychophora, Chilopoda, and Lesser Protostomata*. Wiley-Liss, New York.

D'Haese, C.A. 2004. Phylogénie des hexapodes et implications pour l'hypothèse de leur origine aquatique. Hexapoda phylogeny: implications for an aquatic origin hypothesis. *Journal de la Société de Biologie* 198: 311–321.

Diaz, J.-P., Mani-Ponset, L., Blasco, C., and Connes, R. 2002. Cytological detection of the main phases of lipid metabolism during early post-embryonic development in three teleost species: *Dicentrarchus labrax*, *Sparus aurata* and *Stizostedion lucioperca*. *Aquatic Living Resources* 15: 169–178.

Diggle, P.K. 2002. A developmental morphologist's perspective on plasticity. *Evolutionary Ecology* 16: 267–283.

Di Palma, A. and Alberti, G. 2001. Fine structure of the female genital system in phytoseiid mites with remarks on egg nutrimentary development, sperm-access system, sperm transfer, and capacitation (Acari, Gamasida, Phytoseiidae). *Experimental and Applied Acarology* 25: 525–591.

di-Poï, N., Zákány, J., and Duboule, D. 2007. Distinct roles and regulations for *Hoxd* genes in metanephric kidney development. *PLoS Genetics* 3(12): 2500–2514.

Dittmann, F. 1998. Activation of a Ca++ current coincides with the onset of oocyte differentiation in the larval telotrophic trophocyte-oocyte syncytium of *Dysdercus intermedius*. *Invertebrate Reproduction & Development* 34: 91–96.

Dobens, L.L. and Raftery, L.A. 2000. Integration of epithelial patterning and morphogenesis in *Drosophila* ovarian follicle cells. *Developmental Dynamics* 218: 80–93.

Doe, D.A. 1986. Ultrastructure of the copulatory organ of *Haplopharynx quadristimulus* and its phylogenetic sig-

nificance (Plathelminthes, Haplopharyngida). *Zoomorphology* 106: 163–173.

Dohle, W. 1980. Sind die Myriapoden eine monophyletische Gruppe? *Verhandlungen des naturwissenschaftlichen Verein in Hamburg* 23: 45–104 [in German].

Dohle, W. 1985. Phylogenetic pathways in Chilopoda. *Bijdragen tot de Dierkunde* 55: 55–66.

Dohle, W. 1988. *Myriapoda and the Ancestry of Insects*. Manchester Polytechnic, Manchester.

Dohle, W. 1989. Differences in cell pattern formation in early embryology and their bearing on evolutionary changes in morphology. *Geobios, mémoire speciale* 12: 145–155.

Dohle, W. 1999. The ancestral cleavage pattern of the clitellates and its phylogenetic deviations. *Hydrobiologia* 402: 267–283.

Dohle, W. 2001. Are the insects terrestrial crustaceans? A discussion of some new facts and arguments and the proposal of the proper name 'Tetraconata' for the monophyletic unit Crustacea+Hexapoda. *Annales de la Société Entomologique de France* 37: 85–103.

Dohle, W. 2004. Die Verwandtschaftsbeziehungen der Großgruppen der Deuterostomier: Alternative Hypothesen und ihre Begründungen. *Sitzungsberichte der Gesellschaft Naturforschender Freunde zu Berlin, Neue Folge* 43: 123–162 [in German].

Dohle, W. and Scholtz, G. 1988. Clonal analysis of the crustacean segment: the discordance between genealogical and segmental borders. *Development* 104 (suppl.): 147–160.

Dong, P.D.S., Chu, J., and Panganiban, G. 2001. Proximodistal domain specification and interactions in developing *Drosophila* appendages. *Development* 128: 2365–2372.

Dong, X.-P., Donoghue, P.C.J., Cunningham, J.A., Liu, J.-B., and Cheng, H. 2006. The anatomy, affinity, and phylogenetic significance of *Markuelia*. *Evolution & Development* 7: 468–482.

Donoghue, P.C.J. 2007. Embryonic identity crisis. *Nature* 445: 155–156.

Donoghue, P.C.J. and Purnell, M.A. 2005. Genome duplication, extinction and vertebrate evolution. *Trends in Ecology & Evolution* 20: 312–319.

Donoghue, P.C.J. and Benton, M.J. 2007. Rocks and clocks: calibrating the Tree of Life using fossils and molecules. *Trends in Ecology and Evolution* 22: 424–431.

Donoghue, P.C.J., Purnell, M.A., and Aldridge, R.J. 1998. Conodont anatomy, chordate phylogeny and vertebrate classification. *Lethaia* 31: 211–219.

Donoghue, P.C.J., Forey, P.L., and Aldridge, R.J. 2000. Conodont affinity and chordate phylogeny. *Biological Reviews* 75: 191–251.

Donoghue, P.C.J., Bengtson, S., Dong, X.-P. *et al.* 2006a. Synchrotron X-ray tomographic microscopy of fossil embryos. *Nature* 442: 680–683.

Donoghue, P.C.J., Kouchinski, A., Waloszek, D. *et al.* 2006b. Fossilized embryos are widespread but the record is temporally and taxonomically biased. *Evolution & Development* 8: 232–238.

Donworth, P.J. 1986. A reappraisal and the validation of the species *Dinophilus taeniatus* Harmer, 1889 and the taxonomically significant features in monomorphic dinophilids (Annelida Polychaeta). *Zoologischer Anzeiger* 216: 32–38.

Dopazo, H. and Dopazo, J. 2005. Genome-scale evidence of the nematode-arthropod clade. *Genome Biology* 6: R41.

Dopazo, H., Santoyo, J., and Dopazo, J. 2004. Phylogenomics and the number of characters required for obtaining an accurate phylogeny of eukaryote model species. *Bioinformatics* 20 (suppl. 1): I116–I121.

Dorresteijn, A.W.C., O'Grady, B., Fischer, A., Porchet-Henneré, E., and Boilly-Marer, Y. 1993. Molecular specification of cell lines in the embryo of *Platynereis* (Annelida). *Roux's Archiv of Developmental Biology* 202: 260–269.

Dove, H. and Stollewerk, A. 2003. Comparative analysis of neurogenesis in the myriapod *Glomeris marginata* (Diplopoda) suggests more similarities to chelicerates than to insects. *Development* 130: 2161–2171.

Dowton, M. and Austin, A.D. 1999. Evolutionary dynamics of a mitochondrial rearrangement "hot spot" in Hymenoptera. *Molecular Biology and Evolution* 16: 298–309.

Doyère, L. 1840. Mémoire sur les Tardigrades. *Annales des Sciences Naturelles (Zoologie)* (2)14: 269–361 [in French].

Drew, G.A. 1899. Some observations on the habits, anatomy, and embryology of members of the Protobranchia. *Anatomischer Anzeiger* 15: 493–519.

Drew, G.A. 1901. The life-history of *Nucula delphinodonta* (Mighels). *Quarterly Journal of Microscopical Science* 44: 313–392.

Drobysheva, I.M. 2003. On mitosis in embryos and larvae of the polyclads *Cycloporus japonicus* and *Notoplana humilis* (Plathelminthes) with different types of development. *Zoologicheskii Zhurnal* 82: 1292–1299.

Droser, M.L., Jensen, S. and Gehlig, J.G. 2002. Trace fossils and substrates of the terminal Proterozoic-Cambrian transition: implications for the record of early bilaterians and sediment mixing. *Proceedings of the National Academy of Sciences USA* 99: 12572–12576.

Dubinin, V.B. 1959. Chelicerophorous animals (Chelicerophora W. Dubinin, nom.n.) and their position in the system. *Zoologicheski Zhurnal* 38: 1163–1189 [in Russian].

Duboule, D. 2007. The rise and fall of Hox gene clusters. *Development* 134: 2549–2560.

Dubrulle, J. and Pourquié, O. 2002. From head to tail: links between the segmentation clock and antero-posterior patterning of the embryo. *Current Opinion in Genetics & Development* 12: 519–523.

Duda, T.F. and Palumbi, S.R. 1999. Developmental shifts and species selection in gastropods. *Proceedings of the National Academy of Sciences USA* 96: 10272–10277.

Dumollard, R., Duchen, M. and Carroll, L. 2007. The role of the mitochondrial function in the oocyte and embryo. *Current Topics in Developmental Biology* 77: 21–49.

Dunagan, T.T. and Miller, D.M. 1991. Acanthocephala, pp. 299–332 in F.W. Harrison and E.R. Ruppert (eds), *Microscopic Anatomy of Invertebrates. 4. Aschelminthes.* Wiley-Liss, New York.

Duncan, D.M., Burgess, E.A., and Duncan, I. 1998. Control of distal antennal identity and tarsal development in *Drosophila* by *spineless-aristapedia*, a homolog of the mammalian dioxin receptor. *Genes and Development* 12: 1290–1303.

Dunlop, J.A. 2002. Phylogeny of Chelicerata, pp. 117–141 in J.J. Llorente Bousquets and J.J. Morrone (eds), *Biodiversidad, taxonomía y biogeografía de artrópodos de Mexico: hacia una síntesis de su conocimiento*, vol. 3. Prensas de Ciencias, Universidad Nacional Autónoma de México, Mexico.

Dunlop, J.A. and Arango, C.P. 2005. Pycnogonid affinities: a review. *Journal of Zoological Systematics and Evolutionary Research* 43: 8–21.

Dunn, C.W., Hejnol, A., Matus, D.Q. *et al.* 2008. Broad phylogenomic sampling improves resolution of the animal tree of life. *Nature* 452: 745–749.

Dunwoodie, S.L. 2007. Combinatorial signalling in the heart orchestrates cardiac induction, lineage specification and chamber formation. *Seminars in Cell & Developmental Biology* 18: 54–66.

Dworkin, I.M., Tanda, S., and Larsen, E. 2001. Are entrenched characters developmentally constrained? Creating biramous limbs in an insect. *Evolution & Development* 3: 424–431.

Dzik, J. 1991. Is fossil evidence consistent with traditional views of the early metazoan phylogeny?, pp. 47–56 in A.M. Simonetta and S. Conway Morris (eds), *The Early Evolution of Metazoa and the Significance of Problematic Taxa.* Cambridge University Press, Cambridge.

Dzik, J. 1993. Early metazoan evolution and the meaning of its fossil record, pp. 339–386 in M.K. Hecht (ed.), *Evolutionary Biology*, vol. 27. Plenum Press, New York.

Dzik, J. 1995. *Yunnanozoon* and the ancestry of chordates. *Acta Palaeontologica Polonica* 40: 341–360.

Dzik, J. and Krumbiegel, G. 1989. The oldest 'onychophoran' *Xenusion*: a link connecting phyla? *Lethaia* 22: 169–181.

Eaves, A.A. and Palmer, A.R. 2003. Reproduction: widespread cloning in echinoderm larvae. *Nature* 425: 146.

Eberhard, W.G. 2002. Restraint with constraints: a reply to Wagner and Müller. *Evolution & Development* 4: 7–8.

Eckelbarger, K.J. 2005. Oogenesis and oocytes. *Hydrobiologia* 535/536: 179–198.

Edgecombe, G.D. 2004. Morphological data, extant Myriapoda, and the myriapod stem-group. *Contributions to Zoology* 73: 207–252.

Edgecombe, G.D. and Giribet, G. 2002. Myriapod phylogeny and the relationships of Chilopoda, pp. 143–168 in J. Llorente Bousquets and J.J. Morrone (eds), *Biodiversidad, taxonomía y biogeografía de artrópodos de México: hacia una síntesis de su conocimiento*. Vol. 3. Prensas de Ciencias, Universidad Nacional Autonoma de México, México.

Edgecombe, G.D. and Giribet, G. 2004. Adding mitochondrial sequence data (16S rRNA and citochrome c oxidase subunit I) to the phylogeny of centipedes (Myriapoda: Chilopoda): an analysis of morphology and four molecular loci. *Journal of Zoological Systematics and Evolutionary Research* 42: 89–134.

Edgecombe, G.D. and Giribet, G. 2007. Evolutionary biology of centipedes (Myriapoda: Chilopoda). *Annual Review of Entomology* 52: 151–170.

Edgecombe, G.D., Wilson, G.D.F., Colgan, D.J., Gray, M.R., and Cassis, G. 2000. Arthropod cladistics: combined analysis of histone H3 and U2 sn RNA sequences and morphology. *Cladistics* 16: 204–231.

Eeckhaut, I. and Lanterbecq, D. 2005. Myzostomida: a review of the phylogeny and ultrastructure. *Hydrobiologia* 535/536: 253–275.

Eeckhaut, I., McHugh, D., Mardulyn, P. *et al.* 2000. Myzostomida: a link between trochozoans and flatworms? *Proceedings of the Royal Society of London Series B, Biological Sciences* 267: 1383–1392.

Eernisse, D.J., Albert, J.S., and Anderson, F.E. 1992. Annelida and Arthropoda are not sister taxa: a phylogenetic analysis of spiralian metazoan morphology. *Systematic Biology* 41: 305–330.

Egger, B., Ladurner, P., Nimeth, K., Gschwentner, R., and Rieger, R. 2006. The regeneration capacity of the flatworm *Macrostomum lignano*—on repeated regeneration, rejuvenation, and the minimal size needed for regeneration. *Development, Genes and Evolution* 216: 565–577.

Ehlers, U. 1985. *Das phylogenetische System der Plathelminthes*. G. Fischer, Stuttgart-New York [in German].

Ehlers, U. 1993. Ultrastructure of the spermatozoa of *Halammohydra schulzei* (Cnidaria, Hydrozoa): the significance of acrosomal structures for the systematization of the Eumetazoa. *Microfauna Marina* 8: 115–130.

Ehlers, U. and Sopott-Ehlers, B. 1997. Ultrastructure of the subepidermal musculature of *Xenoturbella bocki*, the adelphotaxon of the Bilateria. *Zoomorphology* 117: 71–79.

Ehlers, U., Ahlrichs, W., Lemburg, C., and Schmidt-Rhaesa, A. 1996. Phylogenetic systematization of the Nemathelminthes (Aschelminthes). *Verhandlungen der Deutschen Zoologischen Gesellschaft* 89: 8.

Ehrenberg, C.G. 1831, 1832, in Hemprich, F.W. 1828–1845. *Symbolae physicae seu icones et descriptiones corporum naturalium novorum aut minus cognitorum quae ex itineribus per Libyam, Aegyptum, Nubiam, Dongalam, Syriam, Arabiam et Habessiniam. Pars Zoologica II(2). Animalia Evertebrata. Phytozoa Turbellaria Africana et Asiatica.* Reimer, Berlin [in Latin].

Eibye-Jacobsen, D. 2004. A reevaluation of *Wiwaxia* and the polychaetes of the Burgess Shale. *Lethaia* 37: 317–335.

Eldredge, N. and Gould, S.J. 1972. Punctuated equilibria: an alternative to phyletic gradualism, pp. 82–115 in T.J.M. Schopf (ed.), *Models in Paleobiology*. Freeman, Cooper and Co., San Francisco, CA.

Ellis, C.H. and Fausto-Sterling, A. 1997. Platyhelminths, the flatworms, pp. 115–130 in S.F. Gilbert and A.M. Raunio (eds), *Embryology: Constructing the Organism*. Sinauer Associates, Sunderland, MA.

Ender, A. and Schierwater, B. 2003. Placozoa are not derived cnidarians: evidence from molecular morphology. *Molecular Biology and Evolution* 20: 130–134.

Engel, M.S. and Grimaldi, D.A. 2004. New light shed on the oldest insect. *Nature* 427: 627–630.

Enghoff, H., Dohle, W., and Blower, J.G. 1993. Anamorphosis in millipedes (Diplopoda). The present state of knowledge and phylogenetic considerations. *Zoological Journal of the Linnean Society* 109: 103–234.

Erasmus, D.A. 1972. *The Biology of Trematodes*. Arnold, London.

Ereskovsky, A.V. 2005. *Comparative Embryology of Sponges*. Sanct-Petersburg University Press, St. Petersburg.

Ereskovsky, A.V. and Boury-Esnault, N. 2002. Cleavage pattern in *Oscarella* species (Porifera, Demospongiae, Homoscleromorpha): transmission of maternal cells and symbiotic bacteria. *Journal of Natural History* 36: 1761–1775.

Ereskovski, A.V. and Dondua, A.K. 2006. The problem of germ layers in sponges (Porifera) and some issues concerning early metazoan evolution. *Zoologischer Anzeiger* 245: 65–76.

Ereskovsky, A.V. and Korotkova, G.P. 1997. The reasons of sponge sexual morphogenesis peculiarities. *Berliner geowissenschaftliche Abhandlungen* 20: 25–33.

Ereskovsky, A.V. and Willenz, P. 2007. *Esperiopsis koltuni* sp. nov. (Demospongiae: Poecilosclerida: Esperiopsidae), a carnivorous sponge from deep water of the Sea of Okhotsk (North Pacific). *Journal of the Marine Biological Association of the United Kingdom* Special Issue 87(6): 1379–1386.

Eriksson, B.J. and Budd, G.E. 2000. Onychophoran cephalic nerves and their bearing on our understanding of head segmentation and stem-group evolution of Arthropoda. *Arthropod Structure and Development* 29: 197–209.

Eriksson, B.J., Tait, N.N., and Budd, G.E. 2003. Head development in the onychophoran *Euperipatoides kanangrensis* with particular reference to the central nervous system. *Journal of Morphology* 255: 1–23.

Eriksson, J. 2003. Evolution and development of the onychophoran head and nervous system. *Acta Universitatis Upsaliensis.* Comprehensive Summaries of Uppsala Dissertations from the Faculty of Science and Technology 833. University of Uppsala, Uppsala.

Erséus, C. 2005. Phylogeny of the oligochaetous Clitellata. *Hydrobiologia* 535/536: 357–372.

Erséus, C. and Kallersjö, M. 2004. 18S rDNA phylogeny of Clitellata (Annelida). *Zoologica Scripta* 33: 187–196.

Erwin, D.H. 2006. The developmental origin of animal bodyplans, pp. 160–198 in S. Xiao and A.J. Kaufman (eds), *Neoproterozoic Geobiology and Paleobiology*, Topics in Geobiology 27. Springer, Dordrecht.

Eschscholtz, J.F. 1829–1833. *Zoologischer Atlas, enthaltend Abbildungen und Beschreibungen neuer Thierarten, während des Flottcapitains von Kotzebue zweiter Reise um die Welt, auf der Russisch-Kaiserlichen Kriegsschlupp Predpriaetië in den Jahren 1823–1826.* Reimer, Berlin [in German].

Extavour, C.G.M. 2008. Urbisexuality: the evolution of bilaterian germ cell specification and reproductive systems, pp. 321–342 in A. Minelli and G. Fusco (eds), *Evolving Pathways: Key Themes in Evolutionary Developmental Biology.* Cambridge University Press, Cambridge.

Extavour, C.G. and Akam, M. 2003. Mechanisms of germ cell specification across the metazoans: epigenesis and preformation. *Development* 130: 5869–5884.

Extavour, C.G., Pang, K., Matus, D.Q., and Martindale, M.Q. 2005. *vasa* and *nanos* expression patterns in a sea anemone and the evolution of bilaterian germ cell specification mechanisms. *Evolution & Development* 7: 201–215.

Fabrizio, J.J., Hime, G., Lemmon, S.K., and Bazinet, C. 1998. Genetic dissection of sperm individualization in *Drosophila melanogaster. Development* 125: 1833–1843.

Fanenbruck, M. and Harzsch, S. 2005. A brain atlas of *Godzilliognomus frondosus* Yager, 1989 (Remipedia, Godzilliidae) and comparison with the brain of *Speleonectes tulumensis* Yager, 1987 (Remipedia, Speleonectidae): implications for arthropod relationships. *Arthropod Structure and Development* 34: 343–378.

Fauchald, K and Rouse, G. 1997. Polychaete systematics: past and present. *Zoologica Scripta* 26: 71–138.

Fautin, D.G. and Mariscal, R.N. 1991. Cnidaria: Anthozoa, pp. 267–358 in F.G. Harrison and J.A. Westfall (eds), *Microscopic Anatomy of Invertebrates*, vol. 2. Alan Liss, New York.

Fedonkin, M. 1994. Vendian body fossils and trace fossils, pp. 370–388 in S. Bengtson (ed.), *Early Life on Earth*, Nobel Symposium no. 84. Columbia University Press, New York.

Fell, H.P. 1948. Echinoderm embryology and the origin of chordates. *Biological Reviews* 23: 81–107.

Fell, P.E. 1974. Porifera, pp. 51–132 in A.C. Giese and J.S. Pearse (eds), *Reproduction of Marine Invertebrates.* Academic Press, New York.

Fell, P.E. 1997. Poriferans, the sponges, pp. 39–54 in S.F. Gilbert and A.M. Raunio (eds), *Embryology: Constructing the Organism.* Sinauer Associates, Sunderland, MA.

Fenton, A., Paterson, S., Viney, M.E., and Gardner, M.P. 2004. Determining the optimal developmental route of *Strongyloides ratti*: an evolutionarily stable strategy approach. *Evolution* 58: 989–1000.

Fernald, R.D. 2006. Casting a genetic light on the evolution of eyes. *Science* 313: 1914–1918.

Ferraguti, M. and Melone, G. 1999. Spermiogenesis in *Seison nebaliae* (Rotifera, Seisonidea): further evidence of a rotifer-acanthocephalan relationship. *Tissue and Cell* 31: 428–440.

Ferrari, F.D. and Grygier, M.J. 2003. Comparative morphology among trunk limbs of *Caenestheriella gifuensis* and of *Leptestheria kawachiensis* (Crustacea: Branchiopoda: Spinicaudata). *Zoological Journal of the Linnean Society* 139: 547–564.

Ferrier, D.E.K., Minguillón, C., Holland, P.W.H., and García-Fernàndez, J. 2000. The amphioxus Hox cluster: deuterostome posterior flexibility and *Hox14. Evolution & Development* 2: 284–293.

Field, K.G., Olsen, G.J., Lane, D.J. *et al.* 1988. Molecular phylogeny of the animal kingdom. *Science* 239: 748–753.

Finkelstein, R. and Boncinelli, E. 1994. From fly head to mammalian forebrain: the story of *otd* and *Otx. Trends in Genetics* 10: 310–315.

Finnerty, J. 1998. Homeoboxes in sea anemones and other nonbilaterian animals: implications for the evolution of the hox cluster and the zootype. *Current Topics in Developmental Biology* 40: 211–254.

Finnerty, J.R. 2003. The origins of axial patterning in the metazoa: how old is bilateral symmetry? *International Journal of Developmental Biology* 47: 523–529.

Finnerty, J.R. and Martindale, M.Q. 1999. Ancient origins of axial patterning genes: Hox genes and ParaHox genes in the Cnidaria. *Evolution & Development* 1: 16–23.

Finnerty, J.R., Pang, K., Burton, P., Paulson, D., and Martindale, M.Q. 2004. Origins of bilateral symmetry: *Hox* and *Dpp* expression in a sea anemone. *Science* 304: 1335–1337.

Flemming, A.J., Shen, Z.Z., Cunha, A., Emmons, S.W. and Leroi, A.M. 2000. Somatic polyploidization and cellular proliferation drive body size evolution in nematodes. *Proceedings of the National Academy of Sciences USA* 97: 5285–5290.

Foddai, D., Bonato, L., Pereira, L.A., and Minelli, A. 2003. Phylogeny and systematics of the Arrupinae (Chilopoda Geophilomorpha Mecistocephalidae) with the description of the dwarfed *Nannarrup hoffmani* n.g. n.sp. *Journal of Natural History* 37: 1247–1267.

Foe, V.E. 1989. Mitotic domains reveal early commitment of cells in *Drosophila* embryos. *Development* 107: 1–22.

Foe, V.E. and Odell, G.M. 1989. Mitotic domains partition fly embryos reflecting early cell biological consequences of determination in progress. *American Zoologist* 29: 617–652.

Foe, V.E., Odell, G.M., and Edgar, B.A. 1993. Mitosis and morphogenesis in the *Drosophila* embryo: point and counterpoint, pp. 149–300 in M. Bate and A. Martinez-Arias (eds), *The Development of Drosophila melanogaster*. Cold Spring Harbor Press, Cold Spring Harbor, NY.

Fontanillas, E., Welch, J.J., Thomas, J.A., and Bromham, M.L. 2007. The influence of body size and net diversification rate on molecular evolution during the radiation of animal phyla. *BMC Evolutionary Biology* 7: 95.

Forgacs, G. and Newman, S.A. 2005. *Biological Physics of the Developing Embryo*. Cambridge Universitry Press, Cambridge.

Foronda, D., Estrada, B., de Navas, L., and Sánchez-Herrero, E. 2006. Requirement of *abdominal-A* and *Abdominal-B* in the developing genitalia of *Drosophila* breaks the posterior downregulation rule. *Development* 133: 117–127.

Fortey, R.A., Briggs, D.E.G., and Wills, M.A. 1996. The Cambrian evolutionary 'explosion': decoupling cladogenesis from morphological disparity. *Biological Journal of the Linnean Society* 57: 13–33.

Franc, A. 1993. Classe des Scyphozoaires, pp. 597–884 in P.P. Grassé (ed.), *Traité de Zoologie* 3(2), Masson, Paris [in French].

Franch-Marro, X., Martín, N., Averof, M., and Casanova, J. 2006. Association of tracheal placodes with leg primordia in *Drosophila* and implications for the origin of insect tracheal systems. *Development* 133: 785–790.

Fransen, M.E. 1988. Coelomic and vascular systems, pp. 199–213 in W. Westheide and C.O. Hermans (eds), *The Ultrastructure of Polychaeta*. Gustav Fischer Verlag, Stuttgart.

Franzén, A. and Afzelius, B.A. 1987. The ciliated epidermis of *Xenoturbella bocki* (Platyhelminthes, Xenoturbellida) with some phylogenetic considerations. *Zoologica Scripta* 16: 9–17.

Fratini, S., Vannini, M., Cannicci, S., and Schubart, C.D. 2005. Tree-climbing mangrove crabs: a case of convergent evolution *Evolutionary Ecology Research* 7: 219–233.

Freeman, G. 1993. Regional specification during embryogenesis in the articulate brachiopod *Terebratalia*. *Developmental Biology* 160: 196–213.

Freeman, G. and Martindale, M.Q. 2002. The origin of mesoderm in phoronids. *Developmental Biology* 252: 301–311.

Freitas, R., Zhang, G.J., and Cohn, M.J. 2006. Evidence that mechanisms of fin development evolved in the midline of early vertebrates. *Nature* 442: 1033–1037.

Frenzel, J. 1892. Untersuchungen über die mikroskopische Fauna Argentiniens. *Archiv für Naturgeschichte* 58: 66–96 [in German].

Frey, H. and Leuckart, R. 1847. *Beträge zur Kenntnis wirbelloser Thiere mit besonderer Berücksichtigung der Fauna des norddeutschen Meeres*. Vieweg & Sohn, Braunschweig [in German].

Frick, J.E. and Ruppert, E.E. 1996. Primordial germ cells of *Synaptula hydriformis* (Holothuroidea, Echinodermata) are epithelial flagellated-collar cells: their primary egg polarity. *Biological Bulletin* 191: 168–177.

Friedrich, M. and Tautz, D. 1995. Ribosomal DNA phylogeny of the major extant arthropod classes and the evolution of myriapods. *Nature* 376: 165–167.

Friedrich, M. and Tautz, D. 2001. Arthropod rDNA phylogeny revisited: a consistency analysis using Monte Carlo simulation, pp. 21–40 in T. Deuve (ed.), *Origin of the Hexapoda*. Annales de la Société Entomologique de France, Nouvelle Série 37. Société Entomologique de France, Paris.

Fryer, G. 1961. The developmental history of *Mutela bourguignati* (Ancey) Bourguignat (Mollusca: Bivalvia). *Philosophical Transactions of the Royal Society of London Series B, Biological Sciences* 244: 259–298.

Fryer, G. 1987 A new classification of the branchiopod Crustacea. *Zoological Journal of the Linnean Society* 91: 357–383.

Fuchs, F. 1914. Die Keimblätterentwicklung von *Cyclops viridis* Jurine. *Zoologische Jahrbücher, Abteilung für Anatomie* 38: 103–156 [in German].

Fukatsu, T. and Nikoh, N. 1998. Two intracellular symbiotic bacteria from the mulberry psyllid *Anomoneura mori* (Insecta, Homoptera). *Applied and Environmental Microbiology* 64: 3599–3606.

Funch, P. 1996. The chordoid larva of *Symbion pandora* (Cycliophora) is a modified trochophore. *Journal of Morphology* 230: 231–263.

Funch, P. and Kristensen, R.M. 1995. Cycliophora is a new phylum with affinities to Entoprocta and Ectoprocta. *Nature* 378: 711–714.

Funch, P. and Kristensen, RM. 1997. Cycliophora, pp. 409–474 in F.W. Harrison and R.M. Woollacott (eds), *Microscopic Anatomy of Invertebrates. Vol. 13. Lophophorates, Entoprocta, and Cycliophora*. Wiley-Liss, New York

Funch, P., Sørensen, M.V., and Obst, M. 2005. On the phylogenetic position of Rotifera—have we come any further? *Hydrobiologia* 546: 11–28.

Furlong, R.F. and Holland, P.W.H. 2002. Bayesian phylogenetic analysis supports monophyly of Ambulacraria and of Cyclostomes. *Zoological Science* 19: 593–599.

Furuya, H., Tsuneki, K., and Khoshida, Y. 1992. Development of the infusoriform embryo of *Dicyema japonicum* (Mesozoa: Dicyemidae). *Biological Bulletin* 183: 248–257.

Furuya, H., Tsuneki. K., and Koshida, Y. 1997. Fine structure of. dicyemid mesozoans, with special reference to cell junctions. *Journal of Morphology* 231: 297–305.

Fürst von Lieven, A. 2005. The embryonic moult in diplogastrids (Nematoda)—homology of developmental stages and heterochrony as a prerequisite for morphological diversity. *Zoologischer Anzeiger* 244: 79–91.

Fusco, G. 2001. How many processes are responsible for phenotypic evolution? *Evolution & Development* 3: 279–286.

Fusco, G. 2005. Trunk segment numbers and sequential segmentation in myriapods. *Evolution & Development* 7: 608–617.

Gad, G. 2005a. A parthenogenetic, simplified adult in the life cycle of *Pliciloricus pedicularis* sp. n. (Loricifera) from the deep sea of the Angola Basin (Atlantic). *Organisms Diversity and Evolution* 5 (suppl. 1): 77–103.

Gad, G. 2005b. Giant Higgins-larvae with paedogenetic reproduction from the deep sea of the Angola Basin—evidence for a new life cycle and for abyssal gigantism in Loricifera? *Organisms Diversity and Evolution* 5: 59–75.

Gaino, E., Burlando, B., and Buffa, P. 1987. Structural and ultrastructural aspects of growth in *Oscarella lobularis* (Porifera, Demospongiae). *Growth* 51: 451–460.

Galaktionov, K.V. and Dobrovolskij, A.A. 2003. *The Biology and Evolution of Trematodes. An Essay on the Biology, Morphology, Life Cycles, and Evolution of Digenetic Trematodes*. Kluwer Academic Publishers, Dordrecht.

Galko, M.J. and Krasnow, M.A. 2004. Cellular and genetic analysis of wound healing in *Drosophila* larvae. *PLoS Biology* 2(8): 1114–1126.

Galliot, B. 2000. Conserved and divergent genes in apex and axis development of cnidarians. *Current Opinion in Genetics & Development* 10: 629–637.

Galliot, B. and Schmid, V. 2002. Cnidarians as a model system for understanding evolution and regeneration. *International Journal of Developmental Biology* 46: 39–48.

Ganot, P., Bouquet, J.-M., and Thompson, E.M. 2006. Comparative organization of follicle, accessory cells and spawning anlagen in dynamic semelparous clutch manipulators, the urochordate Oikopleuridae. *Biology of the Cell* 98: 389–401.

Ganot, P., Kallesøe, T., and Thompson, E.M. 2007. The cytoskeleton organizes germ nuclei with divergent fates and asynchronous cycles in a common cytoplasm during oogenesis in the chordate *Oikopleura*. *Developmental Biology* 302: 577–590.

García-Bellido, A., Ripoll, P., and Morata, G. 1973. Developmental compartmentalisation of the wing disc of *Drosophila*. *Nature New Biology* 245: 251–253.

García-Bellido, D.C. and Collins, D.H. 2004. Moulting arthropod caught in the act. *Nature* 429: 40.

García-Ruiz, J.M., Checa, A., and Rivas, A. 1990. On the origin of ammonite sutures. *Paleobiology* 16: 349–354.

García-Varela, M. and Nadler, S.A. 2006. Phylogenetic relationships among Syndermata inferred from nuclear and mitochondrial gene sequences. *Molecular Phylogenetics and Evolution* 40: 61–72.

Gardiner, S.L. 1992. Polychaeta: general organization, integument, musculature, coelom, and vascular system, pp. 19–52 in F.W. Harrison and S.L. Gardiner (eds), *Microscopic Anatomy of Invertebrates. Vol. 7. Annelida*. Wiley-Liss, New York.

Gardiner, S.L. and Jones, M.L. 1993. Vestimentifera, pp. 371–460 in F.W. Harrison and M.E. Rice (eds), *Microscopic Anatomy of Invertebrates. Vol. 12. Onychophora, Chilopoda, and Lesser Protostomata*. Wiley-Liss, New York.

Garey, J.R. and Schmidt-Rhaesa, A. 1998. The essential role of "minor" phyla in molecular studies of animal evolution. *American Zoologist* 38: 907–917.

Garey, J.R., Near, T.J., Nonnemacher, M.R., and Nadler, S.A. 1996. Molecular evidence for Acanthocephala as a subtaxon of Rotifera. *Journal of Molecular Evolution* 43: 287–292.

Garrone, R. 1998. Evolution of metazoan collagens, pp. 118–137 in W.E.G. Müller (ed.), *Towards the Origin of Metazoa*. Progress in Molecular and Subcellular Biology 21. Springer, Berlin.

Garstang, W. 1894. Preliminary note on a new theory of the phylogeny of the Chordata. *Zoologischer Anzeiger* 17: 122–125.

Garstang, W. 1922. The theory of recapitulation: a critical restatement of the biogenetic law. *Journal of the Linnean Society, Zoology* 35: 81–101.

Gass, G.L. and Bolker, J.A. 2003. Modularity, pp. 260–267 in B.K. Hall and W.M. Olson (eds), *Keywords and Concepts in Evolutionary Developmental Biology*. Harvard University Press, Cambridge, MA.

Gee, H. 2001. Deuterostome phylogeny: the context for the origin and evolution of the vertebrates, pp. 1–14 in P.E. Ahlberg (ed.), *Major Events in Early Vertebrate Evolution: Palaeontology, Phylogeny, Genetics, and Development*. Taylor and Francis, London.

Gegenbaur, C. 1859. *Grundzüge der vergleichenden Anatomie*. W. Engelmann, Leipzig [in German].

Gegenbaur, C. 1870. *Grundzüge der vergleichenden Anatomie*, 2nd edn. W. Engelmann, Leipzig [in German].

Gehring, W.J. 1998. *Master Control Genes in Development and Evolution*. Yale University Press, New Haven.

Gehring, W.J. 2002. The genetic control of eye development and its implications for the evolution of the various eye-types. *International Journal of Developmental Biology* 46: 65–73.

Gehring, W.J. and Ikeo, K. 1999. *Pax6*: mastering eye morphogenesis and eye evolution. *Trends in Genetics* 15: 371–377.

Geoffroy Saint-Hilaire, E. 1807. Considérations sur les pièces de la tête osseuse des animaux vertébrés, et particulièrement sur celles du crâne des oiseaux. *Annales du Muséum d'Histoire naturelle de Paris* 10: 342–365 [in French].

Geoffroy Saint-Hilaire, E. 1822. Considérations générales sur la vertèbre. *Mémoires du Muséum d'Histoire Naturelle, Paris* 9: 89–119 [in French].

Gerhart, J. 2002. Changing the axis changes the perspective. *Developmental Dynamics* 225: 380–383.

Gerhart, J. 2006. The deuterostome ancestor. *Journal of Cellular Physiology* 209: 677–685.

Gerhart, J.C. and Kirschner, M.W. 1997. *Cells, Embryos and Evolution*. Blackwell Science, Boston, MA.

Gerhart, J., Lowe, C., and Kirschner, M. 2005. Hemichordates and the origin of chordates. *Current Opinion in Genetics & Development* 15: 461–467.

Ghiara, G. 1995. Structure of the early vertebrate gastrula and causal factors of morphogenesis and differentiation: a critical survey of the hundred year history of ideas, pp. 199–232 in G. Lanzavecchia, R. Valvassori and M.D. Candia Carnevali (eds), *Body Cavities: Function and Phylogeny*. Mucchi, Modena.

Ghiselin, M.T. 1988. The origin of molluscs in the light of molecular evidence. *Oxford Surveys in Evolutionary Biology* 5: 66–95.

Giard, A. 1877. Sur les Orthonectida, classe nouvelle d'animaux parasites des Échinodermes et des Turbellariés. *Comptes rendus hebdomadaires des séances de l'Académie des Sciences* 85: 812–814 [in French].

Giard, A. 1905. La poecilogonie, pp. 617–646 in *Comptes Rendus di VI Congrès International de Zoologie*, Berne 1904 [in French].

Gibson, G.D. and Chia, F.-S. 1989. Developmental variability (benthic and pelagic) in *Haminaea callidigenita* (Opisthobranchia: Cephalaspidea) is influenced by egg mass jelly. *Biological Bulletin* 176: 103–110.

Gilbert, J.J. 1983. Spermatogenesis and sperm function, pp. 181–193 in K.G. Adiyodi and R.G. Adiyodi (eds), *Reproductive Biology of Invertebrates*, vol. II. Wiley, New York.

Gilbert, S.F. 2003a. Evo-devo, devo-evo, and devgen-popgen. *Biology and Philosophy* 18: 347–353.

Gilbert, S.F. 2003b. The reactive genome, pp. 87–101 in G.B. Müller and S.A. Newman (eds), *Origination of Organismal Form. Beyond the Gene in Developmental and Evolutionary Biology*. MIT Press, Cambridge, MA.

Gillespie, J.M. and Bain, B.A. 2006. Postembryonic development of *Tanystylum bealensis* [sic] (Pycnogonida, Ammotheidae) from Barkley Sound, British Columbia, Canada. *Journal of Zoology* 267: 308–317.

Gillooly, J.F., Allen, A.P., West, G.B., and Brown, J.H. 2005. The rate of DNA evolution: effects of body size and temperature on the molecular clock. *Proceedings of the National Academy of Sciences USA* 102: 140–145.

Giribet, G. 2003. Molecules, development and fossils in the study of metazoan evolution; Articulata versus Ecdysozoa revisited. *Zoology* 106: 303–326.

Giribet, G. and Ribera, C. 1998. The position of arthropods in the animal kingdom: a search for a reliable outgroup for internal arthropod phylogeny. *Molecular Phylogenetics and Evolution* 9: 481–488.

Giribet, G. and Ribera, C. 2000. A review of arthropod phylogeny: new data based on ribosomal DNA sequences and direct character optimization. *Cladistics* 16: 204–231.

Giribet, G. and Wheeler, W. 2002. On bivalve phylogeny: a high-level analysis of the Bivalvia (Mollusca) based on combined morphology and DNA sequence data. *Invertebrate Biology* 121: 271–324.

Giribet, G., Carranza, S., Baguñá, J., Riutort, M., and Ribera, C. 1996. First molecular evidence for the existence of a Tardigrada + Arthropoda clade. *Molecular Biology and Evolution* 13: 76–84.

Giribet, G., Distel, D.L., Polz, M., Sterrer, W., and Wheeler, W.C. 2000. Triploblastic relationships with emphasis on the acoelomates and the position of Gnathostomulida, Cycliophora, Plathelminthes, and Chaetognatha: a combined approach of 18S rDNA sequences and morphology. *Systematic Biology* 49: 539–562.

Giribet, G., Edgecombe, G.D., and Wheeler, W.C. 2001. Arthropod phylogeny based on eight molecular loci and morphology. *Nature* 413: 157–161.

Giribet, G., Edgecombe, G.D., Wheeler, W.C., and Babbitt, C. 2002. Phylogeny and systematic position of Opiliones: a combined analysis of chelicerate relationships using morphological and molecular data. *Cladistics* 18: 5–70.

Giribet, G., Edgecombe, G.D., Carpenter, J.M., D'Haese, C.A., and Wheeler, W.C. 2004a. Is Ellipura monophyletic? A combined analysis of basal hexapod relationships with emphasis on the origin of insects. *Organisms, Diversity and Evolution* 4: 319–340.

Giribet, G., Sørensen, M.V., Funch, P., Kristensen, R.M., and Sterrer, W. 2004b. Investigations into the phylogenetic position of Micrognathozoa using four molecular loci. *Cladistics* 20: 1–13.

Giribet, G., Okusu, A., Lindgren, A.R., Huff, S.W., Schrödl, M., and Nishiguchi, M.K. 2006. Evidence for a clade composed of mollusca with serially repeated structures: monoplacophorans are related to chitons. *Proceedings of the National Academy of Sciences USA* 103: 7723–7728.

Glaessner, M.F. 1984. *The Dawn of Animal Life: a Biohistorical Study*. Cambridge University Press, Cambridge.

Glenner, H. and Høeg, J.T. 1995. A new motile, multicellulars stage involved in host invasion by parasitic barnacles (Rhizocephala). *Nature* 377: 147–150.

Glenny, R.W. and Robertson, H.T. 1990. Fractal properties of pulmonary blood flow: characterization of spatial heterogeneity. *Journal of Applied Physiology* 69: 532–545.

Gliznutsa, L.A. and Dautov, S.S. 2005. Ultrastructural features of embryogenesis of the ophiuroid *Amphipholis kochii* (Lutken, 1872). *Biologiya Mory* 31: 194–201.

Godeaux, J.E.A. 1990. Urochordata—Thaliacea, pp. 453–469 in K.G. Adiyodi and R.G. Adiyodi (eds), *Reproductive Biology of Invertebrates*, vol. IV, part B. Wiley, Chicester.

Godt, D., Couderc, J.L., Cramton, S.E., and Laski, F.A. 1993. Pattern formation in the limbs of *Drosophila*: *bric à brac* is expressed in both a gradient and a wave-like pattern and is required for specification and proper segmentation of the tarsus. *Development* 119: 799–812.

Goette, A. 1899. Über die Entwicklung des knöchernen Rückenschildes (Carapax) der Schildkröten. *Zeitschrift für Wissenschaftliche Zoologie* 66: 407–434 [in German].

Golding, D.W. 1992. Polychaeta: nervous system, pp. 153–179 in F.W. Harrison and S.L. Gardiner (eds), *Microscopic Anatomy of Invertebrates. Vol. 7. Annelida*. Wiley-Liss, New York.

Goldstein, B., Frisse, L.M., and Thomas, K. 1998. Embryonic axis specification in nematodes: evolution of the first step in development. *Current Biology* 8: 157–160.

Goodrich, E.J. 1897. On the relation of the arthropod head to the annelid prostomium. *Quarterly Journal of Microscopic Science* 40: 259–268.

Goodrich, E.S. 1958. *Studies on the Structure and Development of Vertebrates*. Dover, New York.

Gosse, P.H. 1864. The natural history of the hairy-backed animalcules (Chaetonotidae). *The Intellectual Observer* 5: 387–406.

Götze, E. 1938. Bau und Leben von *Caecum glabrum* (Montagu). *Zoologische Jahrbücher, Abteilung für Systematik* 71: 55–122.

Goudeau, M. 1977. Contribution à la biologie d'un crustacé parasite: *Hemioniscus balani* Buchholz, Isopode Epicaride. Nutrition, mues et croissance de la femelle et des embryons. *Cahiers de Biologie Marine* 18: 201–242.

Gould, S.J. 1977. *Ontogeny and Phylogeny*. Belknap Press of Harvard University Press, Cambridge, MA.

Gould, S.J. 2000. Of coiled oysters and big brains: how to rescue the terminology of heterochrony, now gone astray. *Evolution & Development* 2: 241–248.

Gould, S.J. and Vrba, E.S. 1982. Exaptation—a missing term in the science of form. *Paleobiology* 8: 4–15.

Grande, C., Templado, J., Cervera, J.L., and Zardoya, R. 2004a. Molecular phylogeny of Euthyneura (Mollusca: Gastropoda). *Molecular Biology and Evolution* 21: 303–313.

Grande, C., Templado, J., Cervera, J.L., and Zardoya, R. 2004b. Phylogenetic relationships among Opisthobranchia (Mollusca: Gastropoda) based on mitochondrial *cox1*, *trnV*, and *rrnL* genes. *Molecular Phylogenetics and Evolution* 33: 378–388.

Grant, R. 1836–1852. Animal kingdom, pp. 107–118 in R.B. Todd (ed.), *The Cyclopaedia of Anatomy and Physiology*, vol. 1. Sherwood, Gilbert and Piper, London.

Grassé, P.-P. 1970. Embranchement des Myxozoaires, pp. 107–112 in P.-P. Grassé, R.A. Poisson and O. Tuzet (eds), *Précis de Zoologie, Vol. 1. Invertébrés*, 2nd edn. Masson, Paris [in French].

Greene, E. 1999. Phenotypic variation in larval development and evolution: polymorphism, polyphenism, and developmental reaction norms, pp. 379–410 in B.K. Hall and M.H. Wake (eds), *The Origin and Evolution of Larval Forms*. Academic Press, San Diego, CA.

Greenspan, R.J. 2001. The flexible genome. *Nature Reviews Genetics* 2: 383–387.

Gregory, T.R. 2008. *Animal Genome Size Database*. www.genomesize.com.

Gregory, T.R., Hebert, P.D.N., and Kolasa, J. 2000. Evolutionary implications of the relationship between genome size and body size in flatworms and copepods. *Heredity* 84: 201–208.

Gregory, W.K. 1910. The orders of mammals. *Bulletin of the American Museum of Natural History* 27: 1–524.

Grell, K.G. 1971. *Trichoplax adhaerens* und die Entstehung der Metazoen. *Naturwissenschaftliche Rundschau* 24(4): 160–161 [in German].

Grimaldi, D. and Engel, M.S. 2005. *Evolution of the Insects.* Cambridge University Press, Cambridge.

Grimmelikhuijzen, C.J.P. and Westfall, J.A. 1995. The nervous system of cnidarians, pp. 7–24 in O. Breidbach and W. Kutsch (eds), *The Nervous System of Invertebrates: an Evolutionary and Comparative Approach.* Birkhäuser, Basel.

Grishanin, A.K., Khudolii, G.A., Shaikhaev, G.O., Brodskii, V.Y., Makarov, V.B., and Akif'ev, A.P. 1996. Chromatin diminution in *Cyclops kolensis* (Copepoda, Crustacea) is a unique example of genetic engineering in nature. *Russian Journal of Genetics* 32: 424–430.

Grobben, C. 1908. Die systematische Einteilung des Tierreiches. *Verhandlungen der zoologisch-botanischen Gesellschaft in Wien* 58: 491–511 [in German].

Grobben, K. 1910. *Lehrbuch der Zoologie*, 2nd edn. Elwert, Marburg [in German].

Gröger, H., Callaerts, P., Gehring, W.J., and Schmid, V. 1999. Gene duplication and recruitment of a specific tropomyosin into striated muscle cells in the jellyfish *Podocoryne carnea. Journal of Experimental Zoology* 285: 378–386.

Grosberg, R.K. and Strathmann, R.R. 1998. One cell, two cell, red cell, blue cell: the persistence of a unicellular stage in multicellular life histories. *Trends in Ecology and Evolution* 13: 112–116.

Gross, P.S. and Knowlton, R.E. 2002. Morphological variation among larval-postlarval intermediates produced by eyestalk ablation in the snapping shrimp *Alpheus heterochaelis* Say. *Biological Bulletin* 202: 43–52.

Grosshans, H., Johnson, T., Reinert, K.L., Gerstein, M., and Slack, F.J. 2005. The temporal patterning microRNA *let-7* regulates several transcription factors at the larval to adult transition in *C. elegans. Developmental Cell* 8: 321–330.

Grube, E., 1853. Untersuchungen über den Bau von *Peripatus Edwardsii. Archiv für Anatomie, Physiologie und wissenschaftliche Medicin* 20: 322–360 [in German].

Guillaumet, A., Ferdy, J.B., Desmarais, E., Godelle, B., and Crochet, P.-A. 2008. Testing Bergmann's rule in the presence of potentially confounding factors: a case study with three species of *Galerida* larks in Morocco. *Journal of Biogeography* 35: 579–591.

Guralnick, R.P. and Lindberg, D.R. 2001. Reconnecting cell and animal lineages: what do cell lineages tell us about the evolution and development of Spiralia? *Evolution* 55: 1501–1519.

Guralnick, R.P. and Lindberg, D.R. 2002. Cell lineage data and spiralian evolution: a reply to Nielsen and Meier. *Evolution* 56: 2558–2560.

Gurdon, J.B. 1992. The generation of diversity and pattern in animal development. *Cell* 68: 185–199.

Gurney, R. 1942. *Larvae of Decapod Crustacea.* Ray Society, London.

Gustafson, R.G. and Lutz, R.A. 1992. Larval and early post-larval development of the protobranch bivalve *Solemya velum* (Mollusca: Bivalvia). *Journal of the Marine Biological Association of the United Kingdom* 72: 383–402.

Guthrie, S. and Lumsden, A. 1991. Formation and regeneration of rhombomere boundaries in the developing chick hindbrain. *Development* 112: 221–229.

Guthrie, S., Prince, V., and Lumsden, A. 1993. Selective dispersal of avian rhombomere cells in orthotopic and heterotopic grafts. *Development* 118: 527–538.

Haase, A., Stern, M., Wächtler, K., and Bicker, G. 2001. A tissue-specific marker of Ecdysozoa. *Development, Genes and Evolution* 211: 428–433.

Hadfield, M.G. 2000. Why and how marine-invertebrate larvae metamorphose so fast. *Seminars in Cell & Developmental Biology* 11: 437–443.

Hadfield, M.G., Carpizo-Ituarte, E.J., Del Carmen, K., and Nedved, B.T. 2001. Metamorphic competence, a major adaptive convergence in marine invertebrate larvae. *American Zoologist* 41: 1123–1131.

Hadži, J. 1955. K diskusiji o novi sistematiki živalstva (Zur Diskussion über das neue zoologische System). *Razprave, Slovenska Akademija Znanosti in Umetnosti, Razrad za Prirodoslovne in Medicinske Vede, Ljubljana* 3: 175–207 [in Slovenian with German summary].

Hadži, J. 1963. *The Evolution of the Metazoa.* Pergamon Press, Oxford.

Haeckel, E. 1866. *Generelle Morphologie der Organismen. Allgemeine Grundzüge der organischen Formen-Wissenschaft, mechanisch begründet durch die von Charles Darwin reformirte Descendenz-Theorie. Band 1: Allgemeine Anatomie der Organismen.* Georg Reimer, Berlin.

Haeckel, E. 1874. Die Gastrea-Theorie, die phylogenetische Classification des Thierreiches und die Homologie der Keimblätter. *Jenaische Zeitschrift für Naturwissenschaft* 8: 1–55.

Haen, K.M., Lang, B.F., Pomponi, S.A., and Lavrov, D.V. 2007. Glass sponges and bilaterian animals share derived mitochondrial genomic features: a common ancestry or parallel evolution? *Molecular Biology and Evolution* 24: 1518–1527.

Hagadorn, J.W., Waggoner, B.M., and Fedo, C.M. 2000. Early Cambrian Ediacaran-type fossils from California. *Journal of Palaeontology* 74: 731–740.

Hagadorn, J.W., Xiao, S., Donoghue, P.C.J. *et al.* 2006. Cellular and subcellular structure of Neoproterozoic animal embryos. *Science* 314: 2291–2294.

Halanych, K.M. 1995. The phylogenetic position of the pterobranch hemichordates based on 18S rDNA sequence data. *Molecular Phylogenetic and Evolution* 4: 72–76.

Halanych, K.M. 2004. The new view of animal phylogeny. *Annual Review of Ecology Evolution and Systematics* 35: 229–256.

Halanych, K.M., Bacheller, J.D., Aguinaldo, A.M.A., Liva, S.M., Hillis, D.M., and Lake, J.A. 1995. Evidence from 18S ribosomal DNA that the lophophorates are protostome animals. *Science* 267: 1641–1643.

Halanych, K.M., Dahlgren, T.G., and McHugh, D. 2002. Unsegmented annelids? Possible origins of four lophotrochozoan worm taxa. *Integrative and Comparative Biology* 42: 678–684.

Halder, G., Callaerts, P. and Gehring, W.J. 1995. Induction of ectopic eyes by targeted expression of the *eyeless* gene in *Drosophila. Science* 267: 1788–1792.

Hall, B.K. 1998a. *Evolutionary Developmental Biology*, 2nd edn. Chapman & Hall, London.

Hall, B.K. 1998b. Germ layers and the germ-layer theory revisited: primary and secondary germ layers, neural crest as a fourth germ layer, homology, demise of the germ-layer theory. *Evolutionary Biology* 30: 121–186.

Hall, B.K. 1999. *The Neural Crest in Development and Evolution*. Springer, New York.

Hall, B.K. and Wake, M.H. (eds) 1999. *The Origin and Evolution of Larval Forms*. Academic Press, San Diego, CA.

Hall, B.K. and Olson, W.M. (eds) 2003. *Keywords and Concepts in Evolutionary Developmental Biology*. Harvard University Press, Cambridge, MA.

Hallez, P. 1887. *Embryogénie des dendrocoeles d'eau douce.* Paris [in French].

Hamilton, K.A., Nisbet, A.J., Lehane, M.J., Taylor, M.A., and Billingsley, P.F. 2003. A physiological and biochemical model for digestion in the ectoparasitic mite, *Psoroptes ovis* (Acari: Psoroptidae). *International Journal for Parasitology* 33: 773–785.

Hammarsten, O.D. 1918. *Beitrag zur Embryonalentwicklung der* Malacobdella grossa *(Müll.).* Inaugural Dissertation, Stockholm, Uppsala [in German].

Hammersten, O.D. and Runnström, J. 1925. Zur Embryologie von *Acanthochiton discrepans* Brown. *Zoologische Jahrbücher, Abteilung für Anatomie* 47: 261–318 [in German].

Hanelt, B., Van Schyndel, D., Adema, C.M., Lewis, L.A., and Loker, E.S. 1996. The phylogenetic position of *Rhopalura ophiocomae* (Orthonectida) based on 18s ribosomal DNA sequence analysis. *Molecular Biology and Evolution* 13: 1187–1191.

Hanken, J. and Wake, D.B. 1993. Miniaturization of body size: organismal consequences and evolutionary

significance. *Annual Review of Ecology and Systematics* 24: 501–519.

Hanken J. and Wake, D.B. 1998. Biology of tiny animals: systematics of the minute salamanders (*Thorius:* Plethodontidae) from Veracruz and Puebla, Mexico, with descriptions of five new species. *Copeia* 1998: 312–345.

Hansen, T.F. 2006. The evolution of genetic architecture. *Annual Reviews of Ecology, Evolution and Systematics* 37: 123–157.

Hare, W.A. and Owen, W.G. 1998. Effects of bicarbonate versus HEPES buffering on measured properties of neurons in the salamander retina. *Visual Neuroscience* 15: 263–271.

Harris, W.A. 1997. *Pax-6*: where to be conserved is not conservative. *Proceedings of the National Academy of Sciences USA* 94: 2098–2100.

Harrison, F.W. and De Vos, L. 1991. Porifera, pp. 29–89 in F.W. Harrison and J.A. Westfall (eds), *Microscopic Anatomy of Invertebrates. Vol. 2. Placozoa, Porifera, Cnidaria, and Ctenophora*. Wiley-Liss, New York.

Hart, M. 2000. Phylogenetic analyses of mode of larval development. *Seminars in Cell & Developmental Biology* 11: 411–418.

Hartenstein, V. and Ehlers, U. 2000. The embryonic development of the rhabdocoel flatworm *Mesostoma lingua* (Abildgaard, 1798). *Development, Genes and Evolution* 210: 399–415.

Hartenstein, V. and Jones, M. 2003. The embryonic development of the bodywall and nervous system of the cestode flatworm *Hymenolepis diminuta*. *Cell & Tissue Research* 311: 427–435.

Hartmann, H. and Reichert, H. 1998. The genetics of embryonic brain development in *Drosophila. Molecular and Cellular Neuroscience* 12: 194–205.

Hartmann, M. 1904. Fortpflanzungsweisen der Organismen. Neubenennung und Einteilung derselben, erlautert an Protozoen, Volvocineen und Dicyemiden. *Biologisches Centralblatt* 24: 18–32 [in German].

Harvey, A.W., Martin, J.W., and Wetzer, R. 2002. Phylum Arthropoda: Crustacea, pp. 337–369 in C.M. Young (ed.), *Atlas of Marine Invertebrate Larvae*. Academic Press, San Diego, CA.

Harzsch, S. 2003. Ontogeny of the ventral nerve cord in malacostracan crustaceans: a common plan for neuronal development in Crustacea, Hexapoda and other Arthropoda? *Arthropod Structure and Development* 32: 17–37.

Harzsch, S. 2004. Phylogenetic comparison of serotonin-immunoreactive neurons in representatives of the Chilopoda, Diplopoda, and Chelicerata: implications for arthropod relationships. *Journal of Morphology* 259: 198–213.

Harzsch, S. 2006. Neurophylogeny: architecture of the nervous system and a fresh view on arthropod phylogeny. *Integrative and Comparative Biology* 46: 162–194.

Harzsch, S. and Müller, C.H.G. 2007. A new look at the ventral nerve centre of *Sagitta*: implications for the phylogenetic position of Chaetognatha (arrow worms) and the evolution of the bilaterian nervous system. *Frontiers in Zoology* 4: 14.

Harzsch, S., Müller, C.H.G., and Wolf, H. 2005. From variable to constant cell numbers: cellular characteristics of the arthropod nervous system argue against a sister-group relationship of Chelicerata and "Myriapoda" but favour the Mandibulata concept. *Development, Genes and Evolution* 215: 53–68.

Hassanin, A. 2006. Phylogeny of Arthropoda inferred from mitochondrial sequences: strategies for limiting the misleading effects of multiple changes in pattern and rates of substitution. *Molecular Phylogenetics and Evolution* 38: 100–116.

Haszprunar, G. 1988. On the origin and evolution of major gastropod groups, with special reference to the Streptoneura. *Journal of Molluscan Studies* 54: 367–441.

Haszprunar, G. 1996. The Mollusca: coelomate turbellarians or mesenchymate annelids?, pp. 1–28 in J.D. Taylor (ed.), *Origin and Evolutionary Radiation of the Mollusca*. Oxford University Press, Oxford.

Haszprunar, G. 2000. Is the Aplacophora monophyletic? A cladistic point of view. *American Malacological Bulletin* 15: 115–130.

Haszprunar, G. and Wanninger, A. 2000. Molluscan muscle systems in development and evolution. *Journal of Zoological Systematics and Evolutionary Research* 38: 157–163.

Haszprunar, G. and Wanninger, A. 2008. On the fine structure of the creeping larva of *Loxosomella murmanica*: additional evidence for a clade of Kamptozoa (Entoprocta) and Mollusca. *Acta Zoologica* 88: 137–148.

Haszprunar, G., Rieger, R.M., and Schuchert, P. 1991. Extant "Problematica" within or near the Metazoa, pp. 99–105 in A.M. Simonetta and S. Conway Morris (eds), *The Early Evolution of Metazoa and the Significance of Problematic Taxa*. Cambridge University Press, Cambridge.

Haszprunar, G., von Salvini-Plawen, L., and Rieger, R.M. 1995. Larval plankotrophy. *Acta Zoologica* 76: 141–154.

Hatschek, B. 1878. Studien über die Entwicklungsgeschichte der Anneliden. Ein Beitrag zur Morphologie der Bilaterien. *Arbeiten aus dem Zoologischen Institut der Universität Wien* 1: 277–404 [in German].

Hatschek, B. 1888. *Lehrbuch der Zoologie. Eine morphologische Übersicht des Thiereiches zur Einführung in das Studium dieser Wissenschaft*. 1. Gustav Fischer, Jena [in German].

Hausdorf, B., Helmkampf, M., Meyer, A. *et al.* 2007. Spiralian phylogenomics supports the resurrection of Bryozoa comprising Ectoprocta and Entoprocta. *Molecular Biology and Evolution* 24: 2723–2729.

Hayward, D.C., Samuel, G., Pontynen, P.C. *et al.* 2002. Localized expression of a dpp/BMP2/4 ortholog in a coral embryo. *Proceedings of the National Academy of Sciences USA* 99: 8106–8111.

Hayward, D.C., Miller, D.J., and Ball, E.E. 2004. *snail* expression during embryonic development of the coral *Acropora*: blurring the diploblast/triploblast divide? *Developmant Genes and Evolution* 214: 257–260.

Hedges, S.B. and Kumar, S. 2003. Genomic clocks and evolutionary timescales. *Trends in Genetics* 19: 200–206.

Hedges, S.B. and Kumar, S. 2004. Precision of molecular time estimates. *Trends in Genetics* 20: 242–247.

Hejnol, A. and Martindale, M.Q. 2008. Acoel development supports a simple planula-like urbilaterian. *Philosophical Transactions of the Royal Society of London Series B, Biological Sciences* 363: 1493–1501.

Held, L.I. 2002. *Imaginal Discs: the Genetic and Cellular Logic of Pattern Formation*. Cambridge University Press, Cambridge-New York.

Helfenbein, K.G. and Boore, J.L. 2004. The mitochondrial genome of *Phoronis architecta*—comparisons demonstrate that phoronids are lophotrochozoan protostomes. *Molecular Biology and Evolution* 21: 153–157.

Helfenbein, K.G., Fourcade, H.M., Vanjani, R.G., and Boore, J.L. 2004. The mitochondrial genome of *Paraspadella gotoi* is highly reduced and reveals that chaetognaths are a sister group to protostomes *Proceedings of the National Academy of Sciences USA* 101: 10639–10643.

Heming, B.S. 1980. Development of the mouthparts in embryos of *Haplothrips verbasci* (Osborn) (Insects, Thysanoptera, Phlaeothripidae). *Journal of Morphology* 164: 235–263.

Heming, B.S. 2003. *Insect Development and Evolution*. Comstock Publishing Associates/Cornell University Press, Ithaca, NY.

Hendelberg, J. and Åkesson, B. 1988. *Convolutriloba retrogemma* gen. et sp. n., a turbellarian (Acoela, Platyhelminthes) with reversed polarity of reproductive buds. *Fortschritte der Zoologie* 36: 321–327.

Hendelberg, J. and Åkesson, B. 1991. Studies of the budding process in *Convolutriloba retrogemma* (Acoela, Platyhelminthes). *Hydrobiologia* 227: 11–17.

Hennig, W. 1950. *Grundzüge einer Theorie der phylogenetischen Systematik*. Deutscher Zentralverlag, Berlin [in German].

Hennig, W. 1965. Phylogenetic systematics. *Annual Review of Entomology* 10: 97–116.

Hennig, W. 1966. *Phylogenetic Systematics*. University of Illinois Press, Urbana, IL.

Hennig, W. 1969. *Die Stammesgeschichte der Insekten*. Kramer, Frankfurt [in German].

Hennig, W. 1979. *Wirbellose I (ausgenommen Gliedertiere)*. *Taschenbuch der Speziellen Zoologie*, 4th edn, vol. 2. Fischer, Jena [in German].

Henry, J. and Martindale, M.Q. 1997. Nemerteans, the ribbon worms, pp. 151–166 in S.F. Gilbert and A.M. Raunio (eds), *Embryology: Constructing the Organism*. Sinauer Associates, Sunderland, MA.

Henry, J.Q. and Martindale, M.Q. 2004. Inductive interactions and embryonic equivalence groups in a basal metazoan, the ctenophore *Mnemiopsis leidyi*. *Evolution & Development* 6: 17–24.

Henry, J.Q., Martindale, M.Q., and Boyer, B.C. 2000. The unique developmental program of the acoel flatworm, *Neochildia fusca*. *Development Biology* 220: 285–295.

Henry, J.Q., Perry, K.J., and Martindale, M.Q. 2007. Molecular controls of axis specification and cell determination in marine invertebrate embryos and larvae. www.sicb.org/meetings/2007/schedule/abstractdetails

Herlyn, H. and Ehlers, U. 1997. Ultrastructure and function of the pharynx of *Gnathostomula paradoxa* (Gnathostomulida). *Zoomorphology* 117: 135–145.

Herlyn, H., Piskurek, O., Schmitz, J., Ehlers, U., and Zischler, H. 2003. The syndermatan phylogeny and the evolution of acanthocephalan endoparasitism as inferred from 18S rDNA sequences. *Molecular Phylogenetics and Evolution* 26: 155–164.

Hernandez-Nicaise, M.L. 1991. Ctenophora, pp. 359–418 in F.W. Harrison and J.A. Westfall (eds), *Microscopic Anatomy of Invertebrates. Vol. 2. Placozoa, Porifera, Cnidaria and Ctenophora*. Wiley-Liss, New York.

Hernandez-Nicaise, M.L. and Amsellem, J. 1980. Ultrastructure of the giant smooth muscle fiber of the ctenophore *Beroe ovata*. *Journal of Ultrastructural Research* 72: 151–168.

Hérouard, E. 1909. Sur les cycles évolutifs d'un scyphistome. *Comptes Rendus Hebdomadaires des Séances de l'Academie des Sciences, Paris* 148: 320–323 [in French].

Herrera, J.C., McWeeney, S.K., and McEdwards, L.R. 1999. Diversity of energetic strategies among echinoid larvae and the transition from feeding to nonfeeding development. *Oceanologica Acta* 19: 313–321.

Hertel, L.A., Bayne, C.J., and Loker, E.S. 2002. The symbiont *Capsaspora owczarzaki*, nov. gen. nov. sp., isolated from three strains of the pulmonate snail *Biomphalaria glabrata* is related to members of the Mesomycetozoea. *International Journal for Parasitology* 32: 1183–1191.

Hertzler, P.L. and Clark, Jr, W.H. 1992. Cleavage and gastrulation in the shrimp *Sicyonia ingentis*. *Development* 116: 127–140.

Hessling, R. 2002. Metameric organisation of the nervous system in developmental stages of *Urechis caupo* (Echiura) and its phylogenetic implications. *Zoomorphology* 121: 221–234.

Hessling, R. 2003. Novel aspects of the nervous system of *Bonellia viridis* (Echiura) revealed by the combination of immunohistochemistry, confocal laser-scanning microscopy and three-dimensional reconstruction. *Hydrobiologia* 496: 225–239.

Hessling, R. and Westheide, W. 1999. CLSM analysis of development and structure of the central nervous system of *Enchytraeus crypticus* („Oligochaeta", Enchytraeidae). *Zoomorphology* 199: 37–47.

Hessling, R. and Westheide, W. 2002. Are Echiura derived from a segmented ancestor? Immunohistochemical analysis of the nervous system in developmental stages of *Bonellia viridis*. *Journal of Morphology* 252: 100–113.

Hildemann, W.H., Johnson, I.S. and Jokiel, P.L. 1979. Immunocompetence in the lowest metazoan phylum: transplantation immunity in sponge. *Science* 204: 420–422.

Hilken, G. 1998. Vergleich von Tracheensystemen unter phylogenetischem Aspekt. *Verhandlungen des naturwissenschaftlichen Vereins in Hamburg, Neue Folge* 37: 5–94 [in German].

Hill, S.D. and Boyer, B.C. 2001. Phalloidin labeling of developing muscle in embryos of the polychaete *Capitella* sp. I. *Biological Bulletin* 201: 257–258.

Hinman, V.F., O'Brien, E.K., Richards, G.S. and Degnan, B.M. 2003. Expression of anterior Hox genes during larval development of the gastropod *Haliotis asinina*. *Evolution & Development* 5: 508–521.

Ho, S.Y.W. and Jermiin, L.S. 2004. Tracing the decay of the historical signal in biological sequence data. *Systematic Biology* 53: 623–637.

Hodor, P.G. and Ettensohn, C.A. 1998. The dynamics and regulation of mesenchymal cell fusion in the sea urchin embryo. *Developmental Biology* 199: 111–124.

Hoekstra, H.E. and Nachman, M.W. 2003. Different genes underlie adaptive melanism in different populations of rock pocket mice. *Molecular Ecology* 12: 1185–1194.

Holland, L.Z. and Holland, N.D. 1998. Developmental gene expression in amphioxus: new insights into the evolutionary origin of vertebrate brain regions, neural crest, and rostrocaudal segmentation. *American Zoologist* 38: 647–658.

Holland, L.Z., Kene, M., Williams, N.A., and Holland, N.D. 1997. Sequence and embryonic expression of the amphioxus *engrailed* gene (*AmphiEn*): the metameric pattern of transcription resembles that of its segment-polariry homolog in *Drosophila*. *Development* 124: 1723–1732.

Holland, N.D. 1991. Echinodermata: Crinoidea, pp. 247–97 in J.S. Pearse and V.B. Pearse (eds), *Reproduction of Marine Invertebrates, vol. VI, Echinoderms and Lophophorates.* Boxwood Press, Pacific Grove, CA.

Holland, N.D. and Chen, J. 2001. Origin and early evolution of the vertebrates: new insights from advances in molecular biology, anatomy, and palaeontology. *BioEssays* 23: 142–151.

Holland, N.D. and Holland, L.Z. 2006. Stage- and tissue-specific patterns of cell division in embryonic and larval tissues of amphioxus during normal development. *Evolution & Development* 8: 142–149.

Holland, P.W.H. 1988. Homeobox genes and the vertebrate head. *Development* (suppl.) 103: 17–24.

Holland, P.W.H. 1990. Homeobox genes and segmentation: co-option, co-evolution, and convergence. *Seminars in Developmental Biology* 1: 135–145.

Holland, P.W.H. 2001. Beyond the Hox: how widespread is homeobox gene clustering? *Journal of Anatomy* 199: 13–33.

Holland, P., Ingham, P., and Krauss, S. 1992. Mice and flies head to head. *Nature* 358: 627–628.

Holley, S.A., Jackson, P.D., Sasai, Y. *et al.* 1995. A conserved system for dorsal-ventral patterning in insects and vertebrates involving *sog* and *chordin*. *Nature* 376: 249–253.

Holt, R.A., Subramanian, G.M., Halpera, A. *et al.* 2002. The genome sequence of the malaria mosquito *Anopheles gambiae*. *Science* 298: 129–149.

Holterman, M., Van der Wurff, A., Van der Elsen, S. *et al.* 2006. Phylum-wide analysis of SSU rDNA reveals deep phylogenetic relationships among nematodes and accelerated evolution toward crown clades. *Molecular Biology and Evolution* 23: 1792–1800.

Honeybee Genome Sequencing Consortium. 2006. Insights into social insects from the genome of the honeybee *Apis mellifera*. *Nature* 443: 931–949.

Hong, S.G. and Huang, Q. 1999. Studies on spermatogenesis in *Tachypleus tridentatus*: II. The spermiogenesis. *Acta Zoologica Sinica* 45: 252–259.

Horton, A.C., Mahadevan, N.R., Ruvinsky, I., Gibson-Brown, J.J. 2003. Phylogenetic analyses alone are insufficient to determine whether genome duplication(s) occurred during early vertebrate evolution. *Journal of Experimantal Zoology (Molecular Developmental Evolution)* 299B: 41–53.

Hoshiyama, D., Suga, H., Iwabe, N. *et al.* 1998. Sponge Pax cDNA related to *Pax2/5/8* and ancient gene duplications in the Pax family. *Journal of Molecular Evolution* 47: 640–648.

Hou, X.-G. 1987. Early Cambrian large bivalved arthropods from Chengjiang, eastern Yunnan. *Acta Palaeontologica Sinica* 26: 286–298.

Hou, X.-G. and Chen, J.-Y. 1989. Early Cambrian tentacled worm-like animals (*Facivermis* gen. nov.) from Chengjiang, eastern Yunnan. *Acta Palaeontologica Sinica* 28: 32–41.

Hou, X.-G. and Bergström, J. 1995. Cambrian lobopodians—ancestors of extant onychophorans? *Zoological Journal of the Linnean Society* 114: 3–19.

Hou, X. and Bergström, J. 1997. Arthropods of the Lower Cambrian Chengjiang fauna, southwest China. *Fossils & Strata* 45: 1–136.

Hou, X., Ramsköld, L., and Bergström, J. 1991. Composition and preservation of the Chengjiang fauna—a lower Cambrian soft-bodied biota. *Zoologica Scripta* 20: 395–411.

Hou X.-G., Aldridge, R.J., Siveter, D.J., Siveter, D.J., and Feng, X.-H. 2002. New evidence on the anatomy and phylogeny of the earliest vertebrates. *Proceedings of the Royal Society of London Series B, Biological Sciences* 269: 1865–1869.

Hou, X.-G., Aldridge, R.J., Bergström, J., Siveter, D.J., Siveter, D.J., and Feng, X.-H. 2004. *The Cambrian Fossils of Chengjiang, China: the Flowering of Early Animal Life*. Blackwell, Oxford.

Hou, X., Bergström, J. and Yang, J. 2006. Distinguishing anomalocaridids from arthropods and priapulids. *Geological Journal* 41: 259–269.

House, M.R. (ed.) 1979. *The Origin of Major Invertebrate Groups*. The Systematics Association Special Volume no. 12. Academic Press, London.

Houthoofd, W., Willems, M., Vangestel, S., Mertens, C., Bert, W., and Borgonie, G. 2006. Different roads to form the same gut in nematodes. *Evolution & Development* 8: 362–369.

Huber, B.S. 2006. Cryptic female exaggeration: the asymmetric female internal genitalia of *Kaliana yuruani* (Araneae: Pholcidae). *Journal of Morphology* 267: 705–712.

Hueber, S.D., Bezdan, D., Henz, S.R., Blank, M., Wu, H. and Lohmann, I. 2007. Comparative analysis of *Hox* downstream genes in *Drosophila*. *Development* 134: 381–392.

Hughes, C.L. and Kaufman, T.C. 2002a. Exploring myriapod segmentation: the expression patterns of *even-skipped*, *engrailed*, and *wingless* in a centipede. *Developmental Biology* 247: 47–61.

Hughes, C.L. and Kaufman, T.C. 2002b. Hox genes and the evolution of the arthropod body plan. *Evolution & Development* 4: 459–499.

Hughes, C.L. and Kaufman, T.C. 2002c. Exploring the myriapod body plan: expression patterns of the ten Hox genes in a centipede. *Development* 129: 1225–1238.

Hughes, N.C., Minelli, A., and Fusco, G. 2006. The ontogeny of trilobite segmentation: a comparative approach. *Paleobiology* 32: 602–627.

Hughes, R.N., D'Amato, M.E., Bishop, J.D.D. *et al.* 2005. Paradoxical polyembryony? Embryonic cloning in an ancient order of marine bryozoans. *Biology Letters* 1: 178–180.

Hull, D.L. 1988. *Science as a Process: an Evolutionary Account of the Social and Conceptual Development of Science.* University of Chicago Press, Chicago, IL.

Hülskamp, M., Schröder, C, Pfeifle, C., Jäckle, H. and Tautz, D. 1989. Posterior segmentation of the *Drosophila* embryo in the absence of a maternal posterior organizer gene. *Nature* 338: 629–632.

Huxley, J. S. 1932. *Problems of Relative Growth.* Methuen, London,

Hwang, U.W., Friedrich, M., Tautz, D., Park, C.J., and Kim, W. 2001. Mitochondrial protein phylogeny joins myriapods with chelicerates. *Nature* 413: 154–157.

Hyman, L.H. 1940. *The Invertebrates. Volume 1, Protozoa through Ctenophora.* McGraw-Hill, New York.

Hyman, L.H. 1951. *The Invertebrates. Volume 2, Platyhelminthes and Rhynchocoela.* McGraw-Hill, New York.

Hyman, L.H. 1955. *The Invertebrates. Volume 4, Echinodermata.* McGraw-Hill, New York.

Hyman, L.H. 1959. *The Invertebrates. Volume 5, Smaller Coelomate Groups.* McGraw-Hill, New York.

Hyman, L.H. 1967. *The Invertebrates, Volume 6, Mollusca I.* McGraw-Hill, New York.

Ibrahim, I.A. and Gad, A.M. 1975. The occurrence of paedogenesis in *Eristalis* larvae (Diptera: Syrphidae). *Journal of Medical Entomology* 12: 268.

Ikuta, T., Yoshida, N., Satoh, N. and Saiga, H. 2004. *Ciona intestinalis* Hox gene cluster: its dispersed structure and residual collinear expression in development. *Proceedings of the National Academy of Sciences USA* 101: 15118–15123.

Illiger, C. 1811. *Prodromus systematis mammalium et avium additis terminis zoographicis utriusque classis.* C. Salfeld, Berlin [in Latin].

Imai, K.S., Levine, M., Satoh, N., and Satou, Y. 2006. Regulatory blueprint of a chordate embryo. *Science* 312: 1183–1187.

Ingersoll, E.P. and Wilt, H.F. 1998. Matrix metalloproteinase inhibitors disrupt spicule formation by primary mesenchyme cells in the sea urchin embryo. *Developmental Biology* 196: 95–106.

Ingham, P.W. and McMahon, A.P. 2001. Hedgehog signaling in animal development: paradigms and principles. *Genes & Development* 15: 3059–3087.

Inoue, I. 1958. Studies on the life history of *Chordodes japonensis,* a species of Gordiacea. I. The development

and structure of the larva. *Japanese Journal of Zoology* 12: 203–218.

Irimia, M., Maeso, I., Penny, D., García-Fernàndez, J., and Roy, S.W. 2007. Rare coding sequence changes are consistent with Ecdysozoa, not Coelomata. *Molecular Biology and Evolution* 24: 1604–1607.

Irwin, D.M. and Wilson, A.C. 1993. Limitations of molecular methods for establishing the phylogeny of mammals, with special reference to the position of elephants, pp. 257–267 in F.S. Szalay, M.J. Novacek and M.C. McKenna (eds), *Mammal Phylogeny: Placentals.* Springer-Verlag, New York.

Israelsson, O. 1997. *Xenoturbella's* molluscan relatives. *Nature* 390: 32.

Israelsson, O. 1999. New light on the enigmatic *Xenoturbella* (phylum unknown) ontogeny and phylogeny. *Proceedings of the Royal Society of London Series B, Biological Sciences* 266: 835–841.

Ivanov, A.V. 1973. *Trichoplax adhaerens,* a phagocytella-like animal. *Zoologiceskij Zurnal* 52: 1117–1131 [in Russian with English abstract].

Ivanov, A.V. and Mamkaev, Yu.V. 1973. *Turbellarians (Turbellaria). Their Origin and Evolution: a Phylogenetic Essay.* Nauka, Leningrad [in Russian].

Ivanov, P.P. 1937. *General and Comparative Embryology.* Ogiz-Biomedgiz, Moscow [in Russian].

Ivantsov, A.Y. and Fedonkin, M.A. 2001. Locomotion trails of the Ediacara-type organisms preserved with the producer's body fossils, White Sea, Russia. *Geological Associaton of Canada and Mineralogical Association of Canada Abstracts* 26: 69–70.

Iyer, L.M., Aravind, L., Coon, S.L., Klein, D.C., and Koonin, E.V. 2004. Evolution of cell-cell signaling in animals: did late horizontal gene transfer from bacteria have a role? *Trends in Genetics* 20: 292–299.

Jablonski, D. 1993. The tropics as a source of evolutionary novelty through geological time. *Nature* 364: 142–144.

Jablonski, D. 2005. Evolutionary innovations in the fossil record: the intersection of ecology, development, and macroevolution. *Journal of Experimental Zoology (Molecular and Developmental Evolution)* 304B: 504–519.

Jacobs, D.K. and Gates, R.D. 2003. Developmental genes and the reconstruction of metazoan evolution—implications of evolutionary loss, limits on inference of ancestry and type 2 errors. *Integrative and Comparative Biology* 43: 11–18.

Jacobs, D.K., Wray, C.G., Wedeen, C.J., and Kostriken, R. 2000. Molluscan *engrailed* expression, serial organization, and shell evolution. *Evolution & Development* 2: 340–347.

Jacobs, D.K., Hughes, N.C., Fitz-Gibbon, S.T., and Winchell, C.J. 2005. Terminal addition, the Cambrian radiation and the Phanerozoic evolution of bilaterian form. *Evolution & Development* 7: 498–514.

Jacobs, D.K., Nakanishi, N., Yuan, D., Camara, A., Nichols, S.A. and Hartenstein, V. 2007. Evolution of sensory structures in basal metazoa. *Integrative and Comparative Biology* 47: 712–723.

Jaeckle, W.B. and Rice, M.E. 2002. Phylum Sipuncula, pp. 375–396 in C.M. Young (ed.), *Atlas of Marine Invertebrate Larvae*. Academic Press, London.

Jägersten, G. 1955. On the early phylogeny of the Metazoa. The Bilaterogastraea-theory. *Zoologiska Bidrag från Uppsala* 30: 321–354.

Jägersten, G. 1972. *Evolution of the Metazoan Life Cycle. A Comprehensive Theory*. Academic Press, London.

Jaillon, O., Aury, J.-M., Brunet, F. *et al.* 2004. Genome duplication in the teleost fish *Tetraodon nigroviridis* reveals the early vertebrate proto-karyotype. *Nature* 431: 946–957.

Jakob, W., Sagasser, S., Dellaporta, S., Holland, P., Kuhn, K., and Schierwater, B. 2004. The Trox-2 Hox/ParaHox gene of *Trichoplax* (Placozoa) marks an epithelial boundary. *Genes and Evolution* 214: 170–175.

James, B.L. 1964. The life cycle of *Parvatrema homoeotecnum* sp. nov. (Trematoda: Digenea) and a review of the family Gymnophallidae Morozov, 1955. *Parasitology* 54: 1–41.

James, B.L. and Bowers, A. 1967. Reproduction in the daughter sporocyst of *Cercaria bucephalopsis haimeana* (Lacaze-Duthiers 1854) (Bucephalidae) and *Cercaria dichotoma* Lebour, 1911 (non Müller) (Gymnophallidae). *Parasitology* 57: 607–625.

Jamieson, B.G.M. 1992. Oligochaeta, pp. 217–322 in F.W. Harrison and S.L. Gardiner (eds), *Microscopic Anatomy of Invertebrates. Vol. 7. Annelida*. Wiley-Liss, New York.

Janssen, R., Prpic, N.M. and Damen, W.G.M. 2004 Gene expression suggests decoupled dorsal and ventral segmentation in the millipede *Glomeris marginata* (Myriapoda: Diplopoda). *Developmental Biology* 268: 89–104.

Jarman, A.P., Grau, Y., Jan, L.Y., and Jan, Y.N. 1993. *atonal* is a proneural gene that directs chordotonal organ formation in the Drosophila peripheral nervous system. *Cell* 73: 1307–1321.

Jedrzejowska, I. and Kubrakiewicz, J. 2002. Formation and structure of nutritive cords in telotrophic ovarioles of snake flies (Insecta: Raphidioptera). *Folia Histochemica et Cytobiologica* 40: 77–83.

Jefferies, R.P.S. 1986. *The Ancestry of the Vertebrates*. British Museum of Natural History, London.

Jeffery, J.E., Bininda-Emonds, O.R.P., Coates, M.I., and Richardson, M.K. 2002a. Analyzing evolutionary patterns in amniote embryonic development. *Evolution & Development* 4: 292–302.

Jeffery, J.E., Richardson, M.K., Coates, M.I. and Bininda-Emonds, O.R.P. 2002b. Analyzing developmental sequences within a phylogenetic framework, *Systematic Biology* 51: 478–491.

Jeffery, W.R. 2006. Ascidian neural crest-like cells: phylogenetic distribution, relationship to larval complexity, and pigment cell fate. *Journal of Experimental Zoology (Molecular and Developmental Evolution)* 306B: 470–480.

Jeffery, W.R. 2007. Chordate ancestry of the neural crest: new insights from ascidians. *Seminars in Cell and Developmental Biology* 18: 81–91.

Jeffery, W.R., Strickler, A.G., and Yamamoto, Y. 2004. Migratory neural crest-like cells form body pigmentation in a urochordate embryo. *Nature* 431: 696–699.

Jenner, R.A. 1999. Metazoan phylogeny as a tool in evolutionary biolgy: current problems and discrepancies in application. *Belgian Journal of Zoology* 129: 245–262.

Jenner, R.A. 2000. Evolution of animal body plans: the role of metazoan phylogeny at the interface between pattern and process. *Evolution & Development* 2: 208–221.

Jenner, R.A. 2003. Unleashing the force of cladistics? Metazoan phylogenetics and hypothesis testing. *Integrative and Comparative Biology* 43: 207–218.

Jenner, R.A. 2004a. Accepting partnership by submission? Morphological phylogenetics in a molecular millennium. *Systematic Biology* 53: 333–342.

Jenner, R.A. 2004b. The scientific status of metazoan cladistics: why current research practice must change. *Zoologica Scripta* 33: 293–310.

Jenner, R.A. 2004c. Towards a phylogeny of the Metazoa: evaluating alternative phylogenetic positions of Platyhelminthes, Nemertea, and Gnathostomulida, with a critical reappraisal of cladistic characters. *Contributions to Zoology* 73: 3–163.

Jenner, R.A. 2006. Challenging received wisdoms: some contributions of the new microscopy to the new animal phylogeny. *Integrative and Comparative Biology* 46: 93–101.

Jennings, R.M. and Halanych, K.M. 2005. Mitochondrial genomes of *Clymenella torquata* (Maldanidae) and *Riftia pachyptila* (Siboglinidae): evidence for conserved gene order in Annelida. *Molecular Biology and Evolution* 22: 210–222.

Jensen, S., Gehling, J.G., and Droser, M.L. 1998. Ediacara-type fossils in Cambrian sediments. *Nature* 393: 567–569.

Jeram, A.J., Selden, P.A., and Edwards, D. 1990. Land animals in the Silurian: arachnids and myriapods from Shropshire, England. *Science* 250: 658–661.

Jermiin, L.S., Ho, S.Y.W., Ababneh, F., Robinson, J., and Larkum, A.W.D. 2004. The biasing effect of compositional heterogeneity on phylogenetic estimates may be underestimated. *Systematic Biology* 53: 638–643.

Jermiin, L.S., Poladian, L., and Charleston, M.A. 2005. Is the "Big Bang" in animal evolution real? *Science* 310: 1910–1911.

Jiménez-Guri, E., Okamura, B., and Holland, P.W.H. 2007. Origin and evolution of a myxozoan worm. *Integrative and Comparative Biology* 47: 752–758.

Jockusch, E.L., Nulsen, C., Newfeld, S.J., and Nagy, L.M. 2000. Leg development in flies versus grasshoppers: differences in *dpp* expression do not lead to differences in the expression of downstream components of the leg patterning pathway. *Development* 127: 1617–1626.

Joerdens, J., Struck, T., and Purschke, G. 2004. Phylogenetic inference regarding Parergodrilidae and *Hrabeiella periglandulata* ('Polychaeta', Annelida) based on 18S rDNA, 28S rDNA and COI sequences. *Journal of Zoological Systematics and Evolutionary Research* 42: 270–280.

Joffe, B.I. and Kornakova, E.E. 1998. *Notentera ivanovi* Joffe *et al.*, 1997: a contribution to the question of phylogenetic relationships between 'turbellarians' and the parasitic Plathelminthes (Neodermata). *Hydrobiologia* 383: 245–250.

Joffe, B.I., Solovei, I.V., Sewell, K.B. and Cannon, L.R.G. 1995. Organization of the epidermal syncytial mosaic in *Diceratocephala boschmai* (Temnocephalida, Platyhelminthes). *Australian Journal of Zoology* 43: 509–518.

Johansson, K. E. 1937. Über *Lamellisabella zachsi* und ihre systematische Stellung. *Zoologischer Anzeiger* 117: 23–26 [in German].

Johnson, A.D., Drum, M., Bachvarova, R.F., Masi, T., White, M.E., and Crother, B.I. 2003. Evolution of predetermined germ cells in vertebrate embryos: implications for macroevolution. *Evolution & Development* 5: 414–431.

Johnson, C.D. and Stretton, A.O.W. 1980. Neural control of locomotion in *Ascaris*: anatomy, electrophysiology, and biochemistry, pp. 159–95 in B. Zuckerman (ed.), *Nematodes as Biological Models*. Academic Press, New York.

Johnson, M.D., Oldach, D., Delwiche, C.F., and Stoecker, D.K. 2007. Retention of transcriptionally active cryptophyte nuclei by the ciliate *Myrionecta rubra*. *Nature* 445: 426–428.

Johnston, G. 1842. *A History of British Sponges and Lithophytes*. W.H. Lizard, Edinburgh.

Johnston, J.S., Ross, L.D., Beani, L., Hughes, D.P., and Kathirithamby, J. 2004. Tiny genomes and endoreduplication in Strepsiptera. *Insect Molecular Biology* 13: 581–585.

Johnston, L.A. and Gallant, P. 2002. Control of growth and organ size in *Drosophila*. *BioEssays* 24: 54–64.

Jondelius, U., Ruiz-Trillo, I., Baguna, J., and Riutort, M. 2002. The Nemertodermatida are basal bilaterians and not members of the Platyhelminthes. *Zoologica Scripta* 31: 201–215.

Kadner, D. and Stollewerk, A. 2004. Neurogenesis in the chilopod *Lithobius forficatus* suggests more similarities to chelicerates than to insects. *Development, Genes and Evolution* 214: 367–379.

Kai, T. and Spradling, A. 2004. Differentiating germ cells can revert into functional stem cells in *Drosophila melanogaster* ovaries. *Nature* 428: 564–569.

Kalandadze, N.N. and Rautian, S.A. 1992. [The system of mammals and historical zoogeography], pp. 44–152 in O.L. Rossolimo (ed.), *Phylogenetics of Mammals*. Archives of the Zoological Museum, Moscow State University, 29 [in Russian].

Kamm, K., Schierwater, B., Jakob, W., Dellaporta, S.L., and Miller, D.J. 2006. Axial patterning and diversification in the Cnidaria predate the Hox system. *Current Biology* 16: 920–926.

Kammermeier, L. and Reichert, H. 2001. Common developmental genetic mechanisms for patterning invertebrate and vertebrate brains. *Brain Research Bulletin* 55: 675–682.

Kapp, H. 2000. The unique embryology of Chaetognatha. *Zoologischer Anzeiger* 239: 263–266.

Karkach, A.S. 2006. Trajectories and models of individual growth. *Demographic Research* 15: 347–400.

Karuppaswamy, S.A. 1977. Occurrence of β-chitin in the cuticle of a pentastomid *Raillietiella gowrii*. *Experientia* 33: 735–736.

Kasatkina, A.P. and Buryi, G.I. 1996. On the relation of chaetognaths and conodonts. *Albertiana* 18: 21–23.

Kasatkina, A.P. and Buryi, G.I. 1997. Chaetodonta, a new animal superphylum and its position in animal systematics. *Doklady Biological Sciences* 356: 503–505.

Kasatkina, A.P. and Buryi, G.I. 1999. The position of the phyla Chaetognatha and Euconodontophylea in the classification of Metazoa. *Zoosystematica Rossica* 8: 21–26.

Kauffman, S.A. 1993. *Origins of Order: Self-organization and Selection in Evolution*. Oxford University Press, Oxford.

Kawakami, Y., Raya, A., Raya, R.M., Rodríguez-Esteban, C. and Izpisúa Belmonte, J.C. 2005. Retinoic acid signalling links left/right asymmetric patterning and bilaterally symmetric somitogenesis in the zebrafish embryo. *Nature* 435: 165–171.

Keller, E.F. 2002. *Making Sense of Life*. Harvard University Press. Cambridge, MA.

Keller, R. 2006. Mechanisms of elongation in embryogenesis. *Development* 133: 2291–2302.

Kellis, M., Birren, B.W., and Lander, E.S. 2004. Proof and evolutionary analysis of ancient genome duplication in the yeast *Saccharomyces cerevisiae*. *Nature* 428: 617–624.

Kempf, S.C. and Todd, C.D. 1989. Feeding potential in the lecithotrophic larva of *Adalaria proxima* and *Tritonia hombergi*: an evolutionary perspective. *Journal of the Marine Biological Association of the United Kingdom* 69: 659–682.

Kettle, C., Johnstone, J., Jowett, T., Arthur, H., and Arthur, W. 2003. The pattern of segment formation, as revealed by *engrailed* expression, in a centipede with a variable number of segments. *Evolution & Development* 5: 198–207.

Khan, P., Linkhart, B. and Simon, H. 2002. Different regulation of T-box genes *Tbx4* and *Tbx5* during limb development and limb regeneration. *Developmental Biology* 250: 383–392.

Kim, J.H., Kim, W. and Cunningham, C.W. 1999. A new perspective on lower metazoan relationships from 18S rDNA sequences. *Molecular Biology and Evolution* 16: 423–427.

Kimmel, C. B. 1996. Was Urbilateria segmented? *Trends in Genetics* 12: 329–331.

King, N. 2004. The unicellular ancestry of animal development. *Developmental Cell* 7: 313–325.

King, N. and Carroll, S.B. 2001. A receptor tyrosine kinase from choanoflagellates: molecular insights into early animal evolution. *Proceedings of the National Academy of Sciences USA* 98: 15032–7.

King, N., Hittinger, C.T., and Carroll, S.B. 2003. Evolution of key cell signaling and adhesion protein families predates animal origins. *Science* 301: 361–363.

King, P.E. 1973. *Pycnogonids*. Hutchinson, London.

Kinzelbach, R.K. 1971. Strepsiptera (Facherflugler), pp. 1–68 in J.G. Helmcke, D. Starck and H. Wermuth (eds), *Handbuch der Zoologie*, vol. 4, section 2, part 2/24. de Gruyter, Berlin [in German].

Kishimoto, Y., Murate, M., and Sugiyama, T. 1996. Hydra regeneration from recombined ectodermal and endodermal tissue I. Epibolic ectodermal spreading is driven by cell intercalation. *Journal of Cell Science* 109: 763–772.

Kissinger, J.C. and Raff, R.A. 1998. Evolutionary changes in sites and timing of expression of actin genes in the direct- and indirect-developing sea urchins *Heliocidaris erythrogramma* and *H. tuberculata*. *Development, Genes and Evolution* 208: 82–93.

Kjer, K.M. 2004. Aligned 18S and insect phylogeny *Systematic Biology* 53: 506–514.

Kjer, K.M., Carle, F.L., Litman, J., and Ware, J. 2006. A molecular phylogeny of Hexapoda. *Arthropod Systematics & Phylogeny* 64: 35–44.

Klass, K.-D. and Kristensen, N.P. 2001. The ground plan and affinities of hexapods: recent progress and open problems, pp. 265–298 in T. Deuve (ed.), *Origin of the Hexapoda*, vol. 37. Annales de la Société Entomologique de France, Nouvelle Série 37. Société Entomologique de France, Paris.

Klass, K.D., Zompro, O., Kristensen, N.P., and Adis, J. 2002. Mantophasmatodea: a new insect order with extant members in the Afrotropics. *Science* 296: 1456–1459.

Klawe, W.L. and Dickie, L.M. 1957. Biology of the bloodworm, *Glycera dibranchiata* Ehlers, and its relation to the bloodworm of the Maritime Provinces. *Bulletin of the Fisheries Research Board of Canada* 115: 1–37.

Klein, J.T. 1734. *Naturalis dispositio echinodermatum*. Schreiber, Gedani [in Latin].

Klingenberg, C.P. 1998. Heterochrony and allometry: the analysis of evolutionary change in ontogeny. *Biological Reviews* 73: 79–123.

Kluge, A.G. 1999. The science of phylogenetic systematics: explanation, prediction, and test. *Cladistics* 15: 429–435.

Kluge, A.G. 2001. Parsimony with and without scientific justification. *Cladistics* 17: 199–210.

Klussmann-Kolb, A. and Dinapoli, A. 2006. Systematic position of the pelagic Thecosomata and Gymnosomata within Opisthobranchia (Mollusca, Gastropoda)—revival of the Pteropoda. *Journal of Zoological Systematics and Evolutionary Research* 44: 118–129.

Kmita, M., Fraudeau, N., Hérault, Y., and Duboule, D. 2002. A serial deletion/duplication strategy *in vivo* suggests a molecular basis for *Hoxd* genes colinearity in limbs. *Nature* 420: 145–150.

Knauss, E. 1979. Indication of an anal pore in Gnathostomulida. *Zoologica Scripta* 8: 181–186.

Knight, C.G., Patel, M.N., Azevedo, R.B.R. and Leroi, A.M. 2002. A novel mode of ecdysozoan growth in *Caenorhabditis elegans*. *Evolution & Development* 4: 16–27.

Knoll, A.H. and Carroll, S.B. 1999. Early animal evolution: emerging views from comparative biology and geology. *Science* 284: 2129–2137.

Kobayashi, M., Furuya, H., and Holland, P.W.H. 1999. Dicyemids are higher animals. *Nature* 401: 762.

Koch, M. 2003. Monophyly of the Myriapoda? Reliability of current arguments. *African Invertebrates* 44: 137–153.

Koenemann, S., Schram, F.R., Bloechl, A., Iliffe, T.M., Hoenemann, M. and Held, C. 2007. Post-embryonic development of remipede crustaceans. *Evolution & Development* 9: 117–121.

Køie, M. 1982. The redia, cercaria and early stages of *Aporocotyle simplex* Odhner, 1900 (Sanguinicolidae)—a digenetic trematode which has a polychaete annelid as the only intermediate host. *Ophelia* 21: 115–145.

Kojima, T., Sato, M. and Saigo, K. 2000. Formation and specification of distal leg segments in *Drosophila* by dual Bar homeobox genes, *BarH1* and *BarH2*. *Development* 127: 769–778.

Komaru, A., Kawagishi, T., and Konishi, K. 1998. Cytological evidence of spontaneous androgenesis in the freshwater clam *Corbicula leana* Prime. *Development, Genes and Evolution* 208: 46–50.

Konopová, B. and Zrzavý, J. 2005. Ultrastructure, development, and homology of insect embryonic cuticles. *Journal of Morphology* 264: 339–362.

Kontani, K., Moskowitz, I.P.G. and Rothman, J.H. 2005. Repression of cell-cell fusion by components of the *C. elegans* vacuolar ATPase complex. *Developmental Cell* 8: 787–794.

Korn, H. 1982. Annelida, pp. 1–599 in F. Seidel (ed.), *Morphogenese der Tiere. Erste Reihe: Deskriptive Morphogenese*, part 5. H-I. Fischer, Stuttgart [in German].

Kortschak, R.D., Samuel, G., Saint, R., and Miller, D.J. 2003. EST Analysis of the cnidarian *Acropora millepora* reveals extensive gene loss and rapid sequence divergence in the model invertebrates. *Current Biology* 13: 2190–2195.

Kottelat, M., Britz, R., Hui, T.H., and Witte, K.-E. 2006. *Paedocypris*, a new genus of Southeast Asian cyprinid fish with a remarkable sexual dimorphism, comprises the world's smallest vertebrate. *Proceedings of the Royal Society of London Series B, Biological Sciences* 273: 895–899.

Kowalewski, A. 1866. Entwicklungsgeschichte der einfachen Ascidien. *Mémoires de l'Académie Imperiale des Sciences de St. Petersbourg* (7)10: 1–119 [in German].

Kozaric, Z., Kuzir, S., Petrinec, Z., Gjurcevic, E., and Opacak, A. 2006. Histochemical distribution of digestive enzymes in intestine of goldline, *Sarpa salpa* L. 1758. *Journal of Applied Ichthyology* 22: 43–48.

Kozmik, Z. 2005. *Pax* genes in eye development and evolution. *Current Opinion in Genetics & Development* 15: 430–438.

Kozmik, Z., Daube, M., Frei, E. *et al.* 2003. Role of *Pax* genes in eye evolution: a cnidarian *PaxB* gene uniting *Pax2* and *Pax6* functions. *Developmental Cell* 5: 773–785.

Krakauer, D. and Plotkin, J. 2002. Redundancy, antiredundancy, and the robustness of genomes. *Proceedings of the National Academy of Sciences USA* 99: 1405–1409.

Krakauer, D. and Plotkin, J. 2005. Robustness in biological systems: a provisional taxonomy, pp. 71–103 in E. Jen (ed.), *Robust Design: a Repertoire for Biology, Ecology and Engineering*. Oxford University Press, New York.

Kramer, J.M., French, R.P., Park, E.C., and Johnson, J.J. 1990. The *Caenorhabditis elegans rol-6* gene, which interacts with the *sqt-1* collagen gene to determine organismal morphology, encodes a collagen. *Molecular and Cellular Biology* 10: 2081–2089.

Kraus, O. 1998. Phylogenetic relationships between higher taxa of tracheate arthropods, pp. 295–303 in R.A. Fortey and R.H. Thomas (eds), *Arthropod Relationships*. Chapman & Hall, London.

Kraus, O. 2001. "Myriapoda" and the ancestry of Hexapoda, pp. 105–127 in T. Deuve (ed.), *Origin of the Hexapoda*. Annales de la Société Entomologique de France, Nouvelle Série 37. Société Entomologique de France, Paris.

Kraus, O. and Kraus, M. 1994. On "Myriapoda"-Insecta interrelationships, phylogenetic age and primary ecological niches (Arthropoda, Tracheata). *Verhandlungen des naturwissenschaftlichen Vereins in Hamburg, Neue Folge* 34: 5–31.

Krell, F.-T. 1992. Verschmelzung von Antennomeren (Symphysocerie) als Regelfall bei *Temnorhynchus repandus* Burmeister, 1847, sowie phylogenetische, taxonomische, faunistische und nomenklaturische Anmerkungen zu diversen Taxa dieser Gattung (Coleoptera. Scarabaeoidea, Melolonthidae, Dynastinae, Pentodontini). *Deutsche entomologische Zeitschrift, Neue Folge* 39: 295–367.

Kristensen, N.P. 1998. The groundplan and basal diversification of the hexapods, pp. 281–293 in R.A. Fortey and R.H. Thomas (eds), *Arthropod Relationships*, Chapman & Hall, London.

Kristensen, R.M. 1983. Loricifera, a new phylum with Aschelminthes characters from the meiobenthos. *Zeitschrift für Zoologische Systematik und Evolutionsforschung* 21: 163–180.

Kristensen, R.M. 1995. Are Aschelminthes pseudocoelomate or acoelomate?, pp. 41–43 in G. Lanzavecchia, R. Valvassori and M.D. Candia Carnevali (eds), *Body Cavities: Function and Phylogeny*. Mucchi, Modena.

Kristensen, R.M. 2002. An Introduction to Loricifera, Cycliophora, and Micrognathozoa. *Integrative and Comparative Biology* 42: 641–651.

Kristensen, R.M. 2003. Comparative morphology: do the ultrastructural investigations of Loricifera and Tardigrada support the clade Ecdysozoa?, pp. 467–77 in A. Legakis, S. Sfenthourakis, R. Polymeni, and M. Thessalou-Legaki (eds), *The New Panorama of Animal Evolution*. Proceedings of the 18th International Congress of Zoology. Pensoft, Sofia/Moscow.

Kristensen, R.M. and Higgins, R.P. 1991. Kinorhyncha, pp. 377–404 in F.W. Harrison and E.E. Ruppert (eds), *Microscopic Anatomy of Invertebrates, Volume 4. Aschelminthes*, Wiley-Liss, New York.

Kristensen, R.M. and Funch, P. 2000. Micrognathozoa: a new class with complicated jaws like those of Rotifera and Gnathostomulida. *Journal of Morphology* 246: 1–49.

Kropf, C. 1998. Slit sense organs of *Comaroma simonii* Bertkau: a morphological atlas (Araneae, Anapidae), pp. 151–159 in P.A. Selden (ed.), *Proceedings of the 17th European Colloquium of Arachnology, Edinburgh 1997*. British Arachnological Society, Burnham Beeches.

Krug, P.J. 1998. Poecilogony in an estuarine opisthobranch: planktotrophy, lecitotrophy, and mixed clutches in a population of the ascoglossan *Alderia modesta*. *Marine Biology* 132: 483–494.

Kruse, M., Leys, S.P., Müller, I.M., and Müller, W.E.G. 1998. Phylogenetic position of the Hexactinellida within the phylum Porifera based on the amino acid sequence of the protein kinase C from *Rhabdocalyptus dawsoni*. *Journal of Molecular Evolution* 46: 721–728.

Kukalova-Peck, J. 1987. New Carboniferous Diplura, Monura, and Thysanura, the hexapod ground plan, and the role of thoracic side lobes in the origin of wings (Insecta). *Canadian Journal of Zoology* 65: 2327–2345.

Kulakova, M.A., Kostyuchenko, R.P., Andreeva, T.F., and Dondua, A.K. 2002. The *Abdominal-B*-like gene expression during larval development of *Nereis virens* (Polychaeta). *Mechanisms of Development* 115: 177–179.

Kulakova, M., Bakalenko, N., Nivikova, E. *et al.* 2007. Hox gene expression in larval development of the polychaetes *Nereis virens* and *Platynereis dumerilii* (Annelida, Lophotrochozoa). *Development, Genes and Evolution* 217: 39–54.

Kumar, J.P. 2001. Signalling pathways in *Drosophila* and vertebrate retinal development. *Nature Reviews Genetics* 2: 846–857.

Kurland, C.G., Canback, B., and Berg, O.G. 2003. Horizontal gene transfer: a critical view. *Proceedings of the National Academy of Sciences USA* 100: 9658–9662.

Kusche, K., Ruhberg, H. and Burmester, T. 2002. A hemocyanin from the Onychophora and the emergence of respiratory proteins. *Proceedings of the National Academy of Sciences USA* 99: 10545–10548.

Kusche, K., Hembach, A., Hagner-Holler, S., Gebauer, W., and Burmester, T. 2003. Complete subunit sequences, structure and evolution of the 6x6-mer hemocyanin from the common house centipede, *Scutigera coleoptrata*. *European Journal of Biochemistry* 270: 2860–2868.

Labat-Robert, J., Robert, L., Auger, C., Lethias, C., and Garrone, R. 1981. Fibronectine-like protein in Porifera: its role in cell aggregation. *Proceedings of the National Academy of Sciences USA* 78: 6261–6265.

Lacalli, T.C. 1982. The nervous system and ciliary band of Müller's larva. *Proceedings of the Royal Society of London Series B, Biological Sciences* 217: 37–58.

Lacalli, T. 1996. Dorsoventral axis inversion: a phylogenetic perspective. *Bioessays* 18: 251–254.

Lacalli, T. 2002. Vetulicolians—are they deuterostomes? Chordates? *BioEssays* 24: 208–211.

Ladurner, P. and Rieger, R. 2000. Embryonic muscle development of *Convoluta pulchra* (Turbellaria—Acoelomorpha, Platyhelminthes). *Developmental Biology* 222: 359–375.

Lafay, B., Boury-Esnault, N., Vacelet, J., and Christen, R. 1992. An analysis of partial 28S ribosomal RNA sequences suggests early radiations of sponges. *BioSystems* 28: 139–151.

Lake, J.A. 1990. Origin of the Metazoa. *Proceedings of the National Academy of Sciences USA* 87: 763–766.

Lall, S., Grun, D., Krek, A. *et al.* 2006. A genome-wide map of conserved microRNA targets in *C. elegans*. *Current Biology* 16: 460–471.

Lamarck, J.-B. 1801 *Discours d'ouverture du Cours de Zoologie, donné dans le Muséum National d'Histoire Naturelle l'an 8 de la République*. Déterville, Paris [in French].

Lamarck, J.-B. 1809. *Philosophie zoologique*. Dentu, Paris [in French].

Lamarck, J.-B. 1816. *Histoire naturelle des animaux sans vertèbres*, vol. 3. Déterville, Paris [in French].

Lamarck, J.-B. 1933. *The Lamarck manuscripts at Harvard edited by William Morton Wheeler and Thomas Barbour*. Harvard University Press, Cambridge, MA.

Lammert, V. 1991. Gnathostomulida, pp. 19–39 in F.W. Harrison and E.E. Ruppert (eds), *Microscopic Anatomy of Invertebrates. Vol. 4. Aschelminthes*. Wiley-Liss, New York.

Lanfear, R., Thomas, J.A., Welch, J.J., Brey, T., and Bromham, L. 2007. Metabolic rate does not calibrate the molecular clock. *Proceedings of the National Academy of Sciences USA* 104: 15388–15393.

Lang, B.F., O'Kelly, C., Nerad, T., Gray, M.W., and Burger, G. 2002. The closest unicellular relatives of animals. *Current Biology* 12: 1773–1778.

Langton, P.H., Cranston, P.S. and Armitage, P. 1988. The parthenogenetic midge of water supply systems, *Paratanytarsus grimmi* (Schneider) (Diptera: Chironomidae). *Bulletin of Entomological Research* 78: 317–328.

Lankester, E.R. 1873. On the primitive cell layers of the embryo as the basis of genealogical classification of animals, and on the origins of vascular and lymph systems. *Annals and Magazine of Natural History* (4)11: 321–328.

Lankester, E.R. 1877. Notes on the embryology and classification of the animal kingdom: comprising a revision of speculations relative to the origin and significance of germ layers. *Quarterly Journal of Microscopic Science* 17: 399–454.

Lans, D., Wedeen, C.J., and Weisblat, D.A. 1993. Cell lineage analysis of the expression of an *engrailed* homolog in leech embryos. *Development* 117: 857–871.

Lanzavecchia, G., De Eguileor, M., Valvassori, R. and Scarì, G. 1995. Body cavities of Nematomorpha, pp. 45–60 in G. Lanzavecchia, R. Valvassori and M.D. Candia Carnevali (eds), *Body Cavities: Function and Phylogeny*. Mucchi, Modena.

Larink, O. and Westheide, W. 2006. *Coastal Plankton. Photo Guide for European Seas*. Pfeil, München.

Larroux, C., Fahey, B., Liubicich, D. *et al*. 2006. Developmental expression of transcription factor genes in a demosponge: insights into the origin of metazoan multicellularity. *Evolution & Development* 8: 150–173.

Larroux, C., Fahey, B., Degnan, S.M., Adamski, M., and Degnan, B.M. 2007. The NK homeobox gene cluster predates the origin of Hox genes. *Current Biology* 17: 706–710.

Larsen, E. 2003. Genes, cell behaviour, and the evolution of form, pp. 119–131 in G.B. Müller and S.A. Newman (eds), *Origination of Organismal Form. Beyond the Gene in Developmental and Evolutionary Biology*. MIT Press, Cambridge, MA.

Larson, A., Kirk, M.M., and Kirk, D.L. 1992. Molecular phylogeny of the volvocine flagellates. *Molecular Biology and Evolution* 9: 85–105.

Lartillot, N., Le Gouar, M., and Adoutte, A. 2002. Expression patterns of *forkhead* and *goosecoid* homologues in the mollusc *Patella vulgata* support the ancestry of the anterior mesendoderm across Bilateria. *Development, Genes and Evolution* 212: 551–561.

Latreille, P. A. 1829. *Le Règne animal, par M. le baron Cuvier. Nouvelle Edition, revue et augmentée. Tome IV. Crustacés, Arachnides et Partie des Insectes*. Déterville et Crochard, Paris [in French].

Lauterbach, K.-E. 1983. Zum Problem der Monophylie der Crustacea. *Verhandlungen des naturwissenschaftlichen Vereins in Hamburg, Neue Folge* 26: 293–320 [in German].

Lauzon, R.J., Kidder, S.J., and Long, P. 2007. Suppression of programmed cell death regulates the cyclical degeneration of organs in a colonial urochordates. *Developmental Biology* 301: 92–105.

Lavrov, D.V. 2007. Key transitions in animal evolution: a mitochondrial DNA perspective. *Integrative and Comparative Biology* 47: 734–743.

Lavrov, D.V., Brown, W.M., and Boore, J.L. 2004. Phylogenetic position of the Pentastomida and (pan) crustacean relationships. *Proceedings of the Royal Society of London Series B, Biological Sciences* 271: 537–544.

Lavrov, D.V., Forget, L., Kelly, M., and Lang, B.F. 2005. Mitochondrial genomes of two demosponges provide insights into an early stage of animal evolution. *Molecular Biology and Evolution* 22: 1231–1239.

Lee, M.S.Y. 2005. Squamate phylogeny, taxon sampling, and data congruence. *Organisms Diversity and Evolution* 5: 25–45.

Lee, P.N., Callaerts, P., de Couet, H.G., and Martindale, M.Q. 2003. Cephalopod *Hox* genes and the origin of morphological novelties. *Nature* 424: 1061–1065.

Lemaire, P. 2006. How many ways to make a chordate? *Science* 312: 1145–1146.

Lemburg, C. 1995. Ultrastructure of the sense organs and receptor cells of the neck and lorica of *Halicryptus spinulosus* larva (Priapulida). *Microfauna Marina* 10: 7–30.

Lemburg, C. 1999. *Ultrastrukturelle Untersuchungen an den Larven von* Halicryptus spinulosus *und* Priapulus caudatus. *Hypothesen zur Phylogenie der Priapulida und deren Bedeutung für die Evolution der Nemathelminthes*. Cuvillier, Göttingen [in German].

Lengyel, J.A. and Iwaki, D.D. 2002. It takes guts: the *Drosophila* hindgut as a model system for organogenesis. *Developmental Biology* 243: 1–19.

Leuckart, R. 1854. Bericht über die Leistungen in der Naturgeschichte der niederen Tiere während der Jahre 1848–1853. *Archiv für Naturgeschichte* 20: 340–351 [in German].

Leve, C., Gajewski, M., Rohr, K.B., and Tautz, D. 2001. Homologues of *c-hairy1* (*her9*) and *lunatic fringe* in zebrafish are expressed in the developing central nervous system, but not in the presomitic mesoderm. *Development, Genes and Evolution* 211: 493–500.

Lévi, C. 1963. Gastrulation and larval phylogeny in sponges, pp. 375–382 in E.C. Dougherty (ed.), *The Lower Metazoa: Comparative Biology and Phylogeny*. University of California Press, Berkeley, CA.

Lévi, C. 1973. Systématique de la classe des Demospongiaria (Démosponges), pp. 577–632 in P.P. Grassé (ed.), *Traité de Zoologie*, vol. 3(1). Masson, Paris [in French].

Levinton, J., Dubb, L., and Wray, G.A. 2004. Simulations of evolutionary radiations and their application to understanding the probability of a Cambrian explosion. *Journal of Paleontology* 78: 31–38.

Leys, S.P. 2003. The significance of syncytial tissues for the position of the Hexactinellida in the Metazoa. *Integrative and Comparative Biology* 43: 19–27.

Leys, S.P., Cheung, E., and Boury-Esnault, N. 2006. Embryogenesis in the glass sponge *Oopsacas minuta*: formation of syncytia by fusion of blastomeres. *Integrative and Comparative Biology* 46: 104–117.

Leys, S.P. and Degnan, B.M. 2002. Embryogenesis and metamorphosis in a haplosclerid demosponge: gastrulation and transdifferentiation of larval ciliated cells to choanocytes. *Invertebrate Biology* 121: 171–189.

Leys, S.P. and Eerkes-Medrano, D. 2005. Gastrulation in calcareous sponges: in search of Haeckel's Gastraea. *Integrative and Comparative Biology* 45: 342–351.

Leys, S.P. and Ereskovsky, A.V. 2006. Embryogenesis and larval differentiation in sponges. *Canadian Journal of Zoology* 84: 262–287.

Li, C.-W., Chen, J.-Y., and Hua, T.-E. 1998. Precambrian sponges with cellular structures. *Science* 279: 879–882.

Li, H., Zhang, X.-Y. and Wang, A.-T. 2005. Exploration on primordial nervous substances in sponges. *Acta Zoologica Sinica* 51: 1091–1101.

Lichtneckert, R. and Reichert, H. 2005. Insights into the bilaterian brain: conserved genetic patterning mechanisms in insect and vertebrate brain development. *Heredity* 94: 465–477.

Lin, J.-P., Scott, A.C., Li, C.-W. *et al.* 2006. Silicified egg clusters from a Middle Cambrian Burgess Shale-type deposit, Guizhou, south China. *Geology* 34: 1037–1040.

Lin, X., Zheng, X., Xiao, S., and Wang, R. 2004. Phylogeny of the cuttlefishes (Mollusca: Cephalopoda) based on mitochondrial COI and 16S rRNA gene sequence data. *Acta Oceanologica Sinica* 23: 699–707.

Lindberg, D.R. and Guralnick, R.P. 2003. Phyletic patterns of early development in gastropod molluscs. *Evolution & Development* 5: 494–507.

Lindberg, D.R., Ponder, W.F., and Haszprunar, G. 2004. The Mollusca: relationships and patterns from their first half-billion years, pp. 252–278 in J. Cracraft and M.J. Donoghue (eds), *Assembling the Tree of Life*. Oxford University Press, Oxford.

Linder, F. 1952. Contributions to the morphology and taxonomy of the Branchiopoda Notostraca, with special reference to the north American species. *Proceedings of the United States National Museum* 102: 1–69.

Lindgren, A.R., Giribet G., and Nishiguchi, M.K. 2004. A combined approach to the phylogeny of Cephalopoda (Mollusca) *Cladistics* 20: 454–486.

Linnaeus, C. 1758. *Systema Naturae per regna tria Naturae secundum classes, ordines, genera, species, cum characteribus, diferentiis, synonymis, locis. Tomus I. Edition decima, reformata*, Salvius, Holmiae [in Latin].

Lipscomb, D.L., Farris, J.S., Kallersjö, M., and Tehler, A. 1998. Support, ribosomal sequences and the phylogeny of the eukaryotes. *Cladistics* 14: 303–338.

Littlewood, D.T.J. and Olson, P.D. 2001. Small subunit rDNA and the Platyhelminthes: signal, noise, conflict and compromise, pp. 262–78 in D.T.J. Littlewood and R.A. Bray (eds), *Interrelationships of the Platyhelminthes*. Taylor & Francis, London.

Littlewood, D.T.J., Smith, A.B., Clough, K.A., and Emson, R.H. 1997. The interrelationships of the echinoderm classes: morphological and molecular evidence. *Biological Journal of the Linnean Society* 61: 409–438.

Littlewood, D.T.J., Telford, M.J., Clough, K.A., and Rohde, K. 1998. Gnathostomulida—an enigmatic metazoan phylum from both morphological and molecular perspectives. *Molecular Phylogenetics and Evolution* 9: 72–79.

Littlewood, D.T.J., Rohde, K., and Clough, K.A. 1999. The interrelationships of all major groups of Platyhelminthes: phylogenetic evidence from morphology and molecules. *Biological Journal of the Linnean Society* 66: 75–114.

Littlewood, D.T.J., Olson, P.D., Telford, M.J., Herniou, E.A., and Riutort, M. 2001. Elongation factor 1-alpha sequences alone do not assist in resolving the position of the Acoela within the Metazoa. *Molecular Biology and Evolution* 18: 437–442.

Littlewood, D.T.J., Telford, M.J., and Bray, R.A. 2004. Protostomes and Platyhelminthes. The worm's turn, pp. 209–234 in J. Cracraft and M.J. Donoghue (eds), *Assembling the Tree of Life*. Oxford University Press, Oxford.

Locke, J.M. 2000. Ultrastructure of the statocyst of the marine enchytraeid *Grania americana* (Annelida: Clitellata). *Invertebrate Biology* 119: 83–93.

Lockyer, A.E., Olson, P.D., and Littlewood, D.T.J. 2003. Utility of complete large and small subunit rRNA genes in resolving the phylogeny of the Neodermata (Platyhelminthes): Implications and a review of the cercomer theory. *Biological Journal of the Linnean Society* 78: 155–171.

Loesel, R., Nässel, D.R. and Strausfeld, N.J. 2002. Common design in a unique midline neuropil in the brains of arthropods. *Arthropod Structure and Development* 31: 77–91.

Loker, E.S., Adema, C.M., Zhang, S.-M., and Kepler, T.B. 2004. Invertebrate immune systems—not homogeneous, not simple, not well understood. *Immunological Reviews* 198: 10–24.

Long, C.A. 2005. Intricate sutures as fractal curves. *Journal of Morphology* 185: 285–295.

Lorenzen, S. 1985. Phylogenetic aspects of pseudocoelomate evolution, pp. 210–223 in S.C. Morris, J.D. George, R. Gibson and H.M. Platt (eds), *The Origins and Relationships of Lower Invertebrates*. Oxford University Press, Oxford.

Lovén, S. 1874. Études sur les echinoidées. *Kongelige Svenska Vetenskaps-Akademiens Handlinger, new series* 11: 1–91 [in French].

Lowe, C.J. and Wray, G.A. 1997. Radical alterations in the roles of homeobox genes during echinoderm evolution. *Nature* 389: 718–721.

Lowe, C.J., Wu, M., Salic, A. *et al.* 2003. Anteroposterior pattering in hemichordates and the origins of the chordate nervous system. *Cell* 113: 853–865.

Lowe, C.J., Terasaki, M., Wu, M. *et al.* 2006. Dorsoventral patterning in hemichordates: insights into early chordate evolution. *PLoS Biology* 4: 1603–1619.

Luan, Y.-X., Mallatt, J.M., Xie, R.-D., Yang, Y.-M., and Yin, W.-Y. 2005. The phylogenetic positions of three basal-hexapod groups (Protura, Diplura, and Collembola) based on ribosomal RNA gene sequences. *Molecular Biology and Evolution* 22: 1579–1592.

Lucas, W.J., Ding, B. and van der Schoot, C. 1993. Plasmodesmata and the supracellular nature of plants. *New Phytologist* 125: 435–476.

Lüter, C. 2000. The origin of the coelom in Brachiopoda and its phylogenetic significance. *Zoomorphology* 120: 15–28.

Maas, A. and Waloszek, D. 2001. Cambrian derivatives of the early arthropod stem lineage, pentastomids, tardigrades and lobopodians—an 'Orsten' perspective. *Zoologischer Anzeiger* 240: 451–459.

Mabee, P.M. 1989. An empirical rejection of the ontogenetic polarity criterion. *Cladistics* 5: 409–416.

Machida, R. 2006. Evidence from embryology for reconstructing the relationships of hexapod basal clades. *Arthropod Systematics & Phylogeny* 64: 95–104.

Mackie, G.O. and Singla, C.L. 1983. Studies on hexactinellid sponges. I. Histology of *Rhabdocalyptus dawsoni* (Lambe, 1873). *Philosophical Transactions of the Royal Society of London Series B, Biological Sciences* 301: 365–400.

Maduro, M.F. 2006. Endomesoderm specification in *Caenorhabditis elegans* and other nematodes. *BioEssays* 28: 1010–1022.

Maduro, M.F. and Rothman, J.H. 2002. Making worm guts: the gene regulatory network of the *Caenorhabditis elegans* endoderm. *Developmental Biology* 246: 68–85.

Malakhov, V.V. 1980. Cephalorhyncha, a new type of animal kingdom uniting Priapulida, Kinorhyncha, Gordiacea, and a system of Aschelminthes worms. *Zoologicheskii Zhurnal* 59: 485–499.

Malakhov, V.V. 2003. Evolution of nematode embryogenesis. *Journal of Nematology* 35: 351–352.

Malakhov, V.V. and Adrianov, V.A. 1995. *Cephalorhyncha—a new phylum of the Animal Kingdom*. KMK Scientific Press, Moscow.

Maldonado, M. 2004. Choanoflagellates, choanocytes, and animal multicellularity. *Invertebrate Biology* 123: 1–22.

Maldonado, M. and Bergquist, P.R. 2002. Phylum Porifera, pp. 21–50 in C.M. Young (ed.), *Atlas of Marine Invertebrate Larvae*. Academic Press, London.

Mallatt, J. and Winchell, C.J. 2002. Testing the new animal phylogeny: first use of combined large-subunit and small-subunit rRNA to classify the protostomes. *Molecular Biology and Evolution* 19: 289–301.

Mallatt, J. and Giribet, G. 2006. Further use of nearly complete, 28S and 18S rRNA genes to classify Ecdysozoa: 37 more arthropods and a kinorhynch. *Molecular Phylogenetics and Evolution* 40: 772–794.

Mallatt, J. and Winchell, C.J. 2007. Ribosomal RNA genes and deuterostomes phylogeny revisited: more cyclostomes, elasmobranchs, reptiles, and a brittle star. *Molecular Phylogenetics and Evolution* 43: 1005–1022.

Mallatt, J.M., Garey, J.R., and Shultz, J.W. 2004. Ecdysozoan phylogeny and Bayesian inference: first use of nearly complete 28S and 18S rRNA gene sequences to classify the arthropods and their kin. *Molecular Phylogenetics and Evolution* 31: 178–191.

Mamkaev, Y.V. 1995. An elaboration of the evolutionary morphological basis for the systematics of the Plathelminthes. *Hydrobiologia* 305: 15–19.

Mandelbrot, B.B. 1982. *The Fractal Geometry of Nature*. Freeman, New York.

Mann, R.S. and Casares, F. 2002. Signalling legacies. *Nature* 418: 737–9.

Manning, G. and Krasnow, M.A. 1993. Development of the *Drosophila* tracheal system, pp. 609–85 in M. Bate and A. Martinez Arias (eds), *The Development of Drosophila melanogaster*. Cold Spring Harbor Press, Cold Spring Harbor, NY.

Manoukian, A.S. and Krause, H.M. 1992. Concentration-dependent activities of the Even-skipped protein in *Drosophila* embryos. *Genes and Development* 6: 1740–1751.

Manton, S.M. 1973. Arthropod phylogeny—a modern synthesis. *Journal of Zoology* 171: 11–130.

Manton, S.M. 1977. *The Arthropoda: Habits, Functional Morphology, and Evolution*. Clarendon Press, Oxford.

Manton, S.M. and Anderson, D.T. 1979 Polyphyly and the evolution of arthropods, pp. 269–321 in M.R. House (ed.), *The Origin of Major Invertebrate Groups*. Academic Press, London.

Mantovani, B. and Scali, V. 1992. Hybridogenesis and androgenesis in the stick insect *Bacillus rossius-grandii benazzii* (Insecta, Phasmatodea). *Evolution* 46: 783–796.

Manuel, M. 2006. Phylogeny and evolution of calcareous sponges. *Canadian Journal of Zoology* 84: 225–241.

Manuel, M., Borchiellini, C., Alivon, E., Le Parco, Y., Vacelet, J., and Boury-Esnault, N. 2003. Phylogeny and evolution of calcareous sponges: monophyly of Calcinea and Calcaronea, high levels of morphological homoplasy, and the primitive nature of axial symmetry. *Systematic Biology* 52: 311–333.

Manuel, M., Jager, M., Murienne, J., Clabaut, C., and Le Guyader, H. 2006. Hox genes in sea spiders (Pycnogonida) and the homology of arthropod head segments. *Development, Genes and Evolution* 216: 481–491.

Marcaggi, P., Jeanne, M., and Coles, J.A. 2004. Neuronglial trafficking of NH^{4+} and K^+: separate routes of uptake into glial cells of bee retina. *European Journal of Neuroscience* 19: 966–976.

Marcus, E. 1958. On the evolution of the animal phyla. *Quarterly Review of Biology* 33: 24–58.

Margulis, L. 1981. *Symbiosis in Cell Evolution: Microbial Communities in the Archean and Proterozoic Eons.* Freeman, New York.

Mariani, F.V. and Martin, G.R. 2003. Deciphering skeletal patterning: clues from the limb. *Nature* 423: 319–325.

Mark Welch, D.B. 2005. Bayesian and maximum likelihood analyses of rotifer-acanthocephalan relationships. *Hydrobiologia* 546: 47–54.

Marlétaz, F., Martin, E., Perez, Y. *et al.* 2006. Chaetognath phylogenomics: a protostome with deuterostomes-like development. *Current Biology* 16: R577–R578.

Marois, R. and Carew, T.J. 1997. Fine structure of the apical ganglion and its serotonergic cells in the larva of *Aplysia californica*. *Biological Bulletin* 192: 388–398.

Marques, A.C. and Collins, A.G. 2004. Cladistic analysis of Medusozoa and cnidarian evolution. *Invertebrate Biology* 123: 23–42.

Marshall, J.A. and Dixon, K.E. 1978. Cell specialization in the epithelium of the small intestine of feeding *Xenopus laevis* tadpoles. *Journal of Anatomy* **126**: 133–144.

Martí Mus, M. and Bergström, J. 2001. The skeletonmuscular system of hyolithids. *American Zoologist* 41: 1514.

Martin, W., Deusch, O., Stawski, N., Grunheit, N., and Goremykin, V. 2005. Chloroplast genome phylogenetics: why we need independent approaches to plant molecular evolution. *Trends in Plant Science* 10: 203–209.

Martindale, M.Q. 2005. The evolution of metazoan axial properties. *Nature Review Genetics* 6: 917–927.

Martindale, M.Q. and Henry, J.Q. 1995. Modifications of cell fate specification in equal-cleaving nemertean embryos—alternate patterns of spiralian development. *Development* 121: 3175–3185.

Martindale, M.Q. and Henry, J.Q. 1997. Reassessing embryogenesis in the Ctenophora: the inductive role of e1 micromeres in organizing ctene row formation in the "mosaic" embryo, *Mnemiopsis leidyi*. *Development* 124: 1999–2006.

Martindale, M.Q. and Henry, J.Q. 1998. The development of radial and biradial symmetry: the evolution of bilaterality. *American Zoologist* 38: 672–684.

Martindale, M.Q. and Henry, J.Q. 1999. Intracellular fate mapping in a basal metazoan, the ctenophore *Mnemiopsis leidyi*, reveals the origins of mesoderm and the existence of indeterminate cell lineages. *Developmental Biology* 214: 243–257.

Martindale, M.Q., Finnerty, J.R. and Henry, J.Q. 2002. The Radiata and the evolutionary origins of the bilaterian body plan. *Molecular Phylogenetics and Evolution* 24: 358–365.

Martindale, M.Q., Pang, K., and Finnerty, J.R. 2004. Investigating the origins of triploblasty: `mesodermal' gene expression in a diploblastic animal, the sea anemone *Nematostella vectensis* (phylum, Cnidaria; class, Anthozoa). *Development* 131: 2463–2474.

Martinez, D.E., Bridge, D., Masuda-Nakagawa, L.M., and Cartwright, P. 1998. Cnidarian homeoboxes and the zootype. *Nature* 393: 748–749.

Martinez-Arias, A. and Lawrence, P.A. 1985. Parasegments and compartments in the *Drosophila* embryo. *Nature* 313: 639–642.

Martinez-Morales, J.R., Rodrigo, I., and Bovolenta, P. 2004. Eye development: a view from the retina pigmented epithelium. *BioEssays* 26: 766–777.

Maruzzo, D., Bonato, L., Brena, C., Fusco, G., and Minelli, A. 2005. Appendage loss and regeneration in arthropods: a comparative view, pp. 215–245 in S. Koenemann, R. Jenner and R. Vonk (eds), *Crustacea and Arthropod Phylogeny*. Crustacean Issues 16. Taylor and Francis, Boca Raton, FL.

Maslakova, S.A., Martindale, M.Q., and Norenburg, J.L. 2004. Fundamental properties of the spiralian developmental program are displayed by the basal nemertean *Carinoma tremaphoros* (Palaeonemertea, Nemertea). *Developmental Biology* 267: 342–360.

Massagué, J. and Chen, Y.-G. 2000. Controlling TGF-β signaling. *Genes & Development* 14: 627–644.

Mastick, G.S., McKay, R., Oligino, T., Donovan, K., and Lopez, A.J. 1995 Identification of target genes regulated by homeotic proteins in *Drosophila melanogaster* through genetic selection of Ultrabithorax protein-binding sites in yeast. *Genetics* 139: 349–363.

Mathieson, B.R.F. and Lehane, M.J. 2002. Ultrastructure of the alimentary canal of the sheep scab mite, *Psoroptes ovis* (Acari: Psoroptidae). *Veterinary Parasitology* 104: 151–166.

Matsuda, R. 1987. *Animal Evolution in Changing Environment with Special Reference to Abnormal Metamorphosis*. Wiley, New York.

Mattei, X. and Marchand, B. 1987. Les spermatozoïdes des acanthocephales et des myzostomides. Ressemblances et conséquences phyletiques. *Comptes Rendus de l'Academie des Sciences, Paris, Science de la Vie* 305: 525–529 [in French].

Matus, D.Q., Copley, R.R., Dunn, C.W. *et al.* 2006a. Broad taxon and gene sampling indicate that chaetognaths are protostomes. *Current Biology* 16: R575–R576.

Matus, D.Q., Pang, K., Marlow, H., Dunn, C.W., Thomsen, G.H., and Martindale, M.Q. 2006b. Molecular evidence for deep evolutionary roots of bilaterality in animal development. *Proceedings of the National Academy of Sciences USA* 103: 11195–11200.

Mayer, G. 2006. Origin and differentiation of nephridia in the Onychophora provide no support for the Articulata. *Zoomorphology* 125: 1–12.

Mayer, G. and Bartolomaeus, T. 2003. Ultrastructure of the stomochord and the heart-glomerulus complex in *Rhabdopleura compacta* (Pterobranchia): phylogenetic implications. *Zoomorphology* 122: 125–133.

Mayer, G. and Koch, M. 2005. Ultrastructure and fate of the nephridial anlagen in the antennal segment of *Epiperipatus biolleyi* (Onychophora Peripatidae)—evidence for the onychophoran antennae being modified legs. *Arthropod Structure and Development* 34: 471–480.

Mayer, G., Ruhberg, H., and Bartolomaeus, T. 2004. When a epithelium ceases to exist—an ultrastructural study on the fate of the embryonic coelom in *Epiperipatus biolleyi* (Onychophora, Peripatidae). *Acta Zoologica* 85: 163–170.

Maynard Smith, J. and Szathmáry, E. 1995. *The Major Transitions in Evolution*. W.H. Freeman, Oxford.

Maynard Smith, J., Burian, R., Kauffman, S.A. *et al.* 1985. Developmental constraints and evolution. *Quarterly Review of Biology* 60: 265–267.

Mayr, E. 1960. The emergence of evolutionary novelties, pp. 349–380 in S. Tax (ed.), *Evolution after Darwin*, vol. 1. *The Evolution of Life*. University of Chicago Press, Chicago, IL.

Mayr, E. 1969. *Principles of Systematic Zoology*. McGraw-Hill, New York.

Mayr, E. and Ashlock, P.D. 1991. *Principles of Systematic Zoology*, 2nd edn. McGraw-Hill, New York.

Mayr, E., Linsley, E.G., and Usinger, R. 1953. *Methods and Principles of Systematic Zoology*. McGraw-Hill, New York.

Mazet, F. and Shimeld, S.M. 2002. Gene duplication and divergence in the early evolution of vertebrates. *Current Opinions in Genetics & Development* 12: 393–396.

McEdward, L.R. and Janis, D.A. 1997. Relationships among development, ecology and morphology in the evolution of echinoderm larvae and life cycles. *Biological Journal of the Linnean Society* 60: 381–400.

McGhee, Jr, G.R. 1999. *Theoretical Morphology. The Concept and its Applications*. Columbia University Press, New York.

McGhee, Jr, G.R. 2007. *The Geometry of Evolution. Adaptive Landscapes and Theoretical Morphospaces*. Cambridge University Press, Cambridge.

McGhee, J.D. 2000. Homologous tails? Or tales of homology? *BioEssays* 22: 781–785.

McHugh, D. and Rouse, G.W. 1998. Life history evolution of marine invertebrates: new views from phylogenetic systematics. *Trends in Ecology and Evolution* 13: 182–186.

McKenna, M.C. 1975. Toward a phylogenetic classification of the Mammalia, pp. 21–46 in W.P. Luckett and F.S. Szalay (eds), *Phylogeny of the Primates*. Plenum, New York.

McKenna, M.C. and Bell, S.K. 1997. *Classification of Mammals Above the Species Level*. Columbia University Press, New York.

McKinney, M.L. (ed.) 1988. *Heterochrony in Evolution: a Multidisciplinary Approach*. Plenum Press, New York.

McKinney, M.L. and McNamara, K.J. 1991. *Heterochrony: the Evolution of Ontogeny*. Plenum Press, New York.

McLaughlin, P.A. and Lemaitre, R. 1997. Carcinization in the Anomura—fact or fiction? I. Evidence from adult morphology. *Contributions to Zoology* 67: 79–123.

McLaughlin, P.A., Lemaitre, R., and Tudge, C.C. 2004. Carcinization in the Anomura—fact or fiction? II. Evidence from larval, megalopal and early juvenile morphology. *Contributions to Zoology* 73: 165–205.

McNamara, K.J. 1986. A guide to the nomenclature of heterochrony. *Journal of Paleontology* 60: 4–13.

McNamara, K.J. (ed.) 1995. *Evolutionary Change and Heterochrony*. Wiley, Chichester.

McShea, D.W. 1991. Complexity and evolution: what everybody knows. *Biology and Philosophy* 6: 303–324.

McShea, D.W. 1996a. Complexity and homoplasy, pp. 207–25 in M.J. Sanderson and L. Hufford (eds), *Homoplasy: the Recurrence of Similarity in Evolution*. Academic Press, San Diego, CA.

McShea, D.W. 1996b. Metazoan complexity and evolution: is there a trend? *Evolution* 50: 477–492.

McShea, D.W. 2000. Functional complexity in organisms: parts as proxies. *Biology & Philosophy* 15: 641–668.

McShea, D.W. 2001. Parts and integration: the consequences of hierarchy, pp. 27–60 in J. Jackson, S. Lidgard and K. McKinney (eds), *Evolutionary Patterns: Growth, Form, and Tempo in the Fossil Record*, University of Chicago Press, Chicago, IL.

McShea, D.W. 2002. A complexity drain on cells in the evolution of multicellularity. *Evolution* 56: 441–456.

McShea, D.W. 2005. A universal generative tendency towards increased organismal complexity, pp. 435–453 in B. Hallgrímsson and B.K. Hall (eds), *Variation: a Central Concept in Biology*. Elsevier, Amsterdam.

Medina, M., Collins, A.G., Silberman, J.D., and Sogin, M.L. 2001. Evaluating hypotheses of basal animal phylogeny using complete sequences of large and small subunit rRNA. *Proceedings of the National Academy of Sciences USA* 98: 9707–9712.

Meier, P., Finch, A., and Evan, G. 2000. Apoptosis in development. *Nature* 407: 796–801.

Meinertzhagen, I.A. 2005. Eutely, cell lineage, and fate within the ascidian larval nervous system: determinacy or to be determined? *Canadian Journal of Zoology* 83: 184–195.

Meinhardt, H. 2002. The radial-symmetric hydra and the evolution of the bilateral body plan: an old body became a young brain. *BioEssays* **242**: 185–191.

Meinhardt, H. 2006. Primary body axes of vertebrates: generation of a near-Cartesian coordinate system and the role of Spemann-type organizer. *Developmental Dynamics* 235: 2907–2919.

Meldal, B.H.M., Debenham, N.J., De Ley, P., Tandingan De Ley, I., Vanfleteren, J.R., Vierstraete, A.R. *et al.* 2007. An improved molecular phylogeny of the Nematoda with special emphasis on marine taxa. *Molecular Phylogenetics and Evolution* 42: 622–636.

Mellon, D. 1992. Connective tissue and supporting structures, pp. 77–116 in F.W. Harrison and A.G. Humes (eds), *Microscopic Anatomy of Invertebrates. Vol. 10. Decapod Crustacea*. Wiley-Liss, New York.

Melone, G., and Ferraguti, M. 1999. Rotifera, pp. 157–169 in B.G.M. Jamieson (ed.), *Reproductive Biology of Invertebrates. Vol. IX Part A. Progress in Male Gamete Biology*. Oxford and IBH Publishing Co., New Delhi.

Mendel, G. 1866. Versuche über Pflanzenhybriden. *Verhandlungen des naturforschenden Vereins in Brünn* 4: 3–47 [in German].

Mendoza, L., Ajello, L., and Taylor, J.W. 2001. The taxonomic status of *Lacazia loboi* and *Rhinosporidium seeberi* has been finally resolved with use of molecular tools. *Revista Iberoamericana de Micología* 18: 95–98.

Metschnikoff, E. 1865. Ueber einige wenig bekannte niedere Thierformen. *Zeitschrift für wissenschaftliche Zoologie* 15: 450–463 [in German].

Metschnikoff, E. 1886. *Embryologische Studien an Medusen: ein Beitrag zur Genealogie der primitiven Organe*. Hölder, Wien [in German].

Meyer, K. and Bartolomaeus, T., 1996. Ultrastructure and formation of the hooked setae in *Owenia fusiformis* delle Chiaje, 1842: implications for annelid phylogeny. *Canadian Journal of Zoology* 74: 2143–2153.

Michod, R.E. and Roze, D. 2001. Cooperation and conflict in the evolution of multicellularity. *Heredity* 86: 1–7.

Mickoleit, G. 2004. *Phylogenetische Systematik der Wirbeltiere*. Pfeil, München [in German].

Minelli, A. 1975. Cell contacts and pattern formation. *Bollettino di Zoologia* 42: 381–393.

Minelli, A. 1981. Of locomotion in terrestrial planarians. *Bollettino di Zoologia* 48: 41–50.

Minelli, A. 1995. Body cavities and body segmentation: problems of homology and phylogenetic reconstruction, pp. 69–73 in G. Lanzavecchia, R. Valvassori and M.D. Candia Carnevali (eds), *Body Cavities: Function and Phylogeny*. Mucchi, Modena.

Minelli, A. 1996. Segments, body regions and the control of development through time. *Memoirs of the California Academy of Sciences* 20: 55–61.

Minelli, A. 1998. Molecules, developmental modules and phenotypes: a combinatorial approach to homology. *Molecular Phylogenetics and Evolution* 9: 340–347.

Minelli, A. 2000a. Holomeric vs. meromeric segmentation: a tale of centipedes, leeches, and rhombomeres. *Evolution & Development* 2: 35–48.

Minelli, A. 2000b. Limbs and tail as evolutionarily diverging duplicates of the main body axis. *Evolution & Development* 2: 157–165.

Minelli, A. 2000c. The ranks and the names of species and higher taxa, or, a dangerous inertia of the language of natural history, pp. 339–351 in M.T. Ghiselin and A.E. Leviton (eds), *Cultures and Institutions of Natural History. Essays in the History and Philosophy of Science*. California Academy of Sciences, San Francisco, CA.

Minelli, A. 2001. A three-phase model of arthropod segmentation. *Development, Genes and Evolution* 211: 509–521.

Minelli, A. 2003. *The Development of Animal Form: Ontogeny, Morphology, and Evolution*. Cambridge University Press, Cambridge.

Minelli, A. 2004. Bits and pieces. *Science* 306: 1693–1694.

Minelli, A. 2005a. A segmental analysis of the beetle antenna. *Studi Trentini, Acta Biologica* 81 (2004): 91–101.

Minelli, A. 2005b. A morphologist's perspective on terminal growth and segmentation. *Evolution & Development* 7: 568–573.

Minelli, A. 2007. Invertebrate taxonomy and evolutionary developmental biology. *Zootaxa* 1668: 55–60.

Minelli, A. 2009. *Forms of Becoming*. Princeton University Press, Princeton, NJ.

Minelli, A. and Bortoletto, S. 1988. Myriapod metamerism and arthropod segmentation. *Biological Journal of the Linnean Society* 33: 323–343.

Minelli, A. and Fusco, G. 1995. Body segmentation and segment differentiation: the scope for heterochronic change, pp. 49–63 in K.J. McNamara (ed.), *Evolutionary Change and Heterochrony*. Wiley, London.

Minelli, A. and Fusco, G. 2004. Evo-devo perspectives on segmentation: model organisms, and beyond. *Trends in Ecology & Evolution* 19: 432–429.

Minelli, A. and Fusco, G. 2005. Conserved versus innovative features in animal body organization. *Journal of Experimental Zoology (Molecular and Developmental Evolution)* 304B: 520–525.

Minelli, A. and Fusco, G. (eds) 2008. *Evolving Pathways: Key Themes in Evolutionary Developmental Biology*. Cambridge University Press, Cambridge.

Minelli, A., Fusco, G. and Sartori, S. 1991. Self-similarity in biological classification. *BioSystems* 26: 89–97

Minelli, A., Brena, C., Deflorian, G., Maruzzo, D., and Fusco, G. 2006. From embryo to adult—beyond the conventional periodization of arthropod development. *Development, Genes and Evolution* 216: 373–383.

Minelli, A., Negrisolo, E., and Fusco, G. 2007. Reconstructing animal phylogeny in the light of evolutionary developmental biology, pp. 177–190 in T.R. Hodkinson, J.A.N. Parnell and S. Waldren (eds), *Reconstructing the Tree of Life: Taxonomy and Systematics of Species Rich Taxa*. Systematics Association Special Series, vol. 72. Taylor and Francis, CRC Press, Boca Raton, FL.

Miner, B.G., McEdward, L.A., and McEdward, L.R. 2005. The relationship between egg size and the diversity of the facultative feeding period in marine invertebrate larvae. *Journal of Experimental Marine Biology and Ecology* 321: 135–144.

Mineta, K., Nakazawa, M., Cebrià, F., Ikeo, K., Agata, K., and Gojobori, T. 2003. Origin and evolutionary process of the CNS elucidated by comparative genomics analysis of planarian ESTs. *Proceedings of the National Academy of Sciences USA* 100: 7666–7671.

Minguillón, C., Gardeneyes, J., Serra, E. *et al.* 2005. No more than 14: the end of the amphioxus Hox cluster. *International Journal of Biological Sciences* 1: 19–23.

Mirsky, A.E. and Ris, H. 1951. The desoxyribonucleic acid content of animal cells and its evolutionary significance. *Journal of General Physiology* 34: 451–462.

Misof, B., Niehuis, O., Bischoff, I. *et al.* 2007. Towards an 18S phylogeny of hexapods: accounting for group-speficif character covariance in optimized mixed nucleotide/doublet models. *Zoology* 110: 409–429.

Mitman, G. and Fausto-Sterling, A. 1992. Whatever happened to Planaria? C.M. Child and the physiology of inheritance, pp. 172–197 in A.E. Clarke and J.H. Fujimura (eds), *The Right Tool for the Right Job: at Work in Twentieth-Century Life Sciences*. Princeton University Press, Princeton, NJ.

Mittman, B. and Scholtz, G. 2003. Development of the nervous system in the "head" of *Limulus polyphemus* (Chelicerata: Xiphosura): morphological evidence for a correspondence between the segments of the chelicerae and of the (first) antennae of Mandibulata. *Development, Genes and Evolution* 213: 9–17.

Moczek, A.P. 2006. Integrating micro- and macroevolution of development through the study of horned beetles. *Heredity* 97: 168–178.

Moczek, A.P. and Nagy, L.M. 2005. Diverse developmental mechanisms contribute to different levels of diversity in horned beetles. *Evolution & Development* 7: 175–185.

Moczek, A.P., Rose, D., Sewell, W., and Kesselring, B.R. 2006. Conservation, innovation, and the evolution of horned beetle diversity. *Development, Genes and Evolution* 216: 655–665.

Møller, A.P. and Swaddle, J.P. 1997. *Asymmetry, Developmental Stability, and Evolution*. Oxford University Press, Oxford.

Molnár, A., Schwach, F., Studholme, D.J., Thunemann, E.C., and Baulcombe, D.C. 2007. miRNAs control gene expression in the single-cell alga *Chlamydomonas reinhardtii*. *Nature* 447: 1126–1129.

Monteiro, A., Brakefield, P.M., and French, V. 1997. Butterfly eyespots: the genetics and development of the color rings. *Evolution* 51: 1207–1216.

Monteiro, A.S., Schierwater, B., Dellaporta, S.L., and Holland, P.W.H. 2006. A low diversity of ANTP class homeobox genes in Placozoa. *Evolution & Development* 8: 174–182.

Montgelard, C., Catzeflis, F., and Douzery, E. 1997. Phylogenetic relationships of artiodactyls and cetaceans as deduced from the comparison of cytochrome b and 12S rRNA mitochondrial sequences. *Molecular Biology and Evolution* 14: 550–559.

Mooi, R. and David, B. 1997. Skeletal homologies of echinoderms. *Paleontological Society Papers* 3: 305–335.

Moon, R.T., Bowerman, B., Boutros, M., and Perrimon, N. 2002. The promise and perils of Wnt signaling through β-catenin. *Science* 296: 1644–1646.

Moore, J. and Willmer, P. 1997. Convergent evolution in invertebrates. *Biological Reviews* 72: 1–60.

Morris, D.J., Terry, R.S., and Adams, A. 2005. Development and molecular characterisation of the microsporidian *Schroedera airthreyi* n. sp. in a freshwater bryozoan *Plumatella* sp. (Bryozoa: Phylactolaemata). *Journal of Eukaryotic Microbiology* 52: 31–37.

Moshel, S.M., Levine, M., and Collier, J.R. 1998. Shell differentiation and *engrailed* expression in the *Ilyanassa* embryo. *Development, Genes and Evolution* 208: 135–141.

Mossel, E. and Steel, M. 2004. A phase transition from a random cluster model on phylogenetic trees. *Mathematical Biosciences* 187: 189–203.

Mossel, E. and Steel, M. 2005. How much can evolved characters tell us about the tree that generated them?, pp. 384–412 in O. Gascuel (ed.), *Mathematics of Evolution and Phylogeny*. Oxford University Press, Oxford.

Moussian, B., Schwarz, H., Bartoszewski, S., and Nüsslein-Volhard, C. 2005. Involvement of chitin in exoskeleton morphogenesis in *Drosophila melanogaster*. *Journal of Morphology* 264: 117–130.

Mukai, H., Terakado, K., and Reed, C.G. 1997. Bryozoa, pp. 45–206 in F.W. Harrison and R.M. Woollacott (eds), *Microscopic Anatomy of Invertebrates. Vol. 13. Lophophorates, Entoprocta, and Cycliophora*. Wiley-Liss, New York.

Müller, C.H.G., Rosenberg, J., Richter, S., and Meyer-Rochow, V.B. 2003. The compound eye of *Scutigera coleoptrata* (Linnaeus, 1758) (Chilopoda: Notostigmophora): an ultrastructural reinvestigation that adds support to the Mandibulata concept. *Zoomorphology* 122: 191–209.

Müller, G.B. 2003. Homology: the evolution of morphological organization, pp. 51–69 in G.B. Müller and S.A. Newman (eds), *Origination of Organismal Form. Beyond the Gene in Developmental and Evolutionary Biology*, MIT Press, Cambridge, MA.

Müller, G.B. 2008. Evo-devo as a discipline, pp. 5–30 in A. Minelli and G. Fusco (eds), *Evolving Pathways. Key Themes in Evolutionary Developmental Biology*. Cambridge University Press, Cambridge.

Müller, G.B. and Newman, S.A. 1999. Generation, integration, autonomy: three steps in the evolution of homology, pp. 65–73 in G.R. Bock and G. Cardew (eds), *Homology*. Wiley, Chichester.

Müller, G.B. and Newman, S.A. 2003. *Origination of Organismal Form. Beyond the Gene in Developmental and Evolutionary Biology*, MIT Press, Cambridge, MA.

Müller, J. and Reisz, R.R. 2005. Four well-constrained calibration points from the vertebrate fossil record for molecular clock estimates. *BioEssays* 27: 1069–1075.

Müller, M.C.M. 2004. Nerve development, growth and differentiation during regeneration in *Enchytraeus fragmentosus* and *Stylaria lacustris* (Oligochaeta). *Development, Growth and Differentiation* 46: 471–478.

Müller, M.C.M. 2006. Polychaete nervous systems: ground pattern and variations—cLS microscopy and the importance of novel characteristics in phylogenetic analysis. *Integrative and Comparative Biology* 46: 125–133.

Müller, M.C.M. and Westheide, W. 2000. Structure of the nervous system of *Myzostoma cirriferum* (Annelida) as revealed by immunohistochemistry and cLSM analyses. *Journal of Morphology* 245: 87–98.

Müller, M.C.M. and Henning, L. 2004. Ground plan of the polychaete brain: I. Pattern of nerve development during regeneration in *Dorvillea bermudensis* (Dorvilleidae). *Journal of Comparative Neurology* 471: 49–58.

Müller, W.E.G. 1998. Molecular phylogeny of Eumetazoa: experimental evidence for monophyly of animals based on genes in sponges (Porifera). *Progress in Molecular and Subcellular Biology* 19: 89–132.

Müller, W.E.G. 2001. How was metazoan threshold crossed? The hypothetical Urmetazoa. *Comparative Biochemistry and Physiology—Part A: Molecular and Integrative Physiology* 129: 433–460.

Müller, W.E.G. 2003. The origin of metazoan complexity: Porifera as integrated animals. *Integrative and Comparative Biology* 43: 3–10.

Müller, W.E.G. 2005. Spatial and temporal expression patterns in animals, pp. 269–309 in R.A. Myers (ed.), *Encyclopedia of Molecular Cell Biology and Molecular Medicine*, vol. 13. Wiley-VCH, Weinheim.

Müller, W.E.G. and Müller, I.M. 2003a. The urmetazoa: molecular biological studies with living fossils—Porifera, pp. 99–104 in A. Legakis, S. Sfenthourakis, R. Polymeni, and M. Thessalou-Legaki (eds), *The New Panorama of Animal Evolution*. Proceedings of the 18th International Congress of Zoology. Pensoft, Sofia/Moscow.

Müller, W.E.G. and Müller, I.M. 2003b. Origin of the metazoan immune system; identification of the moleculaes and their function in sponges. *Integrative and Comparative Biology* 43: 281–292.

Müller, W.E.G., Blumbach, B., and Müller, I.M. 1999. Evolution of the innate and adaptive immune systems: relationships between potential immune molecules in the lowest metazoan phylum (Porifera) and those in vertebrates. *Transplantation* 68: 1215–1227.

Müller, W.E.G., Wiens, M., Adell, T., Gamulin, V., Schroder, H.C., and Müller, I.M. 2004. Bauplan of Urmetazoa: basis for genetic complexity of Metazoa. *International Review of Cytology* 235: 53–92.

Mumm, J.S. and Kopan, R. 2000. Notch signaling: from the outside in. *Developmental Biology* 228: 151–165.

Mundel, P. 1979. The centipedes (Chilopoda) of the Mazon Creek, pp. 361–378 in M.H. Nitecki (ed.), *Mazon Creek Fossils*. Academic Press, New York

Mundy, N.I., Badcock, N.S., Hart, T., Scribner, K., Janssen, K., and Nadeau, N.J. 2004. Conserved genetic basis of a quantitative plumage trait involved in mate choice. *Science* 303: 1870–1873.

Muricy, G. and Díaz, M.C. 2002. Order Homosclerophorida Dendy, 1905, Family Plakinidae Schulze, 1880, pp. 71–84 in J.N.A. Hooper and R.W.M. van Soest (eds), *Systema Porifera. A Guide to the Classification of Sponges*. Kluwer Academics/Plenum Publishers, New York.

Murphy, W.J., Elzirik, E., Johnson, W.E., Zhang, Y.P., Ryder, O.A., and O'Brien, S.J. 2001. Molecular phylogenetics and the origins of placental mammals. *Nature* 409: 614–618.

Nagy, L.M. and Carroll, S. 1994. Conservation of wingless patterning functions in the short-germ embryos of *Tribolium castaneum*. *Nature* 367: 460–463.

Nam, J. and Nei, M. 2005. Evolutionary change of the numbers of homeobox genes in bilateral animals. *Molecular Biology and Evolution* 22: 2386–2394.

Nardi, F., Carapelli, A., Fanciulli, P.P., Dallai, R., and Frati, F. 2001. The complete mitochondrial DNA sequence of the basal hexapod *Tetrodontophora bielanensis*: evidence for heteroplasmy and tRNA translocations. *Molecular Biology and Evolution* 18: 1293–1304.

Nardi, F., Spinsanti, G., Boore, J.L., Carapelli, A., Dallai, R., and Frati, F. 2003. Hexapod origins: monophyletic or paraphyletic? *Science* 299: 1887–1889.

Near, T.J., Garey, J.R., and Nadler, S.A. 1998. Phylogenetic relationships of the Acanthocephala inferred from 18S ribosomal DNA sequences. *Molecular Phylogenetics and Evolution* 10: 287–298.

Nebelsick, M. 1993. Introvert, mouth cone, and nervous system of *Echinoderes capitatus* (Kinorhyncha, Cyclorhagida) and implications for the phylogenetic relationships of Kinorhyncha. *Zoomorphology* 113: 211–232.

Nederbragt, A.J., van Loon, A.E., and Dictus, W.J.A.G. 2002. Expression of *Patella vulgata* orthologs of *engrailed* and *dpp-BMP2/4* in adjacent domains during molluscan shell development suggests a conserved compartment boundary mechanism. *Developmental Biology* 246: 341–355.

Negrisolo, E., Minelli, A., and Valle, G. 2004. The mitochondrial genome of the house centipede *Scutigera* and the monophyly versus paraphyly of myriapods. *Molecular Biology and Evolution* 21: 770–780.

Nekka, F., Kyriacos, S., Kerrigan, C., and Cartilier, L. 1996. A model of growing vascular structures. *Bulletin of Mathematical Biology* 58: 409–424.

Nelson, G. 1978. Ontogeny, phylogeny, paleontology, and the biogenetic law. *Systematic Zoology* 27: 324–345.

Neuhaus, B., Kristensen, R.M., and Lemburg, C. 1996. Ultrastructure of the cuticle of Nemathelminthes and electron microscopical localization of chitin. *Verhandlungen der Deutschen Zoologischen Gesellschaft* 89: 221.

Neuhaus, B., Bresciani, J., and Peters, W. 1997a. Ultrastructure of the pharyngeal cuticle and lectin labelling with wheat germ agglutinin-gold conjugate indicate chitin in the pharyngeal cuticle of *Oesophagostomum dentatum* (Strongylida, Nematoda). *Acta Zoologica* 78: 205–213.

Neuhaus, B., Kristensen, R.M., and Peters, W. 1997b. Ultrastructure of the cuticle of Loricifera and demonstration of chitin using gold-labelled wheat germ agglutinin. *Acta Zoologica* 78: 215–225.

Neumann, C.J. and Nüsslein-Volhard, C. 2000. Patterning of the zebrafish retina by a wave of *sonic hedgehog* activity. *Science* 289: 2137–2139.

Newman, S.A. 1993. Is segmentation generic? *Bioessays* 15: 277–283.

Newman, S.A. 2003. Hierarchy, pp. 169–174 in B.K. Hall and W.M. Olson (eds), *Keywords and Concepts in Evolutionary Developmental Biology*. Harvard University Press, Cambridge, MA.

Newman, S.A. 2005. The pre-Mendelian, pre-Darwinian world: shifting relations between genetic and epigenetic mechanisms in early multicellular evolution. *Journal of Biosciences* 30: 75–85.

Newman, S.A. and Müller, G.B. 2000. Epigenetic mechanisms of character origination. *Journal of Experimental Zoology (Molecular and Developmental Evolution)* 288: 304–317.

Nezlin, L.P. and Yushin, V.V. 2004. Structure of the nervous system in the tornaria larva of *Balanoglossus proterogonius* (Hemichordata: Enteropneusta) and its phylogenetic implications. *Zoomorphology* 123: 1–13.

Nichols, S.A. 2005. An evaluation of support for order-level monophyly and interrelationships within the class Demospongiae using partial data from the large subunit rDNA and cytochrome oxidase subunit I. *Molecular Phylogenetics and Evolution* 34: 81–96.

Nicol, J.A.C. 1952. Giant axons and synergic contractions in *Branchiomma vesiculosum*. *Journal of Experimental Biology* 28: 22–31.

Nielsen, C. 1971. Entoproct life-cycles and the entoproct/ ectoproct relationship. *Ophelia* 9: 209–341.

Nielsen, C. 1985. Animal phylogeny in the light of the trochaea theory. *Biological Journal of the Linnean Society* 25: 243–299.

Nielsen, C. 1987. Structure and function of metazoan ciliary bands and their phylogenetic significance. *Acta Zoologica* 68: 205–262.

Nielsen, C. 1991. The development of the brachiopod *Crania* (*Neocrania*) *anomala* (O.F. Müller) and its phylogenetic significance. *Acta Zoologica* 72: 7–28.

Nielsen, C. 1995a. *Animal Evolution: Interrelationships of the Living Phyla*. Oxford University Press, Oxford.

Nielsen, C. 1995b. Origin and evolution of animal life cycles. *Biological Reviews* 73: 125–155.

Nielsen, C. 2001. *Animal Evolution: Interrelationships of the Living Phyla*, 2nd edn. Oxford University Press, Oxford.

Nielsen, C. 2002. The phylogenetic position of Entoprocta, Ectoprocta, Phoronida, and Brachiopoda. *Integrative and Comparative Biology* 42: 685–691.

Nielsen, C. 2003a. Defining phyla: morphological and molecular clues to metazoan evolution. *Evolution & Development* 5: 386–393.

Nielsen, C. 2003b. Proposing a solution to the Articulata-Ecdysozoa controversy. *Zoologica Scripta* 32: 475–482.

Nielsen, C. 2004. Trochophora larvae: cell-lineages, ciliary bands, and body regions. 1. Annelida and

Mollusca. *Journal of Experimental Biology (Molecular and Developmental Evolution)* 302B: 35–68.

Nielsen, C. 2005a. Larval and adult brains. *Evolution & Development* 7: 483–489.

Nielsen, C. 2005b. Trochophora larvae and adult body regions in annelids: some conclusions. *Hydrobiologia* 535/536: 23–24.

Nielsen, C. 2008. *Ontogeny of the spiralian brain*, pp. 399–416 in A. Minelli and G. Fusco (eds), *Evolving Pathways: Key Themes in Evolutionary Developmental Biology*. Cambridge University Press, Cambridge.

Nielsen, C. and Meier, R. 2002. What cell lineages tell us about the evolution of Spiralia remains to be seen. *Evolution* 56: 2554–2557.

Nielsen, C. and Nørrevang, A. 1985. The trochaea theory: an example of life cycle phylogeny, pp. 28–41 in S. Conway Morris, J.D. George, R. Gibson and H.M. Platt (eds), *The Origins and Relationships of Lower Invertebrates*. Clarendon Press, Oxford.

Nielsen, M.G., Wilson, K.A., Raff, E.C., and Raff, R.A. 2000. Novel gene expression patterns in hybrid embryos between species with different modes of development. *Evolution & Development* 2: 133–144.

Nieto, M.A. 2002. The Snail superfamily of zinc-finger transcription factors. *Nature Reviews Molecular Cell Biology* 3: 155–166.

Nijhout, H.F. 1990. Metaphors and the role of genes in development. *BioEssays* 12: 441–446.

Nijhout, H.F. and Wheeler, D.E. 1996. Growth models of complex allometries in holometabolous insects. *American Naturalist* 148: 40–56.

Nikolei, E. 1961. Vergleichende Untersuchungen zur Fortpflanzung der heterogenen Gallmücken unter experimentellen Bedingungen. *Zeitschrift für Morphologie und Ökologie der Tiere* 50: 281–329 [in German].

Nilsson, D. 1912. Beiträge zur Kenntnis des Nervensystems der Polychaeten. *Zoologiska Bidrag från Uppsala* 1: 85–161 [in German].

Nilsson, D.-E. and Pelger, S. 1994. A pessimistic estimate of the time required for an eye to evolve. *Proceedings of the Royal Society of London Series B, Biological Sciences* 256: 53–58.

Nishihara, H., Hasegawa, M., and Okada, N. 2006. Pegasoferae, an unexpected mammalian clade revealed by tracking ancient retroposon insertion. *Proceedings of the National Academy of Sciences USA* 103: 9929–9934.

Niswander, L. 2003. Pattern formation: old models out on a limb. *Nature Reviews Genetics* 4: 133–143.

Nitsche, H. 1869. *Beiträge zur Kenntnis der Bryozoen, I. Beobachtungen über die Entwicklungsgeschichte einiger chilostomen Bryozoen*. W. Engelmann, Leipzig [in German].

Niwa, N., Hiromi, Y., and Okabe, M. 2004. A conserved developmental program for sensory organ formation in *Drosophila melanogaster*. *Nature Genetics* 36: 293–297.

Niwa, R. and Slack, F.J. 2007. The evolution of animal microRNA function. *Current Opinion in Geentics and Development* 17: 145–150.

Nohara, M., Nishida, M., Miya, M., and Nishikawa, T. 2005. Evolution of the mitochondrial genome in Cephalochordata as inferred from complete nucleotide sequences from two *Epigonichthys* species. *Journal of Molecular Evolution* 60: 526–537.

Nolo, R., Abbott, L.A., and Bellen, H.J. 2000. Senseless, a Zn finger transcription factor, is necessary and sufficient for sensory organ development in *Drosophila*. *Cell* 102: 349–362.

Norbeck, B.A. and Denburg, J.L. 1991. Pattern formation during insect leg segmentation: studies with a prepattern of a cell surface antigen. *Roux's Archives of Developmental Biology* 199: 476–491.

Nordström, K., Wallén, R., Seymour, J., and Nilsson, D. 2003. A simple visual system without neurons in jellyfish larvae. *Proceedings of the Royal Society of London Series B, Biological Sciences* 270: 2349–2354.

Norén, M. and Jondelius, U. 1997. *Xenoturbella's* molluscan relatives. *Nature* 390: 31–32.

Noto, T. and Endoh, H. 2004. A "chimera" theory on the origin of dicyemid mesozoans: evolution driven by frequent lateral gene transfer from host to parasite. *BioSystems* 73: 73–83.

Nulsen, C. and Nagy, L.M. 1999. The role of wingless in the development of multibranched crustacean limbs. *Developmental Genes and Evolution* 209: 340–348.

Oakley, T.H. 2003. On homology of arthropod compound eyes. *Integrative and Comparative Biology* 43: 522–530.

Obst, M., Funch, P., and Kristensen, R.M. 2006. A new species of Cycliophora from the mouthparts of the American lobster, *Homarus americanus* (Nephropidae, Decapoda). *Organisms Diversity and Evolution* 6: 83–97.

Ockelmann, K.W. and Vahl, O. 1970. On the biology of the polychaete *Glycera alba*, especially its burrowing and feeding. *Ophelia* 8: 275–294.

Oda, H., Wada, H., Tagawa, K. *et al.* 2002. A novel amphioxus cadherin that localizes to epithelial adherens junctions has an unusual domain organization with implications for chordate phylogeny. *Evolution & Development* 4: 426–434.

Ogura, A., Ikeo, K., and Gojobori, T. 2004. Comparative analysis of gene expression for convergent evolution of camera eye between octopus and human. *Genome Research* 14: 1555–1561.

Ohno, S. 1970. *Evolution by Gene Duplication*. Springer, New York.

Okazaki, R.K., Snyder, M.J., Grimm, C.C., and Chang, E.S. 1998. Ecdysteroids in nemerteans: further characterization and identification. *Hydrobiologia* 365: 281–285.

Okuda, S. 1940. Metamorphosis of a pycnogonid parasitic in a hydromedusa. *Journal of the Faculty of Science Hokkaido University Series VI Zoology* 7: 73–86.

Okumura, T., Matsumoto, A., Tanimura, T., and Muratami, R. (2005) An endoderm-specific GATA factor gene, *dGATAe*, is required for the terminal differentiation of the *Drosophila* endoderm. *Developmental Biology* 278: 576–586.

Olson, E.N. 2006. Gene regulatory networks in the evolution and development of the heart. *Science* 313: 1922–1927.

Olson, L.E., Sargis, E.J. and Martin, R.D. 2005. Intraordinal phylogenetics of treeshrews (Mammalia: Scandentia) based on evidence from the mitochondrial 12S rRNA gene. *Molecular Phylogenetics and Evolution* 35: 656–673.

Olsson, L., Ericsson, R. and Cerny, R. 2005. Vertebrate head development: Segmentation, novelties, and homology. *Theory in Biosciences* 124: 145–163.

Orrhage, L. 1995. On the innervation and homologues of the anterior end appendages of the Eunicea (Polychaeta), with a tentative outline of the fundamental constitution of the cephalic nervous system of the polychaetes. *Acta Zoologica* 76: 229–248.

Orrhage, L. and Müller, M.C.M. 2005. Morphology of the nervous system of Polychaeta (Annelida). *Hydrobiologia* 535/536: 79–111.

Oyama, S. 2000. *The Ontogeny of Information*, 2nd edn. Duke University Press, Durham, NC.

Oyama, S., Griffiths, P.E., and Gray, R.D. 2001. *Cycles of Contingency: Developmental Systems and Evolution*. MIT Press, Cambridge, MA.

Packard, A. 1968. Asexual reproduction in *Balanoglossus* (Stomochordata). *Proceedings of the Royal Society of London Series B, Biological Sciences* 171: 261–272.

Page, D.T. 2004. A mode of arthropod brain evolution suggested by *Drosophila* commissure development. *Evolution & Development* 6: 25–31.

Page, L.R. 2002. Ontogenetic torsion in two basal gastropods occurs without shell attachments for larval retractor muscles. *Evolution & Development* 4: 212–222.

Page, L.R. 2003. Gastropod ontogenetic torsion: developmental remnants of an ancient evolutionary change in body plan. *Journal of Experimantal Zoology (Molecular Developmental Evolution)* 297B: 11–26.

Page, L.R. 2006a. Early differentiating neuron in larval abalone (*Haliotis kamtschatkana*) reveals the relationships between ontogenetic torsion and the crossing off the pleuroviscceral nerve cord. *Evolution & Development* 8: 458–467.

Page, L.R. 2006b. Modern insights on gastropod development: reevaluation of the evolution of a novel body plan. *Integrative and Comparative Biology* 46: 134–143.

Pagel, M. 2004. Limpets break Dollo's law. *Trends in Ecology & Evolution* 19: 278–280.

Palmeirim, I., Henrique, D., Ish-Horowicz, D., and Pourquié, O. 1997. Avian *hairy* gene expression identifies a molecular clock linked to vertebrate segmentation and somitogenesis. *Cell* 91: 639–648.

Palmer, A. R. 1996. From symmetry to asymmetry: phylogenetic patterns of asymmetry variation in animals and their evolutionary significance. *Proceedings of the National Academy of Sciences USA* 93: 14279–14286.

Palmer, A.R. 2000. Quasireplication and the contract of error: lessons from sex ratios, heritabilities and fluctuating asymmetry. *Annual Review of Ecology and Systematics* 31: 441–480.

Pancer, Z., Kruse, M., Müller, I.M., and Müller, W.E.G. 1997. On the origin of metazoan adhesion receptors: cloning of integrin α subunit from the sponge *Geodia cydonium*. *Molecular Biology and Evolution* 14: 391–398.

Pandian, T.J. 1994. Arthropoda-Crustacea, pp. 39–166 in K.G. Adiyodi and R.G. Adiyodi (eds), *Reproductive Biology of Invertebrates, Vol. VI, Part B, Asexual Propagation and Reproductive Strategies*. Wiley, Chichester.

Panganiban, G., Irvine, S.M., Lowe, C. *et al.* 1997. The origin and evolution of animal appendages. *Proceedings of the National Academy of Sciences USA* 94: 5162–5166.

Papillon, D., Perez, Y., Fasano, L., Le Parco, Y. and Caubit, X. 2003. Hox gene survey in the chaetognath *Spadella cephaloptera*: evolutionary implications. *Development, Genes and Evolution* 213: 142–148.

Papillon, D., Perez, Y., Caubit, X., and Le Parco, Y. 2004. Identification of chaetognaths as protostomes is supported by the analysis of their mitochondrial genome. *Molecular Biology and Evolution* 21: 2122–2129.

Parsley, R.L. and Zhao, Y.L. 2006. Long stalked eocrinoids in the basal Middle Cambrian Kaiuli Biota, Taijaing County, Guizhou Province, China. *Journal of Paleontology* 80: 1058–1071.

Pasquinelli, A.E., Reinhard, B.J., Slack, F. *et al.* 2000. Conservation of the sequence and temporal expression of *let-7* heterochronic regulatory RNA. *Nature* 408: 86–89.

Pasquinelli, A.E., McCoy, A., Jiménez, E. *et al.* 2003. Expression of the 22 nucleotide *let-7* heterochronic RNA throughout the Metazoa: a role in life history evolution? *Evolution & Development* 5: 372–378.

Passamaneck, Y.J. and Halanych, K.M. 2004. Evidence from *Hox* genes that bryozoans are lophotrochozoans. *Evolution & Development* 6: 275–281.

Passamaneck, Y. and Halanych, K.M. 2006. Lophotrochozoan phylogeny assessed with LSU and SSU data: evidence of lophophorate polyphyly. *Molecular Phylogenetics and Evolution* 40: 20–28.

Passamaneck, Y.J., Schander, C., and Halanych, K.M. 2004. Investigation of molluscan phylogeny using large-subunit and small-subunit nuclear rRNA sequences. *Molecular Phylogenetics and Evolution* 32: 25–38.

Patel, N.H., Condron, B.G., and Zinn, K. 1994. Pair-rule expression patterns of *even-skipped* are found in both short-and long-germ beetles. *Nature* 367: 429–434.

Patterson, C. 1982. Morphology characters and homology, pp. 21–74 in K.A. Joysey and A.E. Friday (eds), *Problems of Phylogenetic Reconstruction*. Academic Press, London.

Patterson, C. (ed.) 1987. *Molecules and Morphology in Evolution: Conflict or Compromise?* Cambridge University Press, Cambridge.

Paululat, A., Holz, A., and Renkawitz-Pohl, R. 1999. Essential genes for myoblast fusion in *Drosophila* embryogenesis. *Mechanisms of Development* 83: 17–26.

Paulus, H.F. 2004. Einiges zur Stammesgeschichte der Spinnentiere (Arthropoda, Chelicerata). *Denisia* 12: 547–574 [in German].

Pawlowski, J., Montoya-Burgos, J.I., Fahrni, J.F., Wuest, J., and Zaninetti, L. 1996. Origin of the Mesozoa inferred from 18S rRNA gene sequences. *Molecular Biology and Evolution* 13: 1128–1132.

Pawson, D.L. 2007. Phylum Echinodermata, pp. 749–764 in Z.-Q. Zhang and W.A. Shear (eds), *Linnaeus Tercentenary: Progress in Invertebrate Taxonomy (Zootaxa 1668)*. Magnolia Press, Auckland.

Pearse, V.B. 2002. Prodigies of propagation: the many modes of clonal replication in boloceroidid sea anemones (Cnidaria, Anthozoa, Actiniaria). *Invertebrate Reproduction and Development* 41: 201–213.

Pearson, J.C., Lemons, D., and McGinnis, W. 2005. Modulating Hox gene functions during animal body patterning. *Nature Reviews Genetics* 6: 893–904.

Peel, A. 2004. The evolution of arthropod segmentation mechanisms. *BioEssays* 26: 1108–1116.

Peel, J.S. 2006. Scaphopodization in Palaeozoic molluscs. *Palaeontology* 49: 1357–1364.

Penners, A. 1922. Die Furchung von *Tubifex rivulorum* Lam. *Zoologische Jahrbücher, Abteilung für Anatomie* 43: 323–368 [in German].

Penners, A. 1930. Entwicklungsgeschichtliche Untersuchungen an marinen Oligochäten. II. Furchung, Keimstreif und Keimbahn von *Pachydrilus (Lumbricillus) lineatus* Müll. *Zeitschrift für Wissenschaftliche Zoologie* 137: 55–119 [in German].

Pentreath, V.W. and Cottrell, G.A. 1970. The blood supply to the central nervous system of *Helix pomatia*. *Zeitschrift für Zellforschung* 111: 160–178.

Perez, Y., Casanova, J.P., and Mazza, J. 2001. Degrees of vacuolation of the absorptive intestinal cells of five *Sagitta* (Chaetognatha) species: possible ecophysiological implications. *Marine Biology* 138: 125–133.

Perseke, M., Hankeln, T., Weich, B., Fritsch, G., Stadler, P.F., Israelsson, O. *et al.* 2007. The mitochondrial DNA of *Xenoturbella bocki*: genomic architecture and phylogenetic analysis. *Theory in Biosciences* 126: 35–42.

Petersen, K. 1979. Development of coloniality in Hydrozoa, pp. 105–139 in G. Larwood and B.R. Rosen (eds), *Biology and Systematics of Colonial Organisms*. Academic Press, London.

Peterson, K.J. 2004. Isolation of Hox and Parahox genes in the hemichordate *Ptychodera flava* and the evolution of deuterostome Hox genes. *Molecular Phylogenetics and Evolution* 31: 1208–1215.

Peterson, K.J. and Davidson, E.H. 2000. Regulatory evolution and the origin of the bilaterians. *Proceedings of the National Academy of Sciences USA* 97: 4430–4433.

Peterson, K.J. and Eernisse, D.J. 2001. Animal phylogeny and the ancestry of bilaterians: inferences from morphology and 18S rDNA gene sequences. *Evolution & Development* 3: 170–205.

Peterson, K.J. and Butterfield, N.J. 2005. Origin of the Eumetazoa: testing ecological predictions of molecular clocks against the Proterozoic fossil record. *Proceedings of the National Academy of Sciences USA* 102: 9547–9552.

Peterson, K.J., Cameron, R.A., and Davidson, E.H. 1997. Set-aside cells in maximal indirect development: evolutionary and developmental significance. *BioEssays* 19: 623–631.

Peterson, K.J., Cameron, R.A., and Davidson, E.H. 2000a. Bilaterian origins: significance of new experimental observations. *Developmental Biology* 219: 1–17.

Peterson, K.J., Irvine, S.Q., Cameron, R.A., and Davidson, E.H. 2000b. Quantitative assessment of Hox complex expression in the indirect development of the polychaete annelid *Chaetopterus sp. Proceedings of the National Academy of Sciences USA* 97: 4487–4492.

Philip, G.K., Creevey, C.J., and McInerney, J.O. 2005. The Opisthokonta and the Ecdysozoa may not be clades: stronger support for the grouping of plant and animal than for animal and fungi and stronger support for the Coelomata than Ecdysozoa. *Molecular Biology and Evolution* 22: 1175–1184.

Philippe, H., Lartillot, N., and Brinkmann, H. 2005. Multigene analyses of bilaterian animals corroborate the monophyly of Ecdysozoa, Lophotrochozoa, and Protostomia. *Molecular Biology and Evolution* 22: 1246–1253.

Piatigorski, J. and Kozmik, Z. 2004. Cubozoan jellyfish: an Evo-Devo model for eyes and other sensory systems. *International Journal of Developmental Biology* 48: 719–729.

Pick, L. 1998. Segmentation: painting stripes from flies to vertebrates. *Developmental Genetics* 23: 1–10.

Pierce, R.J., Wu, W., Hirai, H. *et al.* 2005. Evidence for a dispersed Hox gene cluster in the platyhelminth parasite *Schistosoma mansoni*. *Molecular Biology and Evolution* 22: 2491–2503.

Pierce, S.K., Massey, S.E., Hanten, J.J., and Curtis, N.E. 2003. Horizontal transfer of functional nuclear genes between multicellulars organisms. *Biological Bulletin* 204: 237–240.

Pietsch, A. and Westheide, W. 1987. Protonephridial organs in *Myzostoma cirriferum*. *Acta Zoologica* 68: 195–203.

Pilato, G., Binda, M.G., Biondi, O. *et al.* 2005. The clade Ecdysozoa, perplexities and questions. *Zoologischer Anzeiger* 244: 43–50.

Pilger, J.F. 1993. Echiura, pp. 185–236 in F.W. Harrison and M.E. Rice (eds), *Microscopic Anatomy of Invertebrates. Vol. 12. Onychophora, Chilopoda, and Lesser Protostomata*. Wiley-Liss, New York.

Pilger, J.F. 1997. Sipunculans and echiurans, pp. 167–188 in S.F. Gilbert and A.M. Raunio (eds), *Embryology: Constructing the Organism*. Sinauer Associates, Sunderland, MA.

Pilger, J.F. 2002. Phylum Echiura, pp. 371–373 in C.M. Young (ed.), *Atlas of Marine Invertebrate Larvae*. Academic Press, London.

Piraino, S., Boero, F., Aeschbach, B., and Schmid, V. 1996. Reversing the life cycle: medusae transforming into polyps and cell transdifferentiation in *Turritopsis nutricula* (Cnidaria, Hydrozoa). *Biological Bulletin* 190: 302–312.

Pires-daSilva, A. and Sommer, R.J. 2003. The evolution of signalling pathways in animal development. *Nature Review Genetics* 4: 39–49.

Pisani, D., Poling, L.L., Lyons-Weiler, M., and Hedges, S.B. 2004. The colonization of land by animals: molecular phylogeny and divergence times among arthropods. *BMC Biology* 2: 1–10.

Podar, M., Haddock, S.H.D., Sogin, M.L., and Harbison, G.R. 2001. A molecular phylogenetic framework for the phylum Ctenophora using 18S rRNA genes. *Molecular Phylogenetics and Evolution* 21: 218–230.

Poddubnaya, L.G., Mackiewicz, J.S., Brunanska, M., and Scholz, T. 2005. Ultrastructural studies on the reproductive system of progenetic *Diplocotyle olrikii* (Cestoda,

Spathebothriidea): ovarian tissue. *Acta Parasitologica* 50: 199–207.

Podlasek, C., Houston, J., McKenna, K.E., and McVary, K.T. 2002. Posterior *Hox* gene expression in developing genitalia. *Evolution & Development* 4: 142–163.

Podsiadlowski, L. and Braband, A. 2006. The complete mitochondrial genome of the sea spider *Nymphon gracile* (Arthropoda: Pycnogonida). *BioMed Central Genomics* 7: 284.

Podsiadlowski, L., Kohlhagen, H., and Koch, M. 2007. The complete mitochondrial genome of *Scutigerella causeyae* (Myriapoda: Symphyla) and the phylogenetic position of Symphyla. *Molecular Phylogenetics and Evolution* 45: 251–260.

Podsiadlowski, L., Braband, A., and Mayer, G. 2008. The complete mitochondrial genome of the onychophoran *Epiperipatus biolleyi* reveals a unique transfer RNA set and provides further support for the Ecdysozoa hypothesis. *Molecular Biology and Evolution* 25: 42–51.

Poe, S. and Wiens, J.J. 2000. Character selection and the methodology of morphological phylogenetics, pp. 20–36 in J.J. Wiens (ed.), *Phylogenetic Analysis of Morphological Data*. Smithsonian Institution Press, Washington DC.

Pollock, D.A. and Normark, B.B. 2002. The life cycle of *Micromalthus debilis* LeConte (1878) (Coleoptera: Archostemata: Micromalthidae): historical review and evolutionary perspective. *Journal of Zoological Systematics and Evolutionary Research* 40: 105–112.

Ponder, W.F. and Lindberg, D.R. 1997. Towards a phylogeny of. gastropod molluscs: an analysis using morphological characters. *Zoological Journal of the Linnean Society* 119: 83–265.

Porchet, M., Gaillet, N., Sauber, F., Charlet, M., and Hoffmann, J.A. 1984. Ecdysteroids in annelids, pp. 346–348 in J. Hoffmann and M. Porchet (eds), *Biosynthesis, Metabolism and Mode of Action of Invertebrate Hormones*. Springer, Berlin.

Pourquié, O. 2003. A biochemical oscillator linked to vertebrate segmentation, pp. 183–194 in G.B. Müller and S.A. Newman (eds), *Origination of Organismal Form. Beyond the Gene in Developmental and Evolutionary Biology*. MIT Press, Cambridge, MA.

Powell, R. 2007. Is convergence more than an analogy? Homoplasy and its implications for macroevolutionary predictability. *Biology and Philosophy* 22: 565–578.

Powers, T.P. and Amemiya, C.T. 2004. Evidence for a Hox14 paralog group in vertebrates. *Current Biology* 14: R183–R184.

Presley, R., Horder, T.J., and Slípka, J. 1996. Lancelet development as evidence of ancestral chordate structure. *Israel Journal of Zoology* 42: 97–116.

Priess, J.R. and Hirsch, D.I. 1986. *Caenorhabditis elegans* morphogenesis: the role of the cytoskeleton in elongating the embryo. *Developmental Biology* 117: 156–173.

Prochnik, S.E., Rokhsar, D.S., and Aboobaker, A.A. 2007. Evidence for a microRNA expansion in the bilaterian ancestor. *Development, Genes and Evolution* 217: 73–77.

Prud'homme, B., Gompel, N., Rokas, A. *et al.* 2006. Repeated morphological evolution through cis-regulatory changes in a pleiotropic gene. *Nature* 440: 1050–1053.

Pulquério, M.J.F. and Nichols, R.A. 2007. Dates from the molecular clock: how wrong can we be? *Trends in Ecology & Evolution* 22: 80–84.

Punzo, C., Kurata, S., and Gehring, W.J. 2001. The *eyeless* homeodomain is dispensable for eye development in *Drosophila*. *Genes & Development* 15: 1716–1723.

Purnell, M.A. 1995. Microwear on conodont elements and macrophagy in the first vertebrates. *Nature* 374: 798–800.

Purschke, G. 1996. Echiura (Echiurida), Igelwürmer, pp. 345–349 in W. Westheide and R.M. Rieger (eds), *Spezielle Zoologie. Erster Teil: Einzeller und Wirbellose Tiere.* Gustav Fischer Verlag, Stuttgart [in German].

Purschke, G. 1997. Ultrastructure of nuchal organs in polychaetes (Annelida)—new results and review. *Acta Zoologica* 78: 123–143.

Purschke, G. 1999. Terrestrial polychaetes: models for the evolution of the Clitellata (Annelida)? *Hydrobiologia* 406: 87–99.

Purschke, G. 2000. Sense organs and central nervous system in an enigmatic terrestrial polychaete, *Hrabeiella periglandulata* (Anellida)—implications for anellid evolution. *Invertebrate Biology* 119: 329–341.

Purschke, G. 2002. On the ground pattern of Annelida. *Organisms Diversity and Evolution* 2: 181–196.

Purschke, G. 2003. Is *Hrabeiella periglandulata* (Anellida, "Polychaeta") the sister group of Clitellata? Evidence from a ultrastructural analysis of the dorsal pharynx in *H. periglandulata* and *Enchytraeus minutus* (Anellida, Clitellata). *Zoomorphology* 122: 55–66.

Purschke, G. and Hessling, R. 2002. Analysis of the central nervous system and sense organs in *Potamodrilus fluviatilis* (Anellida: Potamodrilidae). *Zoologischer Anzeiger* 241: 19–35.

Purschke, G., Walfrath, F., and Westheide, W. 1997. Ultrastructure of the nuchal organ and cerebral organ in *Onchnesoma squamatum* (Sipuncula, Phascolionidae). *Zoomorphology* 117: 23–31.

Purschke, G., Hessling, R., and Westheide, W. 2000. The phylogenetic position of the Clitellata and the Echiura—on the problematic assessment of absent characters. *Journal of Zoological Systematics and Evolutionary Research* 38: 165–173.

Purschke, G., Arendt, D., Hausen, H., and Müller, M.C.M. 2006. Photoreceptor cells and eyes in Annelida. *Arthropod Structure and Development* 35: 211–230.

Putnam, N.H., Srivastava, M., Hellsten, U. *et al.* 2007. Sea anemone genome reveals ancestral eu-metazoan gene repertoire and genomic organization. *Science* 317: 86–94.

Pyle, L.J., Narbonne, G.M., Nowlan, G.S., Xiao, S. and James, N.P. 2006. Early Cambrian eggs, embryos, and phosphatic microfossils from Northwestern Canada. *Journal of Paleontology* 80: 811–825.

Qian, Y., Li, G., Zhu, M., Steiner, M. and Erdtmann, B.-D. 2004. Early Cambrian protoconodonts and conodont-like fossils from China: Taxonomic revisions and stratigraphic implications. *Progress in Natural Science* 14: 173–180.

Quast, B. and Bartolomaeus, T. 2001. Ultrastructure and significance of the transitory nephridia in *Erpobdella octoculata* (Hirudinea, Annelida). *Zoomorphology* 120: 205–213.

Rabinowitz, J.S., Chan, X.Y., Kingsley, E.P., Duan, Y., and Lambert, J.D. 2008. nanos is required in somatic blast cell lineages in the posterior of a mollusc embryo. *Current Biology* 18: 331–336.

Raff, E.C., Villinsky, J.T., Turner, F.R., Donoghue, P.C.J. and Graff, R.A. 2006. Experimental taphonomy shows the feasibility of fossil embryos. *Proceedings of the National Academy of Sciences USA* 103: 5846–5851.

Raff, M.C. 1992. Social controls on cell survival and cell death. *Nature* 356: 397–400.

Raff, M.C., Barres, B.A., Burne, J.F., Coles, H.S., Ishizaki, Y., and Jacobson, M.D. 1993. Programmed cell death and the control of cell survival: lessons from the nervous system. *Science* 262: 695–700.

Raff, M.C., Durand, B., and Gao, F.-B. 1998. Cell number control and timing in animal development: the oligodendrocyte cell lineage. *International Journal of Developmental Biology* 42: 263–267.

Raff, R.A. 1999. Cell lineages in larval development and evolution of echinoderms, pp. 255–273 in B.K. Hall and M.H. Wake (eds), *The Origin and Evolution of Larval Forms.* Academic Press, London.

Raff, R.A. and Kaufman, T.C. 1983. *Embryos, Genes, and Evolution.* MacMillan, New York.

Raff, R.A. and Wray, G.A. 1989. Heterochrony: developmental mechanisms and evolutionary results. *Journal of Evolutionary Biology* 2: 409–434.

Raff, R.A. and Byrne, M. 2006. The active evolutionary lives of echinoderm larvae. *Heredity* 97: 244–252.

Raff, R.A., Marshall, C.R., and Turbeville, J.M. 1994. Using DNA sequences to unravel the Cambrian radiation of the animal phyla. *Annual Review of Ecology and Systematics* 25: 351–375.

Rafinesque C. S. 1814. *Précis des découvertes et travaux somiologiques de Mr. C. S. Rafinesque-Schmaltz entre 1800 et 1814; ou choix raisonné de ses principales découvertes en zoologie et en botanique, pour servir d'introduction à ses ouvrages futurs*. Palermo.

Ragan, M.A., Murphy, C.A., and Rand, T.G. 2003. Are Ichthyosporea animals or fungi? Bayesian phylogenetic analysis of elongation factor 1α of *Ichthyophonus irregularis*. *Molecular Phylogenetics and Evolution* 29: 550–562.

Raikova, O.I., Reuter, M., and Justine, J.L. 2001. Contributions to the phylogeny and systematics of the Acoelomorpha, pp. 13–23 in D.T.J. Littlewood and R.A. Bray (eds), *Interrelationships of the Platyhelminthes*. Taylor & Francis, London.

Raimond, R., Marcadé, I., Bouchon, D., Rigaud, T., Bossy, J. and Souty-Grosset, C. 1999. Organization of the large mitochondrial genome in the isopod *Armadillidium vulgare*. *Genetics* 151: 203–210.

Raineri, M. 2000. Early neurogenesis pattern in *Patella coerulea* (Patellogastropoda) and its possible phylogenetic implications. *Malacologia* 42: 131–148.

Raineri, M. 2006. Are protochordates chordates? *Biological Journal of the Linnean Society* 87: 261–284.

Ramsköld, L. and Hou, X.-G. 1991. New Early Cambrian animal and onychophoran affinities of enigmatic metazoans. *Nature* 351: 225–228.

Ramsköld, L. and Chen, J.-Y. 1998. Cambrian lobopodians: morphology and phylogeny, pp. 107–150 in G.D. Edgecombe (ed.), *Arthropod Fossils and Phylogeny*. Columbia University Press, New York.

Rauskolb, C. 2001. The establishment of segmentation in the *Drosophila* leg. *Development* 128: 4511–4521.

Rauskolb, C. and Irvine, K.D. 1999. Notch-mediated segmentation and growth control of the *Drosophila* leg. *Developmental Biology* 210: 339–350.

Rauther, M. 1909. Morphologie und Verwandtschaftsbeziehungen der Nematoden. *Ergebnisse und Fortschritte der Zoologie* 1: 491–596 [in German].

Reber-Müller, S., Streitwolf-Engel, R., Yanze, N. *et al.* 2006. BMP2/4 and BMP5–8 in jellyfish development and transdifferentiation. *International Journal of Developmental Biology* 50: 377–384.

Regier, J.C. and Shultz, J.W. 1997. Molecular phylogeny of the major arthropod groups indicates polyphyly of crustaceans and a new hypothesis for the origin of hexapods. *Molecular Biology and Evolution* 14: 902–913.

Regier, J.C. and Shultz, J.W. 2001. Elongation factor-2: useful gene for arthropod phylogenetics. *Molecular Phylogenetics and Evolution* 20: 136–148.

Regier, J.C., Shultz, J.W., and Kambic, R.E. 2005. Pancrustacean phylogeny: hexapods are terrestrial crustaceans and maxillopods are not monophyletic. *Proceedings of the Royal Society of London Series B, Biological Sciences* 272: 395–401.

Rehkämper, G. and Welsch, U. 1985. On the fine structure of the cerebral ganglion of *Sagitta* (Chaetognatha). *Zoomorphology* 105: 83–89.

Rehorn, K.-P., Thelen, H., Michelson, A.M., and Reuter, R. 1996. A molecular aspect of haematopoiesis and endoderm development common to vertebrates and *Drosophila*. *Development* 122: 4023–4031.

Reichert, H. 2005. A tripartite organization of the urbilaterian brain: developmental genetic evidence from *Drosophila*. *Brain Research Bulletin* 66: 491–494.

Reinhard, W. 1887. Kinorhyncha (Echinoderes) ihr anatomischer Bau. und ihre Stellung irn System. *Zeitschrift für wissenschaftliche Zoologie* 45: 401–467 [in German].

Reisinger, E. 1924. Die Gattung *Rhynchoscolex*. *Zeitschrift für Morphologie und Ökologie der Tiere* 1: 1–37 [in German].

Reisinger, E. 1957. Zur Entwicklungsgeschichte und Entwicklungsmechanik von *Craspedacusta* (Hydrozoa, Limnotrachylina). *Zeitschrift für Morphologie und Ökologie der Tiere* 45: 656–698 [in German].

Reiswig, H.M. and Mackie, G.O. 1983. Studies on hexactinellid sponges: III. The taxonomic status of Hexactinellida within the Porifera. *Philosophical Transactions of the Royal Society of London Series B, Biological Sciences* 301: 419–428.

Reitner, J. and Mehl, D. 1996. Monophyly of the Porifera. *Verhandlungen des naturwissenschaftlichen Vereins in Hamburg, Neue Folge* 36: 5–32.

Reitner, J. and Wörheide, G. 2002. Non-lithistid fossil Demospongiae—Origins of their palaeobiodiversity and highlights in history of preservation, pp. 52–70 in J.N.A. Hooper and R.W.M. van Soest (eds), *Systema Porifera. A Guide to the Classification of Sponges*. Kluwer Academics/Plenum Publishers, New York.

Remak, R. 1855. *Untersuchungen über die Entwicklung der Wirbelthiere*. Reimer, Berlin [in German].

Rensch, B. 1959. *Evolution Above the Species Level*. Methuen, London.

Rentzsch, F., Anton, R., Saina, M., Hammerschmidt, M., Holstein, T.W., and Technau, U. 2006. Asymmetric expression of the Bmp antagonists chordin and gremlin in the sea anemone *Nematostella vectensis*: implications for the evolution of axial patterning. *Developmental Biology* 296: 375–387.

Reuter, M. and Gustafsson, M.K.S. 1995. The flatworm nervous system: pattern and phylogeny, pp. 25–59 in O. Breidbach and W. Kutsch (eds), *The Nervous System of*

Invertebrates: an Evolutionary and Comparative Approach. Birkhäuser, Basel.

Reuter, M. and Halton, D.W. 2001. Comparative neurobiology of Platyhelminthes, pp. 239–349 in D.T.J. Littlewood and R.A. Bray (eds), *Interrelationships of the Platyhelminthes.* Taylor & Francis, London.

Reuter, M. and Kreshchenko, N. 2004. Flatworm asexual multiplication implicates stem cells and regeneration. *Canadian Journal of Zoology* 82: 334–356.

Reuter, M., Mäntylä, K., and Gustafsson, M.K.S. 1998. Organization of the orthogon—main and minor nerve cords. *Hydrobiologia* 383: 175–182.

Reznick, D., Ghalambor, C. and Nunney, L. 2002a. The evolution of senescence in fish. *Mechanisms of Ageing and Development* 123: 773–789.

Reznick, D.N., Mateos, M., and Springe, M.S. 2002b. Independent origins and rapid evolution of the placenta in the fish genus *Poeciliopsis. Science* 298: 1018–1020.

Ricci, C. 1998. Are lemnisci and proboscis present in the Bdelloidea? *Hydrobiologia* 387/388: 93–96.

Rice, M.E. and Pilger, J.F. 1993. Sipuncula, pp. 297–310 in K.G. Adiyodi and R.G. Adiyodi (eds), *Reproductive Biology of Invertebrates,* vol. 6, part A. Oxford and IBH Publ. Co., New Delhi.

Richardson, M.K. 1995. Heterochrony and the phylotypic period. *Developmental Biology* 172: 412–421.

Richardson, M.K. and Chipman, A.D. 2003. Developmental constraints in a comparative framework: a test case using variations in the phalanx number during amniote evolution. *Journal of Experimental Biology (Molecular and Developmental Evolution)* 296B: 8–22.

Richardson, M.K., Jeffery, J.E., and Tabin, C.J. 2004. Proximodistal patterning of the limb: insights from evolutionary morphology. *Evolution & Development* 6: 1–5.

Richter, S. 2002. The Tetraconata concept: hexapod-crustacean relationships and the phylogeny of Crustacea. *Organisms Diversity and Evolution* 2: 217–237.

Richter, S. and Scholtz, G. 2001. Phylogenetic analysis of the Malacostraca (Crustacea). *Journal of Zoological Systematics and Evolutionary Research* 39: 113–136.

Richter, S. and Wirkner, C. 2004. Kontroversen in der phylogenetischen Systematik der Euarthropoda. *Sitzungsberichte der Gesellschaft Naturforschender Freunde zu Berlin, Neue Folge* 43: 73–102 [in German].

Rieger, R.M. 1976. Monociliated epidermal cells in Gastrotricha, significance for concepts of early metazoan evolution. *Zeitschrift für zoologische Systematik und Evolutionsforschung* 14: 198–226.

Rieger, R.M. 1980. A new group of interstitial worms, Lobatocerebridae nov. fam. (Annelida) and its

significance for metazoan phylogeny. *Zoomorphologie* 95: 41–84.

Rieger, R.M. 1984. Evolution of the cuticle in the lower Eumetazoa, pp. 389–399 in J. Bereiter-Hahn, A.G. Maltosky and K. Sylvia-Richards (eds), *Biology of the Integument. Vol. 1. Invertebrates.* Springer, Berlin.

Rieger, R.M. 1985. The phylogenetic status of the acoelomate organization within the Bilateria: a histological perspective, pp. 101–122 in S. Conway Morris, J.D. George, R. Gibson and H.M. Platt (eds), *The Origins and Relationships of Lower Invertebrates.* Clarendon Press, Oxford.

Rieger, R.M. 1986. Über den Ursprung der Bilateria: Die Bedeutung der Ultrastrukturforschung für ein neues Verstehen der Metazoenevolution. *Verhandlungen der Deutschen Zoologischen Gesellschaft* 79: 31–50 [in German].

Rieger, R.M. 1988. Comparative ultrastructure and the Lobatocerebridae: keys to understand the phylogenetic relationship of Annelida and the acoelomates, pp. 373–382 in W. Westheide and C.O. Hermans (eds), *The Ultrastructure of Polychaeta.* Gustav Fischer Verlag, Stuttgart.

Rieger, R.M. 1991. *Jennaria pulchra,* nov. gen. nov. spec., eine den psammobionten Anneliden nahestehende Gattung aus dem Küstengrundwasser von North Carolina. *Berichte des Naturwissenschaftlich-Medizinischen Vereins in Innsbruck* 78: 203–215 [in German].

Rieger, R.M. 1994a. Evolution of the "lower" Metazoa, pp. 475–488 in S. Bengtson (ed.), *Early Life on Earth.* Nobel Symposium no. 84. Columbia University Press, New York.

Rieger, R.M. 1994b. The biphasic life cycle—a central theme of metazoan evolution. *American Zoologist* 34: 484–491.

Rieger, R.M. 2001. Phylogenetic systematics of the Macrostomorpha, pp. 28–38 in D.T.J. Littlewood and R.A. Bray (eds), *Interrelationships of the Platyhelminthes.* Taylor and Francis, London.

Rieger, R.M. 2003. The phenotypic transition from uni- to multicellulars animals, pp. 247–258 in A. Legakis, S. Sfenthourakis, R. Polymeni, and M. Thessalou-Legaki (eds), *The New Panorama of Animal Evolution.* Proceedings of the 18th International Congress of Zoology. Pensoft, Sofia/Moscow.

Rieger, R.M. and Lombardi, J. 1987. Ultrastructure of coelomic lining in echinoderm podia: significance for concepts in the evolution of muscle and peritoneal cells. *Zoomorphology* 107: 191–208.

Rieger, R.M. and Ladurner, P. 2003. The significance of muscle cells for the origin of mesoderm in Bilateria. *Integrative and Comparative Biology* 43: 47–54.

Rieger, R.M. and Purschke, G. 2005. The coelom and the origin of the annelid body plan. *Hydrobiologia* 535/536: 127–137.

Rieger, R.M., Tyler, S., Smith, III, J.P.S., and Rieger, G.E. 1991. Platyhelminthes: Turbellaria, pp. 7–140 in F.W. Harrison (ed.), *Microscopic Anatomy of the Invertebrates (Vol. 3. Platyhelminthes and Nemertinea)*. Wiley-Liss, New York.

Rieppel, O. 1979. Ontogeny and the recognition of primitive character states. *Zeitschrift für zoologische Systematik und Evolutionsforschung* 17: 57–61.

Rieppel, O. 2001. Turtles as hopeful monsters. *BioEssays* 23: 987–991.

Ripper, W. 1931. Versuch einer Kritik der Homologiefrage der Arthropodentracheen. *Zeitschrift für Wissenschaftliche Zoologie* 138: 303–369.

Roberts, T.R. and Warren, T.J. 1994. Observations of fishes and fisheries in southern Laos and northeastern Cambodia, October 1993-Febuary 1994. *Natural History Bulletin of the Siam Society* 42: 87–115.

Robinson, D.R., Wu, Y.M. and Lin, S.F. 2000. The protein tyrosine kinase family of the human genome. *Oncogene* 19: 5548–5557.

Rodríguez, M.Á., Olalla-Tárraga, M.Á., and Hawkins, B.A. 2008. Bergmann's rule and the geography of mammal body size in the Western Hemisphere. *Global Ecology and Biogeography* 17: 274–283.

Rogozin, I.B., Wolf, Y.I., Carmel, L., and Koonin, E.V. 2007. Ecdysozoan clade rejected by genome-wide analysis of rare amino acid replacements. *Molecular Biology & Evolution* 24: 1080–1090.

Rohde, K., Watson, N.A., and Chisholm, L.A. 1999. Ultrastructure of the eyes of the larva of *Neoheterocotyle rhinobatidis* (Platyhelminthes, Monopisthocotylea), and phylogenetic implications. *International Journal for Parasitology* 29: 511–519.

Rohr, K.B., Tautz, D., and Sander, K. 1999. Segmentation gene expression in the mothmidge *Clogmia albipunctata* (Diptera, Psychodidae) and other primitive dipterans. *Development, Genes and Evolution* 209: 145–154.

Rokas, A. and Carroll, S.B. 2006. Bushes in the tree of life. *PLoS Biology* 4: 1899–1904.

Rokas, A., King, N., Finnerty, J., and Carroll, S.B. 2003. Conflicting phylogenetic signals at the base of the metazoan tree. *Evolution & Development* 5: 346–359.

Rokas, A., Krüger, S. and Carroll, S.B. 2005. Animal evolution in the molecular signature of radiations compressed in time. *Science* 310: 1933–1938.

Rot, C., Goldfarb, I., Ilan, M., and Huchon, D. 2006. Putative cross-kingdom horizontal gene transfer in sponge (Porifera) mitochondria. *BMC Evolutionary Biology* 6: 71.

Rouse, G.W. 1999. Trochophore concepts: ciliary bands and the evolution of larvae in spiralian Metazoa. *Biological Journal of the Linnean Society* 66: 411–464.

Rouse, G. and Fitzhugh, K. 1994. Broadcasting fables: is external fertilization really primitive? Sex, size, and larvae in sabellid polychaetes. *Zoologica Scripta* 23: 271–312.

Rouse, G.W. and Fauchald, K. 1995. The articulation of annelids. *Zoologica Scripta* 24: 269–301.

Rouse, G.W. and Fauchald, K. 1997. Cladistics and polychaetes. *Zoologica Scripta* 26: 139–204.

Rouse, G.W. and Pleijel, F. 2001. *Polychaetes*. Oxford University Press, Oxford.

Rouse, G.W., Goffredi, S.K., and Vrijenhoek, R.C. 2004. *Osedax*: bone-eating marine worms with dwarf males. *Science* 305: 668–671.

Rousset, V., Pleijel, F., Rouse, G.W., Erséus, C., and Siddall, M.E. 2007. A molecular phylogeny of annelids. *Cladistics* 23: 41–63.

Rousset, V., Rouse, G.W., Siddall, M.E., Tillier, A., and Pleijel, F. 2004. The phylogenetic position of Siboglinidae (Annelida) inferred from 18S rRNA, 28S rRNA and morphological data. *Cladistics* 20: 518–533.

Roux, W. 1881. *Der Kampf der Teile im Organismus*. Breitkopf und Härtel, Leipzig.

Rudolphi, C.A. 1808–1810. *Entozoorum, sive vermium intestinalium historia naturalis. 1.* Amstelaedami [in Latin].

Ruiz-Trillo, I., Riutort, M., Littlewood, D.T.J., Herniou, E.A., and Baguñá, J. 1999. Acoel flatworms: earliest extant bilaterian metazoans, not members of Platyhelminthes. *Science* 283: 1919–1923.

Ruiz-Trillo, I., Paps, J., Loukota *et al.* 2002. A phylogenetic analysis of myosin heavy chain type II sequences corroborates that Acoela and Nemertodermatida are basal bilaterians. *Proceedings of the National Academy of Sciences USA* 99: 11246–11251.

Ruiz-Trillo, I., Lane, C.E., Archibald, J.M., and Roger, A.J. 2006. Insights into the evolutionary origin and genome architecture of the unicellular opisthokonts *Capsaspora owczarzaki* and *Sphaeroforma arctica*. *Journal of Eukaryotic Microbiology* 53: 379–384.

Ruiz-Trillo, I., Burger, G., Holland, P.W.H. *et al.* 2007. The origins of multicellularity: a multi-taxon genome initiative. *Trends in Genetics* 23: 113–118.

Ruppert, E.E. 1991a. Introduction to the aschelminth phyla: a consideration of mesoderm, body cavities, and cuticles, pp. 1–17 in F.W. Harrison and E.E. Ruppert (eds), *Microscopic Anatomy of Invertebrates. 4. Aschelminthes*. Wiley-Liss, New York.

Ruppert, E.E. 1991b. Gastrotricha, pp. 41–109 in F.W. Harrison and E.E. Ruppert (eds), *Microscopic Anatomy of Invertebrates, Vol. 4, Aschelminthes.* Wiley-Liss, New York.

Ruppert, E.E. 2005. Key characters uniting hemichordates and chordates: homologies or homoplasies? *Canadian Journal of Zoology* 83: 8–23.

Ruppert, E.E. and Carle, K.J. 1983. Morphology of metazoan circulatory systems. *Zoomorphology* 103: 193–208.

Ruppert, E.E. and Schreiner, S.P. 1980. Ultrastructure and potential significance of the cerebral light-refracting bodies of *Stenostomum virginianum* (Turbellaria, Catenulida). *Zoomorphology* 96: 21–31.

Ruppert, E.E. and Smith, P.R. 1988. The functional organization of filtration nephridia. *Biological Reviews* 63: 231–258.

Rusconi, J.C. and Corbin, V. 1999. A widespread and early requirement for a novel Notch function during *Drosophila* embryogenesis. *Developmental Biology* 215: 388–398.

Ruta, M., Wagner, P.J. and Coates, M.I. 2006. Evolutionary patterns in early tetrapods. I. Rapid initial diversification followed by decrease in rates of character change. *Proceedings of the Royal Society of London Series B, Biological Sciences* 273: 2107–2111.

Ruvkun, G. and Giusto, J. 1989. The *Caenorhabditis elegans* heterochronic gene *lin-14* encodes a nuclear protein that forms a temporal developmental switch. *Nature* 338: 313–319.

Ryan, J.F., Mazza, M.E., Pang, K. *et al.* 2007. Pre-bilaterian origins of the Hox cluster and the Hox code: evidence from the sea snemone, *Nematostella vectensis. PLoS ONE* e153, doi:10.1371/journal.pone.000015.

Sabbadin, A., Zaniolo, G., and Majone, F. 1975. Determination of polarity and bilateral symmetry in the ascidian *Botryllus schlosseri. Developmental Biology* 46: 79–87.

Sahli, F. 1990. On post-adult moults in Julida (Myriapoda, Diplopoda). Why do periodomorphosis and intercalaries occur in males?, pp. 135–56 in A. Minelli (ed.), *Proceedings of the 7th International Congress of Myriapodology.* Brill, Leiden.

Salser, S.J. and Kenyon, C. 1994. Patterning *C. elegans*: homeotic cluster genes, cell fates and cell migrations. *Trends in Genetics* 10: 159–164.

Salvini-Plawen, L. 1978. On the origin and evolution of the lower Metazoa. *Zeitschrift für zoologische Systematik und Evolutionsforschung* 16: 40–88.

Samakovlis, C., Hacohen, N., Manning, G., Sutherland, D.C., Guillemin, K., and Krasnow, M.A. 1996. Development of the *Drosophila* tracheal system occurs by a series of morphologically distinct but genetically coupled branching events. *Development* 122: 1395–1407.

Sauber, F., Reuland, M., Berchtold, J.-P. *et al.* 1983. Cycle de mue et ecdystéroïdes chez une sangsue, *Hirudo medicinalis. Comptes Rendus de l'Academie des Sciences, Paris, Science de la Vie* 296: 413–418 [in French].

Sauer, K.P. and Kullmann, H. 2005. Analyse der historisch-ökologischen Ursachen der Evolution der gastroneuralen Metazoa—Testen einer phylogenetischen Hypothese. *Bonner Zoologische Beiträge* 53: 149–163 [in German].

Sawyer, R.T. 1984. Arthropodization in the Hirudinea: evidence for a phylogenetic link with insects and other Uniramia? *Zoological Journal of the Linnean Society* 80: 303–332.

Schaeffer, B. 1987. Deuterostome monophyly and phylogeny. *Evolutionary Biology* 21: 179–235.

Schäfer, W. 1996. Cnidaria, Nesseltiere, pp. 145–81 in W. Westheide and R.M. Rieger (eds), *Spezielle Zoologie. Erster Teil: Einzeller und Wirbellose Tiere.* Gustav Fischer, Stuttgart.

Scharnofske, P. 1984. *Anatomie, Ultrastruktur und Funktion der männlichen Geschlechtsorgane der Dinophilidae und Histriobdellidae (Annelida, Polychaeta).* Thesis, University of Göttingen [in German].

Schauenstein, K., Felsner, P., Rinner, I. *et al.* 2000. In vivo immunomodulation by peripheral adrenergic and cholinergic agonists/antagonists in rat and mouse models. *Annals of the New York Academy of Sciences* 917: 618–627.

Scheltema, A.H. 1993. Aplacophora as progenetic aculiferans and the coelomate origin of mollusks as the sister taxon of Sipuncula. *Biological Bulletin* 184: 57–78.

Scheltema, A.H. 1996. Phylogenetic position of Sipuncula, Mollusca and the progenetic Aplacophora, pp. 53–58 in J.D. Taylor (ed.), *Origin and Evolutionary Radiation of the Mollusca.* Oxford University Press, Oxford.

Schierenberg, E. 2001. Three sons of fortune: early embryogenesis, evolution and ecology of nematodes. *BioEssays* 23: 841–847.

Schierenberg, E. 2005. Unusual cleavage and gastrulation in a freshwater nematode: developmental and phylogenetic implications. *Development, Genes and Evolution* 215: 103–108.

Schierenberg, E. and Schulze, J. 2008. Many roads lead to Rome: different ways to construct a nematode, pp. 261–280 in A. Minelli and G. Fusco (eds), *Evolving Pathways: Key Themes in Evolutionary Developmental Biology.* Cambridge University Press, Cambridge.

Schierwater, B. and Kuhn, K. 1998. Homology of the Hox genes and the zootype concept in early Metazoan evolution. *Molecular Phylogenetics and Evolution* 9: 375–381.

Schleip, W. 1929. *Die Determination der Frühentwicklung.* Akademische Verlagsgesellschaft, Leipzig [in German].

Schlosser, G. 2002. Modularity and the units of evolution. *Theory in Biosciences* 121: 1–80.

Schlosser, G. 2007. How old genes make a new head: redeployment of *Six* and *Eya* genes during the evolution of vertebrate cranial placodes. *Integrative and Comparative Biology* 47: 343–359.

Schlosser, G. and Wagner, G.P. (eds) 2004. *Modularity in Development and Evolution*. University of Chicago Press, Chicago.

Schmalhausen, I.I. 1969. *The Problems of Darwinism*. Nauka, Leningrad [in Russian].

Schmid, V. 1974. Regeneration in medusa buds and medusae of Hydrozoa. *American Zoologist* 14: 773–781.

Schmidt-Ott, U., Gonzalez-Gaitan, M., Jäckle, H., and Technau, G.M. 1994. Number, identity, and sequence of the *Drosophila* head segments as revealed by neural elements and their deletion patterns in mutants. *Proceedings of the National Academy of Sciences USA* 91: 8363–8367.

Schmidt-Ott, U., Gonzalez-Gaitan, M., and Technau, G.M. 1995. Analysis of neural elements in head-mutant *Drosophila* embryos suggests segmental origin of the optic lobes. *Wilhelm Roux's Archives of Developmental Biology* 205: 31–44.

Schmidt-Rhaesa, A. 1996. *Zur Morphologie, Biologie und Phylogenie der Nematomorpha. Untersuchungen an Nectonema munidae und Gordius aquaticus*. Cuvillier Verlag, Göttingen [in German].

Schmidt-Rhaesa, A. 1998. Phylogenetic relationships of the Nematomorpha—a discussion of current hypotheses. *Zoologischer Anzeiger* 236: 203–216.

Schmidt-Rhaesa, A. 2002. *Die Saitenwürmer (Die neue Brehm-Bücherei 632)*. Westarp Wissenschaften, Hohenwarsleben [in German].

Schmidt-Rhaesa, A. 2007. *The Evolution of Organ Systems*. Oxford University Press, Oxford.

Schmidt-Rhaesa, A. and Rothe, B.H. 2006. Postembryonic development of dorsoventral and longitudinal musculature in *Pycnophyes kielensis* (Kinorhyncha, Homalorhagida). *Integrative and Comparative Biology* 46: 144–150.

Schmidt-Rhaesa, A., Bartolomaeus, T., Lemburg, C., Ehlers, U., and Garey, J.R. 1998. The position of the Arthropoda in the phylogenetic system. *Journal of Morphology* 238: 263–285.

Schnabel, R., Bischoff, M., Hintze, A. *et al.* 2006. Global cell sorting in the *C. elegans* embryo defines a new mechanism for pattern formation. *Developmental Biology* 294: 418–431.

Schneider, A. 1873. Untersuchungen über Plathelminthen. *Jahresberichte der Oberhessischen Gesellschaft für Natur- und Heilkunde in Gießen* 14: 69–140 [in German].

Schneider, S.Q., Finnerty, J.R. and Martindale, M.Q. 2003. Protein evolution: structure-function relationships of the oncogene beta-catenin in the evolution of multicellular animals. *Journal of Experimental Zoology (Molecular and Developmental Evolution)* 295B: 25–44.

Schöck, F. and Perrimon, N. 2002. Molecular mechanisms of epithelial morphogenesis. *Annual Reviews Cell and Developmental Biology* 18: 463–493.

Scholtz, G. 1995. Expression of the *engrailed* gene reveals nine putative segment-anlagen in the embryonic pleon of the freshwater crayfish *Cherax destructor* (Crustacea, Malacostraca, Decapoda). *Biological Bulletin* 188: 157–165.

Scholtz, G. 1998. Cleavage, germ band formation and head segmentation: the ground pattern of the Euarthropoda, pp. 317–332 in R.A. Fortey and R.H. Thomas (eds), *Arthropod Relationships*. Systematics Association Special Volume Series 55. Chapman & Hall, London.

Scholtz, G. 2000. Evolution of the nauplius stage in malacostracan crustaceans. *Journal of Zoological Systematics and Evolutionary Research* 38: 175–187.

Scholtz, G. 2002. The Articulata hypothesis—or what is a segment? *Organisms Diversity and Evolution* 2: 197–215.

Scholtz, G. 2003. Is the taxon Articulata obsolete? Arguments in favour of a close relationship between annelids and arthropods, pp. 489–501 in A. Legakis, S. Sfenthourakis, R. Polymeni, and M. Thessalou-Legaki (eds), *The New Panorama of Animal Evolution*. Proceedings of the 18th International Congress of Zoology. Pensoft, Sofia/Moscow.

Scholtz, G. 2004a. Coelenterata versus Acrosomata—zur Position der Rippenquallen (Ctenophora) im phylogenetischen System der Metazoa. *Sitzungsberichte der Gesellschaft Naturforschender Freunde zu Berlin, Neue Folge* 43: 15–33 [in German].

Scholtz, G. 2004b. Baupläne versus ground patterns, phyla versus monophyla: aspects of patterns and processes in evolutionary developmental biology, pp. 3–16 in G. Scholtz (ed.), *Evolutionary Developmental Biology of Crustacea*. A.A. Balkema, Lisse.

Scholtz, G. 2005. Homology and ontogeny: pattern and process in comparative developmental biology. *Theory in Biosciences* 124: 121–143.

Scholtz, G. 2008. On comparisons and causes in evolutionary developmental biology, pp. 144–159 in A. Minelli and G. Fusco (eds), *Evolving Pathways: Key Issues in Evolutionary Developmental Biology*. Cambridge University Press, Cambridge.

Scholtz, G. and Dohle, W. 1996. Cell lineage and cell fate in crustacean embryos—a comparative approach. *International Journal of Developmental Biology* 40: 211–220.

Scholtz, G. and Wolff, C. 2002. Cleavage pattern, gastrulation, and germ disc formation of the amphipod crustacean *Orchestia cavimana*. *Contributions to Zoology* 71: 9–28.

Scholtz, G. and Edgecombe, G.D. 2006. The evolution of arthropod heads: reconciling morphological, developmental and palaeontological evidence. *Development, Genes and Evolution* 216: 395–415.

Scholtz, G. and Kamenz, C. 2006. The book lungs of Scorpiones and Tetrapulmonata (Chelicerata, Arachnida): evidence for homology and a single terrestrialisation event of a common arachnid ancestor. *Zoology* 109: 2–13.

Schram, F.R. 1991. Cladistic analysis of metazoan phyla and the placement of fossil problematica, pp. 35–46 in A.M. Simonetta and S. Conway Morris (eds), *The Early Evolution of Metazoa and the Significance of Problematic Taxa*. Cambridge University Press, Cambridge.

Schram, F.R. and Hof, C.H.J. 1998. Fossils and the interrelationships of major crustacean groups, pp. 233–302 in G.D. Edgecombe (ed.), *Arthropod Fossils and Phylogeny*. Columbia University Press, New York.

Schröter, R.H. 2006. Blown fuse regulates stretching and outgrowth but not myoblast fusion of the circular visceral muscles in *Drosophila*. *Differentiation* 74: 608–621.

Schubert, M., Escriva, H., Xavier-Neto, J. and Laudet, V. 2006. Amphioxus and tunicates as evolutionary model systems. *Trends in Ecology & Evolution* 21: 269–277.

Schuchert, P., Reber-Müller, S. and Schmid, V. 1993. Life stage specific expression of a myosin heavy chain in the hydrozoan *Podocoryne carnea*. *Differentiation* 54: 11–18.

Schulman, B.R., Esquela-Kerscher, A., and Slack, F.J. 2005. Reciprocal expression of *lin-41* and the microRNAs *let-7* and *mir-125* during mouse embryogenesis. *Developmental Dynamics* 234: 1046–1054.

Schulmeister, S. and Wheeler, W. 2004. Comparative and phylogenetic analysis of developmental sequences, *Evolution & Development* 6: 50–57.

Schultz, R.J. 1967. Gynogenesis and triploidy in the viviparous fish *Poeciliopsis*. *Nature* 157: 1564–1567.

Schultze, M.S. 1851. *Beiträge zur Naturgeschichte der Turbellarien*. C.A. Koch Verlag, Greifswald [in German].

Scotland, R.W., Olmstead, R.G. and Bennett, J.R. 2003. Phylogeny reconstruction: the role of morphology. *Systematic Biology* 52: 539–548.

Scott, A. 1941. Reversal of sex production in *Micromalthus*. *Biological Bulletin* 81: 420–431.

Scott, M.P. 1994. Intimations of a creature. *Cell* 79: 1121–1124.

Seaver, E.C. and Kamshige, L.M. 2006. Expression of 'segmentation' genes during larval and juvenile development in the polychaetes *Capitella* sp. I and *H.* [*sic*, pro *Hydroides*] *elegans*. *Developmental Biology* 289: 179–194.

Seaver, E.C., Paulson, D., Irvine, S.Q., and Martindale. M.Q. 2001. The spatial and temporal expression of *Ch-en*, the *engrailed* gene in the polychaete *Chaetopterus* does not support a role in body axis segmentation. *Developmental Biology* 236: 195–209.

Seaver, E.C., Thamm, K., and Hill, S.D. 2005. Growth patterns during segmentation in the two polychaete annelids, *Capitella* sp. I and *Hydroides elegans*: comparison of distinct life history stages. *Evolution & Development* 7: 312–326.

Sebens, K.P. 1987. The ecology of indeterminate growth in animals. *Annual Review of Ecology and Systematics* 18: 371–407.

Sedgwick, A. 1898. *A Student's Textbook of Zoology*. Swan Sonnenschein, London.

Seilacher, A. 1984. Late Precambrian and Early Cambrian metazoa: preservational or real extinction?, pp. 159–168 in H.D. Holland and A.F. Trendall (eds), *Patterns of Change in Earth Evolution*. Springer, Berlin.

Seilacher, A. 1989. Vendozoa: organismic construction in the Proterozoic biosphere. *Lethaia* 22: 229–239.

Seilacher, A. 1994. Early multicellular life: Late Proterozoic fossils and the Cambrian explosion, pp. 389–400 in S. Bengtson (ed.), *Early Life on Earth*. Nobel Symposium no. 84. Columbia University Press, New York.

Seilacher, A. 1997. *Fossil Art*. Royal Tyrrell Museum of Paleontology, Drumheller, Alberta.

Seipel, K. and Schmid, V. 2005. Evolution of striated muscle: jellyfish and the origin of triploblasty. *Developmental Biology* 282: 14–26.

Seipel, K. and Schmid, V. 2006. Mesodermal anatomies in cnidarian polyps and medusae. *International Journal of Developmental Biology* 50: 589–599.

Seipp, S., Schmich, J., and Leitz, T. 2001. Apoptosis—a death-inducing mechanism tightly linked with morphogenesis in *Hydractinia echinata* (Cnidaria, Hydrozoa). *Development* 128: 4891–4898.

Selander, R.B. 1991. Rhipiphoridae (Tenebrionoidea), pp. 509–512 in F.W. Stehr (ed.), *Immature insects*, vol. 2. Kendal-Hunt, Dubuque, IO.

Selleck, M.A. and Stern, C.D. 1991. Fate mapping and cell lineage analysis of Hensen's node in the chick embryo. *Development* 112: 615–626.

Sempere, L.F., Cole, C.N., McPeek, M.A., and Peterson, K.J. 2006. The phylogenetic distribution of metazoan microRNAs: insights into evolutionary complexity and constraint. *Journal of Experimental Zoology (Molecular and Developmental Evolution)* 306B: 575–588.

Sempere, L.F., Martinez, P., Cole, C., Baguñá, J., and Peterson, K.J. 2007. Phylogenetic distribution of microRNAs supports the basal position of acoel flatworms and the polyphyly of Platyhelminthes. *Evolution & Development* 9: 409–415.

Seo, H.-C., Kube, M., Edvardsen, R.B. *et al.* 2001. Miniature genome in the marine chordate *Oikopleura dioica*. *Science* 294: 2506.

Seo, H.C., Edvardsen, R.B., Mæland, A.D. *et al.* 2004. Hox cluster disintegration with persistent anteroposterior order of expression in *Oikopleura dioica*. *Nature* 431: 67–71.

Sewertzoff, A.N. 1931. *Morphologische Gesetzmässigkeiten der Evolution*. Gustav Fischer, Jena [in German].

Shankland, M. 2003. Evolution of body axis segmentation in the bilaterian radiation, pp. 187–195 in A. Legakis, S. Sfenthourakis, R. Polymeni, and M. Thessalou-Legaki (eds), *The New Panorama of Animal Evolution*. Proceedings of the 18th International Congress of Zoology. Pensoft, Sofia/Moscow.

Shao, R., Campbell, N.J.H., Schmidt, E.R., and Barker, S.C. 2001. Increased rate of gene rearrangement in the mitochondrial genomes of three orders of hemipterid insects. *Molecular Biology and Evolution* 18: 1828–1832.

Shear, W.A., Bonamo, P.M., Grierson, J.D., Rolfe, W.D.I., Smith, E.L., and Norton, R.A. 1984. Early land animals in North America. *Science* 224: 492–494.

Shimizu, T., Kitamura, K., Arai, A., and Nakamoto, A. 2001. Pattern formation in embryos of the oligochaete annelid *Tubifex*: cellular basis for segmentation and specification of segmental identity. *Hydrobiologia* 463: 123–131.

Shimizu, H., Takaku, Y., Zhang, X., and Fujisawa, T. 2007. The aboral pore of hydra: evidence that the digestive tract of hydra is a tube, not a sac. *Development, Genes and Evolution* 217: 563–568.

Shimkevich, V.M. 1908. Metorisis as an embryological principle. *Izvestiya imperatorskoi Akademii nauk*, ser. 6, 2 [in Russian].

Shimotori, T. and Goto, T. 2001. Developmental fates of the first four blastomeres of the chaetognath *Paraspadella gotoi*: relationships to protostomes. *Development, Growth and Differentiation* 43: 371–382.

Shinn, G.L. 1997. Chaetognatha, pp. 103–220 in F.W. Harrison and E.E. Ruppert (eds), *Microscopic Anatomy of Invertebrates. Vol. 15. Hemichordata, Chaetognatha, and the Invertebrate Chordates*. Wiley-Liss, New York.

Shu, D.-G., Conway Morris, S., Han, J. *et al.* 2001. Primitive deuterostomes from the Chengjiang Lagerstätte (Lower Cambrian, China). *Nature* 414: 419–424.

Shu, D., Conway Morris, S., Zhang, Z.F. *et al.* 2003. A new species of *Yunnanozoon* with implications for deuterostome evolution. *Science* 299: 1380–1384.

Shu, D.-G., Conway Morris, S., Han, J. *et al.* 2006. Lower Cambrian vendobionts from China and Early diploblast evolution. *Science* 312: 731–734.

Shubin, N.H., Daeschler, E.B., and Coates, M.I. 2004. The early evolution of the tetrapod humerus. *Science* 304: 90–93.

Shull, A.F. 1918. Cell inconstancy in *Hydatina senta*. *Journal of Morphology* 30: 455–464.

Shultz, J.W. 2007. A phylogenetic analysis of the arachnid orders based on morphological characters. *Zoological Journal of the Linnean Society* 150: 221–265.

Shultz, J.W. and Regier, J. C. 2000. Phylogenetic analysis of arthropods using two nuclear protein-encoding genes supports a crustacean + hexapod clade. *Proceedings of Royal Society of London Series B, Biological Sciences* 267: 1011–1019.

Siddall, M.E. and Whiting, M.F. 1999. Long-branch abstractions. *Cladistics* 15: 9–24.

Siddall, M.E., Martin, D.S., Bridge, D., Desser, S.S., and Cone, D.K. 1995. The demise of a phylum of protists: phylogeny of Myxozoa and other parasitic Cnidaria. *Journal of Parasitology* 81: 961–967.

Siddall, M.E., Borda, E., and Rouse, G.W. 2004. Toward a tree of life for Annelida, pp. 237–251 in J. Cracraft and M.J. Donoghue (eds), *Assembling the Tree of Life*. Oxford University Press, Oxford.

Siddall, M.E., Bely, A., and Borda, E. 2006. Hirudinea, pp. 393–429 in G. Rouse and F. Pleijel (eds), *Reproductive Biology and Phylogeny of Annelida*. Science Publishers, Enfield, NH.

Sierwald, P. and Bond, J.E. 2007. Current status of the myriapod class Diplopoda (millipedes): taxonomic diversity and phylogeny. *Annual Review of Entomology* 52: 401–420.

Sierwald, P., Shear, W.A., Shelley, R.M., and Bond, J.E. 2003. Millipede phylogeny revisited in the light of the enigmatic order Siphoniulida. *Journal of Zoological Systematics and Evolutionary Research* 41: 87–99.

Siewing, R. 1969. *Lehrbuch der vergleichenden Entwicklungsgeschichte der Tiere*. Parey, Hamburg [in German].

Signorovitch, A.Y., Dellaporta, S.L., and Buss, L.W. 2005. Molecular signatures for sex in the Placozoa. *Proceedings of the National Academy of Sciences USA* 102: 15518–15522.

Signorovitch, A.Y., Dellaporta, S.L., and Buss, L.W. 2006. Caribbean placozoan phylogeography. *Biological Bulletin* 211: 149–156.

Silver, S.J. and Rebay, I. 2005. Signaling circuitries in development: insights from the retinal determination gene network. *Development* 132: 3–13.

Simonetta, A.M. 2004. Are the traditional classes of arthropods natural ones? Recent advances in palaeontology and some considerations on morphology. *Italian Journal of Zoology* 71: 247–264.

Simpson, G.G. 1945. The principles of classification and a classification of mammals. *Bulletin of the American Museum of Natural History* 85: 1–350.

Simpson, G.G. 1961. *Principles of Animal Taxonomy.* Columbia University Press, New York.

Siveter, D.J., Sutton, M.D., Briggs, D.E.G., and Siveter, D.J. 2004. A Silurian sea spider. *Nature* 431: 978–980.

Slack, J.M.W., Holland, P.W.H. and Graham, C.F. 1993. The zootype and the phylotypic stage. *Nature* 361: 490–492.

Sly, B.J., Snoke, M.S., and Raff, R.A. 2003. Who came first—larvae or adults? Origins of bilaterian metazoan larvae. *International Journal of Developmental Biology* 47: 623–632.

Slyusarev, G.S. 1994. The fine structure of the female *Intoshia variabili* (Alexandrov & Sljusarev) (Mesozoa: Orthonectida). *Acta Zoologica* 75: 311–321.

Slyusarev, G.S. 2000. Fine structure and development of the cuticle of *Intoshia variabili* (Orthonectida). *Acta Zoologica* 81: 1–8.

Slyusarev, G.S. and Kristensen, R. 2003. Fine structure of the ciliated cells and ciliary rootlets of *Intoshia variabilis* (Orthonectida). *Zoomorphology* 122: 33–39.

Small, S. and Levine, M. 1991. The initiation of pair-rule stripes in the *Drosophila* blastoderm. *Current Opinion in Genetics and Development* 1: 255–260.

Smith, A.B. 2005. The pre-radial history of echinoderms. *Geological Journal* 40: 255–280.

Smith, J.P.S. and Tyler, S. 1985. The acoel turbellarians: kingpins of metazoan evolution or a specialized offshoot?, pp. 123–142 in C. Conway Morris, J.D. George, R. Gibson and H.M. Platt (eds), *The Origins and Relationships of Lower Invertebrates.* Oxford University Press, Oxford.

Smith, K.K. 1997. Comparative patterns of craniofacial development in eutherian and metatherian mammals. *Evolution* 51: 1663–1678.

Smith, K.K. 2001. Heterochrony revisited: the evolution of developmental sequences. *Biological Journal of the Linnean Society* 73: 169–186.

Smith, M.J., Arndt, A., Gorski, S., and Fajber, E. 1993. The phylogeny of echinoderm classes based on mitochondrial gene rearrangements. *Journal of Molecular Evolution* 36: 545–554.

Smith, P.R. and Ruppert, E.E. 1988. Nephridia, pp. 231–262 in W. Westheide and C.O. Hermans (eds), *Ultrastructure of Polychaeta (Microfauna Marina 4).* Fischer, Stuttgart.

Smith, P.R., Lombardi, J., and Rieger, R.M. 1986. Ultrastructure of the body cavity lining in a secondary acoelomate, *Microphthalmus* cf. *listensis* Westheide

(Polychaeta: Hesionidae). *Journal of Morphology* 188: 257–271.

Sneath, P.H.A. and Sokal, R.R. 1973. *Numerical Taxonomy.* W.H. Freeman, San Francisco, CA.

Snell, E.A., Furlong, R.F., and Holland, P.W.H. 2001. Hsp70 sequences indicate that choanoflagellates are closely related to animals. *Current Biology* 11: 967–970.

Snodgrass, R.E. 1935. *Principles of Insect Morphology.* McGraw-Hill, NewYork.

Snodgrass, R.E. 1938. Evolution of the Annelida, Onychophora and Arthropoda. *Smithsonian Miscellaneous Collection* 97: 1–159.

Sokal, R.R. and Sneath, P.H.A. 1963. *Principles of Numerical Taxonomy.* W.H. Freeman, San Francisco, CA.

Soldi, R., Ramella, L., Gambi, M.C., Sordino, P., and Sella, G. 1994. Genome size in polychaetes: relationship with body length and life habit, pp. 129–135 in J.-C. Dauvin, L. Laubier and D.J. Reish (eds), *Actes de la 4ième Conférence internationale des Polychètes. Mémoires du Muséum national d'Histoire naturelle, Paris,* 162.

Soliman, M.I. 2003. Ultrastructure of spermatogenesis in the rat lung nematode *Angiostrongylus cantonensis. Journal of the Egyptian-German Society of Zoology* 42(d): 113–124.

Sollas, W.J. 1884. On the development of *Halisarca lobularis* (O. Schmidt). *Quarterly Journal of Microscopical Science* 24: 603–621.

Song, J.L., Wong, J.L., and Wessel, G.M. 2006. Oogenesis: single cell development and differentiation. *Developmental Biology* 300: 385–405.

Sørensen, M.V. 2003. Further structures in the jaw apparatus of *Limnognathia maerski* (Micrognathozoa), with notes on the phylogeny of the Gnathifera. *Journal of Morphology* 255: 131–145.

Sørensen, M.V. and Sterrer, W. 2002. New characters in the gnathostomulid mouth-parts revealed by scanning electron microscopy. *Journal of Morphology* 253: 310–334.

Sørensen, M.V., Funch, P., Willerslev, E., Hansen, A.J. and Olesen, J. 2000. On the phylogeny of the Metazoa in the light of the Cycliophora and Micrognathozoa. *Zoologischer Anzeiger* 239: 297–318.

Soto, A.M. and Sonnenschein, C. 2004. The somatic mutation theory of cancer: growing problems with the paradigm? *BioEssays* 26: 1097–1107.

Spears, T. and Abele, L.G. 1998. Crustacean phylogeny inferred from 18S rDNA, pp. 169–187 in R.A. Fortey and R.H. Thomas (eds), *Arthropod Relationships.* Chapman & Hall, London.

Spears, T., DeBry, R.W., Abele, L.G., and Chodyla, K. 2005. Peracarid monophyly and interordinal phylogeny inferred from nuclear small-subunit ribosomal

DNA sequences (Crustacea: Malacostraca: Peracarida). *Proceedings of the Biological Society of Washington* 118: 117–157.

Spies, R.B. 1977. Reproduction and larval development of *Flabelliderma commensalis* (Moore), pp. 323–345 in D.J. Reish and F. Fauchald (eds), *Essays on Polychaetous Annelids in Memory of Dr. Olga Hartman*. The Allan Hancock Foundation, University of Southern California, Los Angeles, CA.

Spitz, F., Gonzalez F., and Duboule, D. 2003. A gobal control region defines a chromosomal landscape containing the HoxD cluster. *Cell* 113: 405–417.

Sprecher, S.G. and Reichert, H. 2003. The urbilaterian brain: developmental insights into the evolutionary origin of the brain in insects and vertebrates. *Arthropod Structure & Development* 32: 141–156.

Spring J. 2002. Genome duplication strikes back. *Nature Genetics* 31: 128–129.

Spring, J. 2003. Major transitions in evolution by genome fusion: from prokaryotes to eukaryotes, metazoans, bilaterians and vertebrates. *Journal of Structural and Functional Genomics* 3: 19–25.

Spring, J., Yanze, N., Middel, A.M., Stierwald, M., Groeger, H., and Schmid, V. 2000. Ancestral role of the mesoderm specification factor Twist in the life cycle of jellyfish. *Developmental Biology* 228: 363–375.

Spring, J., Yanze, N., Jösch, C., Middel, M., Winninger, B., and Schmid, V. 2002. Conservation of *Brachyury*, *Mef2*, and *Snail* in the myogenic lineage of jellyfish: a connection to the mesoderm of Bilateria. *Developmental Biology* 244: 372–384.

Springer, M.S., Cleven, G., Madsen, O. *et al.* 1997. Endemic African mammals shake the phylogenetic tree. *Nature* 388: 61–64.

Sprinkle, J. 1973. *Morphology and Evolution of Blastozoan Echinoderms*. Special Publication. Harvard University Museum of Comparative Zoology, Cambridge, MA.

Stach, T. and Turbeville, J.M. 2002. Phylogeny of Tunicata inferred from molecular and morphological characters. *Molecular Phylogenetics and Evolution* 25: 408–428.

Stanhope, M.J., Waddell, V.G., Madsen, O. *et al.* 1998. Molecular evidence for multiple origins of Insectivora and for a new order of endemic African insectivore mammals. *Proceedings of the National Academy of Sciences USA* 95: 9967–9972.

Stark, A., Brennecke, J., Bushati, N., Russell, R.B., and Cohen, S.M. 2005. Animal microRNAs confer robustness to gene expression and have a significant impact on 3′UTR evolution. *Cell* 123: 1–14.

Staton, J.L. 2003. Phylogenetic analysis of the mitochondrial cytochrome c oxidase subunit 1 gene from 13

sipunculan genera: intra- and interphylum relationships. *Invertebrate Biology* 122: 252–264.

Stauber, M., Taubert, H. and Schmidt-Ott, U. 2000 Function of *bicoid* and *hunchback* homologs in the basal cyclorrhaphan fly *Megaselia* (Phoridae). *Proceedings of the National Academy of Sciences USA* 97: 10844–10849.

Steenkamp, E.T., Wright, J. and Baldauf, S.L. 2006. The protistan origins of animals and fungi. *Molecular Biology and Evolution* 23: 93–106.

Steenstrup, J.J.S. 1845. *On the Alternation of Generation or the Propagation and Development of Animals through Alternate Generations*. Ray Society, London.

Steinauer, M.L., Nickol, B.B., Broughton, R., and Orti, G. 2005. First sequenced mitochondrial genome from the phylum Acanthocephala (*Leptorhynchoides thecatus*) and its phylogenetic position within Metazoa. *Journal of Molecular Evolution* 60: 706–715.

Steinböck, O. 1963. Origin and affinities of the lower Metazoa: the "aceloid" ancestry of the Eumetazoa, pp. 40–54 in E.C. Dougherty (ed.), *The Lower Metazoa. Comparative Biology and Phylogeny*. University of California Press, Berkeley, CA.

Steinböck, O. and Ausserhofer, B. 1950. Zwei grundverschiedene Entwicklungsabläufe bei einer Art (*Prorhynchus stagnatilis* M. Sch., Turbellaria). *Archiv für Entwicklungsmechanik* 144: 155–177 [in German].

Steiner, G. 2004. Neuralgische Punkte in der Phylogenie der Mollusca. *Sitzungsberichte der Gesellschaft Naturforschender Freunde zu Berlin, Neue Folge* 43: 51–71 [in German].

Steiner, G. and Dreyer, H. 2003. Molecular phylogeny of Scaphopoda (Mollusca) inferred from 18S rDNA sequences: support for a Scaphopoda-Cephalopoda clade. *Zoologica Scripta* 32: 343–356.

Steiner, G. and Salvini-Plawen, L. 2001. *Acaenopax*—polychaete or mollusc? *Nature* 414: 602.

Steiner, M., Zhu, M., Li, G., Qian, Y. and Erdtmann, B.-D. 2004. New Early Cambrian bilaterian embryos and larvae from China. *Geology* 32: 833–836.

Steinmetz, P.R.H., Zelada-Gonzáles, F., Burgtorf, C., Wittbrodt, J. and Arendt, D. 2007. Polychaete trunk neuroectoderm converges and extends by mediolateral cell intercalation. *Proceedings of the National Academy of Sciences USA* 104: 2727–2732.

Sterelny, K. 2001. Niche construction, developmental systems, and the extended replicator, pp. 333–349 in S. Oyama, P.E. Griffiths and R.D. Gray (eds), *Cycles of Contingency. Developmental Systems and Evolution*. MIT Press, Cambridge, MA.

Stern, C.D., Charité, J., Deschamps, J. *et al.* 2006. Head-tail patterning of the vertebrate embryo: one, two or many unresolved problems? *International Journal of Developmental Biology* 50: 3–15.

Sterrer, W., Mainitz, M., and Rieger, R.M. 1985. Gnathostomulida: enigmatic as ever, pp. 181–199 in S. Conway Morris, J.D. George, R. Gibson and H.M. Platt (eds), *The Origins and Relationships of Lower Invertebrates.* Clarendon Press, Oxford.

Stollewerk, A. 2002. Recruitment of cell groups through Delta/Notch signalling during spider neurogenesis. *Development* 129: 5339–5348.

Stollewerk, A. 2008. Evolution of neurogenesis in arthropods, pp. 359–380 in A. Minelli and G. Fusco (eds), *Evolving Pathways: Key Themes in Evolutionary Developmental Biology.* Cambridge University Press, Cambridge.

Stollewerk, A. and Chipman, A.D. 2006. Neurogenesis in myriapods and chelicerates and its importance for understanding arthropod relationships. *Integrative and Comparative Biology* 46: 195–206.

Stollewerk, A., Weller, M. and Tauz, D. 2001. Neurogenesis in the spider *Cupiennius salei. Development* 128: 2673–2688.

Stollewerk, A., Schoppmeier, M., and Damen, W.G.M. 2003. Involvement of *Notch* and *Delta* genes in spider segmentation. *Nature* 423: 863–865.

Stone, J.R. and Hall, B.K. 2004. Latent homologues for the neural crest as an evolutionary novelty. *Evolution & Development* 6: 123–129.

Storch, V. 1991. Priapulida, pp. 333–350 in F.W. Harrison and E.E. Ruppert (eds), *Microscopic Anatomy of Invertebrates. Vol. 4. Aschelminthes.* Wiley-Liss, New York.

Storch, V., Higgins, R.P., and Morse, M.P. 1989. The internal anatomy of *Meiopriapulus fijiensis* (Priapulida). *Transactions of the American Microscopical Society* 108: 245–261.

Stotz, H.H., Augustin, R., Khalturin, K. *et al.* 2003. Novel approaches for the analysis of immune reactions in tunicate and cnidarian model organisms, pp. 127–132 in A. Legakis, S. Sfenthourakis, R. Polymeni, and M. Thessalou-Legaki (eds), *The New Panorama of Animal Evolution.* Proceedings of the 18th International Congress of Zoology. Pensoft, Sofia/Moscow.

Strathmann, R.R. 2000. Functional design in the evolution of embryos and larvae. *Seminars in Cell & Developmental Biology* 11: 395–402.

Strausfeld, N.J., Strausfeld, C.M., Loesel, R., Rowell D., and Stowe S. 2006. Arthropod phylogeny: onychophoran brain organization suggests an archaic relationship with a chelicerate stem lineage. *Proceedings of the Royal Society of London Series B, Biological Sciences* 273: 1857–1866.

Striedter, G.F. 1998. Stepping into the same river twice: homologues as recurring attractors in epigenetic landscapes. *Brain Behavior and Evolution* 52: 218–231.

Strogatz, S. 2003. *Sync: How Order Emerges from Chaos in the Universe, Nature, and Daily Life.* Hyperion, New York.

Struck, T.H. and Fisse, F. 2008. Phylogenetic position of Nemertea derived from phylogenomic data. *Molecular Biology and Evolution* 25: 728–736.

Struck, T.H. and Purschke, G. 2005. The sister group relationship of Aeolosomatidae and Potamodrilidae (Annelida: 'Polychaeta')—a molecular phylogenetic approach based on 18S rDNA and cytochrome oxidase I. *Zoologischer Anzeiger* 243: 281–293.

Struck, T.H., Schult, N., Kusen, T. *et al.* 2007. Annelid phylogeny and the status of Sipuncula and Echiura. *BMC Evolutionary Biology* 7: 57.

Studnička, F. K. 1934. The symplastic state of the tissues of the animal body. *Biological Reviews and Biological Proceedings of the Cambridge Philosophical Society* 9: 263–298.

Sulston, J.E., Schierenberg, E., White, J.G. and Thomson, J.N. 1983. The embryonic cell lineage of the nematode *Caenorhabditis elegans. Developmental Biology* 100: 64–119.

Sumrall, C.D. and Sprinkle, J. 1999. Early ontogeny of the glyptocystitid rhombiferan *Lepadocystis moorei,* pp. 409–414 in M.D. Candia Carnevali and F. Bonasoro (ed.), *Echinoderm Research 1998.* A.A. Balkema, Rotterdam.

Sumrall, C.D. and Wray, G.A. 2007. Ontogeny in the fossil record: diversification of body plans and the evolution of "aberrant" symmetry in Paleozoic echinoderms. *Paleobiology* 33: 149–163.

Sun, W.-G. and Hou, X.-G. 1987. Early Cambrian medusae from Chengjiang, Yunnan, China. *Acta Palaeontologica Sinica* 26: 257–271.

Sutton, M.D., Briggs, D.E.G., Siveter, D.J., and Siveter, D.J. 2001. An exceptionally preserved vermiform mollusc fron the Silurian of England. *Nature* 410: 461–463.

Sweet, W.C. and Donoghue, P.C.J. 2001. Conodonts: past, present, future. *Journal of Paleontology* 75: 1174–1184.

Syed, T. and Schierwater, B. 2002. The evolution of the Placozoa: a new morphological model. *Senckenbergiana Lethaea* 82: 315–324.

Szalay, F.S., Novacek, M.J., and McKenna, M.C. (eds) 1993. *Mammal Phylogeny: Placentals.* Springer-Verlag, New York.

Szaniawski, H. 2002. New evidence for the protoconodont origin of chaetognaths. *Acta Palaeontologica Polonica* 47: 405–419.

Szaniawski, H. 2005. Cambrian chaetognaths recognized in Burgess Shale fossils. *Acta Palaeontologica Polonica* 50: 1–8.

Tabin, C.J., Carroll, S.B. and Panganiban, G. 1999. Out on a limb: parallels in vertebrate and invertebrate limb patterning and the origin of appendages. *American Zoologist* 39: 650–663.

Tagawa, K., Satoh, N. and Humphreys, T. 2001. Molecular studies of hemichordate development: a key to understanding the evolution of bilateral animals and chordates. *Evolution and Development* 3: 443–454.

Takezaki, N., Figueroa, F., Zaleska-Rutczynska, Z., Takahata, N., and Klein, J. 2004. The phylogenetic relationship of tetrapod, coelacanth, and lungfish revealed by the sequences of forty-four nuclear genes. *Molecular Biology and Evolution* 21: 1512–1524.

Tanaka, K. and Truman, J.W. 2007. Molecular patterning mechanisms underlying metamorphosis of the thoracic leg in *Manduca sexta*. *Developmental Biology* 305: 539–560.

Tao, Y. and Schulz, R.A. 2007. Heart development in *Drosophila*. *Seminars in Cell and Developmental Biology* 18: 3–15.

Tarchini, B., Duboule, D. and Kmita, M. 2006. Regulatory constraints in the evolution of the tetrapod limb anterior-posterior polarity. *Nature* 443: 985–988.

Tartar, V. 1961. *The Biology of* Stentor. Pergamon Press, Oxford.

Tattersall, W.M. and Sheppard, E.M. 1934. Observations on the bipinnaria of the asteroid genus *Luidia*, pp. 35–61 in R.J. Daniel (ed.), *James Johnstone Memorial Volume*. University Press of Liverpool, Liverpool.

Taube, E. 1909. Beiträge zur Entwicklungsgeschichte der Euphausiden. I. Die Furchung des Eies bis zur Gastrulation. *Zeitschrift für wissenschaftliche Zoologie* 92: 427–464 [in German].

Tautz, D. 2004. Segmentation. *Developmental Cell* 7: 301–312.

Technau, U. 2001. *Brachyury*, the blastopore and the evolution of the mesoderm. *BioEssays* 23: 788–794.

Technau, U. and Bode, H.R. 1999. *HyBra1*, a *Brachyury* homologue, acts during head formation in *Hydra*. *Development* 126: 999–1010.

Technau, U. and Scholz, C.B. 2003. Origin and evolution of endoderm and mesoderm. *International Journal of Developmental Biology* 47: 531–539.

Technau, U., Miller, M.A., Bridge, D., and Steele, R.E. 2003. Arrested apoptosis of nurse cells during *Hydra* oogenesis and embryogenesis. *Developmental Biology* 260: 191–206.

Technau, U., Rudd, S., Maxwell, P. *et al.* 2005. Maintenance of ancestral complexity and non-metazoan genes in two basal cnidarians. *Trends in Genetics* 21: 633–639.

Telford, M.J. 2000. Turning Hox "signatures" into synapomorphies. *Evolution & Development* 2: 360–364.

Telford, M.J. 2004. The multimeric β-thymosin found in nematodes and arthropods is not a synapomorphy of the Ecdysozoa. *Evolution & Development* 6: 90–94.

Telford, M.J. and Thomas, R.H. 1998. Expression of homeobox genes shows chelicerate arthropods retain their deutocerebral segment. *Proceedings of the National Academy of Sciences USA* 95: 10671–10675.

Telford, M.J. and Littlewood, D.T.J. (eds) 2008. Evolution of the animals: a Linnean tercentenary celebration. *Philosophical Transactions of the Royal Society Series B, Biological Sciences* 363: 1421–1457.

Telford, M.J., Herniou, E.A., Russell, R.B. and Littlewood, D.T.J. 2000. Changes in mitochondrial genetic codes as phylogenetic characters: two examples from the flatworms. *Proceedings of the National Academy of Sciences USA* 97: 11359–11364.

Telford, M.J., Lockyer, A.E., Cartwright-Finch, C. and Littlewood, D.T.J. 2003. Combined large and small subunit ribosomal RNA phylogenies support a basal position of the acoelomorph flatworms. *Proceedings of the Royal Society of London Series B, Biological Sciences* 270: 1077–1083.

Terry, M.D. and Whiting, M.F. 2005. Mantophasmatodea and phylogeny of lower [*sic*] neopterous insects. *Cladistics* 21: 240–257.

Thewissen, J.G.M. and Madar, S.I. 1999. Ankle morphology of the earliest cetaceans and its implications for the phylogeneti relations among ungulates. *Systematic Biology* 48: 21–30.

Thewissen, J.G.M., Madar, S.I., and Hussain, S.T. 1996. *Ambuloketus natans*, an Eocene cetacean (Mammalia) from Pakistan. *Courier des Forschungsinstituts Senckenberg* 191: 1–86.

Thisse, C. and Zon, L.I. 2002. Organogenesis—heart and blood formation from the zebrafish point of view. *Science* 295: 457–462.

Thollesson, M. and Norenburg, J.L. 2003. Ribbon worm relationships: a phylogeny of the phylum Nemertea. *Proceedings of the Royal Society of London Series B, Biological Sciences* 270: 407–415.

Thomas, J.A., Welch, J.J., Woolfit, M., and Bromham, L. 2006. There is no universal molecular clock for invertebrates, but rate variation does not scale with body size. *Proceedings of the National Academy of Sciences USA* 103: 7366–7371.

Thomas, M.B., Freeman, G., and Martin, V.J. 1987. The embryonic origin of neurosensory cells and the role of nerve cells in metamorphosis of *Phialidium gregarium* (Cnidaria, Hydrozoa). *International Journal of Invertebrate Reproduction and Development* 11: 265–287.

Thompson, J.V. 1830a. On the cirripedes or barnacles; demonstrating their deceptive character; the extraordinary metamorphosis they undergo, and the class of animals to which they undisputably belong, pp. 69–85 in J.V. Thomson, *Zoological Researches*. Memoir IV. King and Ridings, Cork.

Thompson, J.V. 1830b. On Polyzoa, a new animal discovered as an inhabitant of some Zoophytes—with a description of the newly instituted genera of *Pedicellaria* and *Vesicularia*, and their species, pp. 89–102 in J.V. Thompson, *Zoological Researches*. Memoir V. King and Ridings, Cork.

Thompson, T.E. 1960. The development of *Neomenia carinata* Tullberg (Mollusca Aplacophora). *Proceedings of the Royal Society of London Series B, Biological Sciences* 153: 263–278.

Thomson, J.M., Newman, M., Parker, J.S., Morin-Kensicki, E.M., Wright, T., and Hammond, S.M. 2006. Extensive post-transcriptional regulation of microRNAs and its implications for cancer. *Genes & Development* 20: 2202–2207.

Thorson, G. 1950. Reproductive and larval ecology of marine bottom invertebrates. *Biological Reviews* 25: 1–45.

Tiegs, O.W. and Manton, S.M. 1958. The evolution of the Arthropoda. *Biological Reviews* 33: 255–337.

Timm, T. 1981. On the origin and evolution of aquatic Oligochaeta. *Eesti NSV Teaduste Akadeemia Toimetised. Biologiiline Seeria* 30: 174–181.

Todaro, M.A., Telford, M.J., Lockyer, A.E., and Littlewood, D.T.J. 2006. Interrelationships of the Gastrotricha and their place among the Metazoa inferred from 18S rRNA genes. *Zoologica Scripta* 35: 251–259.

Torras, R., Yanze, N., Schmid, V., and González-Crespo, S. 2004. *nanos* expression at the embryonic posterior pole and the medusa phase in the hydrozoan *Podocoryne carnea*. *Evolution & Development* 6: 362–371.

Townsend, T.M., Larson, A., Louis, E. and Macey, J.R. 2004. Molecular phylogenetics of Squamata: the position of snakes, amphisbaenians, and dibamids, and the root of the squamate tree. *Systematic Biology* 53: 735–757.

Treccani, L., Khoshnavaz, S., Blank, S. *et al.* 2003. Biomineralizing proteins with emphasis on invertebrate mineralized structures, pp. 289–321 in S.R. Fahnenstock and A. Steibüchl (eds), *Biopolymers*, vol. 8. Wiley-VCH, Weinheim.

Treisman, J.E. 2004. Coming to our senses. *BioEssays* 26: 825–828.

True, J.R. and Carroll, S.B. 2002. Gene co-option in physiological and morphological evolution. *Annual Review of Cell and Developmental Biology* 18: 53–80.

Truman, J.W. and Riddiford, L.M. 1999. The origins of insect metamorphosis. *Nature* 401: 447–452.

Tsuda, M., Sasaoka, Y., Kiso, M. *et al.* 2003. Conserved role of nanos proteins in germ cell development. *Science* 301: 1239–1241.

Turbeville, J.M. 1986. An ultrastructural analysis of coelomogenesis in the hoplonemertine *Prosorhochnus americanus* and the polychaete *Magelona* sp. *Journal of Morphology* 187: 51–60.

Turbeville, J.M. 1991. Nemertinea, pp. 285–328 in F.W. Harrison and B.J. Bogitsh (eds), *Microscopic Anatomy of Invertebrates. Vol. 3. Platyhelminthes and Nemertinea.* Wiley-Liss, New York.

Turbeville, J.M. 1996. Nemertini, Schnurwürmer, pp. 265–275 in W. Westheide and R.M. Rieger (eds), *Spezielle Zoologie. Erster Teil: Einzeller und Wirbellose Tiere.* Gustav Fischer Verlag, Stuttgart.

Turbeville, J.M. and Ruppert, E.E. 1985. Comparative ultrastructure and the evolution of nemertines. *American Zoologist* 25: 53–71.

Turbeville, J.M., Schulz, J.R., and Raff, R.A. 1994. Deuterostome phylogeny and the sister group of the chordates: evidence from molecules and morphology. *Molecular Biology and Evolution* 11: 648–655.

Tuzet, O. 1963. The phylogeny of sponges according to embryological, histological, and serological data, and their affinities with the Protozoa and the Cnidaria, pp. 129–148 in E.C. Dougherty (ed.), *The Lower Metazoa: Comparative Biology and Phylogeny*. University of California Press, Berkeley, CA.

Tuzet, O. 1973. Éponges calcaires, pp. 27–132 in P.-P. Grassé (ed.), *Traité de Zoologie*, vol. 3(1). Masson, Paris.

Tyler, S. 2001. The early worm—origins and relationships of the lower flatworms, pp. 3–12 in D.T.J. Littlewood and R.A. Bray (eds), *Interrelationships of the Platyhelminthes*. Taylor and Francis, London.

Tyler, S. 2002. *Platyhelminthes. The Nature of a Controversial Phylum.* www.developmentalbiology.com/styler/globalworming/platyhelm2002.html

Tyler, S. 2003. Epithelium—the primary building block for metazoan complexity. *Integrative and Comparative Biology* 43: 55–63.

Tyler, S. and Rieger, R.M. 1999. Functional morphology of musculature in the acoelomate worm, *Convoluta pulchra* (Plathelminthes). *Zoomorphology* 119: 127–141.

Tyler, S. and Hoodge, M. 2004. Comparative morphology of the body wall in flatworms (Platyhelminthes). *Canadian Journal of Zoology* 82: 194–210.

Tzetlin, A.B. and Filippova, A.V. 2005. Muscular system in polychaetes (Annelida). *Hydrobiologia* 535/536: 113–126.

Tzetlin, A.B. and Purschke, G. 2006. Fine structure of the pharyngeal apparatus of the pelagosphera larva in *Phascolosoma agassizii* (Sipuncula) and its phylogenetic significance. *Zoomorphology* 125: 109–117.

Tzetlin, A.B., Zhadan, A., Ivanov, I., Müller, M.C.M., and Purschke, G. 2002. On the absence of circular muscle elements in the body wall of *Dysponetus pygmaeus* (Chrysopetalidae, 'Polychaeta', Annelida). *Acta Zoologica* 83: 81–85.

Ullmann, A.J., Lima, C.M.R., Guerrero, F.D., Piesman, J., and Black, W.C. 2005. Genome size and organization in the blacklegged tick, *Ixodes scapularis* and the Southern cattle tick, *Boophilus microplus*. *Insect Molecular Biology* 14: 217–222.

Ulrich, W. 1950. Begriff und Einteilung der Protozoen, pp. 241–250 in H. Grüneberg and W. Ulrich (eds), *Moderne Biologie, Festschrift zum 60. Geburtstag von Hans Nachtsheim*, F.W. Peters, Berlin [in German].

Ulrich, W. 1951. Vorschläge zu einer Revision der Grosseinteilung des Tierreichs. *Verhandlungen der Deutschen Zoologischen Gesellschaft*, Marburg 1950: 215–271 [in German].

Urbach, R. and Technau, G.M. 2003, Segment polarity and DV patterning gene expression reveals segmental organization of the *Drosophila* brain. *Development* 130: 3607–3620.

Urdy, S. and Chirat, R. 2006. Snail shell coiling (re-)evolution and the evo-devo revolution. *Journal of Zoological Systematics and Evolutionary Research* 44: 1–7.

Vaccari, N.E., Edgecombe, G.D. and Escudero, C. 2004. Cambrian origins and affinities of an enigmatic fossil group of arthropods *Nature* 430: 554–557.

Vacelet, J. 1999. Planktonic armoured propagules of the excavating sponge *Alectona* (Porifera: Demospongiae) are larvae: evidence from *Alectona wallichii* and *A. mesatlantica* sp. nov. *Memoirs of the Queensland Museum* 44: 627–642.

Vacelet, J. 2006. New carnivorous sponges (Porifera, Poecilosclerida) collected from manned submersibles in the deep Pacific. *Zoological Journal of the Linnean Society* 148: 553–584.

Vacelet, J. and Boury-Esnault, N. 1995. Carnivorous sponges. *Nature* 373: 333–335.

Valentine, J.W. 2004. *On the Origin of Phyla*. University of Chicago Press, Chicago, IL.

Valentine, J.W. and Collins, A.G. 2000. The significance of moulting in ecdysozoan evolution. *Evolution & Development* 2: 152–156.

Valentine, J.W., Erwin, D.H., and Jablonski, D. 1996. Developmental evolution of metazoan bodyplans: the fossil evidence. *Developmental Biology* 173: 373–381.

Valentine, J.W., Jablonski, D., and Erwin, D.H. 1999. Fossils, molecules and embryos: new perspectives on the Cambrian explosion. *Development* 126: 851–859.

van Beneden, É. 1876. Recherches sur les Dicyémides, survivants actuels d'un embranchement des Mesozoaires.

Bulletin de l'Académie royale de Belgique (2)41: 1160–1205; 42: 35–97 [in French].

van Beneden, É. 1882. Contribution à l'histoire des Dicyémides. *Archives de Biologie, Paris* 3: 197–228 [in French].

van den Biggelaar. J.A.M., van Loon. A.E., and Damen, W.G.M. 1996. Mesentoblast and trochoblast specification in species with spiral cleavage predict their phyletic relations. *Netherlands Journal of Zoology* 46: 8–21.

van den Biggelaar, J.A.M., Dictus, W.J.A.G. and van Loon, A.E. 1997. Cleavage patterns, cell-lineages and cell specification are clues to phyletic lineages in Spiralia. *Seminars in Cell and Developmental Biology* 8: 367–378.

van den Biggelaar, J.A.M., Edsinger-Gonzales, E. and Schram, F. 2002. The improbability of dorso-ventral axis inversion during animal evolution. *Contributions to Zoology* 71: 29–36.

Van de Peer, Y., Taylor, J.S. and Meyer, A. 2003. Are all fishes ancient polyploids? *Journal of Structural and Functional Genomics* 2: 65–73.

Van de Vyver, G. 1993. [Classe des Hydrozoaires] Reproduction sexuée—Embryologie, pp. 417–473 in P.P. Grassé (ed.), *Traité de Zoologie*, 3(2). Masson, Paris [in French].

Van Iten, H., de Leme, J.M., Simoes, M.G., Marques, A.C., and Collins, A.G. 2006. Reassessment of the phylogenetic position of conulariids (?Ediacaran-Triassic) within the subphylum Medusozoa (phylum Cnidaria). *Journal of Systematic Palaeontology* 4: 109–118.

van Loon, A.E. and van den Biggelaar, J.A.M. 1998. Changes in cell lineage specification elucidate evolutionary relations in Spiralia. *Biological Bulletin* 195: 367–369.

Vejdovsky, F. 1886. Zur Morphologie der Gordiiden. *Zeitschrift für wissenschaftliche Zoologie* 43: 369–433 [in German].

Velhagen, Jr, W.A. 1997. Analyzing developmental sequences using sequence units. *Systematic Biology* 46: 204–210.

Verger-Bocquet, M. 1992. Polychaeta: sensory structures, pp. 181–196 in F.W. Harrison and S.L. Gardiner (eds), *Microscopic Anatomy of Invertebrates. Vol. 7. Annelida*. Wiley-Liss, New York.

Verhoeff, K.W. 1923. Periodomorphose. *Zoologischer Anzeiger* 56: 233–238, 241–254 [in German].

Vermeij, G.J. 1996. Animal origins. *Science* 274: 525–526.

Vermeij, G.J. and Dudley, R. 2000. Why are there so few evolutionary transitions between aquatic and terrestrial ecosystems? *Biological Journal of the Linnean Society* 70: 541–554.

Vermot, J. and Pourquié, O. 2005. Retinoic acid coordinates somitogenesis and left/right patterning in vertebrate embryos. *Nature* 435: 215–220.

Vidal, G. 1997. Biodiversity, speciation, and extinction trends of Proterozoic and Cambrian phytoplankton. *Paleobiology* 23: 230–246.

Vidal, N. and Hedges, S.B. 2004. Molecular evidence for a terrestrial origin of snakes. *Proceedings of the Royal Society of London Series B, Biological Sciences* 271 (suppl. 4): S226–S229.

Vilpoux, K. and Waloszek, D. 2003. Larval development and morphogenesis of the sea spider *Pycnogonum litorale* (Stroem, 1762) and the tagmosis of the body of Pantopoda. *Arthropod Structure and Development* 32: 349–383.

Vinogradova, I., Cook, A., and Holden-Dye, L. 2006. The ionic dependence of voltage-activated inward currents in the pharyngeal muscle of *Caenorhabditis elegans*. *Invertebrate Neuroscience* 6: 57–68.

Vinther, J. and Nielsen, C. 2005. The Early Cambrian *Halkieria* is a mollusc. *Zoologica Scripta* 34: 81–89.

Voigt, E. 1938. Ein fossiler Saitenwurm (*Gordius tenuifibrosus*) aus der Eozänen Braunkohle des Geiseltales. *Nova Acta Leopoldina, Neue Folge* 5: 53–56 [in German].

Voigt, O., Collins, A.G., Pearse, V.B. *et al.* 2004. Placozoa: no longer a phylum of one. *Current Biology* 14: R944–R945.

von Baldass, F. 1937. Die Entwicklung von *Holopedium gibberum*. *Zoologische Jahrbücher, Abteilung für Anatomie* 63: 399–454 [in German].

von Baldass, F. 1941. Die Entwicklung von *Daphnia pulex*. *Zoologische Jahrbücher, Abteilung für Anatomie* 67: 1–60 [in German].

von Graff, L. 1882. *Monographie der Turbellarien. 1. Rhabdocoelida*. Wilhelm Engelmann, Leipzig [in German].

von Graff, L. 1905. Marine Turbellarien Orotavas und der Küsten Europas. Teil II. Rhabdocoela. *Zeitschrift für wissenschaftliche Zoologie* 83: 68–154 [in German].

Vonnemann, V., Schroedl, M., Klussmann-Kolb, A., and Waegele, H. 2005. Reconstruction of the phylogeny of the Opisthobranchia (Mollusca: Gastropoda) by means of 18S and 28S rRNA gene sequences *Journal of Molluscan Studies* 71: 113–125.

von Nickisch-Rosenegk, M., Brown, W.M. and Boore, J.L. 2001. Complete sequence of the mitochondrial genome of the tapeworm *Hymenolepis diminuta*: gene arrangements indicate that platyhelminths are eutrochozoans. *Molecular Biology and Evolution* 18: 721–730.

von Salvini-Plaven, L. 1989. Mesoderm heterochrony and metamery in Chordata. *Fortschritte der Zoologie* 35: 213–219.

von Salvini-Plawen, L. 2003. On the phylogenetic significance of the aplacophoran Mollusca. *Iberus* 21: 67–97.

Vorbach, C., Capecchi, M.R. and Penninger, J.M. 2006. Evolution of the mammary gland from the innate immune system? *BioEssays* 28: 606–616.

Wada, H. 2001a. Origin and evolution of the neural crest: a hypothetical reconstruction of its evolutionary history. *Development Growth and Differentiation* 43: 509–520.

Wada, H. 2001b. The origin of the neural crest and insights into evolution of the vertebrate face, pp. 235–240 in H. Sawada, H. Yokosawa and C.C. Lambert (eds), *The Biology of Ascidians*. Springer, Tokyo.

Wada, H. and Satoh, N. 1994. Details of the evolutionary history from invertebrates to vertebrates, as deduced from the sequences of 18S rDNA. *Proceedings of the National Academy of Sciences USA* 91: 1801–1804.

Wada, H., Okuyama, M., Satoh, N., and Zhang, S. 2006. Molecular evolution of fibrillar collagen in chordates, with implications for the evolution of vertebrate skeletons and cordate phylogeny. *Evolution & Development* 8: 370–377.

Waddell, P.J., Cao, Y., Hauf, J. and Hasegawa, M. 1999a. Using novel phylogenetic methods to evaluate mammalian mtDNA, including amino acid-invariant sites-LogDet plus site stripping to detect internal conflicts in the data, with special reference to the positions of hEdgehog, armadillo and elephant. *Systematic Biology* 48: 31–53.

Waddell, P.J., Cao, Y., Hasegawa, M., and Mindell, D.P. 1999b. Assessing the Cretaceous superordinal divergence times within birds and placental mammals by using whole mitochondrial protein sequences and an extended statistical framework. *Systematic Biology* 48: 119–137.

Waddell, P.J., Okada, N. and Hasegawa, M. 1999c. Towards resolving the interordinal relationships of placental mammals. *Systematic Biology* 48: 1–5.

Waddell, P.J., Kishino, H., and Ota, R. 2001. A phylogenetic foundation for comparative mammalian genomics. *Genome Informatics* 12: 141–154.

Wade, C.M., Mordan, P.B., and Naggs, F. 2006. Evolutionary relationships among the pulmonate snails and slugs (Pulmonata, Stylommatophora). *Biological Journal of the Linnean Society* 87: 593–610.

Wägele, H. and Klussmann-Kolb, A. 2005. Opisthobranchia (Mollusca, Gastropoda)—more than just slimy slugs. Shell reduction and its implication on defence and foraging. *Frontiers in Zoology* 2: 3.

Wägele, J.-W., Erikson, T., Lockhart, P., and Misof, B. 1999. The Ecdysozoa: artifact or monophylum? *Journal of Zoological Systematics and Evolutionary Research* 39: 165–176.

Waggoner, B.M. 1996. Phylogenetic hypotheses of the relationships of arthropods to Precambrian and Cambrian problematic fossil taxa. *Systematic Biology* 45: 190–222.

Waggoner, B. 2003. The Ediacaran biotas in space and time. *Integrative and Comparative Biology* 43: 104–113.

Wagner, F. 1890. Zur Kenntnis der ungeschlechtlichen Fortpflanzung von *Microstoma* nebst allegemeinen Bemerkungen über Teilung und Knospung im Tierreich. *Zoologische Jahrbücher, Abtheilung für Anatomie und Ontogenie der Thiere* 4: 349– 423 [in German].

Wagner, G.P. 1989. The biological homology concept. *Annual Review of Ecology and Systematics* 20: 51–69.

Wagner, G.P. 1994. Homology and the mechanisms of development, pp. 273–299 in B.K. Hall (ed.), *Homology: the Hierarchical Basis of Comparative Biology*. Academic Press, San Diego, CA.

Wagner, G.P. 1996. Homologues, natural kinds and the evolution of modularity. *American Zoologist* 36: 36–43.

Wagner, G.P. (ed.) 2001. *The Character Concept in Evolutionary Biology*. Academic Press, San Diego, CA.

Wagner, G.P. and Misof, B.Y. 1993. How can a character be developmentally constrained despite variation in developmental pathways? *Journal of Evolutionary Biology* 6: 449–455.

Wagner, G.P. and Gauthier, J.A. 1999. 1,2,3=2,3,4: a solution to the problem of the homology of the digits in the avian hand. *Proceedings of the National Academy of Sciences USA* 96: 5111–5116.

Wagner, P.J. and Erwin, D.H. 2006. Patterns of convergence in general shell form among Paleozoic gastropods. *Paleobiology* 32: 316–337.

Wagner, G.P., Amemiya, C. and Ruddle F. 2003. Hox cluster duplications and the opportunity for evolutionary novelties. *Proceedings of the National Academy of Sciences USA* **100**: 14603–6.

Wainright, P.O., Hinkle, G., Sogin, M.L., and Stickel, S.K. 1993. The monophyletic origins of the Metazoa: an unexpected evolutionary link with Fungi. *Science* 260: 340–342.

Walcott, C.D. 1911. Middle Cambrian holothurians and medusae. Cambrian geology and paleontology II. *Smithsonian Miscellaneous Collections* 57: 1109–1144.

Wallace, R.L., Ricci, C., and Melone, G. 1996. A cladistic analysis of pseudocoelomate (aschelminth) morphology. *Invertebrate Biology* 115: 104–112.

Wallberg, A., Thollesson, M., Farris, J.S., and Jondelius, U. 2004. The phylogenetic position of the comb jellies (Ctenophora) and the importance of taxonomic sampling. *Cladistics* 20: 558–578.

Wallberg, A, Curini-Galletti, M., Ahmadzadeh, A., and Jondelius, U. 2007. Dismissal of Acoelomorpha: Acoela and Nemertodermatida are separate, early bilaterian clades. *Zoologica Scripta* 36: 509–523.

Waller, T.R. 1998. Origin of the molluscan class Bivalvia and a phylogeny of the major groups, pp. 1–45 in P.A. Johnston and J.W. Haggard (eds), *Bivalves: an Eon of Evolution*. University of Calgary Press, Calgary, Alberta.

Waloszek, D. 1993 The Upper Cambrian *Rehbachiella* and the phylogeny of Branchiopoda and Crustacea. *Fossils & Strata* 32: 1–202.

Waloszek, D. 1999. On the Cambrian diversity of Crustacea, pp. 3–27 in F.R. Schram and J.C. von Vaupel Klein (eds), *Crustaceans and the Biodiversity Crisis*. Proceedings of the Fourth International Crustacean Congress, Amsterdam. Brill, Leiden.

Waloszek, D. 2003a. Cambrian 'Orsten'-type preserved arthropods and the phylogeny of Crustacea, pp. 69–87 in A. Legakis, S. Sfenthourakis, R. Polymeni, and M. Thessalou-Legaki (eds), *The New Panorama of Animal Evolution*. Proceedings of the 18th International Congress of Zoology. Pensoft, Sofia/Moscow.

Waloszek, D. 2003b. The 'Orsten' window: a three-dimensionally preserved Upper Cambrian meiofauna and its contribution to our understanding of the evolution of Arthropoda. *Paleontological Research* 7: 71–88.

Walossek, D. and Müller, K.J. 1994. Pentastomid parasites from the Lower Palaeozoic of Sweden. *Transactions of the Royal Society of Edinburgh, Earth Sciences* 85(1): 1–37.

Walossek, D., Repetski, J.E., and Müller, K.J. 1994. An exceptionally preserved parasitic arthropod, *Heymonsicambria taylori* n. sp. (Arthropoda incertae sedis: Pentastomida), from Cambrian-Ordovician boundary beds of Newfoundland, Canada. *Canadian Journal of Earth Sciences* 31: 1664–1671.

Waloszek, D., Repetski, J.E., and Maas A. 2006. A new Late Cambrian pentastomid and a review of the relationships of this parasitic group. *Transactions of the Royal Society of Edinburgh: Earth Sciences* 96: 163–176.

Waloszek, D., Maas, A., Chen, J., and Stein, M. 2007. Evolution of cephalic feeding structures and the phylogeny of Arthropoda. *Palaeogeography, Palaeoclimatology, Palaeoecology* 254: 273–287.

Walter, M.R. 1987. The timing of major evolutionary innovations from the origin of life to the origins of the Metaphyta and Metazoa: the geological evidence, pp. 15–38 in K.S.W. Campbell and M.F. Day (eds), *Rates of Evolution*. Allen and Unwin, London.

Walthall, W.W. 1995. Repeating patterns of motoneurons in nematodes: the origin of segmentation?, pp. 61–75 in O. Breidbach and W. Kutsch (eds), *The Nervous System of Invertebrates: an Evolutionary and Comparative Approach*. Birkhäuser, Basel.

Wang, X. and Lavrov, D.V. 2007. Mitochondrial genome of the homoscleromorph *Oscarella carmela* (Porifera, Demospongiae) reveals unexpected complexity in the common ancestor of sponges and other animals. *Molecular Biology and Evolution* 24: 363–373.

Wanninger, A. and Haszprunar, G. 2001. The expression of an engrailed protein during embryonic shell formation of the tusk-shell, *Antalis entails* (Mollusca, Scaphopoda). *Evolution & Development* 3: 312–321.

Wanninger, A., Ruthensteiner, B., Lobenwein, S., Salvenmoser, W., Dictus, W.J.A.G., and Haszprunar, G. 1999. Development of the musculature in the limpet *Patella* (Mollusca, Patellogastropoda). *Development, Genes and Evolution* 209: 226–238.

Wanninger, A., Koop, D., Bromham, L., Noonan, E., and Degnan, B.M. 2005. Nervous and muscular system development in *Phascolion strombus*. *Development, Genes and Evolution* 215: 509–518.

Ward, P., Labandeira, C., Laurin, M., and Berner, R.A. 2006. Confirmation of Romer's Gap as a low oxygen interval constraining the timing of initial arthropod and vertebrate terrestrialization. *Proceedings of the National Academy of Sciences USA* 103: 16818–16822.

Watkins, R.F. and Beckenbach, A.T. 1999. Partial sequence of a sponge mitochondrial genome reveals sequence similarity to cnidaria in cytochrome oxidase subunit II and the large ribosomal RNA subunit. *Journal of Molecular Evolution* 48: 542–554.

Webb, M. 1969. *Lamellibrachia barhami*, gen. nov. sp. nov. (Pogonophora), from the Northeast Pacific. *Bulletin of Marine Science* 19: 18–47.

Webster, M. and Zelditch, M.L. 2005. Evolutionary modifications of ontogeny: heterochrony and beyond. *Paleobiology* 31: 354–372.

Wedeen, C.J. 1995. Regionalization and segmentation of the leech. *Journal of Neurobiology* 27: 277–293.

Wedeen, C.J. and Weisblat, D.A. 1991. Segmental expression of an *engrailed*-class gene during early development and neurogenesis in an annelid. *Development* 113: 805–814.

Wedeen, C.J. and Shankland, M. 1997. Mesoderm is required for the formation of a segmented endodermal cell layer in the leech *Helobdella*. *Developmental Biology* 191: 202–214.

Weisblat, D.A. 2006. Germline regeneration: the worm's turn. *Current Biology* 16: R453–R455.

Weisblat, D.A., Gonsalves, F.C., Huang, F.Z., Kang, D., and Song, M.H. 2004. Evolutionary speculations regarding spiral cleavage and segmentation, based on studies of glossiphoniid leech embryos. *Cladistics* 20: 98–99.

Weiss, M.J. 2001. Widespread hermaphroditism in freshwater gastrotrichs. *Invertebrate Biology* 120: 308–341.

Welsch, U. 1995. Evolution of the body cavities in Deuterostomia, pp. 111–134 in G. Lanzavecchia, R. Valvassori and M.D. Candia Carnevali (eds), *Body Cavities: Function and Phylogeny*. Mucchi, Modena.

Werner, B. 1984. 5. Stamm Ctenophora, Rippenquallen, Kammquallen, pp. 306–335 in H.-E. Gruner (ed.), *Lehrbuch der Speziellen Zoologie begründet von A. Kaestner. Band I: Wirbellose Tiere, 2. Teil: Cnidaria, Ctenophora, Mesozoa, Plathelminthes, Nemertini, Entoprocta, Nemathelminthes, Priapulida*. Gustav Fischer, Stuttgart [in German].

Wesley, C.S. 1999. Notch and wingless regulate expression of cuticle patterning genes. *Cell Biology* 19: 5743–5758.

West Eberhard, M.J. 2003. *Developmental Plasticity and Evolution*. Oxford University Press, Oxford.

Westheide, W. 1967. Monographie der Gattungen *Hesionides* Friedrich und *Microphthalmus* Mecznikow (Polychaeta, Hesionidae). *Zeitschrift für Morphologie der Tiere* 61: 1–159 [in German].

Westheide, W. 1982. *Microphthalmus hamosus* sp.n. (Polychaeta, Hesionidae)—an example of evolution leading from the interstitial fauna to a macrofaunal interspecific relationship. *Zoologica Scripta* 11: 189–193.

Westheide, W. 1987. Progenesis as a principle in meiofauna evolution. *Journal of Natural History* 21: 843–854.

Westheide, W. 1990. *Polychaetes: Interstitial Families*. Synopsis of the British Fauna (New Series) No. 44. Universal Book Services/Dr. W. Backhuys, Oegstgeest.

Westheide, W. 2007. Annelida, pp. 374–437 in W. Westheide and R. Rieger (eds), *Spezielle Zoologie. Teil 1: Einzeller und wirbellose Tiere*. Elsevier, Spektrum Akademischer Verlag, Munich [in German].

Westheide, W. and Rieger, R. (eds), 2007. *Spezielle Zoologie. Teil 1: Einzeller und wirbellose Tiere. Zweite Auflage*. Elsevier, Spektrum Akademischer Verlag, Munich [in German].

Weygoldt, P. 1969. *The Biology of Pseudoscorpions*. Harvard University Press, Cambridge, MA.

Weygoldt, P. 1998. Evolution and systematics of the Chelicerata. *Experimental & Applied Acarology* 22: 63–79.

Wheeler, W.C. 1998a. Molecular systematics and arthropods, pp. 9–32 in G.D. Edgecombe (ed.), *Arthropod Fossils and Phylogeny*, Columbia University Press, New York.

Wheeler, W.C. 1998b. Sampling, groundplans, total evidence and the systematics of arthropods, pp. 87–96 in R.A. Fortey and R.H. Thomas (eds), *Arthropod Relationships*. Chapman & Hall, London.

Wheeler, W.C. and Hayashi, C.Y. 1998. The phylogeny of the extant chelicerate orders. *Cladistics* 14: 173–192.

Wheeler, W.C., Schuh, R.T., and Bang, R. 1993. Cladistic relationships among higher groups of Heteroptera. Congruence between morphological and molecular data sets. *Entomologica Scandinavica* 24: 121–137.

Wheeler, W.C., Whiting, M.F., Wheeler, Q.D., and Carpenter, J.M. 2001. The phylogeny of the extant hexapod orders. *Cladistics* 17: 113–169, 403–404.

Whidden, H.P. 2002. Extrinsic snout musculature in Afrotheria and Lipotyphla. *Journal of Mammalian Evolution* 9: 165–187.

Whiting, M.F. 2001. Mecoptera is paraphyletic: multiple genes and a phylogeny for Mecoptera and Siphonaptera. *Zoologica Scripta* 31: 93–104.

Whiting, M.F. and Wheeler, W.C. 1994. Insect homeotic transformation. *Nature* 368: 696.

Whiting, M.F., Carpenter, J.M., Wheeler, Q.D. and Wheeler, W.C. 1997. The Strepsiptera problem: phylogeny of the holometabolous insect orders inferred from 18S and 28S ribosomal DNA sequences and morphology. *Systematic Biology* 46: 1–68.

Whitington, P.M. 1996. Evolution of neural development in the arthropods. *Seminars in Cell and Developmental Biology* 7: 605–614.

Whitington, P.M. and Bacon, J.P. 1998. The organization and development of the arthropod ventral nerve cord: insights into arthropod relationships, pp. 349–367 in R.A. Fortey and R.H. Thomas (eds), *Arthropod Relationships*. Chapman & Hall, London.

Whitten, J.M. 1968. Metamorphic changes in insects, pp. 48–105 in W. Etkin and L.I. Gilbert (eds), *Metamorphosis: a Problem in Developmental Biology*. Appleton-Century-Crofts, New York.

Wienholds, E. and Plasterk, R.H. 2005. MicroRNA function in animal development. *FEBS Letters* 579: 5911–5922.

Wiens, M., Mangoni, A., D'Esposito, M. *et al.* 2003. The molecular basis for the evolution of the metazoan bodyplan: extracellular matrix-mediated morphogenesis in marine demosponges. *Journal of Molecular Evolution* 57 (suppl. 1): S60–S75.

Wiens, J.J., Bonett, R.M., and Chippindale, P.T. 2005. Ontogeny discombobulates phylogeny: paedomorphosis and higher-level salamander relationships. *Systematic Biology* 54: 91–110.

Wiens, J.J., Brandley, M.C., and Reeder, T.W. 2006. Why does a trait evolve multiple times within a clade? Repeated evolution of snakelike body form in squamate reptiles. *Evolution* 60: 123–141.

Wiley, E.O. 1981. *Phylogenetics. The Theory and Practice of Phylogenetic Systematics*. J. Wiley & Sons, New York.

Wilkin, M.B., Becker, M.N., Mulvey, D. *et al.* 2000. *Drosophila dumpy* is a gigantic extracellular protein required to maintain tension at epidermal-cuticle attachment sites. *Current Biology* 10: 559–567.

Wilkins, A.S. 2002. *The Evolution of Developmental Pathways*. Sinauer Associates, Sunderland, MA.

Williams, D.M. and Ebach, M.C. 2007. The shadow of a shade. *Cladistics* 23: 84–89.

Williamson, D.I. 1973. *Amphionides reynaudii* (H. Milne Edwards), representative of a proposed new order of eucaridan Malacostraca. *Crustaceana* 25: 35–50.

Williamson, D.I. 1982. Larval morphology and diversity, pp. 43–110 in L.G. Abele (ed.), *The Biology of Crustacea, Vol. 2. Embryology, Morphology, and Genetics*, Academic Press, New York.

Williamson, D.J. 2006. Hybridization in the evolution of animal form and life-cycle. *Biological Journal of the Linnean Society* 148: 585–602.

Williamson, R. and Chrachri, A. 2004. Cephalopod neural networks. *Neurosignals* 13: 87–98.

Willmann, R. 2001. Die phylogenetischen Beziehungen der Insecta: offene Fragen und Probleme. *Verhandlungen des Westdeutschen Entomologentages 2001*: 1–64 [in German].

Willmann, R. 2003. Phylogenese und System der Insekten, pp. 1–65 in H.H. Dathe (ed.), *Lehrbuch der Speziellen Zoologie, Band I: Wirbellose Tiere, 5, Teil: Insecta*. Gustav Fischer: Spektrum, Heidelberg [in German].

Willmer, E.N. 1960. *Cytology and Evolution*. Academic Press, New York.

Willmer, P. 2003. Convergence and homoplasy in the evolution of organismal form, pp. 33–49 in G.B. Müller and S.A. Newman (eds), *Origination of Organismal Form. Beyond the Gene in Developmental and Evolutionary Biology*. MIT Press, Cambridge, MA.

Wilson, H.M. 2006. Juliformian millipedes from the Lower Devonian of Euramerica: Implications for the timing of millipede cladogenesis in the Paleozoic. *Journal of Paleontology* 80: 638–649.

Wilson, H.V. 1894. Observations on the gemmule and egg development of marine sponges. *Journal of Morphology* 9: 277–406.

Wilson, H.V. 1902. On the asexual origin of the ciliated sponge larva. *American Naturalist* 36: 451–459.

Wilson, H.V. 1907. On some phenomena of coalescence and regeneration in sponges. *Journal of Experimental Zoology* 5: 245–258.

Wilson, K., Cahill, V., Ballment, E., and Benzie, J. 2000. The complete sequence of the mitochondrial genome of the crustacean *Penaeus monodon*: are malacostracan crustaceans more closely related to insects than to branchiopods? *Molecular Biology and Evolution* 17: 863–874.

Wilson, R.A. and Webster, L.A. 1974. Protonephridia. *Biological Reviews* 49: 127–160.

Wilting, J., Buttler, K., Schulte, I., Papoutsi, M., Schweigerer, L., and Männer, J. 2007. The proepicardium delivers hemangioblasts but not lymphangioblasts to the developing heart. *Developmental Biology* 305: 451–459.

Winchell, C.J., Sullivan, J., Cameron, C.B., Swalla, B.J. and Mallatt, J. 2002. Evaluating hypotheses of deuterostome phylogeny and chordate evolution with new LSU and SSU ribosomal DNA data. *Molecular Biology and Evolution* 19: 762–776.

Wingstrand, K.G. 1972. Comparative spermatology of a pentastomid, *Raillietiella hemidactyli*, and a brachiuran crustacean, *Argulus foliaceus*, with a discussion of pentastomid relationships. *Biologiske Skrifter Danske Videnskabernes Selskab* 19(4): 1–72.

Winnepenninckx, B., Backeljau, T. and Kristensen, R.M. 1998a. Relationships of the new phylum Cycliophora. *Nature* 393: 636–638.

Winnepenninckx, B.M.H., van de Peer, Y. and Backeljau, T. 1998b. Metazoan relationships on the basis of 18S rRNA sequences: a few years later... *American Zoologist* 1998: 888–906.

Winsor, M.P. 1976, *Starfish, Jellyfish, and the Order of Life: Issues in Nineteenth-Century Science.* Yale University Press, New Haven, CT.

Wirkner, C.S. and Pass, G. 2002. The circulatory system in Chilopoda: functional morphology and phylogenetic aspects. *Acta Zoologica* 83: 193–202.

Wolf, K. and Markiw, M.E. 1984. Biology contravenes taxonomy in the Myxozoa: new discoveries show alternation of invertebrate and vertebrate hosts. *Science* 225: 1449–1452.

Wolf, Y.I., Rogozin, I.B., and Koonin, E.V. 2004. Coelomata and not ecdysozoa: evidence from genome-wide phylogenetic analysis. *Genome Research* 14: 29–36.

Wollacott, R.M. and Pinto, R.L. 1995. Flagellar basal apparatus and its utility in phylogenetic analysis of the Porifera. *Journal of Morphology* 226: 247–265.

Wolpert, L. 1999. From egg to adult to larva. *Evolution & Development* 1: 3–4.

Wolpert, L., Jessell, T., Lawrence, P., Meyerowitz, E., Robertson, E., and Smith, J. 2007. *Principles of Development.* Oxford Univewrsity Press, Oxford.

Wood, W.B. and Edgar, L.G. 1994. Patterning in the *C. elegans* embryo. *Trends in Genetics* 10: 49–54.

Worsaae, K. and Kristensen, R.M. 2005. Evolution of interstitial Polychaeta (Annelida). *Hydrobiologia* 535/536: 319–340.

Worsaae, K. and Müller, M.C.M. 2004. Nephridial and gonoduct distribution patterns in Nerillidae (Annelida: Polychaeta) examined by tubulin staining and cLSM. *Journal of Morphology* 261: 259–269.

Wray, G.A. 1994. The evolution of cell lineage in echinoderms. *American Zoologist* 34: 353–363.

Wray, G.A., Levinton, J.S., and Shapiro, L.H. 1996. Molecular evidence for deep Precambrian divergences among metazoan phyla. *Science* 274: 568–573.

Wright, K.A. 1991. Nematoda, pp. 111–195 in F.W. Harrison (ed.), *Microscopic Anatomy of the Invertebrates. Vol. 4. Aschelminthes.* Wiley-Liss, New York.

Wright, K.A. 1992. Peripheral sensilla of some lower invertebrates: the Platyhelminthes and Nematoda. *Microscopy Research and Techniques* 22: 285–297.

Wyatt, I.J. 1961. Pupal paedogenesis in the Cecidomyidae (Diptera) I. *Proceedings of the Royal Entomological Society, London* A36: 133–143.

Wyatt, I.J. 1964. Immature stages of Lestremiinae (Diptera: Cecidomyidae) infesting cultivated mushrooms. *Transactions of the Royal Entomological Society, London* 116: 15–27.

Xia, F.-S., Zhang, S.-G., and Wang, Z.-Z. 2007. The oldest bryozoans: new evidence from the Late Tremadocian (Early Ordovician) of East Yangtze Gorges in China. *Journal of Paleontology* 81: 1308–1326.

Xiao, S. and Knoll, A.H. 2000. Phosphatized animal embryos from the Neoproterozoic Doushantuo Formation at Weng'an, Guizhou, South China. *Journal of Paleontology* 74: 767–788.

Xiao, S., Zhang, Y., and Knoll, A.H. 1998. Three-dimensional preservation of algae and animal embryos in a Neoproterozoic phosphorite. *Nature* 391: 553–558.

Xiao, S., Yuan, X., and Knoll, A.H. 2000. Eumetazoan fossils in terminal Proterozoic phosphorites? *Proceedings of the National Academy of Sciences USA* 97: 13684–13689.

Yack, J.E. and Homberg, U. 2003. The functional organization of the nervous system in Lepidoptera, pp. 229–265 in N.P. Kristensen (ed.), *Handbook of Zoology: Lepidoptera, Moths and Butterflies*, vol. 2. W.G. de Gruyter, New York.

Yang, L., Cai, C.-L., Lin, L. *et al.* 2006. Isl1Cre reveals a common Bmp pathway in heart and limb development. *Development* 133: 1575–1585.

Yanze, N., Spring, J., Schmidli, C., and Schmid, V. 2001. Conservation of *Hox/ParaHox*-related genes in the early development of a cnidarian. *Developmental Biology* 236: 89–98.

Yashiro, K., Shiratori, H., and Hamada, H. 2007. Haemodynamics determined by a genetic programme govern asymmetric development of the aortic arch. *Nature* 450: 285–288.

Yeates, G.W. and Boag, B. 2002. Post-embryonic growth of longidorid nematodes. *Nematology* 4: 883–889.

Yochelson, E.L. 1979. Early radiation of Mollusca and mollusc-like groups, pp. 323–358 in M.R. House (ed.), *The Origin of Major Invertebrate Groups.* Academic Press, London.

Young, J.Z. 1939. Fused neurons and synaptic contacts in the giant nerve fibres of cephalopods. *Philosophical*

Transactions of the Royal Society of London Series B, Biological Sciences B229: 465–504.

Young, R.E. and Harman, R.F. 1988. 'Larvae', 'paralarvae' and 'subadult' in cephalopod terminology. *Malacologia* 29: 201–207.

Younossi-Hartenstein, A. and Hartenstein, V. 2000. The embryonic development of the polyclad flatworm *Imogine mcgrathi. Development, Genes and Evolution* 210: 383–398.

Younossi-Hartenstein, A., Tepass, U. and Hartenstein, V. 1993. Embryonic origin of the imaginal discs of the head of *Drosophila melanogaster. Development, Genes and Evolution* 203: 60–73.

Yu, J.-K., Satou, Y., Holland, N.D. *et al.* 2007. Axial patterning in the cephalochordates and the evolution of the organizer. *Nature* 445: 613–617.

Yue, Z. and Bengtson, S. 1999. Embryonic and post-embryonic development of the Early Cambrian cnidarian *Olivooides. Lethaia* 32: 181–195.

Zack, S.P., Penkrot, T.A., Bloch, J.I., and Rose, K.D. 2005. Affinities of 'hyopsodontids' to elephant shrews and a Holarctic origin of Afrotheria. *Nature* 434: 497–501.

Zackson, S.L. 1984. Cell lineage, cell-cell interaction, and segment formation in the ectoderm of a glossiphoniid leech embryo. *Developmental Biology* 104: 143–160.

Zamir, M. 2001. Fractal dimensions and multifractility in vascular branching. *Journal of Theoretical Biology* 212: 183–190.

Zardus, J.D. and Martel, A.L. 2002. Phylum Mollusca: Bivalvia, pp. 289–325 in C.M. Young (ed.), *Atlas of Marine Invertebrate Larvae.* Academic Press, London.

Zelhof, A.C., Hardy, R.W., Becker, A., and Zuker, C.S. 2006. Transforming the architecture of compound eyes. *Nature* 443: 696–699.

Zeng, L. and Swalla, B.J. 2005. Molecular phylogeny of the protochordates: chordate evolution. *Canadian Journal of Zoology* 83: 24–33.

Zhang, J. and Kumar, S. 1997. Detection of convergent and parallel evolution at the amino acid sequence level. *Molecular Biology and Evolution* 14: 527–536.

Zhang, X.-G. and Aldridge, R.J. 2007. Development and diversification of trunk plates of the Lower Cambrian lobopodians. *Palaeontology* 50: 401–415.

Zhang, X.-G., Siveter, D.J., Waloszek, D., and Maas, A. 2007. An epipodite-bearing crown-group crustacean from the Lower Cambrian. *Nature* 449: 595–598.

Zimmer, R.L. 1997. Phoronids, brachiopods, and bryozoans, the lophophorates, pp. 279–305 in S.F. Gilbert and A.M. Raunio (eds), *Embryology: Constructing the Organism.* Sinauer Associates, Sunderland, MA.

Zrzavý, J. 2001. The interrelationships of metazoan parasites: a review of phylum- and higher-level hypotheses from recent morphological and molecular phylogenetic analyses. *Folia Parasitologica* 48: 81–103.

Zrzavý, J. 2003. Gastrotricha and metazoan phylogeny. *Zoologica Scripta* 32: 61–82.

Zrzavý, J. and Štys, P. 1997. The basic body plan of arthropods: insights from evolutionary morphology and developmental biology. *Journal of Evolutionary Biology* 10: 353–367.

Zrzavý, J. and Hypša, V. 2003. Myxozoa, *Polypodium,* and the origin of the Bilateria: the phylogenetic position of "Endocnidozoa" in light of the rediscovery of *Buddenbrockia. Cladistics* 19: 164–169.

Zrzavý, J., Hypša, V. and Vlášková, M., 1998a. Arthropod phylogeny: taxonomic congruence, total evidence and conditional combination approaches to morphological and molecular data sets, pp. 97–107 in R.A. Fortey and R.H. Thomas (eds), *Arthropod Relationships.* Chapman & Hall, London.

Zrzavý, J., Mihulka, S., Kepka, P., Bezdek, A., and Tietz, D. 1998b. Phylogeny of the Metazoa based on morphological and 18S ribosomal DNA evidence. *Cladistics* 14: 249–285.

Zrzavý, J., Hypša, V. and Tietz, D.F. 2001. Myzostomida are not annelids: molecular and morphological support for a clade of animals with anterior sperm flagella. *Cladistics* 17: 170–198.

Zuckerkandl, E. and Pauling, L. 1965. Evolutionary divergence and convergence in proteins, pp. 97–166 in V. Bryson and H.J. Vogel (eds), *Evolving Genes and Proteins.* Academic Press, New York.

Author index

Subject index

life cycle 163
Periodomorphosis 158
Perissodactyla *108*
Peritoneum 215
  ciliated 148
Petalopleura *92*
Petrosiid larva 167
Phagocytellozoa 39
Phagocytosis
  defensive role in sponges 30
Pharate adult 159
Pharyngotremata *39*
Pharynx simplex coronatus 75
Phaselus *168*
Phasmatodea 102
Phenetic school 10
Phenotypic plasticity 237
Pherusa
  gonads 222
Phialidium
  nervous cells from endoderm 143
Philophthalmidae
  miracidium 153
Pholcidae
  asymmetric genitalia 192
Pholidota *108*
Phoronida *39, 41,* 69
  affinities 33
    based on mitochondrial
      genome 18
  body axes 193
  glycocalyx 205
  larva 161, *168*
  larval nervous system 208
  metanephridium 221
  nervous system 207
Phoronidea 78
Phoronis architecta
  mitochondrial genome 79
Phoronozoa 33, *39,* 60, 78–79
  Burgess Shale fossils *27*
  Chengjiang fossils 27
  germ cells formed by
    epigenesis *121*
  larvae 168
  larval nervous system 208
  monociliated cells 205
  regeneration *125*
Photoreceptor 211
  ciliary 212
  phaosomous 212
  rhabdomeric 212
Phragmatopoma lapidosa
  vitellogenesis 121
Phreodrilidae 83
Phthiraptera 102
Phylactometria 98

Phyllodocidae 82
phyllosoma *168*
Phylogenetic
  school 10
  tree 1
  unrooted 1
Phylogenetics 10
Phylogeny 6
  star 14
Phylotypic
  period 240
  stage 50, 240
Phylum 12
Physonectae
  larva *169*
Pikaia 105
Pilidiophora 56, 78
Pilidium 56, 161, *68*
Pinacoderm 32, 47
Pisionidae 82
Pit-1 44
Pitx 15
Pitx2 192
piwi 120
Placenta
  multiple origin of complex kind 6
Placentonema gigantissimum 230
Placode 211
  tracheal 217
Placozoa 22, *39,* 49
  affinities 34
  as Metazoa 30
  asexual reproduction *125*
  lack of extracellular matrix 205
  syncytia 114–115
Placuloid 50
Plagiorchiidae 118
  life cycle 153
  multiple foci of proliferation 153
Plakinidae 48
Plakortis angulospiculatus
  mitochondrial genome 48
Planaria
  negative growth 147
  terrestrial 218
Planipennia
  temporarily blind gut 214
Planktosphaera *169*
Plant lice
  anatomical correlates of
    miniaturization 231
Planula 42, 163, *167*
  photoreceptor 211
  polarity 186
Planuliform larva *168*
Planuloid 'larva' 167
Plasmodesma 114

Plasmodium 113
Plasticity 152, 229
Plathelminthes *39, 41,* 59,
    68, 72, 74, 85
  affinities based on mitochondrial
    genome 18
  longitudinal nerve cords 209
  miRNA 176
  polyphyletic 33
Plathelminthomorpha *39,* 68
  blind gut 214
Platyctenida 50
Platydendron ovale 27
Platygaster
  polyembryony *140*
Platygasteridae
  polyembryony *140*
Platynereis dumerilii
  expression of *Brachyury* 179
  photoreceptors 212
Platyzoa *39,* 59, 74
  blind gut 214
Plecoptera
  aquatic nymph 233
Pleiotropy 172–173
Pleon 150
Plerocercoid *167*
Plesiodactylactis 169
Plesiomorphy 1, 10
Pleurostigmophora 98
Plumatella 116
Pneumodesmus newmani 233
Podocoryne
  carnea
    Brachyury 44
    Mef2 44
    Snail 44
  Hox genes 42
  nanos 129
  polarity of main body axis 186
Podocyst 149
Poecilogony 154
Poeobius
  body axes 194
Pogonophora *39, 41,* 82, 84, 129
  affinities 33
Pollingeria grandis 27
'Polychaeta' 84
  asexual reproduction *125*
  circulatory system 218
  coelom 215
  expression of *engrailed* 180
  food-ova 122
  germ cells formed by
    epigenesis *121*
  larvae 167, 168
  larval vs. adult muscles 219

38808803R00203

Made in the USA
Lexington, KY
27 January 2015